Perspectives in Formal Induction, Revision and Evolution

Perspectives in Formal Induction, Revision and Evolution is a book series focusing on the logics used in computer science and artificial intelligence, including but not limited to formal induction, revision and evolution. It covers the fields of formal representation, deduction, and theories or meta-theories of induction, revision and evolution, where the induction is of the first level, the revision is of the second level, and the evolution is of the third level, since the induction is at the formula stratum, the revision is at the theory stratum, and the evolution is at the logic stratum.

In his book "The Logic of Scientific Discovery", Karl Popper argues that a scientific discovery consists of conjecture, theory, refutation, and revision. Some scientific philosophers do not believe that a reasonable conjecture can come from induction. Hence, induction, revision and evolution have become a new territory for formal exploration. Focusing on this challenge, the perspective of this book series differs from that of traditional logics, which concerns concepts and deduction.

The series welcomes proposals for textbooks, research monographs, and edited volumes, and will be useful for all researchers, graduate students, and professionals interested in the field.

Wei Li · Yuefei Sui

R-Calculus, IV: Propositional Logic

Wei Li
Beihang University
Beijing, China

Yuefei Sui
Institute of Computing Technology
Chinese Academy of Sciences
Beijing, China

ISSN 2731-3689 ISSN 2731-3697 (electronic)
Perspectives in Formal Induction, Revision and Evolution
ISBN 978-981-19-8635-2 ISBN 978-981-19-8633-8 (eBook)
https://doi.org/10.1007/978-981-19-8633-8

Jointly published with Science Press
The print edition is not for sale in China mainland. Customers from China mainland please order the print book from: Science Press.

This Springer imprint is published by the registered company Springer Nature Singapore Pte Ltd.
The registered company address is: 152 Beach Road, #21-01/04 Gateway East, Singapore 189721, Singapore

Preface to the Series

Classical mathematical logics (propositional logic, first-order logic and modal logic) and applied logics (temporal logic, dynamic logic, situation calculus, etc.) concern deduction, a logical process from universal statements to particular statements. Description logics concern concepts which and deduction compose of general logics.

Induction and belief revision are the topics of general logics and philosophical logics, and evolution is a new research area in computer science. To formalize induction, revision and evolution is a goal of this series.

Revision is omnipresent in sciences. A new theory usually is a revision of an old theory or several old theories. Copernicus' heliocentric theory is a revision of the Tychonic system; the theory of relativity is a revision of the classical theory of movement; the quantum theory is a revision of classical mechanics; etc. The AGM postulates is a set of conditions a reasonable revision operator should satisfy. Professor Li proposed a calculus for first-order logic, called R-calculus, which is sound and complete with respect to maximal consistent subsets. R-calculus has several variants which can be used in other logics, such as nonmonotonic logics, can propose new problems in the classical logics, and will be used in bigdata.

Popper proposed in his book The Logic of Scientific Discovery that a scientific discovery consists of four aspects: conjecture, theory, refutation and revision. Some scientific philosophers refuted that a reasonable conjecture should come from induction. Hence, induction, revision and evolution become a new territory to be discovered in a formal way.

An induction process is from particular statements to universal statements. A typical example is the mathematical induction, which is a set of nontrivial axioms in Peano arithmetic which makes Peano arithmetic incomplete with respect to the standard model of Peano arithmetic. A logic for induction is needed to guide data mining in artificial intelligence. Data mining is a canonical induction process, which mines rules from data.

In biology, *Evolution is change in the heritable traits of biological populations over successive generations.* In sciences, Darwin's evolution theory is an evolution of intuitive theories of plants and animals. In logic, an evolution is a generating process of combining two logics into a new logic, where the new logic should have the traits of two logics. Hence, we should define the corresponding heritable traits of logics, sets of logics and sequences of logics. Simply speaking, predicate modal logic is an evolution of propositional modal logic and predicate logic, and there are many new problems in predicate modal logic to be solved, such as the constant domain semantics, the variant domain semantics.

The series should focus on formal representation, deduction, theories or meta-theories of induction, revision and evolution, where the induction is of the first level, the revision is of the second level and the evolution is of the third level, because the induction is at the formula stratum, the revision is at the theory stratum and the evolution is at the logic stratum.

The books in the series differ in level: some are overviews and some highly specialized. Here, the books of overviews are for undergraduated students; and the highly specialized ones are for graduated students and researchers in computer science and mathematical logic.

Beijing, China
October 2016

Wei Li
Jie Luo
Fangming Song
Yuefei Sui
Ju Wang
Wujia Zhu

Preface

Propositional logic is basic. R-calculus is a belief revision operator satisfying AGM postulates. Combining propositional logic and R-calculus produces a new point of view to consider belief revision.

For any $i \in \{0, 1\}$, a theory Γ is $\mathbf{T}^i$-valid if for any assignment v, $v(A) = i$ for some formula $A \in \Gamma$; and is $\mathbf{T}_i$-valid if there is an assignment v such that $v(A) = i$ for each formula $A \in \Gamma$.

For any $Q_1, Q_2 \in \{\mathbf{E}, \mathbf{A}\}$ a sequent $\Gamma \Rightarrow \Delta$ is $\mathbf{G}^{Q_1 Q_2}$-valid if for any assignment v, either $v(A) = 0$ for Q_1-formula $A \in \Gamma$, or $v(B) = 1$ for Q_2-formula $B \in \Delta$; and is $\mathbf{G}_{Q_1 Q_2}$-valid if there is an assignment v such that $v(A) = 1$ for Q_1-formula $A \in \Gamma$, and $v(B) = 1$ for Q_2-formula $B \in \Delta$.

There are variant definitions of the validity of a sequent. For example, for any $Q_1, Q_2 \in \{\mathbf{A}, \mathbf{E}\}$ and $i, j \in \{0, 1\}$, a sequent $\Gamma\Delta$ is $\mathbf{G}^{Q_1 i Q_2 j}$-valid if for any assignment v, either $Q_1 A \in \Gamma(v(A) = i)$ or $Q_2 B \in \Delta(v(B) = j)$; and is $\mathbf{G}_{Q_1 i Q_2 j}$-valid if there is an assignment v such that $Q_1 A \in \Gamma(v(A) = i)$, and $Q_2 B \in \Delta(v(B) = j)$.

The negation of $\mathbf{G}^{Q_1 i Q_2 j}$-validity is $\mathbf{G}_{\bar{Q}_1(1-i)\bar{Q}_2(1-j)}$-validity, where $\bar{\mathbf{E}} = \mathbf{A}$ and $\bar{\mathbf{A}} = \mathbf{E}$.

For any $Q_1, Q_2 \in \{\mathbf{A}, \mathbf{E}\}$ and $i, j \in \{0, 1\}$, there are sound and complete Gentzen deduction systems $\mathbf{G}^{Q_1 Q_2}/_{Q_1 Q_2}$ and $\mathbf{G}^{Q_1 i Q_2 j}/_{Q_1 i Q_2 j}$ for $\mathbf{G}^{Q_1 Q_1}/_{Q_1 Q_2}$- and $\mathbf{G}^{Q_1 i Q_2 j}/_{Q_1 i Q_2 j}$-validity,respectively, which are monotonic in Γ iff $Q_1 = \mathbf{E}$, and nonmonotonic in Δ iff $Q_2 = \mathbf{A}$.

R-calculus is a Gentzen-typed deduction system which is nonmonotonic, and is a concrete belief revision operator which is proved to satisfy AGM postulates and DP postulates. Corresponding to Gentzen deduction system $\mathbf{G}^{Q_1 i Q_2 j}/\mathbf{G}_{Q_1 i Q_2 j}$, there is a sound and complete R-calculus $\mathbf{R}^{Q_1 i Q_2 j}/\mathbf{R}_{Q_1 i Q_2 j}$, respectively.

$\mathbf{R}_{Q_1 i Q_2 j}$ preserves the $\subseteq$-minimal change. Let $\mathbf{R}$, $\mathbf{Q}$, $\mathbf{P}$ denote the $\subseteq$-, $\preceq$- and $\vdash_{\preceq}$-minimal change, respectively. For any $Y_1, Y_2 \in \{\mathbf{R}, \mathbf{Q}\}$, there are sound and complete R-calculi $\mathbf{R}^{Y_1 Q_1 i Y_2 Q_2 j}$ and $\mathbf{R}_{Y_1 Q_1 i Y_2 Q_2 j}$, and R-calculus $\mathbf{R}^{\mathbf{P} Q_1 i \mathbf{P} Q_2 j}/_{\mathbf{P} Q_1 i \mathbf{P} Q_2 j}$ is sound and complete for disjunctive/conjunctive normal sequents, respectively, where sequent $\Gamma \Rightarrow \Delta$ is in disjunctive (conjunctive) normal form if each formula in

Γ is in conjunctive (disjunctive) normal form and each formula in Δ is in conjunctive (disjunctive) normal form.

Applications of R-calculus in $\rightsquigarrow$-propositional logic and logic of supersequents are also given.

Beijing, China
October 2021

Wei Li
Yuefei Sui

Contents

Chapter 1
Introduction

1.1 Propositional Logic

Propositional logic is basic, based on which we developed first-order logic and modal logic, which compose of classical logics. Based on these classical logics, other logics are developed, such as temporal logic, Hoare logic, dynamic logic, computational tree logic, situation calculus, etc.

There are three kinds of the deduction systems in propositional logic:

- Axiomatic deduction system, with three axioms and one deduction rule with which we can deduce all the valid formulas, and with which it is not easy to find a proof in the axiomatic system because so a few axioms are available.
- Gentzen deduction system, with two deduction rules for each logical connectives and one axiom, with which we also can deduce all the valid formulas and a formal proof of a valid formula is easy to be found, just by decomposing formulas into atoms by the deduction rules.
- Between axiomatic deduction system and Gentzen deduction system is natural deduction system, with less deduction rules than the Gentzen one, which is much easier to be mastered than the axiomatic one and harder than the Gentzen one.

To vary propositional logic, a simple way is to use less logical connectives. Traditionally, there are logical connectives in propositional logic:

$$\neg, \wedge, \vee, \rightarrow, \leftrightarrow,$$

where $\{\neg, \wedge\}, \{\neg, \vee\}, \{\neg, \rightarrow\}$ are minimal such that another connective can be represented. Even though logic $L_\wedge$ containing one connective $\{\wedge\}$ is not complete in the expression, it would be interesting to give a sound and complete deduction system for $L_\wedge$. In the second chapter, we will give sound, complete Gentzen deduction systems $\mathbf{G}^\neg$, $\mathbf{G}^\wedge$ and $\mathbf{G}^\vee$ for $L_\neg$, $L_\wedge$ and $L_\vee$, respectively and sound, complete R-calculi $\mathbf{R}^\neg$, $\mathbf{R}^\wedge$ and $\mathbf{R}^\vee$ for sequents. To show how to get a sound and complete Gentzen deduction system from truth-value tables for logical connectives, we will

W. Li and Y. Sui, *R-Calculus, IV: Propositional Logic*,
Perspectives in Formal Induction, Revision and Evolution,
https://doi.org/10.1007/978-981-19-8633-8_1

give sound and complete Gentzen deduction systems $\mathbf{G}^{\neg\to}/\mathbf{G}^{\oplus\otimes}$ to show how we build $\mathbf{G}^{\neg\to}/\mathbf{G}^{\oplus\otimes}$ by truth-value tables for $\neg,\to/\oplus,\otimes$; and sound and complete R-calculi $\mathbf{R}^{\neg\to}/\mathbf{R}^{\oplus\otimes}$ for sequents.

1.2 Variant Deduction Systems

Let Γ, Δ be sets of formulas.

A theory Γ is

- $\mathbf{T}^{\mathtt{t}}$-valid, denoted by $\models^{\mathtt{t}} \Gamma$, if for any assignment v, $v(A) = 1$ for some $A \in \Gamma$;
- $\mathbf{T}^{\mathtt{f}}$-valid, denoted by $\models^{\mathtt{f}} \Delta$, if for any assignment v, $v(B) = 0$ for some $B \in \Delta$;
- $\mathbf{T}_{\mathtt{t}}$-valid, denoted by $\models_{\mathtt{t}} \Gamma$, if there is an assignment v such that $v(A) = 1$ for every $A \in \Gamma$;
- $\mathbf{T}_{\mathtt{f}}$-valid, denoted by $\models_{\mathtt{f}} \Delta$, if there is an assignment v such that $v(B) = 0$ for every $B \in \Delta$.

$$\begin{array}{l}\models^{\mathtt{t}} \Gamma \text{ if } \mathbf{A}v\mathbf{E}A \in \Gamma(v(A) = 1)\\ \models^{\mathtt{f}} \Delta \text{ if } \mathbf{B}v\mathbf{E}B \in \Delta(v(B) = 0)\\ \models_{\mathtt{t}} \Gamma \text{ if } \mathbf{E}v\mathbf{A}A \in \Gamma(v(A) = 1)\\ \models_{\mathtt{f}} \Delta \text{ if } \mathbf{E}v\mathbf{B}B \in \Delta(v(B) = 0).\end{array}$$

There are sound and complete tableau proof systems $\mathbf{T}^{\mathtt{t}}, \mathbf{T}^{\mathtt{f}}, \mathbf{T}_{\mathtt{t}}, \mathbf{T}_{\mathtt{f}}$ for $\mathbf{T}^{\mathtt{t}}, \mathbf{T}^{\mathtt{f}}, \mathbf{T}_{\mathtt{t}}, \mathbf{T}_{\mathtt{f}}$-valid theories Γ, respectively. That is, for any theory Γ,

$$\begin{array}{l}\vdash^{\mathtt{t}} \Gamma \text{ iff } \models^{\mathtt{t}} \Gamma\\ \vdash^{\mathtt{f}} \Delta \text{ iff } \models^{\mathtt{f}} \Delta\\ \vdash_{\mathtt{t}} \Gamma \text{ iff } \models_{\mathtt{t}} \Gamma\\ \vdash_{\mathtt{f}} \Delta \text{ iff } \models_{\mathtt{f}} \Delta.\end{array}$$

A sequent $\Gamma \Rightarrow \Delta$ is

- $\mathbf{G}^{\mathtt{t}}$-valid, denoted by $\models^{\mathtt{t}} \Gamma \Rightarrow \Delta$, if for any assignment v, either $v(A) = 0$ for some $A \in \Gamma$ or $v(B) = 1$ for some $B \in \Delta$;
- $\mathbf{G}^{\mathtt{f}}$-valid, denoted by $\models^{\mathtt{f}} \Gamma \Rightarrow \Delta$, if for any assignment v, either $v(A) = 1$ for some $A \in \Gamma$ or $v(B) = 0$ for some $B \in \Delta$.

  Dually, a co-sequent $\Gamma \mapsto \Delta$ is
- $\mathbf{G}_{\mathtt{t}}$-valid, denoted by $\models_{\mathtt{t}} \Gamma \mapsto \Delta$, if there is an assignment v such that $v(A) = 1$ for every $A \in \Gamma$ and $v(B) = 0$ for every $B \in \Delta$;
- $\mathbf{G}_{\mathtt{f}}$-valid, denoted by $\models_{\mathtt{f}} \Gamma \mapsto \Delta$, if there is an assignment v such that $v(A) = 0$ for every $A \in \Gamma$ and $v(B) = 1$ for every $B \in \Delta$.

$$\begin{array}{l}\models^{\mathtt{t}} \Gamma \Rightarrow \Delta \text{ if } \mathbf{A}v(\mathbf{E}A \in \Gamma(v(A) = 0) \text{ or } \mathbf{E}B \in \Delta(v(B) = 1))\\ \models^{\mathtt{f}} \Gamma \Rightarrow \Delta \text{ if } \mathbf{A}v(\mathbf{E}A \in \Gamma(v(A) = 1) \text{ or } \mathbf{E}B \in \Delta(v(B) = 0))\\ \models_{\mathtt{t}} \Gamma \mapsto \Delta \text{ if } \mathbf{E}v(\mathbf{A}A \in \Gamma(v(A) = 1)\&\mathbf{A}B \in \Delta(v(B) = 0))\\ \models_{\mathtt{f}} \Gamma \mapsto \Delta \text{ if } \mathbf{E}v(\mathbf{A}A \in \Gamma(v(A) = 0)\&\mathbf{A}B \in \Delta(v(B) = 1)).\end{array}$$

$\mathbf{G}^{\mathtt{t}}/\mathbf{G}^{\mathtt{f}}$-validity is complementary to $\mathbf{G}_{\mathtt{t}}/\mathbf{G}_{\mathtt{f}}$-one. There are sound and complete Gentzen deduction systems $\mathbf{G}^{\mathtt{t}}, \mathbf{G}^{\mathtt{f}}, \mathbf{G}_{\mathtt{t}}, \mathbf{G}_{\mathtt{f}}$ for $\mathbf{G}^{\mathtt{t}}, \mathbf{G}^{\mathtt{f}}$-valid sequents $\Gamma \Rightarrow \Delta/\mathbf{G}_{\mathtt{t}}, \mathbf{G}_{\mathtt{f}}$-valid co-sequents $\Gamma \mapsto \Delta$, respectively. That is, for any sequent $\Gamma \Rightarrow \Delta$ and co-sequent $\Gamma \mapsto \Delta$,

$$\begin{array}{l} \vdash^{\mathtt{t}} \Gamma \Rightarrow \Delta \text{ iff } \models^{\mathtt{t}} \Gamma \Rightarrow \Delta \\ \vdash^{\mathtt{f}} \Gamma \Rightarrow \Delta \text{ iff } \models^{\mathtt{f}} \Gamma \Rightarrow \Delta \\ \vdash_{\mathtt{t}} \Gamma \mapsto \Delta \text{ iff } \models_{\mathtt{t}} \Gamma \mapsto \Delta \\ \vdash_{\mathtt{f}} \Gamma \mapsto \Delta \text{ iff } \models_{\mathtt{f}} \Gamma \mapsto \Delta. \end{array}$$

Let $Q_1, Q_2 \in \{\mathbf{A}, \mathbf{E}\}$.

A sequent $\Gamma \Rightarrow \Delta$ is $\mathbf{G}^{Q_1Q_2}$-valid, denoted by $\models^{Q_1Q_2} \Gamma \Rightarrow \Delta$, if for any assignment v, either $Q_1 A \in \Gamma(v(A) = 0)$ or $Q_2 B \in \Delta(v(B) = 1)$.

A co-sequent $\Gamma \mapsto \Delta$ is $\mathbf{G}_{Q_1Q_2}$-valid, denoted by $\models_{Q_1Q_2} \Gamma \mapsto \Delta$, if there is an assignment v such that $Q_1 A \in \Gamma(v(A) = 1)$ and $Q_2 B \in \Delta(v(B) = 0)$.

$$\begin{array}{l} \models^{Q_1Q_2} \Gamma \Rightarrow \Delta \text{ if } \mathbf{A}v(Q_1 A \in \Gamma(v(A) = 0) \text{ or } Q_2 B \in \Delta(v(B) = 1)) \\ \models_{Q_1Q_2} \Gamma \mapsto \Delta \text{ if } \mathbf{E}v(Q_1 A \in \Gamma(v(A) = 1) \& Q_2 B \in \Delta(v(B) = 0)). \end{array}$$

$\mathbf{G}^{Q_1Q_2}$-validity is complementary to $\mathbf{G}_{\overline{Q_1}\,\overline{Q_2}}$, where $\overline{\mathbf{A}} = \mathbf{E}$ and $\overline{\mathbf{E}} = \mathbf{A}$. There are sound and complete Gentzen deduction systems $\mathbf{G}^{Q_1Q_2}/\mathbf{G}_{Q_1Q_2}$ for $\mathbf{G}^{Q_1Q_2}$-valid sequents/$\mathbf{G}_{Q_1Q_2}$-valid co-sequents, respectively, which are monotonic in Γ if and only if $Q_1 = \mathbf{E}$, and nonmonotonic in Δ if and only if $Q_2 = \mathbf{A}$.

Let $i, j \in \{0, 1\}$.

A sequent $\Gamma \Rightarrow \Delta$ is $\mathbf{G}^{Q_1 i Q_2 j}$-valid, denoted by $\models^{Q_1 i Q_2 j} \Gamma \Rightarrow \Delta$, if for any assignment v, either $Q_1 A \in \Gamma(v(A) = i)$ or $Q_2 B \in \Delta(v(B) = j)$.

A co-sequent $\Gamma \mapsto \Delta$ is $\mathbf{G}_{Q_1 i Q_2 j}$-valid, denoted by $\models_{Q_1 i Q_2 j} \Gamma \mapsto \Delta$, if there is an assignment v such that $Q_1 A \in \Gamma(v(A) = i)$ and $Q_2 B \in \Delta(v(B) = j)$.

$$\begin{array}{l} \models^{Q_1 i Q_2 j} \Gamma \Rightarrow \Delta \text{ if } \mathbf{A}v(Q_1 A \in \Gamma(v(A) = i) \text{ or } Q_2 B \in \Delta(v(B) = j)) \\ \models_{Q_1 i Q_2 j} \Gamma \mapsto \Delta \text{ if } \mathbf{E}v(Q_1 A \in \Gamma(v(A) = i) \& Q_2 B \in \Delta(v(B) = j)). \end{array}$$

$\mathbf{G}^{Q_1 i Q_2 j}$-validity is complementary to $\mathbf{G}_{\overline{Q_1}(1-i)\overline{Q_2}(1-j)}$.

There are sound and complete Gentzen deduction systems $\mathbf{G}^{Q_1 i Q_2 j}/\mathbf{G}_{Q_1 i Q_2 j}$ for $\mathbf{G}^{Q_1 i Q_2 j}$-valid sequents/$\mathbf{G}_{Q_1 i Q_2 j}$-valid co-sequents, respectively, which are nonmonotonic in Γ if and only if $Q_1 = \mathbf{A}$, and monotonic in Δ if and only if $Q_2 = \mathbf{E}$. That is, for sequent $\Gamma \Rightarrow \Delta$ and co-sequent $\Gamma \mapsto \Delta$,

$$\begin{array}{l} \vdash^{Q_1 i Q_2 j} \Gamma \Rightarrow \Delta \text{ iff } \models^{Q_1 i Q_2 j} \Gamma \Rightarrow \Delta \\ \vdash_{Q_1 i Q_2 j} \Gamma \mapsto \Delta \text{ iff } \models_{Q_1 i Q_2 j} \Gamma \mapsto \Delta. \end{array}$$

Let $\mathbf{R}, \mathbf{Q}, \mathbf{P}$ denote the $\subseteq$-, $\preceq$- and $\vdash_{\preceq}$-minimal change, respectively. For $Y_1, Y_2 \in \{\mathbf{R}, \mathbf{Q}, \mathbf{P}\}$, there are sound and complete Gentzen deduction systems $\mathbf{G}^{Y_1 Q_1 i Y_2 Q_2 j}/\mathbf{G}_{Y_1 Q_1 i Y_2 Q_2 j}$ for $\mathbf{G}^{Y_1 Q_1 i Y_2 Q_2 j}$-valid sequents/$\mathbf{G}_{Y_1 Q_1 i Y_2 Q_2 j}$-valid co-sequents, respectively, which are nonmonotonic in Γ if and only if $Q_1 = \mathbf{A}$, and monotonic in Δ if and only if $Q_2 = \mathbf{E}$. That is, for sequent $\Gamma \Rightarrow \Delta$ and co-sequent $\Gamma \mapsto \Delta$,

$$\vdash^{Y_1 Q_1 i Y_2 Q_2 j} \Gamma \Rightarrow \Delta \text{ iff } \models^{Q_1 i Y_2 Q_2 j} \Gamma \Rightarrow \Delta$$
$$\vdash_{Y_1 Q_1 i Y_2 Q_2 j} \Gamma \mapsto \Delta \text{ iff } \models_{Y_1 Q_1 i Y_2 Q_2 j} \Gamma \mapsto \Delta.$$

1.3 R-Calculus

The first author (Li 2007) developed a Gentzen-typed deduction system in first-order logic (Alchourrón et al. 1985), called R-calculus, to reduce a configuration $\Delta|\Gamma$ into a consistent theory $\Delta \cup \Theta$, where Θ is a minimal change of Γ by Δ, i.e., a maximal set of Γ which is consistent with Δ, where Δ is a set of atoms, and Γ, Θ are consistent theories of first-order logic. Here, $\Delta|\Gamma$ corresponds to iterating revision $\Gamma \circ \Delta$. Hence, the deduction system gives a concrete revision operator which is shown to satisfy AGM postulates (Alchourrón et al. 1985; Fermé and Hansson 2011).

R-calculus consists of the following rules and axioms:

- Structural rules:

$$\begin{array}{l}
(\text{contraction}^L)\ \Delta, A, A|\Gamma \Rightarrow \Delta, A|\Gamma \\
(\text{contraction}^R)\ \Delta|B, B, \Gamma \Rightarrow \Delta|A, \Gamma \\
(\text{interchange}^L)\ \Delta, A_1, A_2|\Gamma \Rightarrow \Delta, A_2, A_1|\Gamma \\
(\text{interchange}^R)\ \Delta|B_1, B_2, \Gamma \Rightarrow \Delta|B_2, B_1, \Gamma;
\end{array}$$

- R-axiom:

$$(\neg)\ \Delta, \neg A|A, \Gamma \Rightarrow \Delta, \neg A|\Gamma;$$

- R-elimination rule:

$$\frac{\Gamma_1, A \vdash B \quad A \mapsto_T B \quad \Gamma_2, B \vdash C \quad \Delta|C, \Gamma_2 \Rightarrow \Delta|\Gamma_2}{\Delta|\Gamma_1, A, \Gamma_2 \Rightarrow \Delta|\Gamma_1, \Gamma_2};$$

- R-logical deduction rules:

$$\begin{array}{ll}
(\mathbf{R}_1^{\wedge})\ \dfrac{\Delta|B_1, \Gamma \Rightarrow \Delta|\Gamma}{\Delta|B_1 \wedge B_2, \Gamma \Rightarrow \Delta|\Gamma} & (\mathbf{R}_2^{\wedge})\ \dfrac{\Delta|B_2, \Gamma \Rightarrow \Delta|\Gamma}{\Delta|B_1 \wedge B_2, \Gamma \Rightarrow \Delta|\Gamma} \\
(\mathbf{R}^{\vee})\ \dfrac{\Delta|B_1, \Gamma \Rightarrow \Delta|\Gamma \quad \Delta|B_2, \Gamma \Rightarrow \Delta|\Gamma}{\Delta|B_1 \vee B_2, \Gamma \Rightarrow \Delta|\Gamma} & (\mathbf{R}^{\to})\ \dfrac{\Delta|\neg B_1, \Gamma \Rightarrow \Delta|\Gamma \quad \Delta|B_2, \Gamma \Rightarrow \Delta|\Gamma}{\Delta|B_1 \to B_2, \Gamma \Rightarrow \Delta|\Gamma} \\
(\mathbf{R}^{\mathbf{A}})\ \dfrac{\Delta|B(t), \Gamma \Rightarrow \Delta|\Gamma}{\Delta|\mathbf{A}x B(x), \Gamma \Rightarrow \Delta|\Gamma} & (\mathbf{R}^{\mathbf{E}})\ \dfrac{\Delta|B(x), \Gamma \Rightarrow \Delta|\Gamma}{\Delta|\mathbf{E}x B(x), \Gamma \Rightarrow \Delta|\Gamma}
\end{array}$$

where t is a term, and x is a variable not freely occur in Δ and Γ.

The deduction rules in R-calculus are corresponding to those in Gentzen deduction system of first-order logic. In Gentzen deduction system, a sequent $\Gamma \Rightarrow \Delta$ is reduced to atomic sequents $\Gamma' \Rightarrow \Delta'$ by using the rules of the right-hand side and of the left-hand side, where $\Gamma' \Rightarrow \Delta'$ is atomic if Γ', Δ' are sets of atoms and $\Gamma' \Rightarrow \Delta'$ is an axiom if and only if $\Gamma' \cap \Delta' \neq \emptyset$. In R-calculus, a configuration $\Delta|A, \Gamma$ is reduced to literal configurations $\Delta|l, \Gamma$ by using the deduction rules for logical symbols (logical connectives and quantifiers in first-order logic) (Li and Sui 2017a, 2018), where $\Delta|l, \Gamma \Rightarrow \Delta, l|\Gamma$ is an axiom if and only if $\Delta \nvdash \neg l$; otherwise, $\Delta|l, \Gamma \Rightarrow \Delta|\Gamma$, i.e., l is deleted from theory $\{l\} \cup \Gamma$.

1.4 R-Calculi in This Volume

We consider three kinds of R-calculi: one for weak propositional logics, one for propositional logic and one for supersequents and $\rightsquigarrow$-propositional logic.

1.4.1 R-Calculi for Weak Propositional Logics

We consider three weak propositional logics $L^{\neg}, L^{\wedge}, L^{\vee}$, and give sound and complete

- Gentzen deduction systems $\mathbf{G}^{\neg}, \mathbf{G}^{\wedge}, \mathbf{G}^{\vee}$ for $\mathbf{G}^{\neg}, \mathbf{G}^{\wedge}, \mathbf{G}^{\vee}$-valid sequents, respectively,
- R-calculi $\mathbf{R}^{\neg}, \mathbf{R}^{\wedge}, \mathbf{R}^{\vee}$ for $\mathbf{R}^{\neg}, \mathbf{R}^{\wedge}, \mathbf{R}^{\vee}$-valid reductions, respectively;
- Gentzen deduction systems $\mathbf{G}_{\neg}, \mathbf{G}_{\wedge}, \mathbf{G}_{\vee}$ for $\mathbf{G}_{\neg}, \mathbf{G}_{\wedge}, \mathbf{G}_{\vee}$-valid co-sequents, respectively,
- R-calculi $\mathbf{R}_{\neg}, \mathbf{R}_{\wedge}, \mathbf{R}_{\vee}$ for $\mathbf{R}_{\neg}, \mathbf{R}_{\wedge}, \mathbf{R}_{\vee}$-valid co-reductions, respectively.
  Moreover, we consider two propositional logics $L^{\neg\rightarrow}, L^{\oplus\otimes}$, and give sound and complete
- Gentzen deduction systems $\mathbf{G}^{\neg\rightarrow}, \mathbf{G}^{\oplus\otimes}$ for $\mathbf{G}^{\neg\rightarrow}, \mathbf{G}^{\oplus\otimes}$-valid sequents, respectively,
- R-calculi $\mathbf{R}^{\neg\rightarrow}, \mathbf{R}^{\oplus\otimes}$ for $\mathbf{R}^{\neg\rightarrow}, \mathbf{R}^{\oplus\otimes}$-valid sequents, respectively,
- Gentzen deduction systems $\mathbf{G}_{\neg\rightarrow}, \mathbf{G}_{\oplus\otimes}$ for $\mathbf{G}_{\neg\rightarrow}, \mathbf{G}_{\oplus\otimes}$-valid co-sequents, respectively,
- R-calculi $\mathbf{R}_{\neg\rightarrow}, \mathbf{R}_{\oplus\otimes}$ for $\mathbf{R}_{\neg\rightarrow}, \mathbf{R}_{\oplus\otimes}$-valid co-reductions, respectively,

1.4.2 R-Calculi for Tableau Proof/Gentzen Deduction Systems

Corresponding to tableau proof systems and Gentzen deduction systems there are sound and complete R-calculi.

Given a theory Γ and formula A,

- Corresponding to $\mathbf{T}_{\mathtt{t}}$-validity, using Γ to revise A obtains theory Γ, A', denoted by $\models_{\mathtt{t}} \Gamma|A \rightrightarrows \Gamma, A'$, where $A' = A$ iff $\Gamma \cup \{A\}$ is $\mathbf{T}_{\mathtt{t}}$-valid, where $\rightrightarrows$ means "reduced to".
- Corresponding to $\mathbf{T}_{\mathtt{f}}$-validity, using Δ to revise B obtains theory Δ, B', denoted by $\models_{\mathtt{f}} \Delta|B \rightrightarrows \Delta, B'$, where $B' = B$ iff $\Delta \cup \{B\}$ is $\mathbf{T}_{\mathtt{f}}$-valid.
- Corresponding to $\mathbf{T}^{\mathtt{t}}$-validity, using Γ to revise $A \in \Gamma$ obtains theory $\Gamma - \{A'\}$, denoted by $\models^{\mathtt{t}} \Gamma|A \rightrightarrows \Gamma - \{A'\}$, where $A' = A$ iff $\Gamma - \{A\}$ is $\mathbf{T}^{\mathtt{t}}$-valid.
- Corresponding to $\mathbf{T}^{\mathtt{f}}$-validity, using Δ to revise $B \in \Delta$ obtains theory $\Delta - \{B'\}$, denoted by $\models^{\mathtt{f}} \Delta|B \rightrightarrows \Delta - \{B'\}$, where $B' = B$ iff $\Delta - \{B\}$ is $\mathbf{T}^{\mathtt{f}}$-valid.

There are sound and complete R-calculi $\mathbf{S}_{\mathtt{t}}, \mathbf{S}_{\mathtt{f}}, \mathbf{S}^{\mathtt{t}}, \mathbf{S}^{\mathtt{f}}$ for preserving $\mathbf{T}_{\mathtt{t}}, \mathbf{T}_{\mathtt{f}}, \mathbf{T}^{\mathtt{t}}, \mathbf{T}^{\mathtt{f}}$-validities, respectively. That is, for any reduction $\Gamma|A \rightrightarrows \Gamma'/\Delta|B \rightrightarrows \Delta'$,

$$\begin{array}{l} \vdash^{\mathtt{t}} \Gamma|A \rightrightarrows \Gamma' \text{ iff } \models^{\mathtt{t}} \Gamma|A \rightrightarrows \Gamma' \\ \vdash^{\mathtt{f}} \Delta|B \rightrightarrows \Delta' \text{ iff } \models^{\mathtt{f}} \Delta|B \rightrightarrows \Delta' \\ \vdash_{\mathtt{t}} \Gamma|A \rightrightarrows \Gamma' \text{ iff } \models_{\mathtt{t}} \Gamma|A \rightrightarrows \Gamma' \\ \vdash_{\mathtt{f}} \Delta|B \rightrightarrows \Delta' \text{ iff } \models_{\mathtt{f}} \Delta|B \rightrightarrows \Delta'. \end{array}$$

Given a sequent $\Gamma \Rightarrow \Delta$/co-sequent $\Gamma \mapsto \Delta$ and pair (A, B) of formulas,

- Corresponding to $\mathbf{G}_{\mathtt{t}}$-validity, using $\Gamma \mapsto \Delta$ to revise (A, B) obtains a sequent $\Gamma, A' \mapsto \Delta, B'$, denoted by

$$\models_{\mathtt{t}} \Gamma \mapsto \Delta|(A, B) \rightrightarrows \Gamma, A' \mapsto \Delta, B',$$

if $A' = A$ iff $\Gamma, A \mapsto \Delta$ is $\mathbf{G}_{\mathtt{t}}$-valid and $B' = B$ iff $\Gamma, A' \mapsto \Delta, B$ is $\mathbf{G}_{\mathtt{t}}$-valid.
- Corresponding to $\mathbf{G}_{\mathtt{f}}$-validity, using $\Gamma \mapsto \Delta$ to revise (A, B) obtains a sequent $\Gamma, A' \mapsto \Delta, B'$, denoted by

$$\models_{\mathtt{f}} \Gamma \mapsto \Delta|(A, B) \rightrightarrows \Gamma, A' \mapsto \Delta, B',$$

if $A' = A$ iff $\Gamma, A \mapsto \Delta$ is $\mathbf{G}_{\mathtt{f}}$-valid and $B' = B$ iff $\Gamma, A' \mapsto \Delta, B$ is $\mathbf{G}_{\mathtt{f}}$-valid.
- Corresponding to $\mathbf{G}^{\mathtt{t}}$-validity, using $\Gamma \Rightarrow \Delta$ to revise (A, B), where $A \in \Gamma$ and $B \in \Delta$, obtains a sequent $\Gamma - \{A'\} \Rightarrow \Delta - \{B'\}$, denoted by

$$\models^{\mathtt{t}} \Gamma \Rightarrow \Delta|(A, B) \rightrightarrows \Gamma - \{A'\} \Rightarrow \Delta - \{B'\},$$

if $A' = A$ iff $\Gamma - \{A\} \Rightarrow \Delta$ is $\mathbf{G}^{\mathtt{t}}$-valid and $B' = B$ iff $\Gamma - \{A'\} \Rightarrow \Delta - \{B\}$ is $\mathbf{G}^{\mathtt{t}}$-valid.
- Corresponding to $\mathbf{G}^{\mathtt{f}}$-validity, using $\Gamma \Rightarrow \Delta$ to revise (A, B), where $A \in \Gamma$ and $B \in \Delta$, obtains a sequent $\Gamma - \{A'\} \Rightarrow \Delta - \{B'\}$, denoted by

$$\models^{\mathtt{f}} \Gamma \Rightarrow \Delta|(A, B) \rightrightarrows \Gamma - \{A'\} \Rightarrow \Delta - \{B'\},$$

if $A' = A$ iff $\Gamma - \{A\} \Rightarrow \Delta$ is $\mathbf{G}^{\mathtt{f}}$-valid and $B' = B$ iff $\Gamma - \{A'\} \Rightarrow \Delta - \{B\}$ is $\mathbf{G}^{\mathtt{f}}$-valid.

There are sound and complete R-calculi $\mathbf{R}_{\mathtt{t}}, \mathbf{R}_{\mathtt{f}}/\mathbf{R}^{\mathtt{t}}, \mathbf{R}^{\mathtt{f}}$ for preserving $\mathbf{G}_{\mathtt{t}}, \mathbf{G}_{\mathtt{f}}/\mathbf{G}^{\mathtt{t}}, \mathbf{G}^{\mathtt{f}}$-validities, respectively. That is, for any reduction $\Gamma \Rightarrow \Delta|(A, B) \rightrightarrows \Gamma' \Rightarrow \Delta'$/co-reduction $\Gamma \mapsto \Delta|(A, B) \rightrightarrows \Gamma' \mapsto \Delta'$,

$$\begin{array}{l}
\vdash^{\mathtt{t}} \Gamma \Rightarrow \Delta|(A, B) \rightrightarrows \Gamma' \Rightarrow \Delta' \text{ iff } \models^{\mathtt{t}} \Gamma \Rightarrow \Delta|(A, B) \rightrightarrows \Gamma' \Rightarrow \Delta' \\
\vdash^{\mathtt{f}} \Gamma \Rightarrow \Delta|(A, B) \rightrightarrows \Gamma' \Rightarrow \Delta' \text{ iff } \models^{\mathtt{f}} \Gamma \Rightarrow \Delta|(A, B) \rightrightarrows \Gamma' \Rightarrow \Delta' \\
\vdash_{\mathtt{t}} \Gamma \mapsto \Delta|(A, B) \rightrightarrows \Gamma' \mapsto \Delta' \text{ iff } \models_{\mathtt{t}} \Gamma \mapsto \Delta|(A, B) \rightrightarrows \Gamma' \mapsto \Delta' \\
\vdash_{\mathtt{f}} \Gamma \mapsto \Delta|(A, B) \rightrightarrows \Gamma' \mapsto \Delta' \text{ iff } \models_{\mathtt{f}} \Gamma \mapsto \Delta|(A, B) \rightrightarrows \Gamma' \mapsto \Delta'.
\end{array}$$

1.4.3 R-Calculi for Deduction Systems $\mathbf{G}^{Q_1Q_2}/\mathbf{G}_{Q_1Q_2}$

Let $Q_1, Q_2 \in \{\mathbf{A}, \mathbf{E}\}$.

Given a sequent $\Gamma \Rightarrow \Delta$ and pair (A, B) of formulas, the result of $\Gamma \Rightarrow \Delta$ $\mathbf{G}^{Q_1Q_2}$-revising (A, B) is $\Gamma' \Rightarrow \Delta'$, denoted by

$$\models^{Q_1Q_2} \Gamma \Rightarrow \Delta|(A, B) \rightrightarrows \Gamma' \Rightarrow \Delta',$$

if $\Gamma' \Rightarrow \Delta' =$

$$\begin{cases}
\Gamma \pm_1 A \Rightarrow \Delta \pm_2 B & \text{if } \models^{Q_1Q_2} \Gamma \pm_1 A \Rightarrow \Delta \pm_2 B \\
\Gamma \pm_1 A \Rightarrow \Delta & \text{if } \models^{Q_1Q_2} \Gamma \pm_A \Rightarrow \Delta \\
\Gamma \Rightarrow \Delta \pm_2 B & \text{if } \models^{Q_1Q_2} \Gamma \Rightarrow \Delta \pm_2 B \\
\Gamma \Rightarrow \Delta & \text{otherwise,}
\end{cases}$$

where

$$\pm_1 A = \begin{cases} -\{A\} \text{ if } Q_1 = \mathbf{E} \text{ and } A \in \Gamma \\ \cup\{A\} \text{ if } Q_1 = \mathbf{A}, \end{cases}$$
$$\pm_2 B = \begin{cases} -\{B\} \text{ if } Q_2 = \mathbf{E} \text{ and } B \in \Delta \\ \cup\{B\} \text{ if } Q_2 = \mathbf{A}. \end{cases}$$

Given a co-sequent $\Gamma \mapsto \Delta$ and pair (A, B) of formulas, the result of $\Gamma \mapsto \Delta$ $\mathbf{G}_{Q_1Q_2}$-revising (A, B) is $\Gamma' \mapsto \Delta'$, denoted by

$$\models^{Q_1Q_2} \Gamma \mapsto \Delta|(A, B) \rightrightarrows \Gamma' \mapsto \Delta',$$

if $\Gamma' \mapsto \Delta' =$

$$\begin{cases}
\Gamma \pm_1 A \mapsto \Delta \pm_2 B & \text{if } \models_{Q_1Q_2} \Gamma \pm_1 A \mapsto \Delta \pm_2 B \\
\Gamma \pm_1 A \mapsto \Delta & \text{if } \models_{Q_1Q_2} \Gamma \pm_A \mapsto \Delta \\
\Gamma \mapsto \Delta \pm_2 B & \text{if } \models_{Q_1Q_2} \Gamma \mapsto \Delta \pm_2 B \\
\Gamma \mapsto \Delta & \text{otherwise.}
\end{cases}$$

There is a sound and complete R-calculus $\mathbf{R}^{Q_1Q_2}/\mathbf{R}_{Q_1Q_2}$ for preserving $\mathbf{G}^{Q_1Q_2}/\mathbf{G}_{Q_1Q_2}$-validity, respectively. That is, for any reduction $\Gamma\Rightarrow\Delta|(A,B)\rightrightarrows\Gamma'\Rightarrow\Delta'$/co-reduction $\Gamma\mapsto\Delta|(A,B)\rightrightarrows\Gamma'\mapsto\Delta'$,

$$\vdash^{Q_1Q_2}\Gamma\Rightarrow\Delta|(A,B)\rightrightarrows\Gamma'\Rightarrow\Delta' \text{ iff } \models^{Q_1Q_2}\Gamma\Rightarrow\Delta|(A,B)\rightrightarrows\Gamma'\Rightarrow\Delta'$$
$$\vdash_{Q_1Q_2}\Gamma\mapsto\Delta|(A,B)\rightrightarrows\Gamma'\mapsto\Delta' \text{ iff } \models_{Q_1Q_2}\Gamma\mapsto\Delta|(A,B)\rightrightarrows\Gamma'\mapsto\Delta'.$$

1.4.4 R-Calculi for Deduction Systems $\mathbf{G}^{Q_1iQ_2j}/\mathbf{G}_{Q_1iQ_2j}$

Let $Q_1,Q_2\in\{\mathbf{A},\mathbf{E}\}$ and $i,j\in\{0,1\}$.

Given a sequent $\Gamma\Rightarrow_{Q_1iQ_2j}\Delta$ and pair (A,B) of formulas, the result of $\Gamma\Rightarrow\Delta$ $\mathbf{G}^{Q_1iQ_2j}$-revising (A,B) is $\Gamma'\Rightarrow\Delta'$, denoted by

$$\models^{Q_1iQ_2j}\Gamma\Rightarrow\Delta|(A,B)\rightrightarrows\Gamma'\Rightarrow\Delta',$$

if $\Gamma'\Rightarrow\Delta'=$

$$\begin{cases}\Gamma\pm_1A\Rightarrow\Delta\pm_2B & \text{if }\Gamma\pm_1A\Rightarrow\Delta\pm_2B\text{ is }\mathbf{G}^{Q_1iQ_2j}\text{-valid}\\ \Gamma\pm_1A\Rightarrow\Delta & \text{otherwise, if }\Gamma\pm_1A\Rightarrow\Delta\text{ is }\mathbf{G}^{Q_1iQ_2j}\text{-valid}\\ \Gamma\Rightarrow\Delta\pm_2B & \text{otherwise, if }\Gamma\Rightarrow\Delta\pm_2B\text{ is }\mathbf{G}^{Q_1iQ_2j}\text{-valid}\\ \Gamma\Rightarrow\Delta & \text{otherwise,}\end{cases}$$

where

$$\pm_1A=\begin{cases}\cup\{A\} \text{ if } Q_1=\mathbf{A}\\ -\{A\} \text{ if } Q_1=\mathbf{E}\text{ and }A\in\Gamma,\end{cases}$$
$$\pm_2B=\begin{cases}\cup\{B\} \text{ if } Q_2=\mathbf{A}\\ -\{B\} \text{ if } Q_2=\mathbf{E}\text{ and }B\in\Delta.\end{cases}$$

Given a co-sequent $\Gamma\mapsto_{Q_1iQ_2j}\Delta$ and pair (A,B) of formulas, the result of $\Gamma\mapsto\Delta$ $\mathbf{G}_{Q_1iQ_2j}$-revising (A,B) is $\Gamma'\mapsto\Delta'$, denoted by

$$\models^{Q_1iQ_2j}\Gamma\mapsto\Delta|(A,B)\rightrightarrows\Gamma'\mapsto\Delta',$$

if $\Gamma'\mapsto\Delta'=$

$$\begin{cases}\Gamma\pm_1A\mapsto\Delta\pm_2B & \text{if }\Gamma\pm_1A\mapsto\Delta\pm_2B\text{ is }\mathbf{G}_{Q_1iQ_2j}\text{-valid}\\ \Gamma\pm_1A\mapsto\Delta & \text{otherwise, if }\Gamma\pm_1A\mapsto\Delta\text{ is }\mathbf{G}_{Q_1iQ_2j}\text{-valid}\\ \Gamma\mapsto\Delta\pm_2B & \text{otherwise, if }\Gamma\mapsto\Delta\pm_2B\text{ is }\mathbf{G}_{Q_1iQ_2j}\text{-valid}\\ \Gamma\mapsto\Delta & \text{otherwise.}\end{cases}$$

There is a sound and complete R-calculus $\mathbf{R}^{Q_1iQ_2j}/\mathbf{R}_{Q_1iQ_2j}$ for preserving $\mathbf{G}^{Q_1iQ_2j}/\mathbf{G}_{Q_1iQ_2j}$-validity. That is, for any reduction $\Gamma\Rightarrow\Delta|(A,B)\rightrightarrows\Gamma'\Rightarrow\Delta'$/co-reduction $\Gamma\mapsto\Delta|(A,B)\rightrightarrows\Gamma'\mapsto\Delta'$,

$$\vdash^{Q_1iQ_2j} \Gamma \Rightarrow \Delta|(A,B) \rightrightarrows \Gamma' \Rightarrow \Delta' \text{ iff } \models^{Q_1iQ_2j} \Gamma \Rightarrow \Delta|(A,B) \rightrightarrows \Gamma' \Rightarrow \Delta'$$
$$\vdash_{Q_1iQ_2j} \Gamma \mapsto \Delta|(A,B) \rightrightarrows \Gamma' \mapsto \Delta' \text{ iff } \models_{Q_1iQ_2j} \Gamma \mapsto \Delta|(A,B) \rightrightarrows \Gamma' \mapsto \Delta'.$$

1.4.5 R-Calculi for Deduction Systems $\mathbf{G}^{Y_1Q_1iY_2Q_2j}/\mathbf{G}_{Y_1Q_1iY_2Q_2j}$

R-calculus $\mathbf{R}^{Q_1iQ_2j}$ is with respect to the $\subseteq$-minimal change $\mathbf{R}$. There are $\preceq$-minimal change $\mathbf{Q}$ and $\vdash_{\preceq}$-minimal change $\mathbf{P}$, (Li and Sui 2013, 2014) and correspondingly there are R-calculus $\mathbf{R}^{Y_1Q_1iY_2Q_2j}/\mathbf{R}_{Y_1Q_1iY_2Q_2j}$ for $Y_1, Y_2 \in \{\mathbf{R}, \mathbf{Q}, \mathbf{P}\}$.

All these R-calculi $\mathbf{R}^{Y_1Q_1iY_2Q_2j}/\mathbf{R}_{Y_1Q_1iY_2Q_2j}$ are sound and complete with respect to the $\subseteq$-minimal change $\mathbf{R}$, $\preceq$-minimal change $\mathbf{Q}$ and $\vdash_{\preceq}$-minimal change $\mathbf{P}$, respectively. That is, for any reduction $\Gamma \Rightarrow \Delta|(A,B) \rightrightarrows \Gamma' \Rightarrow \Delta'$/co-reduction $\Gamma \mapsto \Delta|(A,B) \rightrightarrows \Gamma' \mapsto \Delta'$,

$$\begin{aligned}\vdash^{Y_1Q_1iY_2Q_2j} &\Gamma \Rightarrow \Delta|(A,B) \rightrightarrows \Gamma' \Rightarrow \Delta' \\ &\text{iff } \models^{Y_1Q_1iY_2Q_2j} \Gamma \Rightarrow \Delta|(A,B) \rightrightarrows \Gamma' \Rightarrow \Delta' \\ \vdash_{Y_1Q_1iY_2Q_2j} &\Gamma \mapsto \Delta|(A,B) \rightrightarrows \Gamma' \mapsto \Delta' \\ &\text{iff } \models_{Y_1Q_1iY_2Q_2j} \Gamma \mapsto \Delta|(A,B) \rightrightarrows \Gamma' \mapsto \Delta'.\end{aligned}$$

About these Gentzen deduction systems and R-calculi, we have the following equivalences:

Gentzen deduction system	R-calculus
$\mathbf{G}^{\mathtt{t}} = \mathbf{G}^{\mathbf{EE}}$	$\mathbf{R}^{\mathtt{t}} = \mathbf{R}^{\mathbf{EE}}$
$\mathbf{G}_{\mathtt{t}} = \mathbf{G}_{\mathbf{AA}}$	$\mathbf{R}_{\mathtt{t}} = \mathbf{R}_{\mathbf{AA}}$
$\mathbf{G}^{\mathtt{f}} = \mathbf{G}^{\mathbf{E1E0}}$	$\mathbf{R}^{\mathtt{f}} = \mathbf{R}^{\mathbf{E1E0}}$
$\mathbf{G}_{\mathtt{f}} = \mathbf{G}_{\mathbf{A0A1}}$	$\mathbf{R}_{\mathtt{f}} = \mathbf{R}_{\mathbf{A0A1}}$
$\mathbf{G}^{Q_1Q_2} = \mathbf{G}^{Q_10Q_21}$	$\mathbf{R}^{Q_1Q_2} = \mathbf{R}^{Q_10Q_21}$
$\mathbf{G}_{Q_1Q_2} = \mathbf{G}_{Q_11Q_20}$	$\mathbf{R}_{Q_1Q_2} = \mathbf{R}_{Q_11Q_20}$
$\mathbf{G}^{Q_1iQ_2j} = \mathbf{G}^{\mathbf{R}Q_1i\mathbf{R}Q_2j}$	$\mathbf{R}^{Q_1iQ_2j} = \mathbf{R}^{\mathbf{R}Q_1i\mathbf{R}Q_2j}$
$\mathbf{G}_{Q_1iQ_2j} = \mathbf{G}_{\mathbf{R}Q_1i\mathbf{R}Q_2j}$	$\mathbf{R}_{Q_1iQ_2j} = \mathbf{R}_{\mathbf{R}Q_1i\mathbf{R}Q_2j}$

1.5 Applications of R-Calculi

We consider two applications of R-calculi to $\leadsto$-propositional logic and supersequents (Li and Sui 2017b; Li et al. 2017a, b), where the former is a logic taking logical connective $\to$ as a nonlogical symbol, and the latter is a pseudo-extension of sequents, that is, a supersequent is equivalent to a sequent.

1.5.1 R-Calculus for $\rightsquigarrow$-Propositional Logic

By taking $\rightarrow$ as a nonlogical symbol (denoted by $\rightsquigarrow$), we can develop $\rightsquigarrow$-propositional logic, and corresponding sound and complete Gentzen deduction systems. The logic can be used in default logic and description logics.

In propositional logic, we have

$$\begin{array}{ll} A_1 \wedge A_2 \rightarrow B \Leftarrow A_1 \rightarrow B \underline{\vee} A_2 \rightarrow B & A_1 \wedge A_2 \not\rightarrow B \equiv A_1 \not\rightarrow B \underline{\wedge} A_2 \not\rightarrow B \\ A_1 \vee A_2 \rightarrow B \equiv A_1 \rightarrow B \underline{\wedge} A_2 \rightarrow B & A_1 \vee A_2 \not\rightarrow B \equiv A_1 \not\rightarrow B \underline{\vee} A_2 \not\rightarrow B \\ A \rightarrow B_1 \wedge B_2 \equiv A \rightarrow B_1 \underline{\wedge} A \rightarrow B_2 & A \not\rightarrow B_1 \wedge B_2 \equiv A \not\rightarrow B_1 \underline{\vee} A \not\rightarrow B_2 \\ A \rightarrow B_1 \vee B_2 \Leftarrow A \rightarrow B_1 \underline{\vee} A \rightarrow B_2; & A \not\rightarrow B_1 \vee B_2 \equiv A \not\rightarrow B_1 \underline{\wedge} A \not\rightarrow B_2; \end{array}$$

where $\underline{\wedge}/\underline{\vee}$ is corresponding to $\wedge/\vee$ in semantics.

Interpreting $A \rightsquigarrow B$ as default $\frac{A:B}{B}$ in default logic (Li et al. 2017a; Reiter 1980), we have

$$\begin{array}{ll} \frac{A_1 \wedge A_2 : B}{B} \Leftarrow \frac{A_1 : B}{B} \underline{\vee} \frac{A_2 : B}{B} & \neg\frac{A_1 \wedge A_2 : B}{B} \equiv \neg\frac{A_1 : B}{B} \underline{\wedge} \neg\frac{A_2 : B}{B} \\ \frac{A_1 \vee A_2 : B}{B} \equiv \frac{A_1 : B}{B} \underline{\wedge} \frac{A_2 : B}{B} & \neg\frac{A_1 \vee A_2 : B}{B} \equiv \neg\frac{A_1 : B}{B} \underline{\vee} \neg\frac{A_2 : B}{B} \\ \frac{A : B_1 \wedge B_2}{B_1 \wedge B_2} \equiv \frac{A : B_1}{B_1} \underline{\wedge} \frac{A : B_2}{B_2} & \neg\frac{A : B_1 \wedge B_2}{B_1 \wedge B_2} \equiv \neg\frac{A : B_1}{B_1} \underline{\vee} \neg\frac{A : B_2}{B_2} \\ \frac{A : B_1 \vee B_2}{B_1 \wedge B_2} \Leftarrow \frac{A : B_1}{B_1} \underline{\vee} \frac{A : B_2}{B_2}; & \neg\frac{A : B_1 \vee B_2}{B_1 \wedge B_2} \equiv \neg\frac{A : B_1}{B_1} \underline{\wedge} \neg\frac{A : B_2}{B_2}. \end{array}$$

Interpreting $A \rightsquigarrow B$ as the subsumption statement $C \sqsubseteq D$ in description logics, we have

$$\begin{array}{ll} C_1 \sqcap C_2 \sqsubseteq D \Leftarrow C_1 \sqsubseteq D \underline{\vee} C_2 \sqsubseteq D & C_1 \sqcap C_2 \not\sqsubseteq D \equiv C_1 \not\sqsubseteq D \underline{\wedge} C_2 \not\sqsubseteq D \\ C_1 \sqcup C_2 \sqsubseteq D \equiv C_1 \sqsubseteq D \underline{\wedge} C_2 \sqsubseteq D & C_1 \sqcup C_2 \not\sqsubseteq D \equiv C_1 \not\sqsubseteq D \underline{\vee} C_2 \not\sqsubseteq D \\ C \sqsubseteq D_1 \sqcap D_2 \equiv C \sqsubseteq D_1 \underline{\wedge} C \sqsubseteq D_2 & C \not\sqsubseteq D_1 \sqcap D_2 \equiv C \not\sqsubseteq D_1 \underline{\vee} C \not\sqsubseteq D_2 \\ C \sqsubseteq D_1 \sqcup D_2 \Leftarrow C \sqsubseteq D_1 \underline{\vee} C \sqsubseteq D_2; & C \not\sqsubseteq D_1 \sqcup D_2 \equiv C \not\sqsubseteq D_1 \underline{\wedge} C \not\sqsubseteq D_2. \end{array}$$

For R-calculus of $\rightsquigarrow$-propositional logic, we consider three kinds of minimal changes:

(i) Subset-minimal ($\subseteq$-minimal) change.
(ii) Pseudo-subformula-minimal ($\preceq$-minimal) change, where $\preceq$ is the pseudo-subformula relation, just as the subformula relation $\leq$, where a formula A is a pseudo-subformula of B if eliminating some subformulas in B results in A.
(iii) Deduction-based minimal ($\vdash_{\preceq}$-minimal) change, where a theory Θ is a $\vdash_{\preceq}$-minimal change of Γ by γ (denoted by $\models_{\mathrm{U}} \Gamma|\gamma \Rightarrow \Theta$), if $\Theta \preceq \Gamma \cup \{\gamma\}$ is consistent, and for any theory Ξ with $\Theta \prec \Xi \preceq \Gamma, \gamma$ either $\Xi \vdash \Theta$ and $\Theta \vdash \Xi$, or Ξ is inconsistent.

For $* \in \{\mathtt{t}, \mathtt{f}\}$, there are tableau proof systems $\mathbf{T}^*_{\rightsquigarrow}/\mathbf{T}^{\rightsquigarrow}_*$, Gentzen deduction systems $\mathbf{G}^*_{\rightsquigarrow}/\mathbf{G}^{\rightsquigarrow}_*$ and R-calculi $\mathbf{R}^{\rightsquigarrow}_{\mathtt{t}}, \mathbf{Q}^{\rightsquigarrow}_{\mathtt{t}}, \mathbf{P}^{\rightsquigarrow}_{\mathtt{t}}$ for tableau proof systems and $\mathbf{R}_{\mathtt{t}}, \mathbf{Q}_{\mathtt{t}}, \mathbf{P}_{\mathtt{t}}$ for Gentzen deduction systems sound and complete with respect to $\subseteq$-, $\preceq$-,$\vdash$- and $\vdash_{\preceq}$-minimal changes, respectively. That is, for any theory Γ,

$$\vdash^*_{\rightsquigarrow} \Gamma \text{ iff } \models^*_{\rightsquigarrow} \Gamma$$
$$\vdash^{\rightsquigarrow}_* \Gamma \text{ iff } \models^{\rightsquigarrow}_* \Gamma,$$

and for any sequent $\Gamma \Rightarrow \Delta$/co-sequent $\Gamma \mapsto \Delta$,

$$\vdash^*_{\rightsquigarrow} \Gamma \Rightarrow \Delta \text{ iff } \models^*_{\rightsquigarrow} \Gamma \Rightarrow \Delta$$
$$\vdash^{\rightsquigarrow}_* \Gamma \mapsto \Delta \text{ iff } \models^{\rightsquigarrow}_* \Gamma \mapsto \Delta,$$

and for any $Y \in \{\mathbf{R}, \mathbf{Q}, \mathbf{P}\}$ and any reduction $\Gamma|A \rightrightarrows \Gamma'/\Gamma|A \rightrightarrows \Gamma'$,

$$\vdash^{\mathtt{t}}_Y \Gamma|A \rightrightarrows \Gamma' \text{ iff } \models^{\mathtt{t}}_Y \Gamma|A \rightrightarrows \Gamma'$$
$$\vdash^Y_{\mathtt{t}} \Gamma|A \rightrightarrows \Gamma' \text{ iff } \models^Y_{\mathtt{t}} \Gamma|A \rightrightarrows \Gamma',$$

and any reduction $\Gamma \Rightarrow \Delta|(A, B) \rightrightarrows \Gamma' \Rightarrow \Delta'/\Gamma \mapsto \Delta|(A, B) \rightrightarrows \Gamma' \mapsto \Delta'$,

$$\vdash^{\mathtt{t}}_Y \Gamma \Rightarrow \Delta|(A, B) \rightrightarrows \Gamma' \Rightarrow \Delta' \text{ iff } \models^{\mathtt{t}}_Y \Gamma \Rightarrow \Delta|(A, B) \rightrightarrows \Gamma' \Rightarrow \Delta'$$
$$\vdash^Y_{\mathtt{t}} \Gamma \mapsto \Delta|(A, B) \rightrightarrows \Gamma' \Rightarrow \Delta' \text{ iff } \models^Y_{\mathtt{t}} \Gamma \mapsto \Delta|(A, B) \rightrightarrows \Gamma' \Rightarrow \Delta'.$$

1.5.2 R-Calculi for Supersequents

A supersequent δ is of form $\Gamma|\Delta \Rightarrow \Sigma|\Pi$, where $\Gamma, \Delta, \Sigma, \Pi$ are sets of formulas. δ is valid, denoted by $\models^+ \delta$, if for any assignment v, either $v(A) = 0$ for some $A \in \Gamma$, or $v(B) = 1$ for some $B \in \Delta$, or $v(C) = 1$ for some $C \in \Sigma$, or $v(D) = 0$ for some $D \in \Pi$. That is, both each formula in Γ having truth-value 1 and each formula in Δ having truth-value 0 imply either some formula in Σ has truth-value 1 or some formula in Π has truth-value 0.

A co-supersequent δ is of form $\Gamma|\Delta \mapsto \Sigma|\Pi$. δ is valid, denoted by $\models_- \delta$, if there is an assignment v such that $v(A) = 1$ for each $A \in \Gamma$, $v(B) = 0$ for each $B \in \Delta$, $v(C) = 0$ for each $C \in \Sigma$ and $v(D) = 1$ for each $D \in \Pi$.

$\Gamma\|\Delta \Rightarrow \Sigma\|\Pi$	$\mathtt{f}\|\mathtt{t} \Rightarrow \mathtt{t}\|\mathtt{f}$
$\Gamma\|\Delta \mapsto \Sigma\|\Pi$	$\mathtt{t}\|\mathtt{f} \mapsto \mathtt{f}\|\mathtt{t}$

The validity of $\Gamma|\Delta \Rightarrow \Sigma|\Pi/\Gamma|\Delta \mapsto \Sigma|\Pi$ is equivalent to that of $\Gamma, \Pi \Rightarrow \Delta, \Sigma/\Gamma, \Pi \mapsto \Delta, \Sigma$.

There is a sound and complete Gentzen deduction system $\mathbf{G}^+/\mathbf{G}_-$ for supersequents/co-supersequents, respectively. That is, for any supersequent/co-supersequent δ,

$$\vdash^+ \Gamma|\Delta \Rightarrow \Sigma|\Pi \text{ iff } \models^+ \Gamma|\Delta \Rightarrow \Sigma|\Pi$$
$$\vdash_- \Gamma|\Delta \mapsto \Sigma|\Pi \text{ iff } \models_- \Gamma|\Delta \mapsto \Sigma|\Pi.$$

A supersequent δ is reduced to sequents:

$$\begin{array}{l} \Gamma \Rightarrow \Sigma,\ \mathbf{G}^{\mathtt{ft}} \\ \Gamma \Rightarrow \Pi,\ \mathbf{G}^{\mathtt{ff}} \\ \Delta \Rightarrow \Sigma,\ \mathbf{G}^{\mathtt{tt}} \\ \Delta \Rightarrow \Pi,\ \mathbf{G}^{\mathtt{tf}}, \end{array}$$

and A co-supersequent δ is reduced to co-sequents:

$$\begin{array}{l} \Gamma \mapsto \Sigma,\ \mathbf{G}_{\mathtt{tf}} \\ \Gamma \mapsto \Pi,\ \mathbf{G}_{\mathtt{tt}} \\ \Delta \mapsto \Sigma,\ \mathbf{G}_{\mathtt{ff}} \\ \Delta \mapsto \Pi,\ \mathbf{G}_{\mathtt{ft}}. \end{array}$$

Correspondingly there are sound and complete R-calculi $\mathbf{R}^+/\mathbf{R}_-$ for supersequents/co-supersequents, which are reduced to the following sound and complete R-calculi for sequents:

$$\begin{array}{l} \Gamma \Rightarrow \Sigma|(A, C) \rightrightarrows \Gamma' \Rightarrow \Sigma',\ \mathbf{R}^{\mathtt{ft}} \\ \Gamma \Rightarrow \Pi|(A, D) \rightrightarrows \Gamma' \Rightarrow \Pi',\ \mathbf{R}^{\mathtt{ff}} \\ \Delta \Rightarrow \Sigma|(B, C) \rightrightarrows \Delta' \Rightarrow \Sigma',\ \mathbf{R}^{\mathtt{tt}} \\ \Delta \Rightarrow \Pi|(B, D) \rightrightarrows \Delta' \Rightarrow \Pi',\ \mathbf{R}^{\mathtt{tf}}, \end{array}$$

and

$$\begin{array}{l} \Gamma \mapsto \Sigma|(A, C) \rightrightarrows \Gamma' \mapsto \Sigma',\ \mathbf{R}_{\mathtt{tf}} \\ \Gamma \mapsto \Pi|(A, D) \rightrightarrows \Gamma' \mapsto \Pi',\ \mathbf{R}_{\mathtt{tt}} \\ \Delta \mapsto \Sigma|(B, C) \rightrightarrows \Delta' \mapsto \Sigma',\ \mathbf{R}_{\mathtt{ff}} \\ \Delta \mapsto \Pi|(B, D) \rightrightarrows \Delta' \mapsto \Pi',\ \mathbf{R}_{\mathtt{ft}}. \end{array}$$

1.6 Notations

All the deduction systems and R-calculi in this book are listed in the following table:

- For theories we have the following systems:

Deduction systems	R-calculi
$\mathbf{T}^{\mathtt{t}}, \mathbf{T}^{\mathtt{f}}/\mathbf{T}_{\mathtt{t}}, \mathbf{T}_{\mathtt{f}}$	$\mathbf{S}^{\mathtt{t}}, \mathbf{S}^{\mathtt{f}}/\mathbf{S}_{\mathtt{t}}, \mathbf{S}_{\mathtt{f}}$
$\mathbf{T}^{\mathtt{t}}_{\rightsquigarrow}, \mathbf{T}^{\mathtt{f}}_{\rightsquigarrow}/\mathbf{T}^{\rightsquigarrow}_{\mathtt{t}}, \mathbf{T}^{\rightsquigarrow}_{\mathtt{f}}$	$\mathbf{R}^{\mathtt{t}}_{\rightsquigarrow}, \mathbf{R}^{\mathtt{f}}_{\rightsquigarrow}$
	$\mathbf{R}^{\rightsquigarrow}_{\mathtt{t}}, \mathbf{Q}^{\rightsquigarrow}_{\mathtt{t}}, \mathbf{P}^{\rightsquigarrow}_{\mathtt{t}}$

- For sequents we have the following systems:

Deduction systems	R-calculi
$\mathbf{G}^{\neg}, \mathbf{G}^{\wedge}, \mathbf{G}^{\vee}$	$\mathbf{R}^{\neg}, \mathbf{R}^{\wedge}, \mathbf{R}^{\vee}$
$\mathbf{G}_{\neg}, \mathbf{G}^{\wedge}, \mathbf{G}_{\vee}$	$\mathbf{R}_{\neg}, \mathbf{R}_{\wedge}, \mathbf{R}_{\vee}$
$\mathbf{G}^{\neg\to}, \mathbf{G}^{\oplus\otimes}$	$\mathbf{R}^{\neg\to}, \mathbf{R}^{\oplus\otimes}$
$\mathbf{G}_{\neg\to}, \mathbf{G}_{\oplus\otimes}$	$\mathbf{R}_{\neg\to}, \mathbf{R}_{\oplus\otimes}$
$\mathbf{G}^{\mathtt{t}}, \mathbf{G}^{\mathtt{f}}/\mathbf{G}_{\mathtt{t}}, \mathbf{G}_{\mathtt{f}}$	$\mathbf{R}^{\mathtt{t}}, \mathbf{R}^{\mathtt{f}}/\mathbf{R}_{\mathtt{t}}, \mathbf{R}_{\mathtt{f}}$
$\mathbf{G}^{Q_1Q_2}/\mathbf{G}_{Q_1Q_2}$	$\mathbf{R}^{Q_1Q_2}/\mathbf{R}_{Q_1Q_2}$
$\mathbf{G}^{Q_1iQ_2j}/\mathbf{G}_{Q_1iQ_2j}$	$\mathbf{R}^{Q_1iQ_2j}/\mathbf{R}_{Q_1iQ_2j}$
$\mathbf{G}^{Y_1Q_1iY_2Q_2j}/\mathbf{G}_{Y_1Q_1iY_2Q_2j}$	$\mathbf{R}^{Y_1Q_1iY_2Q_2j}/\mathbf{R}_{Y_1Q_1iY_2Q_2j}$
$\mathbf{G}^{+}/\mathbf{G}_{-}$	$\mathbf{R}^{+}/\mathbf{R}_{-}$
$\mathbf{G}^{\mathtt{ff}}, \mathbf{G}^{\mathtt{ff}}$	$\mathbf{R}^{\mathtt{ff}}, \mathbf{R}^{\mathtt{ff}}$
$\mathbf{G}_{\mathtt{ff}}, \mathbf{G}_{\mathtt{ff}}$	$\mathbf{R}_{\mathtt{ff}}, \mathbf{R}_{\mathtt{ff}}$
$\mathbf{G}^{\mathtt{t}}_{\leadsto}$	$\mathbf{R}_{\mathtt{t}}, \mathbf{Q}_{\mathtt{t}}, \mathbf{P}_{\mathtt{t}}$

Our notation is standard (W. Li 2010; Mendelson 1964; Takeuti 1987) and given as needed. We use Δ, Γ, Θ to denote theories; A, B, C formulas in propositional logic, and $\Delta|\Gamma$ a configuration which is equivalent to $\Gamma \circ \Delta$ in AGM's belief revision. In the meta-language, we use $\sim$, &, or, **A**, **E** to express $\neg, \wedge, \vee, \forall, \exists$, respectively.

We use $\dfrac{\left[\begin{array}{l}\delta_1\\ \delta_2\end{array}\right.}{\delta}$ to denote that δ_1 and δ_2 imply δ; and $\dfrac{\left\{\begin{array}{l}\delta_1\\ \delta_2\end{array}\right.}{\delta}$ to denote that δ_1 implies δ and δ_2 implies δ. In proof tree of completeness theorem, we use $\left[\begin{array}{l}\delta_1\\ \delta_2\end{array}\right.$ to denote that δ_1 and δ_2 are at a same subnode of the tree; and $\left\{\begin{array}{l}\delta_1\\ \delta_2\end{array}\right.$ to denote that δ_1 and δ_2 are at different subnodes of the tree.

Hence, rules

$$
(\wedge_1^L)\ \frac{\Gamma, A_1 \Rightarrow \Delta}{\Gamma, A_1 \wedge A_2 \Rightarrow \Delta} \quad (\wedge^R)\ \frac{\begin{array}{c}\Gamma \Rightarrow B_1, \Delta\\ \Gamma \Rightarrow B_2, \Delta\end{array}}{\Gamma \Rightarrow B_1 \wedge B_2, \Delta}
$$

$$
(\wedge_2^L)\ \frac{\Gamma, A_2 \Rightarrow \Delta}{\Gamma, A_1 \wedge A_2 \Rightarrow \Delta}
$$

in traditional Gentzen deduction system are represented in this book as following two rules

$$
(\vee^L)\ \frac{\left\{\begin{array}{l}\Gamma, A_1 \Rightarrow \Delta\\ \Gamma, A_2 \Rightarrow \Delta\end{array}\right.}{\Gamma, A_1 \wedge A_2 \Rightarrow \Delta} \quad (\vee^R)\ \frac{\left[\begin{array}{l}\Gamma \Rightarrow B_1, \Delta\\ \Gamma \Rightarrow B_2, \Delta\end{array}\right.}{\Gamma \Rightarrow B_1 \wedge B_2, \Delta}
$$

In a proof tree ξ of completeness theorem, $A_1 \wedge A_2$ and $B_1 \wedge B_2$ have the following subnodes, respectively:

$$\left[\begin{array}{l}\Gamma_1, A_1 \Rightarrow \Delta_1\\ \Gamma_1, A_2 \Rightarrow \Delta_1\end{array}\right. \text{if } \Gamma_1, A_1 \wedge A_2 \Rightarrow \Delta_1 \in \xi$$

$$\left\{\begin{array}{l}\Gamma_1 \Rightarrow B_1, \Delta_1\\ \Gamma_1 \Rightarrow B_2, \Delta_1\end{array}\right. \text{if } \Gamma_1 \Rightarrow B_1 \wedge B_2, \Delta_1 \in \xi$$

Let Γ, Δ be sets of literals. We define

$$\begin{aligned}\text{incon}(\Gamma) &\text{ iff } \mathbf{E}l(l, \neg l \in \Gamma)\\ \text{con}(\Gamma) &\text{ iff } \sim \mathbf{E}l(l, \neg l \in \Gamma)\\ \text{val}(\Delta) &\text{ iff } \mathbf{E}l(l, \neg l \in \Delta)\\ \text{inval}(\Delta) &\text{ iff } \sim \mathbf{E}l(l, \neg l \in \Delta).\end{aligned}$$

Notice that incon(Γ) iff val(Γ) and con(Γ) iff inval(Γ).

References

Alchourrón, C.E., Gärdenfors, P., Makinson, D.: On the logic of theory change: partial meet contraction and revision functions. J. Symbolic Logic **50**, 510–530 (1985)

Fermé, E., Hansson, S.O.: AGM 25 years, twenty-five years of research in belief change. J. Philos. Logic **40**, 295–331 (2011)

Li, W.: Mathematical logic, foundations for information science. In: Progress in Computer Science and Applied Logic, vol. 25, Birkhäuser (2010)

Li, W.: R-calculus: an inference system for belief revision. Comput. J. **50**, 378–390 (2007)

Li, W., Sui, Y.: The sound and complete R-calculi with respect to pseudo-revision and pre-revision. Int. J. Intell. Sci. **3**, 110–117 (2013)

Li, W., Sui, Y.: An **R**-calculus for the propositional logic programming. In: Proceedings of International Conference on Computer Science and Information Technology, pp. 863–870 (2014)

Li, W., Sui, Y.: The R-calculus and the finite injury priority method. J. Comput. **12**, 127–134 (2017)

Li, W., Sui, Y.: The B_4-valued propositional logic with unary logical connectives $\sim_1 / \sim_2 / \neg$. Front. Comput. Sci. **11**, 887–894 (2017)

Li, W., Sui, Y.: Multisequent Gentzen deduction systems for B_2^2-valued first-order logic. Artif. Intell. Res. **7**, 53 (2018)

Li, W., Sui, Y.: A computational framework for Karl Popper's logic of scientific discovery. Sci. China Inf. Sci. **61**, 042101:1–042101:10 (2018)

Li, W., Sui, Y., Wang, Y.: The propositional normal default logic and the finite/infinite injury priority method. Sci. China Inf. Sci. **60**, 092107 (2017)

Li, W., Sui, Y., Wang, Y.: The lattice-modalized propositional logic: distributivity and modularity of the Gentzen deduction system GL. J. Intell. Fuzzy Syst. **33**, 733–740 (2017)

Mendelson, E.: Introduction to Mathematical Logic. Wadsworth & Brooks/Cole Advanced Books & Software, Monterey, California (1964)

Reiter, R.: A logic for default reasoning. Artif. Intell. **13**, 81–132 (1980)

Takeuti, G.: Proof theory. In: Barwise, J. (ed.) Handbook of Mathematical Logic. Studies in Logic and the Foundations of Mathematics. North-Holland, Amsterdam, NL (1987)

Chapter 2
R-Calculus for Simplified Propositional Logics

Assume that the logical language of propositional logic contains three logical symbols: $\neg, \wedge, \vee$. There is a classical Gentzen deduction system **G** for propositional logic which is sound and complete with respect to classical semantics (Li 2010; Mendelson 1964; Takeuti 1987).

This chapter firstly will give three logical sublanguages of propositional logic, where each language contains only one connective; and correspondingly there are

- three weak propositional logics;
- three Gentzen deduction systems which are sound and complete with respect to the classical interpretations of the corresponding logical connectives, respectively; and
- three R-calculi which are sound and complete with respect to the $\subseteq$-minimal change (Li 2007; Li and Sui 2013).

Therefore, soundness and completeness are related only to the semantics of the concerned logics, no matter whether the logical language is complete with the semantics, such as the expressive completeness of $L = \{\neg, \wedge, \vee\}$ in classical propositional logic, where for any n-ary truth-value function $f : \{0, 1\}^n \to \{0, 1\}$, there is a formula A in L such that for any assignment v and any $a_1, \ldots, a_n \in \{0, 1\}$, if $v(p_1) = a_1, \ldots, v(p_n) = a_n$ then $f(a_1, \ldots, a_n) = v(A)$.

There are three ways to deduce valid formulas in a logic: axiomatic systems, natural deduction systems and Gentzen-type deduction systems, where the latter are based on the semantics of logical connectives in the logic to be formalized.

For each logical connective $x \in \{\neg, \wedge, \vee\}$, a Gentzen-type deduction system $\mathbf{G}^x$ will be given so that

- soundness theorem is true, that is, for any sequent $\Gamma \Rightarrow \Delta$, if $\vdash_{\mathbf{G}^x} \Gamma \Rightarrow \Delta$ then $\models_{\mathbf{G}^x} \Gamma \Rightarrow \Delta$; and
- completeness theorem is true, that is, for any sequent $\Gamma \Rightarrow \Delta$, if $\models_{\mathbf{G}^x} \Gamma \Rightarrow \Delta$ then $\vdash_{\mathbf{G}^x} \Gamma \Rightarrow \Delta$;

W. Li and Y. Sui, *R-Calculus, IV: Propositional Logic*,
Perspectives in Formal Induction, Revision and Evolution,
https://doi.org/10.1007/978-981-19-8633-8_2

and two R-calculi $\mathbf{R}^x$ and $\mathbf{R}_x$ are given which are sound and complete with respect to $\subseteq$-minimal change.

Moreover, we will give two logical languages of propositional logic, which contain

- logical connectives $\{\neg, \rightarrow\}$ and $\{\neg, \oplus, \otimes\}$; and
- two Gentzen deduction systems $\mathbf{G}^{\neg\rightarrow}$, $\mathbf{G}^{\oplus\otimes}$ are built which are sound and complete with respect to semantics of logical connectives, and
- four R-calculi $\mathbf{R}^{\neg\rightarrow}$, $\mathbf{R}^{\oplus\otimes}/\mathbf{R}_{\neg\rightarrow}$, $\mathbf{R}_{\oplus\otimes}$ are given which are sound and complete with respect to $\subseteq$-minimal change (Li and Sui 2014).

2.1 Weak Propositional Logics

We consider three weak propositional logics containing logical connective $\{\neg\}$, $\{\wedge\}$ and $\{\vee\}$, respectively, and give sound and complete Gentzen deduction systems $\mathbf{G}^{\neg}, \mathbf{G}^{\wedge}, \mathbf{G}^{\vee}/\mathbf{G}_{\neg}, \mathbf{G}_{\wedge}, \mathbf{G}_{\vee}$ and R-calculi $\mathbf{R}^{\neg}, \mathbf{R}^{\wedge}, \mathbf{R}^{\vee}/\mathbf{R}_{\neg}, \mathbf{R}_{\wedge}, \mathbf{R}_{\vee}$.

The logical language of weak propositional logic $\mathbf{G}^x$ contains the following symbols:

- variables: $p_0, p_1, \ldots$;
- unary connective: x.

Formulas are defined as follows:

$$A ::= p|xA/A_1xA_2.$$

Let v be an assignment, a function from variables to $\{0, 1\}$.

The interpretation $v(A)$ of formulas A in v is:

$$v(A) = \begin{cases} v(p) & \text{if } A = p \\ 1 - v(A_1) & \text{if } A = \neg A_1 \text{ and } x = \neg \\ \min\{v(A_1), v(A_2)\} & \text{if } A = A_1 \wedge A_2 \text{ and } x = \wedge \\ \max\{v(A_1), v(A_2)\} & \text{if } A = A_1 \vee A_2 \text{ and } x = \vee \end{cases}$$

and A is satisfied in v, denoted by $v \models A$, if $v(A) = 1$.

2.1.1 *Gentzen Deduction System* $\mathbf{G}^{\neg}$

Given a sequent $\delta = \Gamma \Rightarrow \Delta$, we say that v satisfies δ, denoted by $v \models^{\neg} \delta$, if $v \models \Gamma$ implies $v \models \Delta$, where $v \models \Gamma$ if for each $A \in \Gamma$, $v \models A$; and $v \models \Delta$ if for some $B \in \Delta$, $v \models B$.

A sequent δ is valid, denoted by $\models^{\neg} \delta$, if for any assignment v, $v \models^{\neg} \delta$.

Gentzen deduction system $\mathbf{G}^{\neg}$ contains the following axiom and deduction rules.

- **Axiom**:

$$(\mathrm{A}^{\neg})\ \frac{\Gamma \cap \Delta \neq \emptyset}{\Gamma \Rightarrow \Delta},$$

where Γ, Δ are sets of variables.
- **Deduction rules:**

$$(\neg^L)\ \frac{\Gamma \Rightarrow A, \Delta}{\Gamma, \neg A \Rightarrow \Delta}\ (\neg^R)\ \frac{\Gamma, B \Rightarrow \Delta}{\Gamma \Rightarrow \neg B, \Delta}.$$

Definition 2.1.1 A sequent $\Gamma \Rightarrow \Delta$ is provable in $\mathbf{G}^{\neg}$, denoted by $\vdash^{\neg} \Gamma \Rightarrow \Delta$, if there is a sequence $\{\Gamma_1 \Rightarrow \Delta_1, \ldots, \Gamma_n \Rightarrow \Delta_n\}$ of sequents such that $\Gamma_n \Rightarrow \Delta_n = \Gamma \Rightarrow \Delta$, and for each $1 \leq i \leq n$, $\Gamma_i \Rightarrow \Delta_i$ is an axiom or is deduced from the previous sequents by one of deduction rules in $\mathbf{G}^{\neg}$.

Theorem 2.1.2 (Soundness theorem) *Given a sequent* $\Gamma \Rightarrow \Delta$, *if* $\vdash^{\neg} \Gamma \Rightarrow \Delta$ *then* $\models^{\neg} \Gamma \Rightarrow \Delta$.

Proof We prove that each axiom is valid and each deduction rule preserves the validity.

$(\mathrm{A}^{\neg})$ Assume that $p \in \Gamma \cap \Delta$. Then, for any assignment v, either $v(p) = 0$ or $v(p) = 1$. Hence, $v \models \Gamma$ or $v \models \Delta$.

$(\neg^L)$ Assume that for any assignment v, $v \models \Gamma$ implies $v \models A, \Delta$. For any assignment v, assume that $v \models \neg A, \Gamma$. Then, $v \models \Gamma$, and by induction assumption, $v \models A, \Delta$. Since $v \not\models A$, $v \models \Delta$.

$(\neg^R)$ Assume that for any assignment v, $v \models \Gamma, B$ implies $v \models \Delta$. For any assignment v, assume that $v \models \Gamma$. If $v \not\models B$ then $v \models \neg B$, and $v \models \neg B, \Delta$; otherwise, $v \models \Gamma, B$ and by induction assumption, $v \models \Delta$, and hence $v \models \neg B, \Delta$. □

Theorem 2.1.3 (Completeness theorem) *Given a sequent* $\Gamma \Rightarrow \Delta$, *if* $\models^{\neg} \Gamma \Rightarrow \Delta$ *then* $\vdash^{\neg} \Gamma \Rightarrow \Delta$.

Proof Let $\delta = \Gamma \Rightarrow \Delta$. Define a linear tree, called the reduction tree for δ, denoted by $T(\delta)$, from which we can obtain either a proof of δ or a show of the nonvalidity of δ.

This reduction tree $T(\delta)$ for δ contains a sequent at each node and is constructed in stages as follows.

Stage 0: $T_0(\delta) = \{\delta\}$.

Stage $k (k > 0)$: $T_k(\delta)$ is defined by cases.

Case 0. If Γ, Δ have no logical connectives in them, write nothing above $\Gamma \Rightarrow \Delta$.

Case 1. Every topmost sequent $\Gamma \Rightarrow \Delta$ in $T_{k-1}(\delta)$ has a common formula in Γ and Δ. Then, stop.

Case 2. Not case 1. $T_k(\delta)$ is defined as follows. Let $\Gamma \Rightarrow \Delta$ be any topmost sequent of the tree which has been defined by stage $k - 1$.

Subcase $(\neg^L)$. Let $\neg A$ be the formula in Γ whose outmost logical symbol is $\neg$, and to which no reduction has been applied in previous stages. Then, write down

$$\Gamma \Rightarrow A, \Delta$$

above $\Gamma \Rightarrow \Delta$. We say that a $(\neg^L)$ reduction has been applied to $\neg A$.

Subcase $(\neg^R)$. Let $\neg B$ be the formula in Δ whose outmost logical symbol is $\neg$, and to which no reduction has been applied in previous stages. Then, write down

$$\Gamma, B \Rightarrow \Delta$$

above $\Gamma \Rightarrow \Delta$. We say that a $(\neg^R)$ reduction has been applied to $\neg B$.

Given a sequent δ, if $T(\delta)$ is ended with a sequent containing common formulas, then it is a routine to construct a proof of δ.

Otherwise, $T(\delta) = \delta_1, \ldots, \delta_n$, there is no rule applicable for δ_n and $\delta_n = \Gamma_n \Rightarrow \Delta_n$ has no common formulas in Γ_n and Δ_n.

Let

$$\cup\Gamma = \{A \in \Gamma_i : \Gamma_i \Rightarrow \Delta_i \in T(\delta)\},$$
$$\cup\Delta = \{B \in \Delta_i : \Gamma_i \Rightarrow \Delta_i \in T(\delta)\}.$$

We define an assignment in which every statement $A \in \cup\Gamma$ is true and every statement in $B \in \cup\Delta$ is false.

Define v such that for any variable p,

$$v(p) = 1 \text{ iff } p \in \cup\Gamma.$$

By induction on the structure of formulas, we prove that if $A \in \cup\Gamma$ then $v \models A$; and if $B \in \cup\Delta$ then $v \not\models B$.

Case $A = \neg A_1 \in \cup\Gamma$. Assume that $\delta' \in T(\delta)$ is the first one such that $A \in \delta'$. Then, there are a $\delta'' \in T(\delta)$ and formula sets Γ, Δ' such that $\delta' = \Gamma' \Rightarrow \neg A_1, \Delta'$ and $\delta'' = \Gamma', A \Rightarrow \Delta'$. By induction assumption, $v \not\models \Gamma', A \Rightarrow \Delta$, and so $v \not\models \Gamma' \Rightarrow \neg A, \Delta'$.

Case $B = \neg B_1 \in \cup\Delta$. Assume that $\delta' \in T(\delta)$ is the first one such that $B \in \delta'$. Then, there are a $\delta'' \in T(\delta)$ and formula sets Γ', Δ' such that $\delta' = \Gamma' \Rightarrow \neg B_1, \Delta'$ and $\delta'' = \Gamma', B \Rightarrow \Delta'$. By induction assumption, $v \not\models \Gamma', B \Rightarrow \Delta'$, and $v \not\models \Gamma' \Rightarrow \neg B, \Delta'$.

This completes the proof. □

Notice that each formula A is of form $(\neg)^i p$, where p is a variable.

2.1.2 *Gentzen Deduction System* $\mathbf{G}_\neg$

Given a co-sequent $\delta = \Gamma \mapsto \Delta$, we say that δ is valid, denoted by $v \models_\neg \delta$, if there is an assignment v such that for each $A \in \Gamma$, $v(A) = 1$, and for each $B \in \Delta$, $v(B) = 0$.

Gentzen deduction system $\mathbf{G}_\neg$ contains the following axiom and deduction rules.

- **Axiom**:

$$(\mathtt{A}_{\neg})\ \frac{\Gamma \cap \Delta = \emptyset}{\Gamma \mapsto \Delta},$$

where Γ, Δ are sets of variables.

- **Deduction rules**:

$$(\neg^L)\ \frac{\Gamma \mapsto A, \Delta}{\Gamma, \neg A \mapsto \Delta}\ (\neg^R)\ \frac{\Gamma, B \mapsto \Delta}{\Gamma \mapsto \neg B, \Delta}$$

Definition 2.1.4 A co-sequent $\Gamma \mapsto \Delta$ is provable in $\mathbf{G}_{\neg}$, denoted by $\vdash_{\neg} \Gamma \mapsto \Delta$, if there is a sequence $\{\Gamma_1 \mapsto \Delta_1, \ldots, \Gamma_n \mapsto \Delta_n\}$ of co-sequents such that $\Gamma_n \mapsto \Delta_n = \Gamma \mapsto \Delta$, and for each $1 \le i \le n$, $\Gamma_i \mapsto \Delta_i$ is an axiom or is deduced from the previous co-sequents by one of the deduction rules in $\mathbf{G}_{\neg}$.

Theorem 2.1.5 (Soundness and completeness theorem) *For any co-sequent* $\Gamma \mapsto \Delta$,

$$\vdash_{\neg} \Gamma \mapsto \Delta \textit{ iff } \models_{\neg} \Gamma \mapsto \Delta.$$

□

2.1.3 Gentzen Deduction System $\mathbf{G}^{\wedge}$

A sequent δ is valid, denoted by $\models^{\wedge} \delta$, if for any assignment v, $v \models \Gamma$ implies $v \models \Delta$, where $v \models \Gamma$ if for each $A \in \Gamma$, $v(A) = 1$; and $v \models \Delta$ if for some $B \in \Delta$, $v(B) = 1$.

Gentzen deduction system $\mathbf{G}^{\wedge}$ contains the following

- **Axiom**:

$$(\mathtt{A}^{\wedge})\ \frac{\Gamma \cap \Delta \neq \emptyset}{\Gamma \Rightarrow \Delta},$$

where Γ, Δ are sets of variables.

- **Deduction rules**:

$$(\wedge^L)\ \frac{\begin{cases}\Gamma, A_1 \Rightarrow \Delta \\ \Gamma, A_2 \Rightarrow \Delta\end{cases}}{\Gamma, A_1 \wedge A_2 \Rightarrow \Delta}\ (\wedge^R)\ \frac{\begin{bmatrix}\Gamma \Rightarrow B_1, \Delta \\ \Gamma \Rightarrow B_2, \Delta\end{bmatrix}}{\Gamma \Rightarrow B_1 \wedge B_2, \Delta}$$

where $\frac{\begin{cases}\delta_1 \\ \delta_2\end{cases}}{\delta}$ means that either δ_1 or δ_2 implies δ; and $\frac{\begin{bmatrix}\delta_1 \\ \delta_2\end{bmatrix}}{\delta}$ means that δ_1 and δ_2 imply δ.

Definition 2.1.6 A sequent $\Gamma \Rightarrow \Delta$ is provable in $\mathbf{G}^{\wedge}$, denoted by $\vdash^{\wedge} \Gamma \Rightarrow \Delta$, if there is a sequence $\{\Gamma_1 \Rightarrow \Delta_1, \ldots, \Gamma_n \Rightarrow \Delta_n\}$ of sequents such that $\Gamma_n \Rightarrow \Delta_n = \Gamma \Rightarrow \Delta$, and for each $1 \le i \le n$, $\Gamma_i \Rightarrow \Delta_i$ is an axiom or is deduced from the previous sequents by one of the deduction rules in $\mathbf{G}^{\wedge}$.

Theorem 2.1.7 (Soundness theorem) *For any sequent* $\Gamma \Rightarrow \Delta$, *if* $\vdash^{\wedge} \Gamma \Rightarrow \Delta$ *then* $\models^{\wedge} \Gamma \Rightarrow \Delta$.

Proof We prove that each axiom is valid and each deduction rule preserves the validity.

($\mathtt{A}^{\wedge}$) Assume that $p \in \Gamma \cap \Delta$. Then, for any assignment v, either $v(p) = 0$ or $v(p) = 1$, i.e., either $v \models \Gamma$ or $v \models \Delta$.

($\wedge^L$) Assume that for any assignment v, $v \models \Gamma, A_1$ implies $v \models \Delta$. For any assignment v, assume that $v \models A_1 \wedge A_2, \Gamma$. Then, $v \models A_1, \Gamma$, and by induction assumption, $v \models \Delta$.

($\wedge^R$) Assume that for any assignment v, $v \models \Gamma$ implies $v \models B_1, \Delta$; and $v \models \Gamma$ implies $v \models B_2, \Delta$. For any assignment v, assume that $v \models \Gamma$. If $v \not\models B_1$ or $v \not\models B_2$ then $v \models \Delta$, which implies $v \models B_1 \wedge B_2, \Delta$; otherwise, $v \models B_1 \wedge B_2$, and $v \models B_1 \wedge B_2, \Delta$. □

Theorem 2.1.8 (Completeness theorem) *For any sequent* $\Gamma \Rightarrow \Delta$, *if* $\models^{\wedge} \Gamma \Rightarrow \Delta$ *then* $\vdash^{\wedge} \Gamma \Rightarrow \Delta$.

Proof Given a sequent $\Delta \Rightarrow \Gamma$, we construct a tree T such that either

(i) For each branch ξ of T, a sequent $\Delta' \Rightarrow \Gamma'$ at the leaf of ξ is an axiom, or
(ii) There is an assignment v such that $v \not\models \Delta \Rightarrow \Gamma$.

T is constructed as follows:

- The root of T is $\Delta \Rightarrow \Gamma$;
- For a node ξ, if each sequent $\Delta' \Rightarrow \Gamma'$ at ξ is atomic then the node is a leaf.
- Otherwise, ξ has the direct children nodes containing the following sequents:

$$\begin{cases} \Gamma_1, A_1, A_2 \Rightarrow \Delta_1 & \text{if } \Gamma_1, A_1 \wedge A_2 \Rightarrow \Delta_1 \in \xi \\ \begin{cases} \Gamma_1 \Rightarrow B_1, \Delta_1 \\ \Gamma_1 \Rightarrow B_2, \Delta_1 \end{cases} & \text{if } \Gamma_1 \Rightarrow B_1 \wedge B_2, \Delta_1 \in \xi \end{cases}$$

where $\left[\begin{matrix}\delta_1 \\ \delta_2\end{matrix}\right.$ means that δ_1 and δ_2 are at a same subnode of the tree; and $\left\{\begin{matrix}\delta_1 \\ \delta_2\end{matrix}\right.$ means that δ_1 and δ_2 are at different subnodes of the tree. Sometimes we use $\left[\begin{matrix}\Gamma_1, A_1 \Rightarrow \Delta_1 \\ \Gamma_1, A_2 \Rightarrow \Delta_1\end{matrix}\right.$ to denote $\Gamma_1, A_1, A_2 \Rightarrow \Delta_1$.

Lemma 2.1.9 *If for each branch* $\xi \subseteq T$, *there is a sequent* $\Delta' \Rightarrow \Gamma' \in \xi$ *which is an axiom in* $\mathbf{G}^{\wedge}$ *then* T *is a proof tree of* $\Delta \Rightarrow \Gamma$.

Proof By the definition of T, T is a proof tree of $\Delta \Rightarrow \Gamma$. □

Lemma 2.1.10 *If there is a branch* $\xi \subseteq T$ *such that each sequent* $\Delta' \Rightarrow \Gamma' \in \xi$ *is not an axiom in* $\mathbf{G}^{\wedge}$ *then there is an assignment* v *such that* $v \not\models \Delta \Rightarrow \Gamma$.

Proof Let γ be the set of all the atomic sequents in ξ.

Let

$$\Theta^L = \bigcup_{\Gamma' \Rightarrow \Delta' \in \gamma} \Gamma',$$
$$\Theta^R = \bigcup_{\Gamma' \Rightarrow \Delta' \in \gamma} \Delta'.$$

Define an assignment v as follows: for any variable p,

$$v(p) = \begin{cases} 1 \text{ if } p \in \Theta^L \\ 0 \text{ otherwise.} \end{cases}$$

Then, v is well-defined and $v \not\models \Gamma'' \Rightarrow \Delta''$, where $\Gamma'' \Rightarrow \Delta''$ is the sequent at the leaf node of ξ.

We proved by induction on ξ that each sequent $\Delta' \Rightarrow \Gamma' \in \xi$ is not satisfied by v.

Case $\Gamma' \Rightarrow \Delta' = \Gamma_2, A_1 \wedge A_2 \Rightarrow \Delta' \in \xi$. Then $\Gamma' \Rightarrow \Delta'$ has a child node $\in \xi$ containing sequent $\Gamma_2, A_1, A_2 \Rightarrow \Delta'$. By induction assumption, $v \not\models \Gamma_2, A_1, A_2 \Rightarrow \Delta'$. Hence, $v \not\models \Gamma_2, A_1 \wedge A_2 \Rightarrow \Delta'$.

Case $\Gamma' \Rightarrow \Delta' = \Gamma' \Rightarrow B_1 \wedge B_2, \Delta_2 \in \xi$. Then $\Gamma' \Rightarrow \Delta'$ has a child node $\in \xi$ containing sequent $\Gamma' \Rightarrow B_i, \Delta_2$. By induction assumption, $v \not\models \Gamma' \Rightarrow B_i, \Delta_2$. Hence, $v \not\models \Gamma' \Rightarrow B_1 \wedge B_2, \Delta_2$. □

Notice that each formula A is of form $p_1 \wedge \cdots \wedge p_n$.

2.1.4 Gentzen Deduction System $\mathbf{G}_\wedge$

Given a co-sequent $\delta = \Gamma \mapsto \Delta$, we say that δ is valid, denoted by $\models_\wedge \delta$, if there is an assignment v such that for each $A \in \Gamma$, $v(A) = 1$, and for each $B \in \Delta$, $v(B) = 0$.

Gentzen deduction system $\mathbf{G}_\wedge$ contains the following

- **Axiom**:

$$(\mathtt{A}_\wedge)\ \frac{\Gamma \cap \Delta = \emptyset}{\Gamma \mapsto \Delta},$$

  where Γ, Δ are sets of variables.
- **Deduction rules**:

$$(\wedge^L)\ \frac{\left[\begin{array}{l}\Gamma, A_1 \mapsto \Delta \\ \Gamma, A_2 \mapsto \Delta\end{array}\right.}{\Gamma, A_1 \wedge A_2 \mapsto \Delta} \quad (\wedge^R)\ \frac{\left\{\begin{array}{l}\Gamma \mapsto B_1, \Delta \\ \Gamma \mapsto B_2, \Delta\end{array}\right.}{\Gamma \mapsto B_1 \wedge B_2, \Delta}$$

Definition 2.1.11 A co-sequent $\Gamma \mapsto \Delta$ is provable in $\mathbf{G}_\wedge$, denoted by $\vdash_\wedge \Gamma \mapsto \Delta$, if there is a sequence $\{\Gamma_1 \mapsto \Delta_1, \ldots, \Gamma_n \mapsto \Delta_n\}$ of co-sequents such that $\Gamma_n \mapsto \Delta_n = \Gamma \mapsto \Delta$, and for each $1 \leq i \leq n$, $\Gamma_i \mapsto \Delta_i$ is an axiom or is deduced from the previous co-sequents by one of the deduction rules in $\mathbf{G}_\wedge$.

Theorem 2.1.12 (Soundness and completeness theorem) *For any co-sequent* $\Gamma \mapsto \Delta$,

$$\vdash_\wedge \Gamma \mapsto \Delta \textit{ if and only if } \models_\wedge \Gamma \mapsto \Delta.$$ □

2.1.5 Gentzen Deduction System $\mathbf{G}^{\vee}$

A sequent $\delta = \Gamma \Rightarrow \Delta$ is $\mathbf{G}^{\vee}$-valid, denoted by $\models^{\vee} \delta$, if for any assignment v, either $v \models \Gamma$ or $v \models \Delta$, where $v \models \Gamma$ if for some $A \in \Gamma$, $v(A) = 0$; and $v \models \Delta$ if for some $B \in \Delta$, $v(B) = 1$.

Gentzen deduction system $\mathbf{G}^{\vee}$ contains the following

- **Axiom**:

$$(\mathtt{A}^{\vee}) \frac{\Gamma \cap \Delta \neq \emptyset}{\Gamma \Rightarrow \Delta},$$

where Γ, Δ are sets of variables.

- **Deduction rules**:

$$(\vee^L) \frac{\left[\begin{array}{l}\Gamma, A_1 \Rightarrow \Delta \\ \Gamma, A_2 \Rightarrow \Delta\end{array}\right.}{\Gamma, A_1 \vee A_2 \Rightarrow \Delta} \quad (\vee^R) \frac{\left\{\begin{array}{l}\Gamma \Rightarrow B_1, \Delta \\ \Gamma \Rightarrow B_2, \Delta\end{array}\right.}{\Gamma \Rightarrow B_1 \vee B_2, \Delta.}$$

Definition 2.1.13 A sequent $\Gamma \Rightarrow \Delta$ is provable in $\mathbf{G}^{\vee}$, denoted by $\vdash^{\vee} \Gamma \Rightarrow \Delta$, if there is a sequence $\{\Gamma_1 \Rightarrow \Delta_1, \ldots, \Gamma_n \Rightarrow \Delta_n\}$ of sequents such that $\Gamma_n \Rightarrow \Delta_n = \Gamma \Rightarrow \Delta$, and for each $1 \leq i \leq n$, $\Gamma_i \Rightarrow \Delta_i$ is an axiom or is deduced from the previous sequents by one of the deduction rules in $\mathbf{G}^{\vee}$.

Theorem 4.2 (Soundness theorem) For any sequent $\Gamma \Rightarrow \Delta$, if $\vdash^{\vee} \Gamma \Rightarrow \Delta$ then $\models^{\vee} \Gamma \Rightarrow \Delta$.

Proof We prove that each axiom is valid, and each deduction rule preserves the validity.

$(\mathtt{A}^{\vee})$ Assume that $p \in \Gamma \cap \Delta \neq \emptyset$. Then, for any assignment v, either $v(p) = 0$ or $v(p) = 1$, i.e., either $v \models \Gamma$, or $v \models \Delta$.

$(\vee^L)$ Assume that for any assignment v, $v \models \Gamma, A_1$ implies $v \models \Delta$; and $v \models A_2, \Gamma$ implies $v \models \Delta$. Then, for any assignment v, assume that $v \models \Gamma, A_1 \vee A_2$. Then, $v \models A_1 \vee A_2$. If $v \models A_1$ then $v \models A_1, \Gamma$, and by induction assumption, $v \models \Delta$; and if $v \models A_2$ then $v \models A_2, \Gamma$, and by induction assumption, $v \models \Delta$.

$(\vee^R)$ Assume that for any assignment v, $v \models \Gamma$ implies $v \models B_i, \Delta$. Then, for any assignment v, assume that $v \models \Gamma$. Then, $v \models B_i, \Delta$, and $v \models B_1 \vee B_2, \Delta$. □

Theorem 2.1.14 (Completeness theorem) *For any sequent* $\Gamma \Rightarrow \Delta$, *if* $\models^{\vee} \Gamma \Rightarrow \Delta$ *then* $\vdash^{\vee} \Gamma \Rightarrow \Delta$. □

Proof Given a sequent $\Delta \Rightarrow \Gamma$, we construct a tree T such that either

(i) For each branch ξ of T, a sequent $\Delta' \Rightarrow \Gamma'$ at the leaf of ξ is an axiom.
(ii) There is an assignment v such that $v \not\models \Delta \Rightarrow \Gamma$.

T is constructed as follows:

- The root of T is $\Delta \Rightarrow \Gamma$.
- For a node ξ, if each sequent $\Delta' \Rightarrow \Gamma'$ at ξ is atomic then the node is a leaf.
- Otherwise, ξ has the direct children nodes containing the following sequents:

$$\left\{\begin{array}{ll}\left\{\begin{array}{l}\Gamma_1, A_1 \Rightarrow \Delta_1 \\ \Gamma_1, A_2 \Rightarrow \Delta_1\end{array}\right. & \text{if } \Gamma_1, A_1 \vee A_2 \Rightarrow \Delta_1 \in \xi \\ \left[\begin{array}{l}\Gamma_1 \Rightarrow B_1, \Delta_1 \\ \Gamma_1 \Rightarrow B_2, \Delta_1\end{array}\right. & \text{if } \Gamma_1 \Rightarrow B_1 \vee B_2, \Delta_1 \in \xi\end{array}\right.$$

Lemma 2.1.15 *If for each branch $\xi \subseteq T$, there is a sequent $\Delta' \Rightarrow \Gamma' \in \xi$ which is an axiom in $\mathbf{G}^{\wedge}$ then T is a proof tree of $\Delta \Rightarrow \Gamma$.*

Proof By the definition of T, T is a proof tree of $\Delta \Rightarrow \Gamma$. □

Lemma 2.1.16 *If there is a branch $\xi \subseteq T$ such that each sequent $\Delta' \Rightarrow \Gamma' \in \xi$ is not an axiom in $\mathbf{G}^{\wedge}$ then there is an assignment v such that $v \not\models \Delta \Rightarrow \Gamma$.*

Proof Let γ be the set of all the atomic sequents in ξ.

Let

$$\begin{array}{l}\Theta^L = \bigcup_{\Gamma' \Rightarrow \Delta' \in \gamma} \Gamma', \\ \Theta^R = \bigcup_{\Gamma' \Rightarrow \Delta' \in \gamma} \Delta'.\end{array}$$

Define an assignment v as follows: for any variable p,

$$v(p) = \begin{cases} 1 \text{ if } p \in \Theta^L \\ 0 \text{ otherwise.}\end{cases}$$

Then, v is well-defined and $v \not\models \Gamma'' \Rightarrow \Delta''$, where $\Gamma'' \Rightarrow \Delta''$ is the sequent at the leaf node of ξ.

We proved by induction on ξ that each sequent $\Delta' \Rightarrow \Gamma' \in \xi$ is not satisfied by v.

Case $\Gamma' \Rightarrow \Delta' = \Gamma_2, A_1 \vee A_2 \Rightarrow \Delta' \in \xi$. Then $\Gamma' \Rightarrow \Delta'$ has children nodes $\in \xi$ containing sequent $\Gamma_2, A_i \Rightarrow \Delta'$. By induction assumption, $v \not\models \Gamma_2, A_i \Rightarrow \Delta'$. Hence, $v \not\models \Gamma_2, A_1 \vee A_2 \Rightarrow \Delta'$.

Case $\Gamma' \Rightarrow \Delta' = \Gamma' \Rightarrow B_1 \vee B_2, \Delta_2 \in \xi$. Then $\Gamma' \Rightarrow \Delta'$ has children nodes $\in \xi$ containing sequent $\Gamma' \Rightarrow B_1, B_2, \Delta_2$. By induction assumption, $v \not\models \Gamma' \Rightarrow B_1, B_2, \Delta_2$. Hence, $v \not\models \Gamma' \Rightarrow B_1 \vee B_2, \Delta_2$. □

Notice that each formula A is of form $p_1 \vee \cdots \vee p_n$.

2.1.6 Gentzen Deduction System $\mathbf{G}_{\vee}$

Given a co-sequent $\delta = \Gamma \mapsto \Delta$, we say that δ is $\mathbf{G}_{\vee}$-valid, denoted by $\models_{\vee} \delta$, if there is an assignment v such that for each $A \in \Gamma$, $v(A) = 1$, and for each $B \in \Delta$, $v(B) = 0$.

Gentzen deduction system $\mathbf{G}_\vee$ contains the following

- **Axiom**:

$$(\mathbb{A}_\vee)\ \frac{\Gamma \cap \Delta = \emptyset}{\Gamma \mapsto \Delta},$$

where Γ, Δ are sets of variables.
- **Deduction rules**:

$$(\vee^L)\ \frac{\begin{cases}\Gamma, A_1 \mapsto \Delta\\ \Gamma, A_2 \mapsto \Delta\end{cases}}{\Gamma, A_1 \vee A_2 \mapsto \Delta}\quad (\vee^R)\ \frac{\left[\begin{array}{l}\Gamma \mapsto B_1, \Delta\\ \Gamma \mapsto B_2, \Delta\end{array}\right.}{\Gamma \mapsto B_1 \vee B_2, \Delta}$$

Definition 2.1.17 A co-sequent $\Gamma \mapsto \Delta$ is provable in $\mathbf{G}_\vee$, denoted by $\vdash_\vee \Gamma \mapsto \Delta$, if there is a sequence $\{\Gamma_1 \mapsto \Delta_1, \ldots, \Gamma_n \mapsto \Delta_n\}$ of co-sequents such that $\Gamma_n \mapsto \Delta_n = \Gamma \mapsto \Delta$, and for each $1 \leq i \leq n$, $\Gamma_i \mapsto \Delta_i$ is an axiom or is deduced from the previous co-sequents by one of the deduction rules in $\mathbf{G}_\vee$.

Theorem 2.1.18 (Soundness and completeness theorem) *For any co-sequent* $\Gamma \mapsto \Delta$,

$$\vdash_\vee \Gamma \mapsto \Delta \text{ if and only if } \models_\vee \Gamma \mapsto \Delta.$$

□

2.2 R-Calculus for Weak Propositional Logics

Correspondingly we have three R-calculi $\mathbf{R}^\neg, \mathbf{R}^\wedge, \mathbf{R}^\vee$ for sequents and $\mathbf{R}_\neg$, **ikm**, $\mathbf{R}_\wedge, \mathbf{R}_\vee$ for co-sequents.

Let $x \in \{\neg, \wedge, \vee\}$.

A sequent $\Gamma \Rightarrow \Delta$ revises a formula pair (A, B), where $A \in \Gamma$ and $B \in \Delta$, and obtains a sequent $\Gamma' \Rightarrow \Delta'$, denoted by $\models^x \Gamma \Rightarrow \Delta|(A, B) \rightrightarrows \Gamma' \Rightarrow \Delta'$, if

$$\Gamma' = \begin{cases}\Gamma - \{A\} & \text{if } \models^x \Gamma - \{A\} \Rightarrow \Delta\\ \Gamma & \text{otherwise;}\end{cases}$$
$$\Delta' = \begin{cases}\Delta - \{B\} & \text{if } \models^x \Gamma' \Rightarrow \Delta - \{B\}\\ \Gamma & \text{otherwise.}\end{cases}$$

A co-sequent $\Gamma \mapsto \Delta$ revises a formula pair (A, B) and obtains a co-sequent $\Gamma' \mapsto \Delta'$, denoted by $\models_x \delta = (A, B)|\Gamma \mapsto \Delta \rightrightarrows \Gamma' \mapsto \Delta'$, if

$$\Gamma' = \begin{cases}\Gamma, A & \text{if } \models^x \Gamma, A \mapsto \Delta\\ \Gamma & \text{otherwise;}\end{cases}$$
$$\Delta' = \begin{cases}\Delta, B & \text{if } \models^x \Gamma' \mapsto \Delta, B\\ \Gamma & \text{otherwise.}\end{cases}$$

2.2.1 *R-Calculus* $\mathbf{R}^{\neg}$

Let $A \in \Gamma$ and $B \in \Delta$. A reduction δ is valid, denoted by $\models^{\neg} \delta$, if

$$\Gamma' = \begin{cases} \Gamma - \{A\} & \text{if } \models^{\neg} \Gamma - \{A\} \Rightarrow \Delta \\ \Gamma & \text{otherwise} \end{cases}$$
$$\Delta' = \begin{cases} \Delta - \{B\} & \text{if } \models^{\neg} \Gamma' \Rightarrow \Delta - \{B\} \\ \Delta & \text{otherwise.} \end{cases}$$

R-calculus $\mathbf{R}^{\neg}$ consists of the following axioms and deduction rules: let $A \in \Gamma$ and $B \in \Delta$.

- **Axioms**:

$$(\mathrm{A}_0^L)\ \frac{\sim \mathbf{E}p' \neq p(p' \in \Gamma \cap \Delta)}{\Gamma \Rightarrow \Delta|(p,q) \rightrightarrows \Gamma \Rightarrow \Delta|q} \quad (\mathrm{A}_-^L)\ \frac{\mathbf{E}p' \neq p(p' \in \Gamma \cap \Delta)}{\Gamma \Rightarrow \Delta|(p,q) \rightrightarrows \Gamma - \{p\} \Rightarrow \Delta|q}$$
$$(\mathrm{A}_0^R)\ \frac{\sim \mathbf{E}q' \neq q(q' \in \Gamma \cap \Delta)}{\Gamma' \Rightarrow \Delta|q \rightrightarrows \Gamma' \Rightarrow \Delta} \quad (\mathrm{A}_-^R)\ \frac{\mathbf{E}q' \neq q(q' \in \Gamma \cap \Delta)}{\Gamma' \Rightarrow \Delta|q \rightrightarrows \Gamma' \Rightarrow \Delta - \{q\}}$$

  where Γ, Δ are sets of variables, $p \in \Gamma$ and $q \in \Delta$.
- **Deduction rules**:

$$(\neg_0^L)\ \frac{\Gamma \Rightarrow \Delta|(\lambda, A; B) \rightrightarrows \Gamma \Rightarrow \Delta|B}{\Gamma \Rightarrow \Delta|(\neg A, B) \rightrightarrows \Gamma \Rightarrow \Delta|B}$$
$$(\neg_-^L)\ \frac{\Gamma \Rightarrow \Delta|(\lambda, A; B) \rightrightarrows \Gamma \Rightarrow \Delta - \{A\}|B}{\Gamma \Rightarrow \Delta|(\neg A, B) \rightrightarrows \Gamma - \{\neg A\} \Rightarrow \Delta|B}$$
$$(\neg_0^R)\ \frac{\Gamma' \Rightarrow \Delta|(B, \lambda) \rightrightarrows \Gamma' \Rightarrow \Delta}{\Gamma' \Rightarrow \Delta|\neg B \rightrightarrows \Gamma' \Rightarrow \Delta}$$
$$(\neg_-^R)\ \frac{\Gamma' \Rightarrow \Delta|(B, \lambda) \rightrightarrows \Gamma' - \{B\} \Rightarrow \Delta}{\Gamma' \Rightarrow \Delta|\neg B \rightrightarrows \Gamma' \Rightarrow \Delta - \{\neg B\}}$$

  where $\Gamma \Rightarrow \Delta|(\lambda, A; B)$ denotes $(\Gamma \Rightarrow \Delta|(\lambda, A)|B$.

Definition 2.2.1 A reduction δ is provable in $\mathbf{R}^{\neg}$, denoted by $\vdash^{\neg} \delta$, if there is a sequence $\{\delta_1, \ldots, \delta_n\}$ of reductions such that $\delta_n = \delta$, and for each $1 \leq i \leq n$, δ_i is an axiom or is deduced from the previous reductions by one of the deduction rules in $\mathbf{R}^{\neg}$.

Theorem 2.2.2 (Soundness and completeness theorem) *For any reduction* $\delta = (A, B)|\Gamma \Rightarrow \Delta \rightrightarrows \Gamma' \Rightarrow \Delta'$,

$$\vdash^{\neg} \delta \textit{ iff } \models^{\neg} \delta.$$

□

2.2.2 *R-Calculus* $\mathbf{R}_\neg$

A reduction δ is valid, denoted by $\models_\neg \delta$, if

$$\Gamma' = \begin{cases} \Gamma, A \text{ if } \Gamma, A \mapsto \Delta \text{ is } \mathbf{G}_\neg\text{-valid} \\ \Gamma \quad \text{otherwise} \end{cases}$$
$$\Delta' = \begin{cases} \Delta, B \text{ if } \Gamma' \mapsto \Delta, B \text{ is } \mathbf{G}_\neg\text{-valid} \\ \Delta \quad \text{otherwise.} \end{cases}$$

R-calculus $\mathbf{R}_\neg$ consists of the following axioms and deduction rules:

- **Axioms**:

$$(\mathrm{A}_L^0)\ \frac{p \in \Delta}{\Gamma \mapsto \Delta|(p,q) \rightrightarrows \Gamma \mapsto \Delta|q} \quad (\mathrm{A}_L^+)\ \frac{p \notin \Delta}{\Gamma \mapsto \Delta|(p,q) \rightrightarrows \Gamma, p \mapsto \Delta|q}$$
$$(\mathrm{A}_R^0)\ \frac{q \in \Gamma'}{\Gamma' \mapsto \Delta|q \rightrightarrows \Gamma' \mapsto \Delta} \quad (\mathrm{A}_R^+)\ \frac{q \notin \Gamma'}{\Gamma' \mapsto \Delta|q \rightrightarrows \Gamma' \mapsto \Delta, q}$$

  where Γ, Δ are sets of variables, and p, q are atomic.
- **Deduction rules**:

$$(\neg_L^0)\ \frac{\Gamma \mapsto \Delta|(\lambda, A; B) \rightrightarrows \Gamma \mapsto \Delta|B}{\Gamma \mapsto \Delta|(\neg A, B) \rightrightarrows \Gamma \mapsto \Delta|B}$$
$$(\neg_L^+)\ \frac{\Gamma \mapsto \Delta|(\lambda, A; B) \rightrightarrows \Gamma \mapsto \Delta, A|B}{\Gamma \mapsto \Delta|(\neg A, B) \rightrightarrows \Gamma, \neg A \mapsto \Delta|B}$$
$$(\neg_R^0)\ \frac{|\Gamma' \mapsto \Delta|(B, \lambda) \rightrightarrows \Gamma' \mapsto \Delta}{\Gamma' \mapsto \Delta|\neg B \rightrightarrows \Gamma' \mapsto \Delta}$$
$$(\neg_R^+)\ \frac{\Gamma' \mapsto \Delta|(B, \lambda) \rightrightarrows \Gamma', B \mapsto \Delta}{\Gamma' \mapsto \Delta|\neg B \rightrightarrows \Gamma' \mapsto \Delta, \neg B}$$

  where $(\lambda, A; B)|\Gamma \mapsto \Delta$ is $B|((\lambda, A)|\Gamma \mapsto \Delta)$.

Definition 2.2.3 A reduction δ is provable in $\mathbf{R}_\neg$, denoted by $\vdash^\neg \delta$, if there is a sequence $\{\delta_1, \ldots, \delta_n\}$ of reductions such that $\delta_n = \delta$, and for each $1 \leq i \leq n$, δ_i is an axiom or is deduced from the previous reductions by one of the deduction rules in $\mathbf{R}_\neg$.

Theorem 2.2.4 (Soundness and completeness theorem) *For any reduction* $\delta = (A, B)|\Gamma \mapsto \Delta \rightrightarrows \Gamma' \mapsto \Delta'$,

$$\vdash_\neg \delta \textit{ iff } \models_\neg \delta.$$

□

2.2.3 *R-Calculus* $\mathbf{R}^{\wedge}$

Let $A \in \Gamma$ and $B \in \Delta$. A reduction δ is valid, denoted by $\models^{\wedge} \delta$, if

$$\Gamma' = \begin{cases} \Gamma - \{A\} \text{ if } \models^{\wedge} \Gamma - \{A\} \Rightarrow \Delta \\ \Gamma \quad\quad \text{otherwise} \end{cases}$$
$$\Delta' = \begin{cases} \Delta - \{B\} \text{ if } \models^{\wedge} \Gamma' \Rightarrow \Delta - \{B\} \\ \Delta \quad\quad \text{otherwise.} \end{cases}$$

R-calculus $\mathbf{R}^{\wedge}$ consists of the following axioms and deduction rules:

- **Axiom**:

$$(\mathrm{A}_0^L)\ \frac{\sim \mathbf{E}p' \neq p(p' \in \Gamma \cap \Delta)}{\Gamma \Rightarrow \Delta|(p,q) \rightrightarrows \Gamma \Rightarrow \Delta|q} \quad (\mathrm{A}_-^L)\ \frac{\mathbf{E}p' \neq p(p' \in \Gamma \cap \Delta)}{\Gamma \Rightarrow \Delta|(p,q) \rightrightarrows \Gamma - \{p\} \Rightarrow \Delta|q}$$
$$(\mathrm{A}_0^R)\ \frac{\sim \mathbf{E}q' \neq q(q' \in \Gamma' \cap \Delta)}{\Gamma' \Rightarrow \Delta|q \rightrightarrows \Gamma' \Rightarrow \Delta} \quad (\mathrm{A}_-^R)\ \frac{\mathbf{E}q' \neq q(q' \in \Gamma' \cap \Delta)}{\Gamma' \Rightarrow \Delta|q \rightrightarrows \Gamma' \Rightarrow \Delta - \{q\}}$$

where Γ, Δ are sets of variables, and p, q are variables.

- **Deduction rules**:

$$(\wedge_0^L)\ \frac{\begin{bmatrix} \Gamma \Rightarrow \Delta|(A_1, B) \rightrightarrows \Gamma \Rightarrow \Delta|B \\ \Gamma \Rightarrow \Delta|(A_2, B) \rightrightarrows \Gamma \Rightarrow \Delta|B \end{bmatrix}}{\Gamma \Rightarrow \Delta|(A_1 \wedge A_2, B) \rightrightarrows \Gamma \Rightarrow \Delta|B}$$
$$(\wedge_-^L)\ \frac{\begin{cases} \Gamma \Rightarrow \Delta|(A_1, B) \rightrightarrows \Gamma - \{A_1\} \Rightarrow \Delta|B \\ \Gamma \Rightarrow \Delta|(A_2, B) \rightrightarrows \Gamma - \{A_2\} \Rightarrow \Delta|B \end{cases}}{\Gamma \Rightarrow \Delta|(A_1 \wedge A_2, B) \rightrightarrows B|\Gamma - \{A_1 \wedge A_2\} \Rightarrow \Delta}$$
$$(\wedge_0^R)\ \frac{\begin{bmatrix} \Gamma' \Rightarrow \Delta|B_1 \rightrightarrows \Gamma' \Rightarrow \Delta \\ \Gamma' \Rightarrow \Delta - \{B_1\}|B_2 \rightrightarrows \Gamma' \Rightarrow \Delta - \{B_1\} \end{bmatrix}}{\Gamma' \Rightarrow \Delta|B_1 \wedge B_2 \rightrightarrows \Gamma' \Rightarrow \Delta}$$
$$(\wedge_-^R)\ \frac{\begin{cases} \Gamma' \Rightarrow \Delta|B_1 \rightrightarrows \Gamma' \Rightarrow \Delta - \{B_1\} \\ \Gamma' \Rightarrow \Delta - \{B_1\}|B_2 \rightrightarrows \Gamma' \Rightarrow \Delta - \{B_1, B_2\} \end{cases}}{\Gamma' \Rightarrow \Delta|B_1 \wedge B_2 \rightrightarrows \Gamma' \Rightarrow \Delta - \{B_1 \wedge B_2\}}$$

Definition 2.2.5 A reduction δ is provable in $\mathbf{R}^{\wedge}$, denoted by $\vdash^{\wedge} \delta$, if there is a sequence $\{\delta_1, \ldots, \delta_n\}$ such that $\delta_n = \delta$, and for each $1 \leq i \leq n$, δ_i is an axiom or is deduced from the previous reductions by one of the deduction rules in $\mathbf{R}^{\wedge}$.

Theorem 2.2.6 (Soundness and completeness theorem) *For any reduction* $\delta = (A, B)|\Gamma \Rightarrow \Delta \rightrightarrows \Gamma' \Rightarrow \Delta'$,

$$\vdash^{\wedge} \delta \textit{ iff } \models^{\wedge} \delta.$$

□

2.2.4 *R-Calculus* $\mathbf{R}_\wedge$

A reduction δ is valid, denoted by $\models_\wedge \delta$, if

$$\Gamma' = \begin{cases} \Gamma, A \text{ if } \Gamma, A \mapsto \Delta \text{ is } \mathbf{G}_\wedge\text{-valid} \\ \Gamma \quad \text{otherwise} \end{cases}$$
$$\Delta' = \begin{cases} \Delta, B \text{ if } \Gamma' \mapsto \Delta, B \text{ is } \mathbf{G}_\wedge\text{-valid} \\ \Delta \quad \text{otherwise.} \end{cases}$$

R-calculus $\mathbf{R}_\wedge$ consists of the following axioms and deduction rules:

- **Axioms**:

$$(\mathrm{A}_0^L)\ \frac{p \in \Delta}{\Gamma \mapsto \Delta|(p,q) \rightrightarrows \Gamma \mapsto \Delta|q} \qquad (\mathrm{A}_+^L)\ \frac{p \notin \Delta}{\Gamma \mapsto \Delta|(p,q) \rightrightarrows \Gamma, p \mapsto \Delta|q}$$
$$(\mathrm{A}_0^R)\ \frac{q \in \Gamma'}{\Gamma' \mapsto \Delta|q \rightrightarrows \Gamma' \mapsto \Delta} \qquad (\mathrm{A}_+^R)\ \frac{q \notin \Gamma'}{\Gamma' \mapsto \Delta|q \rightrightarrows \Gamma' \mapsto \Delta, q}$$

where Γ, Δ are sets of variables and p, q are variables.

- **Deduction rules**:

$$(\wedge_0^L)\ \frac{\begin{cases} \Gamma \mapsto \Delta|(A_1, B) \rightrightarrows \Gamma \mapsto \Delta|B \\ \Gamma, A_1 \mapsto \Delta|(A_2, B) \rightrightarrows \Gamma, A_1 \mapsto \Delta|B \end{cases}}{\Gamma \mapsto \Delta|(A_1 \wedge A_2, B) \rightrightarrows \Gamma \mapsto \Delta|B}$$
$$(\wedge_+^L)\ \frac{\begin{bmatrix} \Gamma \mapsto \Delta|(A_1, B) \rightrightarrows \Gamma, A_1 \mapsto \Delta|B \\ \Gamma, A_1 \mapsto \Delta|(A_2, B) \rightrightarrows \Gamma, A_1, A_2 \mapsto \Delta|B \end{bmatrix}}{\Gamma \mapsto \Delta|(A_1 \wedge A_2, B) \rightrightarrows \Gamma, A_1 \wedge A_2 \mapsto \Delta|B}$$
$$(\wedge_0^R)\ \frac{\begin{bmatrix} \Gamma' \mapsto \Delta|B_1 \rightrightarrows \Gamma' \mapsto \Delta \\ \Gamma' \mapsto \Delta|B_2 \rightrightarrows \Gamma' \mapsto \Delta \end{bmatrix}}{\Gamma' \mapsto \Delta|B_1 \wedge B_2 \rightrightarrows \Gamma' \mapsto \Delta}$$
$$(\wedge_+^R)\ \frac{\begin{cases} \Gamma' \mapsto \Delta|B_1 \rightrightarrows \Gamma' \mapsto \Delta, B_1 \\ \Gamma' \mapsto \Delta|B_2 \rightrightarrows \Gamma' \mapsto \Delta, B_2 \end{cases}}{\Gamma' \mapsto \Delta|B_1 \wedge B_2 \rightrightarrows \Gamma' \mapsto \Delta, B_1 \wedge B_2}$$

Definition 2.2.7 A reduction δ is provable in $\mathbf{R}_\wedge$, denoted by $\vdash_\wedge \delta$, if there is a sequence $\{\delta_1, \ldots, \delta_n\}$ of reductions such that $\delta_n = \delta$, and for each $1 \leq i \leq n$, δ_i is an axiom or is deduced from the previous reductions by one of the deduction rules in $\mathbf{R}_\wedge$.

Theorem 2.2.8 (Soundness and completeness theorem) *For any reduction* $\Gamma \mapsto \Delta|(A, B) \rightrightarrows \Gamma' \mapsto \Delta'$,

$$\vdash_\wedge \Gamma \mapsto \Delta|(A, B) \rightrightarrows \Gamma' \mapsto \Delta' \textit{ iff } \models_\wedge \Gamma \mapsto \Delta|(A, B) \rightrightarrows \Gamma' \mapsto \Delta'.$$

□

2.2.5 *R-Calculus* $\mathbf{R}^{\vee}$

Let $A \in \Gamma$ and $B \in \Delta$. A reduction δ is valid, denoted by $\models^{\vee} \delta$, if

$$\Gamma' = \begin{cases} \Gamma - \{A\} \text{ if } \models^{\vee} \Gamma - \{A\} \Rightarrow \Delta \\ \Gamma \quad\quad \text{otherwise} \end{cases}$$
$$\Delta' = \begin{cases} \Delta - \{B\} \text{ if } \models^{\vee} \Gamma' \Rightarrow \Delta - \{B\} \\ \Delta \quad\quad \text{otherwise.} \end{cases}$$

R-calculus $\mathbf{R}^{\vee}$ consists of the following axioms and deduction rules:

- **Axioms**:

$$(\mathbb{A}^L_-)\ \frac{\sim \mathbf{E}p' \neq p(p' \in \Gamma \cap \Delta)}{\Gamma \Rightarrow \Delta|(p,q) \rightrightarrows \Gamma \Rightarrow \Delta|q} \quad (\mathbb{A}^L_0)\ \frac{\mathbf{E}p' \neq p(p' \in \Gamma \cap \Delta)}{\Gamma \Rightarrow \Delta|(p,q) \rightrightarrows \Gamma - \{p\} \Rightarrow \Delta|q}$$
$$(\mathbb{A}^R_-)\ \frac{\sim \mathbf{E}q' \neq q(q' \in \Gamma \cap \Delta)}{\Gamma' \Rightarrow \Delta|q \rightrightarrows \Gamma' \Rightarrow \Delta} \quad (\mathbb{A}^R_0)\ \frac{\mathbf{E}q' \neq q(q' \in \Gamma \cap \Delta)}{\Gamma' \Rightarrow \Delta|q \rightrightarrows \Gamma' \Rightarrow \Delta - \{q\}}$$

where Γ, Δ are sets of variables, and p, q are variables.

- **Deduction rules**:

$$(\vee^L_-)\ \frac{\begin{bmatrix} \Gamma \Rightarrow \Delta|(A_1, B) \rightrightarrows \Gamma \Rightarrow \Delta|B \\ \Gamma - \{A_1\} \Rightarrow \Delta|(A_2, B) \rightrightarrows \Gamma - \{A_1\} \Rightarrow \Delta|B \end{bmatrix}}{\Gamma \Rightarrow \Delta|(A_1 \vee A_2, B) \rightrightarrows \Gamma \Rightarrow \Delta|B}$$
$$(\vee^L_0)\ \frac{\begin{cases} \Gamma \Rightarrow \Delta|(A_1, B) \rightrightarrows \Gamma - \{A_1\} \Rightarrow \Delta|B \\ \Gamma - \{A_1\} \Rightarrow \Delta(A_2, B) \rightrightarrows \Gamma - \{A_1, A_2\} \Rightarrow \Delta|B \end{cases}}{\Gamma \Rightarrow \Delta|(A_1 \vee A_2, B) \rightrightarrows \Gamma - \{A_1 \vee A_2\} \Rightarrow \Delta|B}$$
$$(\wedge^R_-)\ \frac{\begin{bmatrix} \Gamma' \Rightarrow \Delta|B_1 \rightrightarrows \Gamma' \Rightarrow \Delta \\ \Gamma' \Rightarrow \Delta|B_2 \rightrightarrows \Gamma' \Rightarrow \Delta \end{bmatrix}}{\Gamma' \Rightarrow \Delta|B_1 \vee B_2 \rightrightarrows \Gamma' \Rightarrow \Delta}$$
$$(\vee^R_0)\ \frac{\begin{cases} \Gamma' \Rightarrow \Delta|B_1 \rightrightarrows \Gamma' \Rightarrow \Delta - \{B_1\} \\ \Gamma' \Rightarrow \Delta|B_2 \rightrightarrows \Gamma' \Rightarrow \Delta - \{B_2\} \end{cases}}{\Gamma' \Rightarrow \Delta|B_1 \vee B_2 \rightrightarrows \Gamma' \Rightarrow \Delta - \{B_1 \vee B_2\}}$$

Definition 2.2.9 A reduction δ is provable in $\mathbf{R}^{\vee}$, denoted by $\vdash^{\wedge} \delta$, if there is a sequence $\{\delta_1, \ldots, \delta_n\}$ of reductions such that $\delta_n = \delta$, and for each $1 \leq i \leq n$, δ_i is an axiom or is deduced from the previous reductions by one of the deduction rules in $\mathbf{R}^{\vee}$.

Theorem 2.2.10 (Soundness and completeness theorem) *For any reduction* $\Gamma \Rightarrow \Delta|(A, B) \rightrightarrows \Gamma' \Rightarrow \Delta'$,

$$\vdash^{\vee} \Gamma \Rightarrow \Delta|(A, B) \rightrightarrows \Gamma' \Rightarrow \Delta' \textit{ iff } \models^{\vee} \Gamma \Rightarrow \Delta|(A, B) \rightrightarrows \Gamma' \Rightarrow \Delta'.$$

□

2.2.6 *R-Calculus* $\mathbf{R}_\vee$

A reduction $\delta = \Gamma \mapsto \Delta|(A, B) \rightrightarrows \Gamma' \mapsto \Delta'$ is valid, denoted by $\models_\vee \delta$, if

$$\Gamma' = \begin{cases} \Gamma, A \text{ if } \Gamma, A \mapsto \Delta \text{ is } \mathbf{G}_\vee\text{-valid} \\ \Gamma \quad \text{otherwise} \end{cases}$$
$$\Delta' = \begin{cases} \Delta, B \text{ if } \Gamma' \mapsto \Delta, B \text{ is } \mathbf{G}_\vee\text{-valid} \\ \Delta \quad \text{otherwise.} \end{cases}$$

R-calculus $\mathbf{R}_\vee$ consists of the following axioms and deduction rules:

- **Axioms**:

$$(\mathrm{A}_0^L)\ \frac{p \in \Delta}{\Gamma \mapsto \Delta|(p, q) \rightrightarrows \Gamma \mapsto \Delta|q} \quad (\mathrm{A}_+^L)\ \frac{p \notin \Delta}{\Gamma \mapsto \Delta|(p, q) \rightrightarrows \Gamma, p \mapsto \Delta|q}$$
$$(\mathrm{A}_0^R)\ \frac{q \in \Gamma'}{\Gamma' \mapsto \Delta|q \rightrightarrows \Gamma' \mapsto \Delta} \quad (\mathrm{A}_+^R)\ \frac{q \notin \Gamma'}{\Gamma' \mapsto \Delta|q \rightrightarrows \Gamma' \mapsto \Delta, q}$$

where Γ, Δ are sets of variables, where p, q are variables.
- **Deduction rules**:

$$(\vee_0^L)\ \frac{\left[\begin{array}{l}\Gamma \mapsto \Delta|(A_1, B) \rightrightarrows \Gamma \mapsto \Delta|B \\ \Gamma \mapsto \Delta|(A_2, B) \rightrightarrows \Gamma \mapsto \Delta|B\end{array}\right.}{\Gamma \mapsto \Delta|(A_1 \vee A_2, B) \rightrightarrows \Gamma \mapsto \Delta|B}$$
$$(\vee_+^L)\ \frac{\left\{\begin{array}{l}\Gamma \mapsto \Delta|(A_1, B) \rightrightarrows \Gamma, A_1 \mapsto \Delta|B \\ \Gamma \mapsto \Delta|(A_2, B) \rightrightarrows \Gamma, A_2 \mapsto \Delta|B\end{array}\right.}{\Gamma \mapsto \Delta|(A_1 \vee A_2, B) \rightrightarrows \Gamma, A_1 \vee A_2 \mapsto \Delta|B}$$
$$(\vee_0^R)\ \frac{\left\{\begin{array}{l}\Gamma' \mapsto \Delta|B_1 \rightrightarrows \Gamma' \mapsto \Delta \\ \Gamma' \mapsto \Delta, B_1|B_2 \rightrightarrows \Gamma' \mapsto \Delta, B_1\end{array}\right.}{\Gamma' \mapsto \Delta|B_1 \vee B_2 \rightrightarrows \Gamma' \mapsto \Delta}$$
$$(\vee_+^R)\ \frac{\left[\begin{array}{l}\Gamma' \mapsto \Delta|B_1 \rightrightarrows \Gamma' \mapsto \Delta, B_1 \\ \Gamma' \mapsto \Delta, B_1|B_2 \rightrightarrows \Gamma' \mapsto \Delta, B_1, B_2\end{array}\right.}{\Gamma' \mapsto \Delta|B_1 \vee B_2 \rightrightarrows \Gamma' \mapsto \Delta, B_1 \vee B_2}$$

Definition 2.2.11 A reduction δ is provable in $\mathbf{R}_\vee$, denoted by $\vdash_\vee \delta$, if there is a sequence $\{\delta_1, \ldots, \delta_n\}$ of reductions such that $\delta_n = \delta$, and for each $1 \leq i \leq n$, δ_i is an axiom or is deduced from the previous reductions by one of the deduction rules in $\mathbf{R}_\vee$.

Theorem 2.2.12 (Soundness and completeness theorem) *For any reduction* $\delta = \Gamma \mapsto \Delta|(A, B) \rightrightarrows \Gamma' \mapsto \Delta'$,

$$\vdash_\vee \delta \textit{ iff } \models_\vee \delta.$$

□

2.3 Variant Propositional Logics

We consider two variant propositional logics

- containing logical connectives $\neg, \rightarrow$ or $\neg, \oplus, \otimes$, and
- give Gentzen deduction systems $\mathbf{G}^{\neg\rightarrow}$, $\mathbf{G}^{\oplus\otimes}$ for sequents and $\mathbf{G}_{\neg\rightarrow}$, $\mathbf{G}_{\oplus\otimes}$ for co-sequents, and
- R-calculi $\mathbf{R}^{\neg\rightarrow}$, $\mathbf{R}^{\oplus\otimes}$ for sequents and $\mathbf{R}_{\neg\rightarrow}$, $\mathbf{R}_{\oplus\otimes}$ for co-sequents.

2.3.1 *Gentzen Deduction System* $\mathbf{G}^{\neg\rightarrow}$

The logical language of propositional logic $\mathbf{G}^{\neg\rightarrow}$ contains the following symbols:

- variables: $p_0, p_1, \ldots$;
- unary connective: $\neg$;
- binary connective: $\rightarrow$.

Formulas are defined as follows:

$$A ::= p|\neg A_1|A_1 \rightarrow A_2.$$

Let v be an assignment, a function from variables to $\{0, 1\}$.
The interpretation $v(A)$ of formula A in assignment v is defined as follows:

$$v(A) = \begin{cases} v(p) & \text{if } A = p \\ 1 - v(A_1) & \text{if } A = \neg A_1 \\ \max\{v(\neg A_1), v(A_2)\} & \text{if } A = A_1 \rightarrow A_2 \end{cases}$$

That is, the truth tables for logical connectives $\neg$ and $\rightarrow$ are given as follows:

A	$\neg A$
1	0
0	1

$A_1 \rightarrow A_2$	1	0
1	1	0
0	1	1

Given two sets Γ, Δ of formulas, define

$$v(\Gamma) = \min\{v(A) : A \in \Gamma\},$$
$$v(\Delta) = \max\{v(B) : B \in \Delta\}.$$

Given a sequent $\delta = \Gamma \Rightarrow \Delta$, we say that v satisfies δ, denoted by $v \models \delta$, if $v(\Gamma) = 1$ implies $v(\Delta) = 1$.

A sequent δ is valid, denoted by $\models^{\neg\rightarrow} \delta$, if for any assignment v, $v \models \delta$.

Gentzen deduction system $\mathbf{G}^{\neg\rightarrow}$ consists of the following axiom and deduction rules:

- **Axiom**:

$$(\mathtt{A}^{\neg\rightarrow})\ \frac{\Gamma \cap \Delta \neq \emptyset}{\Gamma \Rightarrow \Delta},$$

  where Γ, Δ are sets of variables.
- **Deduction rules**:

$$\begin{array}{ll}(\neg^L)\ \dfrac{\Gamma \Rightarrow A, \Delta}{\Gamma, \neg A \Rightarrow \Delta} & (\neg^R)\ \dfrac{\Gamma, B \Rightarrow \Delta}{\Gamma \Rightarrow \neg B, \Delta} \\ (\rightarrow^L)\ \dfrac{\left[\begin{array}{l}\Gamma, \neg A_1 \Rightarrow \Delta \\ \Gamma, A_2 \Rightarrow \Delta\end{array}\right.}{\Gamma, A_1 \rightarrow A_2 \Rightarrow \Delta} & (\rightarrow^R)\ \dfrac{\left\{\begin{array}{l}\Gamma \Rightarrow \neg B_1, \Delta \\ \Gamma \Rightarrow B_2, \Delta\end{array}\right.}{\Gamma \Rightarrow B_1 \rightarrow B_2, \Delta.}\end{array}$$

Definition 2.3.1 A sequent $\Gamma \Rightarrow \Delta$ is provable, denoted by $\vdash^{\neg\rightarrow} \Gamma \Rightarrow \Delta$, if there is a sequence $\{\Gamma_1 \Rightarrow \Delta_1, \ldots, \Gamma_n \Rightarrow \Delta_n\}$ of sequents such that $\Gamma_n \Rightarrow \Delta_n = \Gamma \Rightarrow \Delta$, and for each $1 \leq i \leq n$, $\Gamma_i \Rightarrow \Delta_i$ is an axiom or is deduced from the previous sequents by one of the deduction rules in $\mathbf{G}^{\neg\rightarrow}$.

Theorem 2.3.2 (Soundness theorem) *For any sequent* $\Gamma \Rightarrow \Delta$, *if* $\vdash^{\neg\rightarrow} \Gamma \Rightarrow \Delta$ *then* $\models^{\neg\rightarrow} \Gamma \Rightarrow \Delta$.

Proof We prove that each axiom is valid and each deduction rule preserves the validity.

$(\mathtt{A}^{\neg\rightarrow})$ Assume that $p \in \Gamma \cap \Delta$. Then, for any assignment v, either $v(p) = 0$ or $v(p) = 1$, i.e., either $v \models \Gamma$ or $v \models \Delta$.

$(\neg^L)$ Assume that for any assignment v, $v \models \Gamma$ implies $v \models A, \Delta$. Then, for any assignment v, assume that $v \models \Gamma, \neg A$. $v \models \Gamma$, and by induction assumption, $v \models A, \Delta$. Because $v \models \neg A$, $v \models \Delta$.

$(\neg^R)$ Assume that for any assignment v, $v \models \Gamma, B$ implies $v \models \Delta$. Then, for any assignment v, assume that $v \models \Gamma$. If $v \models \neg B$ then $v \models \neg B, \Delta$; otherwise, $v \models \Gamma, B$, and by induction assumption, $v \models \Delta$, which implies $v \models \neg B, \Delta$.

$(\rightarrow^L)$ Assume that for any assignment v,

$$\begin{array}{l}v \models \Gamma, A_2 \text{ implies } v \models \Delta \\ v \models \Gamma, \neg A_1 \text{ implies } v \models \Delta.\end{array}$$

Then, for any assignment v, assume that $v \models \Gamma, A_1 \rightarrow A_2$. Then, either $v \models A_2$ or $v \models \neg A_1$. For each case, we have that $v \models \Delta$.

$(\rightarrow^R)$ Assume that for any assignment v, either $v \models \Gamma$ implies $v \models \neg B_1, \Delta$; or $v \models \Gamma$ implies $v \models B_2, \Delta$. Then, for any assignment v, assume that $v \models \Gamma$. If $v \models \Delta$ then $v \models B_1 \rightarrow B_2, \Delta$; otherwise, by induction assumption, either $v \models \neg B_1$, or $v \models B_2$, which implies $v \models B_1 \rightarrow B_2$, i.e., $v \models B_1 \rightarrow B_2, \Delta$. □

Theorem 2.3.3 (Completeness theorem) *For any sequent* $\Gamma \Rightarrow \Delta$, *if* $\models^{\neg\rightarrow} \Gamma \Rightarrow \Delta$ *then* $\vdash^{\neg\rightarrow} \Gamma \Rightarrow \Delta$.

Proof Let $\delta = \Gamma \Rightarrow A$. We will define a tree, called the reduction tree for δ, denoted by $T(\delta)$, from which we can obtain either a proof of δ or a show of the nonvalidity of δ.

This reduction tree $T(\delta)$ for δ contains a sequent at each node and is constructed in stages as follows.

Stage 0: $T_0(\delta) = \{\delta\}$.

Stage $k(k > 0)$: $T_k(\delta)$ is defined by cases.

Case 0. If Γ and Δ have any statement in common, write nothing above $\Gamma \Rightarrow \Delta$.

Case 1. Every topmost sequent $\Gamma \Rightarrow \Delta$ in $T_{k-1}(\delta)$ has the same statement in Γ and Δ. Then, stop.

Case 2. Not case 1. $T_k(\delta)$ is defined as follows. Let $\Gamma \Rightarrow \Delta$ be any topmost sequent of the tree which has been defined by stage $k - 1$.

Subcase $(\neg^L)$. Let $\neg A_1, \ldots, \neg A_n$ be all the formulas in Γ whose outmost logical symbol is $\neg$, and to which no reduction has been applied in previous stages. Then, write down

$$\Gamma \Rightarrow A_1, \ldots, A_n, \Delta$$

above $\Gamma \Rightarrow \Delta$. We say that a $(\neg^L)$ reduction has been applied to $\neg A_1, \ldots, \neg A_n$.

Subcase $(\neg^R)$. Let $\neg B_1, \ldots, \neg B_n$ be all the formulas in Δ whose outmost logical symbol is $\neg$, and to which no reduction has been applied in previous stages. Then, write down

$$\Gamma, B_1, \ldots, B_n \Rightarrow \Delta$$

above $\Gamma \Rightarrow \Delta$. We say that a $(\neg^R)$ reduction has been applied to $\neg B_1, \ldots, \neg B_n$.

Subcase $(\rightarrow^L)$. Let $A_1^1 \rightarrow A_2^1, \ldots, A_1^n \rightarrow A_2^n$ be all the statements in Γ whose outermost logical symbol is $\rightarrow$, and to which no reduction has been applied in previous stages by any $(\rightarrow^L)$. Then, write down all sequents of the form

$$\Gamma, E_1, \ldots, E_n \Rightarrow \Delta$$

above $\Gamma \Rightarrow \Delta$, where E_i is either $\neg A_1^i$ or A_2^i. We say that a $(\rightarrow^L)$ reduction has been applied to $A_1^1 \rightarrow A_2^1, \ldots, A_1^n \rightarrow A_2^n$.

Subcase $(\rightarrow^R)$. Let $B_1^1 \rightarrow B_2^1, \ldots, B_1^n \rightarrow B_2^n$ be all the statements in Δ whose outermost logical symbol is $\rightarrow$, and to which no reduction has been applied in previous stages by any $(\rightarrow^R)$. Then, write down

$$\Gamma \Rightarrow \neg B_1^1, B_1^1, \ldots, \neg B_1^n, B_2^n, \Delta$$

above $\Gamma \Rightarrow \Delta$. We say that a $(\rightarrow^R)$ reduction has been applied to $B_1^1 \rightarrow B_2^1, \ldots, B_1^n \rightarrow B_2^n$.

So the collection of those sequents which are obtained by the above reduction process, together with the partial order obtained by this process, is the reduction tree for δ, denoted by $T(\delta)$.

A sequence $\delta_0, \ldots$ of sequents in $T(\delta)$ is a branch if $\delta_0 = \delta$, δ_{i+1} is immediately above δ_i.

Given a sequent δ, if each branch of $T(\delta)$ is ended with a sequent containing common formulas, then it is a routine to construct a proof of δ.

Otherwise, there is a branch $\sigma = \delta_1, \ldots, \delta_n$ of $T(\delta)$ such that there is no rule applicable for δ_n and $\delta_n = \Gamma_n \Rightarrow \Delta_n$ has no common statements in Γ_n and Δ_n.

Let

$$\cup\Gamma = \{A \in \Gamma_i : \Gamma_i \Rightarrow \Delta_i \in \sigma\},$$
$$\cup\Delta = \{B \in \Delta_i : \Gamma_i \Rightarrow \Delta_i \in \sigma\}.$$

We define an assignment in which every formula $A \in \cup\Gamma$ is true and every formula in $B \in \cup\Delta$ is false.

Define v such that for any variable p,

$$v(p) = 1 \text{ iff } p \in \cup\Gamma.$$

By the induction on the structure of formulas, we prove that $v(A) = 1$ if $A \in \cup\Gamma$ and $v(B) = 0$ if $B \in \cup\Delta$.

Case $A = p \in \cup\Gamma$ is atomic. By the definition of v, $v(p) = 1$, i.e., $v(A) = 1$.

Case $B = q \in \cup\Delta$ is atomic. By the definition of v, $v(q) = 0$, i.e., $v(B) = 0$.

Case $A = \neg A_1 \in \cup\Gamma$. Let β be the least-length segment of σ such that $\beta = \Gamma', \neg A \Rightarrow \Delta'$ for some Γ' and Δ'. Then, there is a segment γ of σ such that β is a segment of γ and $\gamma = \Gamma' \Rightarrow A, \Delta'$. By induction assumption, $v \models \Gamma'$ and $v \not\models A, \Delta'$, equivalently, $v \models \Gamma', \neg A$ and $v \not\models \Delta'$.

Case $B = \neg B_1 \in \cup\Delta$. Let β be the least-length segment of σ such that $\beta = \Gamma', \neg B \Rightarrow \Delta'$ for some Γ' and Δ'. Then, there is a segment γ of σ such that β is a segment of γ and $\gamma = \Gamma', B \Rightarrow \Delta'$. By induction assumption, $v \models \Gamma', B$ and $v \not\models \Delta'$, equivalently, $v \models \Gamma'$ and $v \not\models \neg B, \Delta'$.

Case $A = A_1 \rightarrow A_2 \in \cup\Gamma$. Let β be the least-length segment of σ such that $\beta = \Gamma', A_1 \rightarrow A_2 \Rightarrow \Delta'$ for some Γ' and Δ'. Then, there is a segment γ of σ such that β is a segment of γ and γ is one of the following forms:

$$\Gamma', \neg A_1 \Rightarrow \Delta',$$
$$\Gamma', A_2 \Rightarrow \Delta'.$$

By induction assumption, either $v \models \Gamma', \neg A_1$ or $v \models \Gamma', A_2$, and $v \not\models \Delta'$. Then, by definition of satisfaction, $v \models \Gamma', A_1 \rightarrow A_2$ and $v \not\models \Delta'$.

Case $B = B_1 \rightarrow B_2 \in \cup\Delta$. Let β be the least-length segment of σ such that $\beta = \Gamma' \Rightarrow B_1 \rightarrow B_2, \Delta'$. Then, there is a segment γ of σ such that β is a segment of γ and

$$\gamma = \Gamma' \Rightarrow \neg B_1, B_2, \Delta'.$$

By the induction assumption, $v \models \Gamma'$ and $v \not\models \neg B_1, B_2, \Delta'$, i.e., $v \not\models B_1 \rightarrow B_2, \Delta'$.

This completes the proof. □

2.3.2 *Gentzen Deduction System* $\mathbf{G}_{\neg\to}$

A co-sequent $\Gamma \mapsto \Delta$ is valid, denoted by $\models_{\neg\to} \Gamma \mapsto \Delta$, if there is an assignment v such that for each formula $A \in \Gamma$, $v(A) = 1$, and for each formula $B \in \Delta$, $v(B) = 0$.

Gentzen deduction system $\mathbf{G}_{\neg\to}$ consists of the following axiom and deduction rules:

- **Axiom**:

$$(\mathrm{A}_{\neg\to})\ \frac{\Gamma \cap \Delta = \emptyset}{\Gamma \mapsto \Delta},$$

where Γ, Δ are sets of variables.

- **Deduction rules**:

$$\begin{array}{ll}(\neg^L)\ \dfrac{\Gamma \mapsto A, \Delta}{\Gamma, \neg A \mapsto \Delta} & (\neg^R)\ \dfrac{\Gamma, B \mapsto \Delta}{\Gamma \mapsto \neg B, \Delta} \\ (\to^L)\ \dfrac{\begin{cases}\Gamma, \neg A_1 \mapsto \Delta \\ \Gamma, A_2 \mapsto \Delta\end{cases}}{\Gamma, A_1 \to A_2 \mapsto \Delta} & (\to^R)\ \dfrac{\begin{bmatrix}\Gamma \mapsto \neg B_1, \Delta \\ \Gamma \mapsto B_2, \Delta\end{bmatrix}}{\Gamma \mapsto B_1 \to B_2, \Delta.}\end{array}$$

Definition 2.3.4 A co-sequent $\Gamma \mapsto \Delta$ is provable in $\mathbf{G}_{\neg\to}$, denoted by $\vdash_{\neg\to} \Gamma \mapsto \Delta$, if there is a sequence $\{\Gamma_1 \mapsto \Delta_1, \ldots, \Gamma_n \mapsto \Delta_n\}$ of co-sequents such that $\Gamma_n \mapsto \Delta_n = \Gamma \mapsto \Delta$, and for each $1 \leq i \leq n$, $\Gamma_i \mapsto \Delta_i$ is an axiom or is deduced from the previous co-sequents by one of the deduction rules in $\mathbf{G}_{\neg\to}$.

Theorem 2.3.5 (Soundness and completeness theorem) *For any co-sequent* $\Gamma \mapsto \Delta$,

$$\vdash_{\neg\to} \Gamma \mapsto \Delta \text{ if and only if } \models_{\neg\to} \Gamma \mapsto \Delta. \qquad \square$$

2.3.3 *Gentzen Deduction System* $\mathbf{G}^{\otimes\oplus}$

Traditionally there are two kinds of the semantics for the logical connectives $\vee$ and $\wedge$: the inclusive one and exclusive one, where in the exclusive one, $\vee$ and $\wedge$ have the following truth tables:

$A_1 \vee A_2$	1	0
1	0	1
0	1	0

$A_1 \wedge A_2$	1	0
1	1	0
0	0	0

and in the inclusive one, $\vee$ and $\wedge$ have the following ones:

$A_1 \vee A_2$	1	0
1	1	1
0	1	0

$A_1 \wedge A_2$	1	0
1	1	0
0	0	0

We consider two connectives which truth tables are given as follows:

$A_1 \oplus A_2$	1 0
1	0 1
0	1 0

$A_1 \otimes A_2$	1 0
1	1 0
0	0 1

The logical language of propositional logic $\mathbf{G}^{\oplus\otimes}$ contains the following symbols:

- variables: $p_0, p_1, \ldots$;
- unary connective: $\neg$;
- binary connective: $\oplus$, $\otimes$.

Formulas are defined as follows:

$$A ::= p|\neg A_1|A_1 \oplus A_2|A_1 \otimes A_2.$$

Let v be an assignment, a function from variables to $\{0, 1\}$.
The truth-value of formula A in an assignment v is defined as follows:

$$v(A) = \begin{cases} v(p) & \text{if } A = p \\ 1 - v(A_1) & \text{if } A = \neg A_1 \\ f_\oplus(v(A_1), v(A_2)) & \text{if } A = A_1 \oplus A_2 \\ f_\otimes(v(A_1), v(A_2)) & \text{if } A = A_1 \otimes A_2 \end{cases}$$

where $f_\oplus, f_\otimes : \{0, 1\}^2 \to \{0, 1\}$ are defined as follows:

A	$\neg A$
1	0
0	1

$f_\oplus$	1 0
1	0 1
0	1 0

$f_\otimes$	1 0
1	1 0
0	0 1

Given two sets Γ, Δ of formulas, define

$$v(\Gamma) = \min\{v(A) : A \in \Gamma\},$$
$$v(\Delta) = \max\{v(B) : B \in \Delta\}.$$

Given a sequent $\delta = \Gamma \Rightarrow \Delta$, we say that v satisfies δ, denoted by $v \models \delta$, if $v(\Gamma) = 1$ implies $v(\Delta) = 1$.

A sequent δ is valid, denoted by $\models^{\oplus\otimes} \delta$, if for any assignment v, $v \models \delta$.

Gentzen deduction system $\mathbf{G}^{\oplus\otimes}$ consists of the following axiom and deduction rules.

- **Axiom**:

$$(\mathrm{A}^{\oplus\otimes})\ \frac{\mathrm{incon}(\Gamma) \text{ or } \mathrm{val}(\Delta) \text{ or } \Gamma \cap \Delta \neq \emptyset}{\Gamma \Rightarrow \Delta}$$

where Γ, Δ are sets of literals.

- **Deduction rules**:

$$(\neg\neg^L)\ \frac{\Gamma, A \Rightarrow \Delta}{\Gamma, \neg\neg A \Rightarrow \Delta} \qquad (\neg\neg^R)\ \frac{\Gamma \Rightarrow B, \Delta}{\Gamma \Rightarrow \neg\neg B, \Delta}$$

$$(\oplus^L)\ \frac{\left\{\begin{array}{l}\left[\begin{array}{l}\Gamma, \neg A_1 \rightarrow \Delta\\ \Gamma, \neg A_2 \Rightarrow \Delta\end{array}\right.\\ \left[\begin{array}{l}\Gamma, A_1 \rightarrow \Delta\\ \Gamma, A_2 \Rightarrow \Delta\end{array}\right.\end{array}\right.}{\Gamma, A_1 \oplus A_2 \Rightarrow \Delta} \qquad (\oplus^R)\ \frac{\left\{\begin{array}{l}\left[\begin{array}{l}\Gamma \Rightarrow B_1, \Delta\\ \Gamma \Rightarrow \neg B_2, \Delta\end{array}\right.\\ \left[\begin{array}{l}\Gamma \Rightarrow \neg B_1, \Delta\\ \Gamma \Rightarrow B_2, \Delta\end{array}\right.\end{array}\right.}{\Gamma \Rightarrow B_1 \oplus B_2, \Delta}$$

$$(\neg\oplus^L)\ \frac{\left\{\begin{array}{l}\left[\begin{array}{l}\Gamma, \neg A_1 \rightarrow \Delta\\ \Gamma, A_2 \Rightarrow \Delta\end{array}\right.\\ \left[\begin{array}{l}\Gamma, A_1 \rightarrow \Delta\\ \Gamma, \neg A_2 \Rightarrow \Delta\end{array}\right.\end{array}\right.}{\Gamma, \neg(A_1 \oplus A_2) \Rightarrow \Delta} \qquad (\neg\oplus^R)\ \frac{\left\{\begin{array}{l}\left[\begin{array}{l}\Gamma \Rightarrow \neg B_1, \Delta\\ \Gamma \Rightarrow \neg B_2, \Delta\end{array}\right.\\ \left[\begin{array}{l}\Gamma \Rightarrow B_1, \Delta\\ \Gamma \Rightarrow B_2, \Delta\end{array}\right.\end{array}\right.}{\Gamma \Rightarrow \neg(B_1 \oplus B_2), \Delta}$$

and

$$(\otimes^L)\ \frac{\left\{\begin{array}{l}\left[\begin{array}{l}\Gamma, \neg A_1 \rightarrow \Delta\\ \Gamma, A_2 \Rightarrow \Delta\end{array}\right.\\ \left[\begin{array}{l}\Gamma, A_1 \rightarrow \Delta\\ \Gamma, \neg A_2 \Rightarrow \Delta\end{array}\right.\end{array}\right.}{\Gamma, A_1 \otimes A_2 \Rightarrow \Delta} \qquad (\otimes^R)\ \frac{\left\{\begin{array}{l}\left[\begin{array}{l}\Gamma \Rightarrow B_1, \Delta\\ \Gamma \Rightarrow B_2, \Delta\end{array}\right.\\ \left[\begin{array}{l}\Gamma \Rightarrow \neg B_1, \Delta\\ \Gamma \Rightarrow \neg B_2, \Delta\end{array}\right.\end{array}\right.}{\Gamma \Rightarrow B_1 \otimes B_2, \Delta}$$

$$(\neg\otimes^L)\ \frac{\left\{\begin{array}{l}\left[\begin{array}{l}\Gamma, \neg A_1 \rightarrow \Delta\\ \Gamma, \neg A_2 \Rightarrow \Delta\end{array}\right.\\ \left[\begin{array}{l}\Gamma, A_1 \rightarrow \Delta\\ \Gamma, A_2 \Rightarrow \Delta\end{array}\right.\end{array}\right.}{\Gamma, \neg(A_1 \otimes A_2) \Rightarrow \Delta} \qquad (\neg\otimes^R)\ \frac{\left\{\begin{array}{l}\left[\begin{array}{l}\Gamma \Rightarrow \neg B_1, \Delta\\ \Gamma \Rightarrow B_2, \Delta\end{array}\right.\\ \left[\begin{array}{l}\Gamma \Rightarrow B_1, \Delta\\ \Gamma \Rightarrow \neg B_2, \Delta\end{array}\right.\end{array}\right.}{\Gamma \Rightarrow \neg(B_1 \otimes B_2), \Delta}$$

Definition 2.3.6 A sequent $\Gamma \Rightarrow \Delta$ is provable in $\mathbf{G}^{\oplus\otimes}$, denoted by $\vdash^{\oplus\otimes} \Gamma \vdash \Delta$, if there is a sequence $\{\Gamma_1 \Rightarrow \Delta_1, \ldots, \Gamma_n \Rightarrow \Delta_n\}$ of sequents such that $\Gamma_n \Rightarrow \Delta_n = \Gamma \Rightarrow \Delta$, and for each $1 \leq i \leq n$, $\Gamma_i \Rightarrow \Delta_i$ is an axiom or is deduced from the previous sequents by one of the deduction rules in $\mathbf{G}^{\oplus\otimes}$.

Theorem 2.3.7 (Soundness theorem) *For any sequent* $\Gamma \Rightarrow \Delta$, *if* $\vdash^{\oplus\otimes} \Gamma \Rightarrow \Delta$ *then* $\models^{\oplus\otimes} \Gamma \Rightarrow \Delta$.

□

Proof We prove that each axiom is valid and each deduction rule preserves the validity.

($\mathrm{A}^{\oplus\otimes}$) Assume that $\mathrm{incon}(\Gamma)$. Then, there is a literal l such that $l, \neg l \in \Gamma$. For any assignment v, either $v(l) = 0$ or $v(\neg l) = 0$, i.e., $v \models \Gamma$.

Assume that $l \in \Gamma \cap \Delta$. Then, for any assignment v, either $v(l) = 0$ or $v(l) = 1$, i.e., either $v \models \Gamma$ or $v \models \Delta$.

($\neg^L$) Assume that for any assignment v, $v \models \Gamma$ implies $v \models A, \Delta$. Then, for any assignment v, assume that $v \models \Gamma, \neg A$. By induction assumption, $v \models A, \Delta$. Because $v \models \neg A$, $v \models \Delta$.

($\neg^R$) Assume that for any assignment v, $v \models \Gamma, B$ implies $v \models \Delta$. Then, for any assignment v, assume that $v \models \Gamma$. If $v \models \neg B$ then $v \models \neg B, \Delta$; otherwise, $v \models \Gamma, B$, and by induction assumption, $v \models \Delta$, which implies $v \models \neg B, \Delta$.

($\oplus^L$) Assume that for any assignment v, either

$$v \models \Gamma, \neg A_1 \text{ implies } v \models \Delta$$
$$v \models \Gamma, \neg A_2 \text{ implies } v \models \Delta$$

or

$$v \models \Gamma, A_1 \text{ implies } v \models \Delta$$
$$v \models \Gamma, A_2 \text{ implies } v \models \Delta.$$

Then, for any assignment v, assume that $v \models \Gamma, A_1 \oplus A_2$. Then, either $v \models \neg A_1$ and $v \models \neg A_2$, or $v \models A_1$ and $v \models A_2$. Hence, either $v \models \Gamma, \neg A_1$ and $v \models \Gamma, \neg A_2$, or $v \models \Gamma, A_1$ and $v \models \Gamma, A_2$. By induction assumption, $v \models \Delta$.

($\oplus^R$) Assume that for any assignment v, either

$$v \models \Gamma \text{ implies } v \models B_1, \Delta$$
$$v \models \Gamma \text{ implies } v \models \neg B_2, \Delta,$$

or

$$v \models \Gamma \text{ implies } v \models \neg B_1, \Delta$$
$$v \models \Gamma \text{ implies } v \models B_2, \Delta,$$

Then, for any assignment v, assume that $v \models \Gamma$. If either $v \models B_1$ and $v \models \neg B_2$, or $v \models \neg B_1$ and $v \models B_2$ then $v \models B_1 \oplus B_2, \Delta$; otherwise, by induction assumption, $v \models \Delta$, which implies $v \models B_1 \oplus B_2, \Delta$. ($\otimes^L$) Assume that for any assignment v, either

$$v \models \Gamma, \neg A_1 \text{ implies } v \models \Delta$$
$$v \models \Gamma, A_2 \text{ implies } v \models \Delta,$$

or

$$v \models \Gamma, A_1 \text{ implies } v \models \Delta$$
$$v \models \Gamma \neg A_2 \text{ implies } v \models \Delta.$$

Then, for any assignment v, assume that $v \models \Gamma, A_1 \otimes A_2$. Then, either $v \models \neg A_1$ and $v \models A_2$, or $v \models A_1$ or $v \models \neg A_2$, i.e., either $v \models \Gamma, \neg A_1$ and $v \models \Gamma, A_2$, or $v \models \Gamma, A_1$ and $v \models \Gamma, \neg A_2$. By induction assumption, $v \models \Delta$.

($\otimes^R$) Assume that for any assignment v, either

$$v \models \Gamma \text{ implies } v \models B_1, \Delta$$
$$v \models \Gamma \text{ implies } v \models B_2, \Delta,$$

or

$$\begin{array}{l} v \models \Gamma \text{ implies } v \models \neg B_1, \Delta \\ v \models \Gamma \text{ implies } v \models \neg B_2, \Delta, \end{array}$$

Then, for any assignment v, assume that $v \models \Gamma$. If either $v \models B_1$ and $v \models B_2$, or $v \models \neg B_1$ and $v \models \neg B_2$ then $v \models B_1 \otimes B_2$, and $v \models B_1 \otimes B_2, \Delta$; otherwise, by induction assumption, $v \models \Delta$, which implies $v \models B_1 \otimes B_2, \Delta$.

Similar for other cases. □

Theorem 2.3.8 (Completeness theorem) *For any sequent* $\Gamma \Rightarrow \Delta$, *if* $\models^{\oplus\otimes} \Gamma \Rightarrow \Delta$ *then* $\vdash^{\oplus\otimes} \Gamma \Rightarrow \Delta$.

Proof Given a sequent $\Delta \Rightarrow \Gamma$, we construct a tree T such that either

(i) For each branch ξ of T, a sequent $\Delta' \Rightarrow \Gamma'$ at the leaf of ξ is an axiom.
(ii) There is an assignment v such that $v \not\models \Delta \Rightarrow \Gamma$.

T is constructed as follows:

- The root of T is $\Delta \Rightarrow \Gamma$.
- For a node ξ, if each sequent $\Delta' \Rightarrow \Gamma'$ at ξ is literal then the node is a leaf.
- Otherwise, ξ has the direct children nodes containing the following sequents:

$$\left\{\begin{array}{ll} \Gamma_1, A \Rightarrow \Delta_1 & \text{if } \Gamma_1, \neg\neg A \Rightarrow \Delta_1 \in \xi \\ \Gamma_1 \Rightarrow B, \Delta_1 & \text{if } \Gamma_1 \Rightarrow \neg\neg B, \Delta_1 \in \xi \\ \left[\begin{array}{l} \left\{\begin{array}{l} \Gamma_1, \neg A_1 \Rightarrow \Delta_1 \\ \Gamma_1, \neg A_2 \Rightarrow \Delta_1 \\ \Gamma_1, A_1 \Rightarrow \Delta_1 \\ \Gamma_1, A_2 \Rightarrow \Delta_1 \end{array}\right. \\ \left\{\begin{array}{l} \Gamma_1 \Rightarrow B_1, \Delta_1 \\ \Gamma_1 \Rightarrow \neg B_2, \Delta_1 \\ \Gamma_1 \Rightarrow B_1, \Delta_1 \\ \Gamma_1 \Rightarrow \neg B_2, \Delta_1 \end{array}\right. \\ \left\{\begin{array}{l} \Gamma_1, \neg A_1 \Rightarrow \Delta_1 \\ \Gamma_1, A_2 \Rightarrow \Delta_1 \\ \Gamma_1, A_1 \Rightarrow \Delta_1 \\ \Gamma_1, \neg A_2 \Rightarrow \Delta_1 \end{array}\right. \\ \left\{\begin{array}{l} \Gamma_1 \Rightarrow \neg B_1, \Delta_1 \\ \Gamma_1 \Rightarrow \neg B_2, \Delta_1 \\ \Gamma_1 \Rightarrow B_1, \Delta_1 \\ \Gamma_1 \Rightarrow B_2, \Delta_1 \end{array}\right. \end{array}\right. & \begin{array}{l} \text{if } \Gamma_1, A_1 \oplus A_2 \Rightarrow \Delta_1 \in \xi \\ \\ \\ \\ \text{if } \Gamma_1 \Rightarrow B_1 \oplus B_2, \Delta_1 \in \xi \\ \\ \\ \\ \text{if } \Gamma_1, \neg(A_1 \oplus A_2) \Rightarrow \Delta_1 \in \xi \\ \\ \\ \\ \text{if } \Gamma_1 \Rightarrow \neg(B_1 \oplus B_2), \Delta_1 \in \xi \end{array} \end{array}\right.$$

and

$$\left\{\begin{array}{ll}
\left[\begin{cases}\Gamma_1, \neg A_1 \Rightarrow \Delta_1 \\ \Gamma_1, A_2 \Rightarrow \Delta_1 \\ \Gamma_1, A_1 \Rightarrow \Delta_1 \\ \Gamma_1, \neg A_2 \Rightarrow \Delta_1\end{cases}\right. & \text{if } \Gamma_1, A_1 \otimes A_2 \Rightarrow \Delta_1 \in \xi \\
\left[\begin{cases}\Gamma_1 \Rightarrow B_1, \Delta_1 \\ \Gamma_1 \Rightarrow B_2, \Delta_1 \\ \Gamma_1 \Rightarrow \neg B_1, \Delta_1 \\ \Gamma_1 \Rightarrow \neg B_2, \Delta_1\end{cases}\right. & \text{if } \Gamma_1 \Rightarrow B_1 \otimes B_2, \Delta_1 \in \xi \\
\left[\begin{cases}\Gamma_1, \neg A_1 \Rightarrow \Delta_1 \\ \Gamma_1, \neg A_2 \Rightarrow \Delta_1 \\ \Gamma_1, A_1 \Rightarrow \Delta_1 \\ \Gamma_1, A_2 \Rightarrow \Delta_1\end{cases}\right. & \text{if } \Gamma_1, \neg(A_1 \otimes A_2) \Rightarrow \Delta_1 \in \xi \\
\left[\begin{cases}\Gamma_1 \Rightarrow \neg B_1, \Delta_1 \\ \Gamma_1 \Rightarrow B_2, \Delta_1 \\ \Gamma_1 \Rightarrow B_1, \Delta_1 \\ \Gamma_1 \Rightarrow \neg B_2, \Delta_1\end{cases}\right. & \text{if } \Gamma_1 \Rightarrow \neg(B_1 \otimes B_2), \Delta_1 \in \xi
\end{array}\right.$$

Lemma 2.3.9 *If for each branch $\xi \subseteq T$, there is a sequent $\Delta' \Rightarrow \Gamma' \in \xi$ which is an axiom in $\mathbf{G}^{\oplus\otimes}$ then T is a proof tree of $\Delta \Rightarrow \Gamma$.*

Proof By the definition of T, T is a proof tree of $\Delta \Rightarrow \Gamma$ in $\mathbf{G}^{\oplus\otimes}$. □

Lemma 2.3.10 *If there is a branch $\xi \subseteq T$ such that each sequent $\Delta' \Rightarrow \Gamma' \in \xi$ is not an axiom in $\mathbf{G}^{\oplus\otimes}$ then there is an assignment v such that $v \not\models \Delta \Rightarrow \Gamma$.*

Proof Let γ be the set of all the literal sequents in ξ.

Let

$$\begin{array}{l}\Theta^L = \bigcup_{\Gamma' \Rightarrow \Delta' \in \gamma} \Gamma', \\ \Theta^R = \bigcup_{\Gamma' \Rightarrow \Delta' \in \gamma} \Delta'.\end{array}$$

Define an assignment v as follows: for any variable p,

$$v(p) = \begin{cases} 1 \text{ if } p \in \Theta^L \text{ or } \neg p \in \Theta^R \\ 0 \text{ otherwise.} \end{cases}$$

Then, v is well-defined and $v \not\models \Gamma'' \Rightarrow \Delta''$, where $\Gamma'' \Rightarrow \Delta''$ is a sequent at the leaf node of ξ.

We proved by induction on ξ that each sequent $\Delta' \Rightarrow \Gamma' \in \xi$ is not satisfied by v.

Case $\Gamma' \Rightarrow \Delta' = \Gamma_2, \neg A \Rightarrow \Delta' \in \xi$. Then $\Gamma' \Rightarrow \Delta'$ has children nodes $\in \xi$ containing sequent $\Gamma_2 \Rightarrow A, \Delta'$. By induction assumption, $v \not\models \Gamma_2 \Rightarrow A, \Delta'$. Hence, $v \not\models \Gamma_2, \neg A \Rightarrow \Delta'$.

Case $\Gamma' \Rightarrow \Delta' = \Gamma' \Rightarrow \neg B, \Delta_2 \in \xi$. Then $\Gamma' \Rightarrow \Delta'$ has children nodes $\in \xi$ containing sequent $\Gamma', B \Rightarrow \Delta_2$. By induction assumption, $v \not\models \Gamma', B \Rightarrow \Delta_2$. Hence, $v \not\models \Gamma' \Rightarrow \neg B, \Delta_2$.

Case $\Gamma' \Rightarrow \Delta' = \Gamma_2, A_1 \oplus A_2 \Rightarrow \Delta' \in \xi$. Then $\Gamma' \Rightarrow \Delta'$ has children nodes $\in \xi$ containing sequents either $\Gamma_2, \neg A_1 \Rightarrow \Delta'$ or $\Gamma_2, \neg A_2 \Rightarrow \Delta'$; and either $\Gamma_2, A_1 \Rightarrow$

Δ' or $\Gamma_2, A_2 \Rightarrow \Delta'$. By induction assumption, either $v \not\models \Gamma_2, \neg A_1 \Rightarrow \Delta'$ or $v \not\models \Gamma_2, \neg A_2 \Rightarrow \Delta'$, and either $v \not\models \Gamma_2, A_1 \Rightarrow \Delta'$ or $v \not\models \Gamma_2, A_2 \Rightarrow \Delta'$. Hence, $v \not\models \Gamma_2, A_1 \oplus A_2 \Rightarrow \Delta'$.

Case $\Gamma' \Rightarrow \Delta' = \Gamma' \Rightarrow B_1 \oplus B_2, \Delta_2 \in \xi$. Then $\Gamma' \Rightarrow \Delta'$ has children nodes $\in \xi$ containing sequents either $\Gamma' \Rightarrow B_1, \Delta_2$ or $\Gamma' \Rightarrow \neg B_2, \Delta_2$, and either $\Gamma' \Rightarrow \neg B_1, \Delta_2$ or $\Gamma' \Rightarrow B_2, \Delta_2$. By induction assumption, either $v \not\models \Gamma' \Rightarrow B_1, \Delta_2$ or $v \not\models \Gamma' \Rightarrow \neg B_2, \Delta_2$; and either $v \not\models \Gamma' \Rightarrow \neg B_1, \Delta_2$ or $v \not\models \Gamma' \Rightarrow B_2, \Delta_2$. Hence, $v \not\models \Gamma' \Rightarrow B_1 \oplus B_2, \Delta_2$.

Case $\Gamma' \Rightarrow \Delta' = \Gamma_2, A_1 \otimes A_2 \Rightarrow \Delta' \in \xi$. Then $\Gamma' \Rightarrow \Delta'$ has children nodes $\in \xi$ containing sequents either $\Gamma_2, \neg A_1 \Rightarrow \Delta'$ or $\Gamma_2, A_2 \Rightarrow \Delta'$; and either $\Gamma_2, A_1 \Rightarrow \Delta'$ or $\Gamma_2, \neg A_2 \Rightarrow \Delta'$. By induction assumption, either $v \not\models \Gamma_2, \neg A_1 \Rightarrow \Delta'$ or $v \not\models \Gamma_2, A_2 \Rightarrow \Delta'$, and either $v \not\models \Gamma_2, A_1 \Rightarrow \Delta'$ or $v \not\models \Gamma_2, \neg A_2 \Rightarrow \Delta'$. Hence, $v \not\models \Gamma_2, A_1 \otimes A_2 \Rightarrow \Delta'$.

Case $\Gamma' \Rightarrow \Delta' = \Gamma' \Rightarrow B_1 \otimes B_2, \Delta_2 \in \xi$. Then $\Gamma' \Rightarrow \Delta'$ has children nodes $\in \xi$ containing sequents either $\Gamma' \Rightarrow B_1, \Delta_2$ or $\Gamma' \Rightarrow B_2, \Delta_2$, and either $\Gamma' \Rightarrow \neg B_1, \Delta_2$ or $\Gamma' \Rightarrow \neg B_2, \Delta_2$. By induction assumption, either $v \not\models \Gamma' \Rightarrow B_1, \Delta_2$ or $v \not\models \Gamma' \Rightarrow B_2, \Delta_2$; and either $v \not\models \Gamma' \Rightarrow \neg B_1, \Delta_2$ or $v \not\models \Gamma' \Rightarrow \neg B_2, \Delta_2$. Hence, $v \not\models \Gamma' \Rightarrow B_1 \otimes B_2, \Delta_2$.

Similar for other cases. □

2.3.4 Gentzen Deduction System $\mathbf{G}_{\oplus\otimes}$

A co-sequent $\Gamma \mapsto \Delta$ is valid, denoted by $\models_{\oplus\otimes} \Gamma \mapsto \Delta$, if there is an assignment v such that for every $A \in \Gamma$, $v(A) = 1$, and for every $B \in \Delta$, $v(B) = 0$.

Gentzen deduction system $\mathbf{G}_{\oplus\otimes}$ consists of the following axiom and deduction rules.

- **Axiom**:

$$(\mathbb{A}_{\oplus\otimes})\ \frac{\mathrm{con}(\Gamma)\&\mathrm{inval}(\Delta)\&\Gamma \cap \Delta = \emptyset}{\Gamma \mapsto \Delta}$$

  where Γ, Δ are sets of literals.

- **Deduction rules**:

$$(\neg^L)\ \frac{\Gamma \mapsto A, \Delta}{\Gamma, \neg A \mapsto \Delta} \qquad (\neg^R)\ \frac{\Gamma, A \mapsto \Delta}{\Gamma \mapsto \neg A, \Delta}$$

$$(\oplus^L)\ \frac{\left\{\begin{array}{l}\left[\begin{array}{l}\Gamma, \neg A_1 \rightarrow \Delta\\ \Gamma, A_2 \mapsto \Delta\end{array}\right.\\ \left[\begin{array}{l}\Gamma, A_1 \rightarrow \Delta\\ \Gamma, \neg A_2 \mapsto \Delta\end{array}\right.\end{array}\right.}{\Gamma, A_1 \oplus A_2 \mapsto \Delta} \qquad (\oplus^R)\ \frac{\left\{\begin{array}{l}\left[\begin{array}{l}\Gamma \mapsto \neg B_1, \Delta\\ \Gamma \mapsto \neg B_2, \Delta\end{array}\right.\\ \left[\begin{array}{l}\Gamma \mapsto B_1, \Delta\\ \Gamma \mapsto B_2, \Delta\end{array}\right.\end{array}\right.}{\Gamma \mapsto B_1 \oplus B_2, \Delta}$$

$$(\neg\oplus^L)\ \frac{\left\{\begin{array}{l}\left[\begin{array}{l}\Gamma, \neg A_1 \rightarrow \Delta\\ \Gamma, \neg A_2 \mapsto \Delta\end{array}\right.\\ \left[\begin{array}{l}\Gamma, A_1 \rightarrow \Delta\\ \Gamma, A_2 \mapsto \Delta\end{array}\right.\end{array}\right.}{\Gamma, \neg(A_1 \oplus A_2) \mapsto \Delta} \qquad (\neg\oplus^R)\ \frac{\left\{\begin{array}{l}\left[\begin{array}{l}\Gamma \mapsto \neg B_1, \Delta\\ \Gamma \mapsto B_2, \Delta\end{array}\right.\\ \left[\begin{array}{l}\Gamma \mapsto B_1, \Delta\\ \Gamma \mapsto \neg B_2, \Delta\end{array}\right.\end{array}\right.}{\Gamma \mapsto \neg(B_1 \oplus B_2), \Delta}$$

and

$$(\otimes^L)\ \frac{\left\{\begin{array}{l}\left[\begin{array}{l}\Gamma, \neg A_1 \rightarrow \Delta \\ \Gamma, \neg A_2 \mapsto \Delta\end{array}\right. \\ \left[\begin{array}{l}\Gamma, A_1 \rightarrow \Delta \\ \Gamma, A_2 \mapsto \Delta\end{array}\right.\end{array}\right.}{\Gamma, A_1 \otimes A_2 \mapsto \Delta} \qquad (\otimes^R)\ \frac{\left\{\begin{array}{l}\left[\begin{array}{l}\Gamma \mapsto \neg B_1, \Delta \\ \Gamma \mapsto B_2, \Delta\end{array}\right. \\ \left[\begin{array}{l}\Gamma \mapsto B_1, \Delta \\ \Gamma \mapsto \neg B_2, \Delta\end{array}\right.\end{array}\right.}{\Gamma \mapsto B_1 \otimes B_2, \Delta}$$

$$(\neg\otimes^L)\ \frac{\left\{\begin{array}{l}\left[\begin{array}{l}\Gamma, \neg A_1 \rightarrow \Delta \\ \Gamma, A_2 \mapsto \Delta\end{array}\right. \\ \left[\begin{array}{l}\Gamma, A_1 \rightarrow \Delta \\ \Gamma, \neg A_2 \mapsto \Delta\end{array}\right.\end{array}\right.}{\Gamma, \neg(A_1 \otimes A_2) \mapsto \Delta} \qquad (\neg\otimes^R)\ \frac{\left\{\begin{array}{l}\left[\begin{array}{l}\Gamma \mapsto \neg B_1, \Delta \\ \Gamma \mapsto \neg B_2, \Delta\end{array}\right. \\ \left[\begin{array}{l}\Gamma \mapsto B_1, \Delta \\ \Gamma \mapsto B_2, \Delta\end{array}\right.\end{array}\right.}{\Gamma \mapsto \neg(B_1 \otimes B_2), \Delta}$$

Definition 2.3.11 A co-sequent $\Gamma \mapsto \Delta$ is provable in $\mathbf{G}_{\oplus\otimes}$, denoted by $\vdash_{\oplus\otimes} \Gamma \vdash \Delta$, if there is a sequence $\{\Gamma_1 \mapsto \Delta_1, \ldots, \Gamma_n \mapsto \Delta_n\}$ of co-sequents such that $\Gamma_n \mapsto \Delta_n = \Gamma \mapsto \Delta$, and for each $1 \leq i \leq n$, $\Gamma_i \mapsto \Delta_i$ is an axiom or is deduced from the previous co-sequents by one of the deduction rules in $\mathbf{G}_{\oplus\otimes}$.

Theorem 2.3.12 (Soundness and completeness theorem) *For any co-sequent* $\Gamma \mapsto \Delta$,

$$\vdash_{\oplus\otimes} \Gamma \mapsto \Delta \text{ if and only if } \models_{\oplus\otimes} \Gamma \mapsto \Delta.$$

□

2.4 R-Calculi for Variant Propositional Logics

Let $x \in \{\{\neg, \rightarrow\}, \{\neg, \oplus, \otimes\}\}$.

A sequent $\Gamma \Rightarrow \Delta$ revises a formula pair (A, B) and obtains a sequent $\Gamma' \Rightarrow \Delta'$, denoted by $\models^x \Gamma \Rightarrow \Delta|(A, B) \rightrightarrows \Gamma' \Rightarrow \Delta'$, if

$$\Gamma' = \begin{cases} \Gamma - \{A\} \text{ if } \models^x \Gamma - \{A\} \Rightarrow \Delta \\ \Gamma \qquad \text{otherwise;} \end{cases}$$
$$\Delta' = \begin{cases} \Delta - \{B\} \text{ if } \models^x \Gamma' \Rightarrow \Delta - \{B\} \\ \Gamma \qquad \text{otherwise.} \end{cases}$$

A co-sequent $\Gamma \mapsto \Delta$ revises a formula pair (A, B) and obtains a sequent $\Gamma' \mapsto \Delta'$, denoted by $\models_x \Gamma \mapsto \Delta|(A, B) \rightrightarrows \Gamma' \mapsto \Delta'$,

$$\Gamma' = \begin{cases} \Gamma, A \text{ if } \models_x \Gamma, A \Rightarrow \Delta \\ \Gamma \qquad \text{otherwise;} \end{cases}$$
$$\Delta' = \begin{cases} \Delta, B \text{ if } \models_x \Gamma' \Rightarrow \Delta, B \\ \Gamma \qquad \text{otherwise.} \end{cases}$$

We will give R-calculi $\mathbf{R}^{\neg\rightarrow}$, $\mathbf{R}^{\oplus\otimes}$ for sequents and $\mathbf{R}_{\neg\rightarrow}$, $\mathbf{R}_{\oplus\otimes}$ for co-sequents, which are sound and complete, that is,

$$\vdash^x \Gamma \Rightarrow \Delta|(A, B) \rightrightarrows \Gamma' \Rightarrow \Delta' \text{ iff } \models^x \Gamma \Rightarrow \Delta|(A, B) \rightrightarrows \Gamma' \Rightarrow \Delta'$$
$$\vdash_x \Gamma \mapsto \Delta|(A, B) \rightrightarrows \Gamma' \mapsto \Delta' \text{ iff } \models_x \Gamma \mapsto \Delta|(A, B) \rightrightarrows \Gamma' \mapsto \Delta'.$$

2.4.1 *R-Calculus* $\mathbf{R}^{\neg\rightarrow}$

Let $A \in \Gamma$ and $B \in \Delta$. A reduction $\delta = \Gamma \Rightarrow \Delta|(A, B) \rightrightarrows \Gamma' \Rightarrow \Delta'$ is $\mathbf{R}^{\neg\rightarrow}$-valid, denoted by $\models^{\neg\rightarrow} \delta$, if

$$\Gamma' = \begin{cases} \Gamma - \{A\} & \text{if } \models^{\neg\rightarrow} \Gamma - \{A\} \Rightarrow \Delta \\ \Gamma & \text{otherwise} \end{cases}$$
$$\Delta' = \begin{cases} \Delta - \{B\} & \text{if } \models^{\neg\rightarrow} \Gamma' \Rightarrow \Delta - \{B\} \\ \Delta & \text{otherwise.} \end{cases}$$

R-calculus $\mathbf{R}^{\neg\rightarrow}$ consists of the following axioms and deduction rules:

- **Axioms**:

$$(\mathtt{A}_0^L)\ \frac{\sim \mathbf{E}l' \neq l(l', \neg l' \in \Gamma \text{ or } l', \neg l' \in \Delta \text{ or } l' \in \Gamma \cap \Delta)}{\Gamma \Rightarrow \Delta|(l, m) \rightrightarrows \Gamma \Rightarrow \Delta|m}$$
$$(\mathtt{A}_-^L)\ \frac{\mathbf{E}l' \neq l(l', \neg l' \in \Gamma \text{ or } l', \neg l' \in \Delta \text{ or } l' \in \Gamma \cap \Delta)}{\Gamma \Rightarrow \Delta|(l, m) \rightrightarrows \Gamma - \{l\} \Rightarrow \Delta|m}$$
$$(\mathtt{A}_0^R)\ \frac{\sim \mathbf{E}m' \neq m(m', \neg m' \in \Gamma \text{ or } m', \neg m' \in \Delta \text{ or } m' \in \Gamma \cap \Delta)}{\Gamma' \Rightarrow \Delta|m \rightrightarrows \Gamma' \Rightarrow \Delta}$$
$$(\mathtt{A}_-^R)\ \frac{\mathbf{E}m' \neq m(m', \neg m' \in \Gamma \text{ or } m', \neg m' \in \Delta \text{ or } m' \in \Gamma \cap \Delta)}{\Gamma' \Rightarrow \Delta|m \rightrightarrows \Gamma' \Rightarrow \Delta - \{m\}}$$

  where Γ, Δ are sets of literals and l, m are literals.
- **Deduction rules**:

$$(\neg\neg_0^L)\ \frac{\Gamma \Rightarrow \Delta|(A, B) \rightrightarrows \Gamma \Rightarrow \Delta|B}{\Gamma \Rightarrow \Delta|(\neg\neg A, B) \rightrightarrows \Gamma \Rightarrow \Delta|B}$$
$$(\neg\neg_-^L)\ \frac{\Gamma \Rightarrow \Delta|(A, B) \rightrightarrows \Gamma - \{A\} \Rightarrow \Delta|B}{\Gamma \Rightarrow \Delta|(\neg\neg A, B) \rightrightarrows \Gamma - \{\neg\neg A\} \Rightarrow \Delta|B}$$
$$(\neg\neg_0^R)\ \frac{\Gamma' \Rightarrow \Delta|B \rightrightarrows \Gamma' \Rightarrow \Delta}{\Gamma' \Rightarrow \Delta|\neg\neg B \rightrightarrows \Gamma' \Rightarrow \Delta}$$
$$(\neg\neg_-^R)\ \frac{\Gamma' \Rightarrow \Delta|B \rightrightarrows \Gamma' \Rightarrow \Delta - \{B\}}{\Gamma \Rightarrow \Delta|\neg\neg B \rightrightarrows \Gamma \Rightarrow \Delta - \{\neg\neg B\}}$$

and

$$(\to_0^L)\ \frac{\left[\begin{array}{l}\Gamma \Rightarrow \Delta|(\neg A_1, B) \rightrightarrows \Gamma \Rightarrow \Delta|B \\ \Gamma - \{\neg A_1\} \Rightarrow \Delta|(A_2, B) \rightrightarrows \Gamma - \{\neg A_1\} \Rightarrow \Delta|B\end{array}\right.}{\Gamma \Rightarrow \Delta|(A_1 \to A_2, B) \rightrightarrows \Gamma \Rightarrow \Delta|B}$$

$$(\to_-^L)\ \frac{\left\{\begin{array}{l}\Gamma \Rightarrow \Delta|(\neg A_1, B) \rightrightarrows \Gamma - \{\neg A_1\} \Rightarrow \Delta|B \\ \Gamma - \{\neg A_1\} \Rightarrow \Delta|(A_2, B) \rightrightarrows \Gamma - \{\neg A_1, A_2\} \Rightarrow \Delta|B\end{array}\right.}{\Gamma \Rightarrow \Delta|(A_1 \to A_2, B) \rightrightarrows \Gamma - \{A_1 \to A_2\} \Rightarrow \Delta|B}$$

$$(\to_0^R)\ \frac{\left[\begin{array}{l}\Gamma' \Rightarrow \Delta|\neg B_1 \rightrightarrows \Gamma' \Rightarrow \Delta \\ \Gamma' \Rightarrow \Delta|B_2 \rightrightarrows \Gamma' \Rightarrow \Delta\end{array}\right.}{\Gamma' \Rightarrow \Delta|B_1 \to B_2 \rightrightarrows \Gamma' \Rightarrow \Delta}$$

$$(\to_-^R)\ \frac{\left\{\begin{array}{l}\Gamma' \Rightarrow \Delta|\neg B_1 \rightrightarrows \Gamma' \Rightarrow \Delta - \{\neg B_1\} \\ \Gamma' \Rightarrow \Delta|B_2 \rightrightarrows \Gamma' \Rightarrow \Delta - \{B_2\}\end{array}\right.}{\Gamma' \Rightarrow \Delta|B_1 \to B_2 \rightrightarrows \Gamma' \Rightarrow \Delta - \{B_1 \to B_2\}}$$

and

$$(\neg\to_0^L)\ \frac{\left[\begin{array}{l}\Gamma \Rightarrow \Delta|(A_1, B) \rightrightarrows \Gamma \Rightarrow \Delta|B \\ \Gamma \Rightarrow \Delta|(\neg A_2, B) \rightrightarrows \Gamma \Rightarrow \Delta|B\end{array}\right.}{\Gamma \Rightarrow \Delta|(\neg(A_1 \to A_2), B) \rightrightarrows \Gamma \Rightarrow \Delta|B}$$

$$(\neg\to_-^L)\ \frac{\left\{\begin{array}{l}\Gamma \Rightarrow \Delta|(A_1, B) \rightrightarrows \Gamma - \{A_1\} \Rightarrow \Delta|B \\ \Gamma \Rightarrow \Delta|(\neg A_2, B) \rightrightarrows \Gamma - \{\neg A_2\} \Rightarrow \Delta|B\end{array}\right.}{\Gamma \Rightarrow \Delta|(\neg(A_1 \to A_2), B) \rightrightarrows \Gamma - \{\neg(A_1 \to A_2)\} \Rightarrow \Delta|B}$$

$$(\neg\to_0^R)\ \frac{\left\{\begin{array}{l}\Gamma' \Rightarrow \Delta|B_1 \rightrightarrows \Gamma' \Rightarrow \Delta \\ \Gamma' \Rightarrow \Delta - \{B_1\}|\neg B_2 \rightrightarrows \Gamma' \Rightarrow \Delta - \{B_1\}\end{array}\right.}{\Gamma' \Rightarrow \Delta|(\neg(A_1 \to A_2), B) \rightrightarrows \Gamma' \Rightarrow \Delta|B}$$

$$(\neg\to_-^R)\ \frac{\left[\begin{array}{l}\Gamma' \Rightarrow \Delta|B_1 \rightrightarrows \Gamma' \Rightarrow \Delta - \{B_1\} \\ \Gamma' \Rightarrow \Delta - \{B_1\}|\neg B_2 \rightrightarrows \Gamma' \Rightarrow \Delta - \{B_1, \neg B_2\}\end{array}\right.}{\Gamma' \Rightarrow \Delta|\neg(B_1 \to B_2) \rightrightarrows \Gamma' \Rightarrow \Delta - \{\neg(B_1 \to B_2)\}}$$

Definition 2.4.1 A reduction δ is provable in $\mathbf{R}^{\neg\to}$, denoted by $\vdash^{\neg\to} \delta$, if there is a sequence $\{\delta_1, \ldots, \delta_n\}$ of reductions such that $\delta_n = \delta$, and for each $1 \le i \le n$, δ_i is an axiom or is deduced from the previous reductions by one of the deduction rules in $\mathbf{R}^{\neg\to}$.

Theorem 2.4.2 (Soundness and completeness theorem) *For any reduction* $\delta = (A, B)|\Gamma \Rightarrow \Delta \rightrightarrows \Gamma' \Rightarrow \Delta'$,

$$\vdash^{\neg\to} \delta \textit{ iff } \models^{\neg\to} \delta.$$

□

2.4.2 R-Calculus $\mathbf{R}_{\neg\to}$

A reduction $\delta = \Gamma \mapsto \Delta|(A, B) \rightrightarrows \Gamma' \mapsto \Delta'$ is $\mathbf{R}_{\neg\to}$-valid, denoted by $\models_{\neg\to} \delta$, if

$$\Gamma' = \begin{cases} \Gamma, A \text{ if } \Gamma, A \mapsto \Delta \text{ is } \mathbf{G}_{\neg\rightarrow}\text{-valid} \\ \Gamma \quad \text{otherwise} \end{cases}$$
$$\Delta' = \begin{cases} \Delta, B \text{ if } \Gamma' \mapsto \Delta, B \text{ is } \mathbf{G}_{\neg\rightarrow}\text{-valid} \\ \Delta \quad \text{otherwise.} \end{cases}$$

R-calculus $\mathbf{R}_{\neg\rightarrow}$ consists of the following axioms and deduction rules:

- **Axioms**:

$$(\mathtt{A}_0^L)\ \frac{\neg l \in \Gamma \text{ or } l \in \Delta}{\Gamma \mapsto \Delta|(l, m) \rightrightarrows \Gamma \mapsto \Delta|m}$$
$$(\mathtt{A}_+^L)\ \frac{\neg l \notin \Gamma \& l \notin \Delta}{\Gamma \mapsto \Delta|(l, m) \rightrightarrows \Gamma, l \mapsto \Delta|m}$$
$$(\mathtt{A}_0^R)\ \frac{m \in \Gamma' \text{ or } \neg m \in \Delta}{\Gamma' \mapsto \Delta|m \rightrightarrows \Gamma' \mapsto \Delta}$$
$$(\mathtt{A}_+^R)\ \frac{m \notin \Gamma' \& \neg m \notin \Delta}{\Gamma' \mapsto \Delta|m \rightrightarrows \Gamma' \mapsto \Delta, m}$$

where Γ, Δ are sets of literals, and l, m are literals.

- **Deduction rules**:

$$(\neg\neg_0^L)\ \frac{\Gamma \mapsto \Delta|(A, B) \rightrightarrows \Gamma \mapsto \Delta|B}{\Gamma \mapsto \Delta|(\neg\neg A, B) \rightrightarrows \Gamma \mapsto \Delta|B}$$
$$(\neg\neg_+^L)\ \frac{\Gamma \mapsto \Delta|(A, B) \rightrightarrows \Gamma, A \mapsto \Delta|B}{\Gamma \mapsto \Delta|(\neg\neg A, B) \rightrightarrows \Gamma, \neg\neg A \mapsto \Delta|B}$$
$$(\neg\neg_0^R)\ \frac{\Gamma' \mapsto \Delta|B \rightrightarrows \Gamma' \mapsto \Delta}{\Gamma' \mapsto \Delta|\neg\neg B \rightrightarrows \Gamma' \mapsto \Delta}$$
$$(\neg\neg_+^R)\ \frac{\Gamma' \mapsto \Delta|B \rightrightarrows \Gamma' \mapsto \Delta, B}{\Gamma \mapsto \Delta|\neg\neg B \rightrightarrows \Gamma \mapsto \Delta, \neg\neg B}$$

and

$$(\rightarrow_0^L)\ \frac{\begin{cases} \Gamma \mapsto \Delta|(\neg A_1, B) \rightrightarrows \Gamma \mapsto \Delta|B \\ \Gamma, \neg A_1 \mapsto \Delta|(A_2, B) \rightrightarrows \Gamma, \neg A_1 \mapsto \Delta|B \end{cases}}{\Gamma \mapsto \Delta|(A_1 \rightarrow A_2, B) \rightrightarrows \Gamma \mapsto \Delta|B}$$
$$(\rightarrow_+^L)\ \frac{\left[\begin{array}{l} \Gamma \mapsto \Delta|(\neg A_1, B) \rightrightarrows \Gamma, \neg A_1 \mapsto \Delta|B \\ \Gamma, \neg A_1 \mapsto \Delta|(A_2, B) \rightrightarrows \Gamma, \neg A_1, A_2 \mapsto \Delta|B \end{array}\right.}{\Gamma \mapsto \Delta|(A_1 \rightarrow A_2, B) \rightrightarrows \Gamma, A_1 \rightarrow A_2 \mapsto \Delta|B}$$
$$(\rightarrow_0^R)\ \frac{\left[\begin{array}{l} \Gamma' \mapsto \Delta|\neg B_1 \rightrightarrows \Gamma' \mapsto \Delta \\ \Gamma' \mapsto \Delta|B_2 \rightrightarrows \Gamma' \mapsto \Delta \end{array}\right.}{\Gamma' \mapsto \Delta|B_1 \rightarrow B_2 \rightrightarrows \Gamma' \mapsto \Delta}$$
$$(\rightarrow_+^R)\ \frac{\begin{cases} \Gamma' \mapsto \Delta|\neg B_1 \rightrightarrows \Gamma' \mapsto \Delta, \neg B_1 \\ \Gamma' \mapsto \Delta|B_2 \rightrightarrows \Gamma' \mapsto \Delta, B_2 \end{cases}}{\Gamma' \mapsto \Delta|B_1 \rightarrow B_2 \rightrightarrows \Gamma' \mapsto \Delta, B_1 \rightarrow B_2}$$

and

$$(\neg \to^L_0)\ \frac{\left[\begin{array}{l}\Gamma \mapsto \Delta|(A_1, B) \rightrightarrows \Gamma \mapsto \Delta|B \\ \Gamma \mapsto \Delta|(\neg A_2, B) \rightrightarrows \Gamma \mapsto \Delta|B\end{array}\right.}{\Gamma \mapsto \Delta|(\neg(A_1 \to A_2), B) \rightrightarrows \Gamma \mapsto \Delta|B}$$

$$(\neg \to^L_+)\ \frac{\left\{\begin{array}{l}\Gamma \mapsto \Delta|(A_1, B) \rightrightarrows \Gamma, A_1 \mapsto \Delta|B \\ \Gamma \mapsto \Delta|(\neg A_2, B) \rightrightarrows \Gamma, \neg A_2 \mapsto \Delta|B\end{array}\right.}{\Gamma \mapsto \Delta|(\neg(A_1 \to A_2), B) \rightrightarrows \Gamma, \neg(A_1 \to A_2) \mapsto \Delta|B}$$

$$(\neg \to^R_0)\ \frac{\left\{\begin{array}{l}\Gamma' \mapsto \Delta|B_1 \rightrightarrows \Gamma' \mapsto \Delta \\ \Gamma' \mapsto \Delta, B_1|\neg B_2 \rightrightarrows \Gamma' \mapsto \Delta, B_1\end{array}\right.}{\Gamma' \mapsto \Delta|\neg(B_1 \to B_2) \rightrightarrows \Gamma' \mapsto \Delta}$$

$$(\neg \to^R_+)\ \frac{\left[\begin{array}{l}\Gamma' \mapsto \Delta|B_1 \rightrightarrows \Gamma' \mapsto \Delta, B_1 \\ \Gamma' \mapsto \Delta, B_1|\neg B_2 \rightrightarrows \Gamma' \mapsto \Delta, B_1, \neg B_2\end{array}\right.}{\Gamma' \mapsto \Delta|\neg(B_1 \to B_2) \rightrightarrows \Gamma' \mapsto \Delta, \neg(B_1 \to B_2)}$$

Definition 2.4.3 A reduction δ is provable in $\mathbf{R}_{\neg\to}$, denoted by $\vdash_{\wedge\to} \delta$, if there is a sequence $\{\delta_1, \ldots, \delta_n\}$ of reductions such that $\delta_n = \delta$, and for each $1 \leq i \leq n, \delta_i$ is an axiom or is deduced from the previous reductions by one of the deduction rules in $\mathbf{R}_{\neg\to}$.

Theorem 2.4.4 (Soundness and completeness theorem) *For any reduction* $\delta = \Gamma \mapsto \Delta|(A, B) \rightrightarrows \Gamma' \mapsto \Delta'$,

$$\vdash_{\neg\to} \delta \text{ iff } \models_{\neg\to} \delta.$$

□

2.4.3 *R-Calculus* $\mathbf{R}^{\oplus\otimes}$

Let $A \in \Gamma$ and $B \in \Delta$. A reduction $\delta = \Gamma \Rightarrow \Delta|(A, B) \rightrightarrows \Gamma' \Rightarrow \Delta'$ is $\mathbf{R}^{\oplus\otimes}$-valid, denoted by $\models^{\oplus\otimes} \delta$, if

$$\Gamma' = \begin{cases}\Gamma - \{A\} & \text{if } \models^{\oplus\otimes} \Gamma - \{A\} \Rightarrow \Delta \\ \Gamma & \text{otherwise}\end{cases}$$
$$\Delta' = \begin{cases}\Delta - \{B\} & \text{if } \models^{\oplus\otimes} \Gamma' \Rightarrow \Delta - \{B\} \\ \Delta & \text{otherwise.}\end{cases}$$

R-calculus $\mathbf{R}^{\oplus\otimes}$ consists of the following axioms and deduction rules:

- **Axioms**:

$$(\mathtt{A}_0^L)\ \frac{\sim \mathbf{E}l' \neq l(l', \neg l' \in \Gamma \text{ or } l', \neg l' \in \Delta \text{ or } l' \in \Gamma \cap \Delta)}{\Gamma \Rightarrow \Delta|(l, m) \rightrightarrows \Gamma \Rightarrow \Delta|m}$$

$$(\mathtt{A}_-^L)\ \frac{\mathbf{E}l' \neq l(l', \neg l' \in \Gamma \text{ or } l', \neg l' \in \Delta \text{ or } l' \in \Gamma \cap \Delta)}{\Gamma \Rightarrow \Delta|(l, m) \rightrightarrows \Gamma - \{l\} \Rightarrow \Delta|m}$$

$$(\mathtt{A}_0^R)\ \frac{\sim \mathbf{E}m' \neq m(m', \neg m' \in \Gamma \text{ or } m', \neg m' \in \Delta \text{ or } m' \in \Gamma \cap \Delta)}{\Gamma' \Rightarrow \Delta|m \rightrightarrows \Gamma' \Rightarrow \Delta}$$

$$(\mathtt{A}_-^R)\ \frac{\mathbf{E}m' \neq m(m', \neg m' \in \Gamma \text{ or } m', \neg m' \in \Delta \text{ or } m' \in \Gamma \cap \Delta)}{\Gamma' \Rightarrow \Delta|m \rightrightarrows \Gamma' \Rightarrow \Delta - \{m\}}$$

where Γ, Δ are sets of literals and l, m are literals.

- **Deduction rules**:

$$(\neg\neg_0^L)\ \frac{\Gamma \Rightarrow \Delta|(A, B) \rightrightarrows \Gamma \Rightarrow \Delta|B}{\Gamma \Rightarrow \Delta|(\neg\neg A, B) \rightrightarrows \Gamma \Rightarrow \Delta|B}$$

$$(\neg\neg_-^L)\ \frac{\Gamma \Rightarrow \Delta|(A, B) \rightrightarrows \Gamma - \{A\} \Rightarrow \Delta|B}{\Gamma \Rightarrow \Delta|(\neg\neg A, B) \rightrightarrows \Gamma - \{\neg\neg A\} \Rightarrow \Delta|B}$$

$$(\neg\neg_0^R)\ \frac{\Gamma' \Rightarrow \Delta|B \rightrightarrows \Gamma' \Rightarrow \Delta}{\Gamma' \Rightarrow \Delta|\neg\neg B \rightrightarrows \Gamma' \Rightarrow \Delta}$$

$$(\neg\neg_-^R)\ \frac{\Gamma' \Rightarrow \Delta|B \rightrightarrows \Gamma' \Rightarrow \Delta - \{B\}}{\Gamma \Rightarrow \Delta|\neg\neg B \rightrightarrows \Gamma \Rightarrow \Delta - \{\neg\neg B\}}$$

and

$$(\oplus_0^L)\ \frac{\left[\begin{array}{l}\left\{\begin{array}{l}\Gamma \Rightarrow \Delta|(\neg A_1, B) \rightrightarrows \Gamma \Rightarrow \Delta|B \\ \Gamma - \{\neg A_1\} \Rightarrow \Delta|(\neg A_2, B) \rightrightarrows \Gamma - \{\neg A_1\} \Rightarrow \Delta|B\end{array}\right. \\ \left\{\begin{array}{l}\Gamma \Rightarrow \Delta|(A_1, B) \rightrightarrows \Gamma \Rightarrow \Delta|B \\ \Gamma - \{A_1\} \Rightarrow \Delta|(A_2, B) \rightrightarrows \Gamma - \{A_1\} \Rightarrow \Delta|B\end{array}\right.\end{array}\right.}{\Gamma \Rightarrow \Delta|(A_1 \oplus A_2, B) \rightrightarrows \Gamma \Rightarrow \Delta|B}$$

$$(\oplus_-^L)\ \frac{\left\{\begin{array}{l}\left[\begin{array}{l}\Gamma \Rightarrow \Delta|(\neg A_1, B) \rightrightarrows \Gamma - \{\neg A_1\} \Rightarrow \Delta|B \\ \Gamma - \{\neg A_1\} \Rightarrow \Delta|(\neg A_2, B) \rightrightarrows \Gamma - \{\neg A_1, \neg A_2\} \Rightarrow \Delta|B\end{array}\right. \\ \left[\begin{array}{l}\Gamma \Rightarrow \Delta|(A_1, B) \rightrightarrows \Gamma - \{A_1\} \Rightarrow \Delta|B \\ \Gamma - \{A_1\} \Rightarrow \Delta|(A_2, B) \rightrightarrows \Gamma - \{A_1, A_2\} \Rightarrow \Delta|B\end{array}\right.\end{array}\right.}{\Gamma \Rightarrow \Delta|(A_1 \oplus A_2, B) \rightrightarrows \Gamma - \{A_1 \oplus A_2\} \Rightarrow \Delta|B}$$

$$(\oplus_0^R)\ \frac{\left[\begin{array}{l}\left\{\begin{array}{l}\Gamma' \Rightarrow \Delta|\neg B_1 \rightrightarrows \Gamma' \Rightarrow \Delta \\ \Gamma' \Rightarrow \Delta - \{\neg B_1\}|B_2 \rightrightarrows \Gamma' \Rightarrow \Delta - \{\neg B_1\}\end{array}\right. \\ \left\{\begin{array}{l}\Gamma' \Rightarrow \Delta|B_1 \rightrightarrows \Gamma' \Rightarrow \Delta \\ \Gamma' \Rightarrow \Delta - \{B_1\}|\neg B_2 \rightrightarrows \Gamma' \Rightarrow \Delta - \{B_1\}\end{array}\right.\end{array}\right.}{\Gamma' \Rightarrow \Delta|B_1 \oplus B_2 \rightrightarrows \Gamma' \Rightarrow \Delta}$$

$$(\oplus_-^R)\ \frac{\left\{\begin{array}{l}\left[\begin{array}{l}\Gamma' \Rightarrow \Delta|\neg B_1 \rightrightarrows \Gamma' \Rightarrow \Delta - \{\neg B_1\} \\ \Gamma' \Rightarrow \Delta - \{\neg B_1\}|B_2 \rightrightarrows \Gamma' \Rightarrow \Delta - \{\neg B_1, B_2\}\end{array}\right. \\ \left[\begin{array}{l}\Gamma' \Rightarrow \Delta|B_1 \rightrightarrows \Gamma' \Rightarrow \Delta - \{B_1\} \\ \Gamma' \Rightarrow \Delta - \{B_1\}|\neg B_2 \rightrightarrows \Gamma' \Rightarrow \Delta - \{B_1, \neg B_2\}\end{array}\right.\end{array}\right.}{\Gamma' \Rightarrow \Delta|B_1 \oplus B_2 \rightrightarrows \Gamma' \Rightarrow \Delta - \{B_1 \oplus B_2\}}$$

and

$$(\neg\oplus_0^L)\ \frac{\left[\begin{cases}\Gamma\Rightarrow\Delta|(\neg A_1,B)\rightrightarrows\Gamma\Rightarrow\Delta|B\\ \Gamma-\{\neg A_1\}\Rightarrow\Delta|(A_2,B)\rightrightarrows\Gamma-\{\neg A_1\}\Rightarrow\Delta|B\end{cases}\quad\begin{cases}\Gamma\Rightarrow\Delta|(A_1,B)\rightrightarrows\Gamma\Rightarrow\Delta|B\\ \Gamma-\{A_1\}\Rightarrow\Delta|(\neg A_2,B)\rightrightarrows\Gamma-\{A_1\}\Rightarrow\Delta|B\end{cases}\right]}{\Gamma\Rightarrow\Delta|(\neg(A_1\oplus A_2),B)\rightrightarrows\Gamma\Rightarrow\Delta|B}$$

$$(\neg\oplus_-^L)\ \frac{\left\{\begin{array}{l}\left[\begin{array}{l}\Gamma\Rightarrow\Delta|(\neg A_1,B)\rightrightarrows\Gamma-\{\neg A_1\}\Rightarrow\Delta|B\\ \Gamma-\{\neg A_1\}\Rightarrow\Delta|(A_2,B)\rightrightarrows\Gamma-\{\neg A_1,A_2\}\Rightarrow\Delta|B\end{array}\right.\\ \left[\begin{array}{l}\Gamma\Rightarrow\Delta|(A_1,B)\rightrightarrows\Gamma,A_1\Rightarrow\Delta|B\\ \Gamma-\{A_1\}\Rightarrow\Delta|(\neg A_2,B)\rightrightarrows\Gamma-\{A_1,\neg A_2\}\Rightarrow\Delta|B\end{array}\right.\end{array}\right.}{\Gamma\Rightarrow\Delta|(\neg(A_1\oplus A_2),B)\rightrightarrows\Gamma-\{\neg(A_1\oplus A_2)\}\Rightarrow\Delta|B}$$

$$(\neg\oplus_0^R)\ \frac{\left[\begin{cases}\Gamma'\Rightarrow\Delta|\neg B_1\rightrightarrows\Gamma'\Rightarrow\Delta\\ \Gamma'\Rightarrow\Delta-\{\neg B_1\}|\neg B_2\rightrightarrows\Gamma'\Rightarrow\Delta-\{\neg B_1\}\end{cases}\quad\begin{cases}\Gamma'\Rightarrow\Delta|B_1\rightrightarrows\Gamma'\Rightarrow\Delta\\ \Gamma'\Rightarrow\Delta-\{B_1\}|B_2\rightrightarrows\Gamma'\Rightarrow\Delta-\{B_1\}\end{cases}\right]}{\Gamma'\Rightarrow\Delta|\neg(B_1\oplus B_2)\rightrightarrows\Gamma'\Rightarrow\Delta}$$

$$(\neg\oplus_-^R)\ \frac{\left\{\begin{array}{l}\left[\begin{array}{l}\Gamma'\Rightarrow\Delta|\neg B_1\rightrightarrows\Gamma'\Rightarrow\Delta-\{\neg B_1\}\\ \Gamma'\Rightarrow\Delta-\{\neg B_1\}|\neg B_2\rightrightarrows\Gamma'\Rightarrow\Delta-\{\neg B_1,\neg B_2\}\end{array}\right.\\ \left[\begin{array}{l}\Gamma'\Rightarrow\Delta|B_1\rightrightarrows\Gamma'\Rightarrow\Delta-\{B_1\}\\ \Gamma'\Rightarrow\Delta-\{B_1\}|B_2\rightrightarrows\Gamma'\Rightarrow\Delta-\{B_1,B_2\}\end{array}\right.\end{array}\right.}{\Gamma'\Rightarrow\Delta|\neg(B_1\oplus B_2)\rightrightarrows\Gamma'\Rightarrow\Delta-\{\neg(B_1\oplus B_2)\}}$$

and

$$(\otimes_0^L)\ \frac{\left[\begin{cases}\Gamma\Rightarrow\Delta|(\neg A_1,B)\rightrightarrows\Gamma\Rightarrow\Delta|B\\ \Gamma-\{\neg A_1\}\Rightarrow\Delta|(A_2,B)\rightrightarrows\Gamma-\{\neg A_1\}\Rightarrow\Delta|B\end{cases}\quad\begin{cases}\Gamma\Rightarrow\Delta|(A_1,B)\rightrightarrows\Gamma\Rightarrow\Delta|B\\ \Gamma-\{A_1\}\Rightarrow\Delta|(\neg A_2,B)\rightrightarrows\Gamma-\{A_1\}\Rightarrow\Delta|B\end{cases}\right]}{\Gamma\Rightarrow\Delta|(A_1\otimes A_2,B)\rightrightarrows\Gamma\Rightarrow\Delta|B}$$

$$(\otimes_-^L)\ \frac{\left\{\begin{array}{l}\left[\begin{array}{l}\Gamma\Rightarrow\Delta|(\neg A_1,B)\rightrightarrows\Gamma-\{\neg A_1\}\Rightarrow\Delta|B\\ \Gamma-\{\neg A_1\}\Rightarrow\Delta|(A_2,B)\rightrightarrows\Gamma-\{\neg A_1,A_2\}\Rightarrow\Delta|B\end{array}\right.\\ \left[\begin{array}{l}\Gamma\Rightarrow\Delta|(A_1,B)\rightrightarrows\Gamma-\{A_1\}\Rightarrow\Delta|B\\ \Gamma-\{A_1\}\Rightarrow\Delta|(\neg A_2,B)\rightrightarrows\Gamma-\{A_1,\neg A_2\}\Rightarrow\Delta|B\end{array}\right.\end{array}\right.}{\Gamma\Rightarrow\Delta|(A_1\otimes A_2,B)\rightrightarrows\Gamma-\{A_1\otimes A_2\}\Rightarrow\Delta|B}$$

$$(\otimes_0^R)\ \frac{\left[\begin{cases}\Gamma'\Rightarrow\Delta|\neg B_1\rightrightarrows\Gamma'\Rightarrow\Delta\\ \Gamma'\Rightarrow\Delta-\{\neg B_1\}|\neg B_2\rightrightarrows\Gamma'\Rightarrow\Delta-\{\neg B_1\}\end{cases}\quad\begin{cases}\Gamma'\Rightarrow\Delta|B_1\rightrightarrows\Gamma'\Rightarrow\Delta\\ \Gamma'\Rightarrow\Delta-\{B_1\}|B_2\rightrightarrows\Gamma'\Rightarrow\Delta-\{B_1\}\end{cases}\right]}{\Gamma'\Rightarrow\Delta|B_1\otimes B_2\rightrightarrows\Gamma'\Rightarrow\Delta}$$

$$(\otimes_-^R)\ \frac{\left\{\begin{array}{l}\left[\begin{array}{l}\Gamma'\Rightarrow\Delta|\neg B_1\rightrightarrows\Gamma'\Rightarrow\Delta-\{\neg B_1\}\\ \Gamma'\Rightarrow\Delta-\{\neg B_1\}|\neg B_2\rightrightarrows\Gamma'\Rightarrow\Delta-\{\neg B_1,\neg B_2\}\end{array}\right.\\ \left[\begin{array}{l}\Gamma'\Rightarrow\Delta|B_1\rightrightarrows\Gamma'\Rightarrow\Delta-\{B_1\}\\ \Gamma'\Rightarrow\Delta-\{B_1\}|B_2\rightrightarrows\Gamma'\Rightarrow\Delta-\{B_1,B_2\}\end{array}\right.\end{array}\right.}{\Gamma'\Rightarrow\Delta|B_1\otimes B_2\rightrightarrows\Gamma'\Rightarrow\Delta-\{B_1\otimes B_2\}}$$

and

$$(\neg\otimes_0^L)\ \frac{\left[\begin{cases}\Gamma \Rightarrow \Delta|(A_1, B) \rightrightarrows \Gamma \Rightarrow \Delta|B \\ \Gamma - \{A_1\} \Rightarrow \Delta|(A_2, B) \rightrightarrows \Gamma - \{A_1\} \Rightarrow \Delta|B\end{cases} \\ \begin{cases}\Gamma \Rightarrow \Delta|(\neg A_1, B) \rightrightarrows \Gamma \Rightarrow \Delta|B \\ \Gamma - \{\neg A_1\} \Rightarrow \Delta|(\neg A_2, B) \rightrightarrows \Gamma - \{\neg A_1\} \Rightarrow \Delta|B\end{cases}\right.}{\Gamma \Rightarrow \Delta|(\neg(A_1 \otimes A_2), B) \rightrightarrows \Gamma \Rightarrow \Delta|B}$$

$$(\neg\otimes_-^L)\ \frac{\begin{cases}\left[\begin{array}{l}\Gamma \Rightarrow \Delta|(A_1, B) \rightrightarrows \Gamma - \{A_1\} \Rightarrow \Delta|B \\ \Gamma - \{A_1\} \Rightarrow \Delta|(A_2, B) \rightrightarrows B|\Gamma - \{A_1, A_2\} \Rightarrow \Delta|B\end{array}\right. \\ \left[\begin{array}{l}\Gamma \Rightarrow \Delta|(\neg A_1, B) \rightrightarrows \Gamma - \{\neg A_1\} \Rightarrow \Delta|B \\ \Gamma - \{\neg A_1\} \Rightarrow \Delta|(\neg A_2, B) \rightrightarrows \Gamma - \{\neg A_1, \neg A_2\} \Rightarrow \Delta|B\end{array}\right.\end{cases}}{\Gamma \Rightarrow \Delta|(\neg(A_1 \otimes A_2, B) \rightrightarrows \Gamma - \{\neg(A_1 \otimes A_2)\} \Rightarrow \Delta|B}$$

$$(\neg\otimes_0^R)\ \frac{\left[\begin{cases}\Gamma' \Rightarrow \Delta|B_1 \rightrightarrows \Gamma' \Rightarrow \Delta \\ \Gamma' \Rightarrow \Delta - \{B_1\}|\neg B_2 \rightrightarrows \Gamma' \Rightarrow \Delta - \{B_1\}\end{cases} \\ \begin{cases}\Gamma' \Rightarrow \Delta|\neg B_1 \rightrightarrows \Gamma' \Rightarrow \Delta \\ \Gamma' \Rightarrow \Delta - \{\neg B_1\}|B_2 \rightrightarrows \Gamma' \Rightarrow \Delta - \{\neg B_1\}\end{cases}\right.}{\Gamma' \Rightarrow \Delta|\neg(B_1 \otimes B_2) \rightrightarrows \Gamma' \Rightarrow \Delta}$$

$$(\neg\otimes_-^R)\ \frac{\begin{cases}\left[\begin{array}{l}\Gamma' \Rightarrow \Delta|B_1 \rightrightarrows \Gamma' \Rightarrow \Delta - \{B_1\} \\ \Gamma' \Rightarrow \Delta - \{B_1\}|\neg B_2 \rightrightarrows \Gamma' \Rightarrow \Delta - \{B_1, \neg B_2\}\end{array}\right. \\ \left[\begin{array}{l}\Gamma' \Rightarrow \Delta|\neg B_1 \rightrightarrows \Gamma' \Rightarrow \Delta - \{\neg B_1\} \\ \Gamma' \Rightarrow \Delta - \{\neg B_1\}|B_2 \rightrightarrows \Gamma' \Rightarrow \Delta - \{\neg B_1, B_2\}\end{array}\right.\end{cases}}{\Gamma' \Rightarrow \Delta|\neg(B_1 \otimes B_2) \rightrightarrows \Gamma' \Rightarrow \Delta - \{\neg(B_1 \otimes B_2)\}}$$

Definition 2.4.5 A reduction δ is provable in $\mathbf{R}^{\oplus\otimes}$, denoted by $\vdash^{\oplus\otimes} \delta$, if there is a sequence $\{\delta_1, \ldots, \delta_n\}$ of reductions such that $\delta_n = \delta$, and for each $1 \leq i \leq n$, δ_i is an axiom or is deduced from the previous reductions by one of the deduction rules in $\mathbf{R}^{\oplus\otimes}$.

Theorem 2.4.6 (Soundness and completeness theorem) *For any reduction* $\delta = (A, B)|\Gamma \Rightarrow \Delta \rightrightarrows \Gamma' \Rightarrow \Delta'$,

$$\vdash^{\oplus\otimes} \delta \textit{ iff } \models^{\oplus\otimes} \delta.$$

□

2.4.4 *R-Calculus* $\mathbf{R}_{\oplus\otimes}$

A reduction $\delta = (A, B)|\Gamma \mapsto \Delta \rightrightarrows \Gamma' \mapsto \Delta'$ is $\mathbf{R}_{\oplus\otimes}$-valid, denoted by $\models_{\oplus\otimes} \delta$, if

$$\Gamma' = \begin{cases}\Gamma, A \text{ if } \Gamma, A \mapsto \Delta \text{ is } \mathbf{G}^{\oplus\otimes}\text{-valid} \\ \Gamma \quad \text{otherwise}\end{cases}$$
$$\Delta' = \begin{cases}\Delta, B \text{ if } \Gamma' \mapsto \Delta, B \text{ is } \mathbf{G}^{\oplus\otimes}\text{-valid} \\ \Delta \quad \text{otherwise.}\end{cases}$$

R-calculus $\mathbf{R}_{\oplus\otimes}$ consists of the following axioms and deduction rules:

- **Axioms**:

$$(\mathtt{A}_0^L)\ \frac{\neg l \in \Gamma \text{ or } l \in \Delta}{\Gamma \mapsto \Delta|(l,m) \rightrightarrows \Gamma \mapsto \Delta|m}$$

$$(\mathtt{A}_+^L)\ \frac{\neg l \notin \Gamma \& l \notin \Delta}{\Gamma \mapsto \Delta|(l,m) \rightrightarrows \Gamma, l \mapsto \Delta|m}$$

$$(\mathtt{A}_0^R)\ \frac{m \in \Gamma' \text{ or } \neg m \in \Delta}{\Gamma' \mapsto \Delta|m \rightrightarrows \Gamma' \mapsto \Delta}$$

$$(\mathtt{A}_+^R)\ \frac{m \notin \Gamma' \& \neg m \notin \Delta}{\Gamma' \mapsto \Delta|m \rightrightarrows \Gamma' \mapsto \Delta, m}$$

where Γ, Δ are sets of literals, and l, m are literals.

- **Deduction rules**:

$$(\neg\neg_0^L)\ \frac{\Gamma \mapsto \Delta|(A,B) \rightrightarrows \Gamma \mapsto \Delta|B}{\Gamma \mapsto \Delta|(\neg\neg A,B) \rightrightarrows \Gamma \mapsto \Delta|B}$$

$$(\neg\neg_+^L)\ \frac{\Gamma \mapsto \Delta|(A,B) \rightrightarrows \Gamma, A \mapsto \Delta|B}{\Gamma \mapsto \Delta|(\neg\neg A,B) \rightrightarrows \Gamma, \neg\neg A \mapsto \Delta|B}$$

$$(\neg\neg_0^R)\ \frac{\Gamma' \mapsto \Delta|B \rightrightarrows \Gamma' \mapsto \Delta}{\Gamma' \mapsto \Delta|\neg\neg B \rightrightarrows \Gamma' \mapsto \Delta}$$

$$(\neg\neg_+^R)\ \frac{\Gamma' \mapsto \Delta|B \rightrightarrows \Gamma' \mapsto \Delta, B}{\Gamma \mapsto \Delta|\neg\neg B \rightrightarrows \Gamma \mapsto \Delta, \neg\neg B}$$

and

$$(\oplus_0^L)\ \frac{\left[\begin{array}{l}\left\{\begin{array}{l}\Gamma \mapsto \Delta|(\neg A_1,B) \rightrightarrows \Gamma \mapsto \Delta|B\\ \Gamma, \neg A_1 \mapsto \Delta|(\neg A_2,B) \rightrightarrows \Gamma, \neg A_1 \mapsto \Delta|B\end{array}\right.\\ \left\{\begin{array}{l}\Gamma \mapsto \Delta|(A_1,B) \rightrightarrows \Gamma \mapsto \Delta|B\\ \Gamma, A_1 \mapsto \Delta|(A_2,B) \rightrightarrows \Gamma, A_1 \mapsto \Delta|B\end{array}\right.\end{array}\right.}{\Gamma \mapsto \Delta|(A_1 \oplus A_2,B) \rightrightarrows \Gamma \mapsto \Delta|B}$$

$$(\oplus_+^L)\ \frac{\left\{\begin{array}{l}\left[\begin{array}{l}\Gamma \mapsto \Delta|(\neg A_1,B) \rightrightarrows \Gamma, \neg A_1 \mapsto \Delta|B\\ \Gamma, \neg A_1 \mapsto \Delta|(\neg A_2,B) \rightrightarrows \Gamma, \neg A_1, \neg A_2 \mapsto \Delta|B\end{array}\right.\\ \left[\begin{array}{l}\Gamma \mapsto \Delta|(A_1,B) \rightrightarrows \Gamma, A_1 \mapsto \Delta|B\\ \Gamma, \neg A_1 \mapsto \Delta|(A_2,B) \rightrightarrows \Gamma, A_1, A_2 \mapsto \Delta|B\end{array}\right.\end{array}\right.}{\Gamma \mapsto \Delta|(A_1 \oplus A_2,B) \rightrightarrows \Gamma, A_1 \oplus A_2 \mapsto \Delta|B}$$

$$(\oplus_0^R)\ \frac{\left[\begin{array}{l}\left\{\begin{array}{l}\Gamma' \mapsto \Delta|\neg B_1 \rightrightarrows \Gamma' \mapsto \Delta\\ \Gamma' \mapsto \Delta, \neg B_1|B_2 \rightrightarrows \Gamma' \mapsto \Delta, \neg B_1\end{array}\right.\\ \left\{\begin{array}{l}\Gamma' \mapsto \Delta|B_1 \rightrightarrows \Gamma' \mapsto \Delta\\ \Gamma' \mapsto \Delta, B_1|\neg B_2 \rightrightarrows \Gamma' \mapsto \Delta, B_1\end{array}\right.\end{array}\right.}{\Gamma' \mapsto \Delta|B_1 \oplus B_2 \rightrightarrows \Gamma' \mapsto \Delta}$$

$$(\oplus_+^R)\ \frac{\left\{\begin{array}{l}\left[\begin{array}{l}\Gamma' \mapsto \Delta|\neg B_1 \rightrightarrows \Gamma' \mapsto \Delta, \neg B_1\\ \Gamma' \mapsto \Delta, \neg B_1|B_2 \rightrightarrows \Gamma' \mapsto \Delta, \neg B_1, B_2\end{array}\right.\\ \left[\begin{array}{l}\Gamma' \mapsto \Delta|B_1 \rightrightarrows \Gamma' \mapsto \Delta, B_1\\ \Gamma' \mapsto \Delta, B_1|\neg B_2 \rightrightarrows \Gamma' \mapsto \Delta, B_1, \neg B_2\end{array}\right.\end{array}\right.}{\Gamma' \mapsto \Delta|B_1 \oplus B_2 \rightrightarrows \Gamma' \mapsto \Delta, B_1 \oplus B_2}$$

and

$$(\neg\oplus_0^L)\ \frac{\left[\begin{array}{l}\left\{\begin{array}{l}\Gamma \mapsto \Delta|(\neg A_1, B) \rightrightarrows \Gamma \mapsto \Delta|B\\ \Gamma, \neg A_1 \mapsto \Delta|(A_2, B) \rightrightarrows \Gamma, \neg A_1 \mapsto \Delta|B\end{array}\right.\\ \left\{\begin{array}{l}\Gamma \mapsto \Delta|(A_1, B) \rightrightarrows \Gamma \mapsto \Delta|B\\ \Gamma, A_1 \mapsto \Delta|(\neg A_2, B) \rightrightarrows \Gamma, A_1 \mapsto \Delta|B\end{array}\right.\end{array}\right.}{\Gamma \mapsto \Delta|(\neg(A_1 \oplus A_2), B) \rightrightarrows \Gamma \mapsto \Delta|B}$$

$$(\neg\oplus_+^L)\ \frac{\left\{\begin{array}{l}\left[\begin{array}{l}\Gamma \mapsto \Delta|(\neg A_1, B) \rightrightarrows \Gamma, \neg A_1 \mapsto \Delta|B\\ \Gamma, \neg A_1 \mapsto \Delta|(A_2, B) \rightrightarrows \Gamma, \neg A_1, A_2 \mapsto \Delta|B\end{array}\right.\\ \left[\begin{array}{l}\Gamma \mapsto \Delta|(A_1, B) \rightrightarrows \Gamma, A_1 \mapsto \Delta|B\\ \Gamma, A_1 \mapsto \Delta|(\neg A_2, B) \rightrightarrows \Gamma, A_1, \neg A_2 \mapsto \Delta|B\end{array}\right.\end{array}\right.}{\Gamma \mapsto \Delta|(\neg(A_1 \oplus A_2), B) \rightrightarrows B|\Gamma, \neg(A_1 \oplus A_2) \mapsto \Delta}$$

$$(\neg\oplus_0^R)\ \frac{\left[\begin{array}{l}\left\{\begin{array}{l}\Gamma' \mapsto \Delta|\neg B_1 \rightrightarrows \Gamma' \mapsto \Delta\\ \Gamma' \mapsto \Delta, \neg B_1|\neg B_2 \rightrightarrows \Gamma' \mapsto \Delta, \neg B_1\end{array}\right.\\ \left\{\begin{array}{l}\Gamma' \mapsto \Delta|B_1 \rightrightarrows \Gamma' \mapsto \Delta\\ \Gamma' \mapsto \Delta, B_1|B_2 \rightrightarrows \Gamma' \mapsto \Delta, B_1\end{array}\right.\end{array}\right.}{\Gamma' \mapsto \Delta|\neg(B_1 \oplus B_2) \rightrightarrows \Gamma' \mapsto \Delta}$$

$$(\neg\oplus_+^R)\ \frac{\left\{\begin{array}{l}\left[\begin{array}{l}\Gamma' \mapsto \Delta|\neg B_1 \rightrightarrows \Gamma' \mapsto \Delta, \neg B_1\\ \Gamma' \mapsto \Delta, \neg B_1|\neg B_2 \rightrightarrows \Gamma' \mapsto \Delta, \neg B_1, \neg B_2\end{array}\right.\\ \left[\begin{array}{l}\Gamma' \mapsto \Delta|B_1 \rightrightarrows \Gamma' \mapsto \Delta, B_1\\ \Gamma' \mapsto \Delta, B_1|B_2 \rightrightarrows \Gamma' \mapsto \Delta, B_1, B_2\end{array}\right.\end{array}\right.}{\Gamma' \mapsto \Delta|\neg(B_1 \oplus B_2) \rightrightarrows \Gamma' \mapsto \Delta, \neg(B_1 \oplus B_2)}$$

and

$$(\otimes_0^L)\ \frac{\left[\begin{array}{l}\left\{\begin{array}{l}\Gamma \mapsto \Delta|(\neg A_1, B) \rightrightarrows \Gamma \mapsto \Delta|B\\ \Gamma, \neg A_1 \mapsto \Delta|(A_2, B) \rightrightarrows \Gamma, \neg A_1 \mapsto \Delta|B\end{array}\right.\\ \left\{\begin{array}{l}\Gamma \mapsto \Delta|(A_1, B) \rightrightarrows \Gamma \mapsto \Delta|B\\ \Gamma, A_1 \mapsto \Delta|(\neg A_2, B) \rightrightarrows \Gamma, A_1 \mapsto \Delta|B\end{array}\right.\end{array}\right.}{\Gamma \mapsto \Delta|(A_1 \otimes A_2, B) \rightrightarrows \Gamma \mapsto \Delta|B}$$

$$(\otimes_+^L)\ \frac{\left\{\begin{array}{l}\left[\begin{array}{l}\Gamma \mapsto \Delta|(\neg A_1, B) \rightrightarrows \Gamma, \neg A_1 \mapsto \Delta|B\\ \Gamma, \neg A_1 \mapsto \Delta|(A_2, B) \rightrightarrows \Gamma, \neg A_1, A_2 \mapsto \Delta|B\end{array}\right.\\ \left[\begin{array}{l}\Gamma \mapsto \Delta|(A_1, B) \rightrightarrows \Gamma, A_1 \mapsto \Delta|B\\ \Gamma, A_1 \mapsto \Delta|(\neg A_2, B) \rightrightarrows \Gamma, A_1, \neg A_2 \mapsto \Delta|B\end{array}\right.\end{array}\right.}{\Gamma \mapsto \Delta|(A_1 \otimes A_2, B) \rightrightarrows \Gamma, A_1 \otimes A_2 \mapsto \Delta|B}$$

$$(\otimes_0^R)\ \frac{\left[\begin{array}{l}\left\{\begin{array}{l}\Gamma' \mapsto \Delta|\neg B_1 \rightrightarrows \Gamma' \mapsto \Delta\\ \Gamma' \mapsto \Delta, \neg B_1|\neg B_2 \rightrightarrows \Gamma' \mapsto \Delta, \neg B_1\end{array}\right.\\ \left\{\begin{array}{l}\Gamma' \mapsto \Delta|B_1 \rightrightarrows \Gamma' \mapsto \Delta\\ \Gamma' \mapsto \Delta, B_1|B_2 \rightrightarrows \Gamma' \mapsto \Delta, B_1\end{array}\right.\end{array}\right.}{\Gamma' \mapsto \Delta|B_1 \otimes B_2 \rightrightarrows \Gamma' \mapsto \Delta}$$

$$(\otimes_+^R)\ \frac{\left\{\begin{array}{l}\left[\begin{array}{l}\Gamma' \mapsto \Delta|\neg B_1 \rightrightarrows \Gamma' \mapsto \Delta, \neg B_1\\ \Gamma' \mapsto \Delta, \neg B_1|\neg B_2 \rightrightarrows \Gamma' \mapsto \Delta, \neg B_1, \neg B_2\end{array}\right.\\ \left[\begin{array}{l}\Gamma' \mapsto \Delta|B_1 \rightrightarrows \Gamma' \mapsto \Delta, B_1\\ \Gamma' \mapsto \Delta, B_1|B_2 \rightrightarrows \Gamma' \mapsto \Delta, B_1, B_2\end{array}\right.\end{array}\right.}{\Gamma' \mapsto \Delta|B_1 \otimes B_2 \rightrightarrows \Gamma' \mapsto \Delta, B_1 \otimes B_2}$$

and

$$
(\neg\otimes_0^L)\ \frac{\left[\begin{array}{l}\left\{\begin{array}{l}\Gamma \mapsto \Delta|(A_1, B) \rightrightarrows \Gamma \mapsto \Delta|B\\ \Gamma, A_1 \mapsto \Delta|(A_2, B) \rightrightarrows \Gamma, A_1 \mapsto \Delta|B\end{array}\right.\\ \left\{\begin{array}{l}\Gamma \mapsto \Delta|(\neg A_1, B) \rightrightarrows \Gamma \mapsto \Delta|B\\ \Gamma, \neg A_1 \mapsto \Delta|(\neg A_2, B) \rightrightarrows \Gamma, \neg A_1 \mapsto \Delta|B\end{array}\right.\end{array}\right.}{\Gamma \mapsto \Delta|(\neg(A_1 \otimes A_2), B) \rightrightarrows \Gamma \mapsto \Delta|B}
$$

$$
(\neg\otimes_+^L)\ \frac{\left\{\begin{array}{l}\left[\begin{array}{l}\Gamma \mapsto \Delta|(A_1, B) \rightrightarrows \Gamma, A_1 \mapsto \Delta|B\\ \Gamma, A_1 \mapsto \Delta|(A_2, B) \rightrightarrows \Gamma, A_1, A_2 \mapsto \Delta|B\end{array}\right.\\ \left[\begin{array}{l}\Gamma \mapsto \Delta|(\neg A_1, B) \rightrightarrows \Gamma, \neg A_1 \mapsto \Delta|B\\ \Gamma, \neg A_1 \mapsto \Delta|(\neg A_2, B) \rightrightarrows \Gamma, \neg A_1, \neg A_2 \mapsto \Delta|B\end{array}\right.\end{array}\right.}{\Gamma \mapsto \Delta|(\neg(A_1 \otimes A_2), B) \rightrightarrows \Gamma, \neg(A_1 \otimes A_2) \mapsto \Delta|B}
$$

$$
(\neg\otimes_0^R)\ \frac{\left[\begin{array}{l}\left\{\begin{array}{l}\Gamma' \mapsto \Delta|B_1 \rightrightarrows \Gamma' \mapsto \Delta\\ \Gamma' \mapsto \Delta, B_1|\neg B_2 \rightrightarrows \Gamma' \mapsto \Delta, B_1\end{array}\right.\\ \left\{\begin{array}{l}\Gamma' \mapsto \Delta|\neg B_1 \rightrightarrows \Gamma' \mapsto \Delta\\ \Gamma' \mapsto \Delta, \neg B_1|B_2 \rightrightarrows \Gamma' \mapsto \Delta, \neg B_1\end{array}\right.\end{array}\right.}{\Gamma' \mapsto \Delta|\neg(B_1 \otimes B_2) \rightrightarrows \Gamma' \mapsto \Delta}
$$

$$
(\neg\otimes_+^R)\ \frac{\left\{\begin{array}{l}\left[\begin{array}{l}\Gamma' \mapsto \Delta|B_1 \rightrightarrows \Gamma' \mapsto \Delta, B_1\\ \Gamma' \mapsto \Delta, B_1|\neg B_2 \rightrightarrows \Gamma' \mapsto \Delta, B_1, \neg B_2\end{array}\right.\\ \left[\begin{array}{l}\Gamma' \mapsto \Delta|\neg B_1 \rightrightarrows \Gamma' \mapsto \Delta, \neg B_1\\ \Gamma' \mapsto \Delta, \neg B_1|B_2 \rightrightarrows \Gamma' \mapsto \Delta, \neg B_1, B_2\end{array}\right.\end{array}\right.}{\Gamma' \mapsto \Delta|\neg(B_1 \otimes B_2) \rightrightarrows \Gamma' \mapsto \Delta, \neg(B_1 \otimes B_2)}
$$

Definition 2.4.7 A reduction δ is provable in $\mathbf{R}_{\oplus\otimes}$, denoted by $\vdash_{\oplus\otimes} \delta$, if there is a sequence $\{\delta_1, \ldots, \delta_n\}$ of reductions such that $\delta_n = \delta$, and for each $1 \leq i \leq n$, δ_i is an axiom or is deduced from the previous reductions by one of the deduction rules in $\mathbf{R}_{\oplus\otimes}$.

Theorem 2.4.8 (Soundness and completeness theorem) *For any reduction* $\delta = (A, B)|\Gamma \mapsto \Delta \rightrightarrows \Gamma' \mapsto \Delta'$,

$$\vdash_{\oplus\otimes} \delta \textit{ iff } \models_{\oplus\otimes} \delta. \qquad \square$$

References

Li, W.: Mathematical logic, foundations for information science. In: Progress in Computer Science and Applied Logic, vol. 25, Birkhäuser (2010)

Li, W.: R-calculus: an inference system for belief revision. Comput. J. **50**, 378–390 (2007)

Li, W., Sui, Y.: The sound and complete *R*-calculi with respect to pseudo-revision and pre-revision. Int. J. Intell. Sci. **3**, 110–117 (2013)

Li, W., Sui, Y.: An **R**-calculus for the propositional logic programming. In: Proceedings of International Conference on Computer Science and Information Technology, pp. 863–870 (2014)

Mendelson, E.: Introduction to Mathematical Logic. Wadsworth & Brooks/Cole Advanced Books & Software, Monterey, California (1964)

Takeuti, G.: Proof theory. In: Barwise, J. (ed.), Handbook of Mathematical Logic. Studies in Logic and the Foundations of Mathematics. North-Holland, Amsterdam, NL (1987)

Chapter 3
R-Calculi for Tableau/Gentzen Deduction Systems

A logic consists of a logical language, syntax and semantics, where the logical language specifies what symbols can be used in the logic, and the symbols are decomposed into two classes: logical and nonlogical, where the logical symbols are the ones used in each language of the logic, and the nonlogical symbols are those which are different for different logical languages of the logic; the syntax specifies what strings of symbols are meaningful (formulas) in logic, and the semantics specifies the truth-values of formulas under an assignment (or a model). In propositional logic, there is no nonlogical symbol, because variables are logical. Hence, there is only one logical language of propositional logic.

There are several choices of a set of logical connectives:

$$\{\neg, \wedge\}, \{\neg, \vee\}, \{\neg, \rightarrow\}, \{\neg, \wedge, \vee\}, \{\neg, \wedge, \vee, \rightarrow, \leftrightarrow\}.$$

We choose $\{\neg, \wedge, \vee\}$ as the set of logical connectives, and other connectives are defined as follows.

$$\begin{aligned}
&A \rightarrow B = \neg A \vee B \\
&A \leftrightarrow B = (A \rightarrow B) \wedge (B \rightarrow A) = (\neg A \vee B) \wedge (\neg B \vee A).
\end{aligned}$$

The semantics of propositional logic is defined by assignments, functions from variables to the $\{0, 1\}$-values.

Li (2007) proposed a belief revision operator called R-calculus **R** in first-order logic, which satisfies the AGM postulates and is nonmonotonic (Clark 1987; Ginsberg 1987; Reiter 1980). These deduction rules in R-calculus are reduced to the following two deduction rules:

$$(0)\ \frac{\Delta \vdash \neg A}{\Delta | A \Rightarrow \Delta} \qquad (+)\ \frac{\Delta \nvdash \neg A}{\Delta | A \Rightarrow \Delta, A},$$

W. Li and Y. Sui, *R-Calculus, IV: Propositional Logic*,
Perspectives in Formal Induction, Revision and Evolution,
https://doi.org/10.1007/978-981-19-8633-8_3

where (0) says that if A is inconsistent with Δ then the result of Δ revising A is Δ; and if A is consistent with Δ then the result of Δ revising A is $\Delta \cup \{A\}$.

Therefore, R-calculus is monotonic by $(-)$ and nonmonotonic by rule $(+)$.

A sequent $\Gamma \Rightarrow \Delta$ is valid (Li 2010; Takeuti and Barwise 1987) if for any assignment v, v satisfying each formula in Γ implies v satisfying some formula in Δ, equivalently, either v satisfies the negation of some formula in Γ or v satisfies some formula in Δ.

Dually, a co-sequent $\Gamma \mapsto \Delta$ is valid if there is an assignment v such that v satisfies each formula in Γ and the negation of each formula in Δ.

$$\begin{array}{l} \models \Gamma \Rightarrow \Delta \text{ iff } \mathbf{A}v(\mathbf{E}A \in \Gamma(v(A) = 0) \vee \mathbf{E}B \in \Delta(v(B) = 1)) \\ \models \Gamma \mapsto \Delta \text{ iff } \mathbf{E}v(\mathbf{A}A \in \Gamma(v(A) = 1) \& \mathbf{A}B \in \Delta(v(B) = 0)). \end{array}$$

When Δ is empty, sequent $\Gamma \Rightarrow$ is valid iff Γ is $\mathtt{f}$-valid; and when Γ is empty, sequent $\Rightarrow \Delta$ is valid iff Δ is $\mathtt{t}$-valid, where

$$\begin{array}{l} \Gamma \text{ is } \mathtt{f}\text{-valid iff } \mathbf{A}v\mathbf{E}A \in \Gamma(v(A) = 0) \\ \Delta \text{ is } \mathtt{t}\text{-valid iff } \mathbf{A}v\mathbf{E}B \in \Delta(v(B) = 1). \end{array}$$

When Δ is empty, co-sequent $\Gamma \mapsto$ is valid iff Γ is $\mathtt{t}$-consistent; and when Γ is empty, co-sequent $\mapsto \Delta$ is valid iff Δ is $\mathtt{f}$-consistent.

$$\begin{array}{l} \Gamma \text{ is } \mathtt{t}\text{-consistent iff } \mathbf{E}v\mathbf{A}A \in \Gamma(v(A) = 1) \\ \Delta \text{ is } \mathtt{f}\text{-consistent iff } \mathbf{E}v\mathbf{A}B \in \Delta(v(B) = 0). \end{array}$$

$\mathtt{t}\mathrm{val}(\Gamma)$, $\mathtt{f}\mathrm{val}(\Gamma)$ are monotonic in Γ; and $\mathtt{tcon}(\Gamma)$, $\mathtt{fcon}(\Gamma)$ are nonmonotonic in Γ. That is,

$$\begin{array}{l} \Gamma \subseteq \Gamma' \ \& \ \mathtt{f}\mathrm{val}(\Gamma) \Rightarrow \mathtt{f}\mathrm{val}(\Gamma') \\ \Delta \subseteq \Delta' \ \& \ \mathtt{t}\mathrm{val}(\Delta) \Rightarrow \mathtt{t}\mathrm{val}(\Delta') \\ \Gamma' \subseteq \Gamma \ \& \ \mathtt{tcon}(\Gamma') \Rightarrow \mathtt{tcon}(\Gamma) \\ \Delta' \subseteq \Delta \ \& \ \mathtt{fcon}(\Delta') \Rightarrow \mathtt{fcon}(\Delta) \end{array}$$

There are sound and complete Gentzen deduction systems $\mathbf{G}$ and $\mathbf{G}'$ for sequents and co-sequents, that is:

$$\begin{array}{l} \vdash \Gamma \Rightarrow \Delta \text{ iff } \models \Gamma \Rightarrow \Delta \\ \vdash \Gamma \mapsto \Delta \text{ iff } \models \Gamma \mapsto \Delta \end{array}$$

where $\mathbf{G}$ is monotonic and $\mathbf{G}'$ nonmonotonic (Cao et al. 2016; Reiter 1980).

A sequent $\Gamma \Rightarrow \Delta$ is decomposed into two sequents $\Gamma \Rightarrow$ and $\Rightarrow \Delta$; and a co-sequent $\Gamma \mapsto \Delta$ into two co-sequents $\Gamma \mapsto$ and $\mapsto \Delta$. For these sequents and co-sequents, we have four tableau proof systems in propositional logic (Hähnle 2001; Malinowski 2009; Urquhart and Gabbay 2001):

- $\mathbf{T}^{\mathtt{f}} : \Gamma \Rightarrow$ is $\mathbf{T}^{\mathtt{f}}$-provable iff Γ is $\mathtt{f}$-valid;
- $\mathbf{T}^{\mathtt{t}} : \Rightarrow \Delta$ is $\mathbf{T}^{\mathtt{t}}$-provable iff Δ is $\mathtt{t}$-valid;
- $\mathbf{T}_{\mathtt{t}} : \Gamma \mapsto$ is $\mathbf{T}_{\mathtt{t}}$-provable iff Γ is $\mathtt{t}$-consistent;
- $\mathbf{T}_{\mathtt{f}} : \mapsto \Delta$ is $\mathbf{T}_{\mathtt{f}}$-provable iff $\mapsto \Delta$ is $\mathtt{f}$-consistent.

That is,

$$\begin{array}{rcl} \vdash^{\mathtt{f}} \Gamma \Rightarrow & \text{iff} & \models^{\mathtt{f}} \Gamma \Rightarrow \\ \vdash^{\mathtt{t}} \Rightarrow \Delta & \text{iff} & \models^{\mathtt{t}} \Rightarrow \Delta \\ \vdash_{\mathtt{t}} \Gamma \mapsto & \text{iff} & \models_{\mathtt{t}} \Gamma \mapsto \\ \vdash_{\mathtt{f}} \mapsto \Delta & \text{iff} & \models_{\mathtt{f}} \mapsto \Delta, \end{array}$$

where $\vdash^{\mathtt{f}} / \vdash^{\mathtt{t}}$ is complementary to $\vdash_{\mathtt{t}} / \vdash_{\mathtt{f}}$, respectively.

G is complementary to $\mathbf{G}'$. Generally, we define

$$\begin{array}{l} \mathbf{G}^{\mathtt{t}} : \models^{\mathtt{t}} \Gamma \Rightarrow \Delta \text{ iff } \mathbf{A}v(\mathbf{E}A \in \Gamma(v(A) = 0) \text{ or } \mathbf{E}B \in \Delta(v(B) = 1)) \\ \mathbf{G}^{\mathtt{f}} : \models^{\mathtt{f}} \Gamma \Rightarrow \Delta \text{ iff } \mathbf{A}v(\mathbf{E}A \in \Gamma(v(A) = 1) \text{ or } \mathbf{E}B \in \Delta(v(B) = 0)) \\ \mathbf{G}_{\mathtt{t}} : \models_{\mathtt{t}} \Gamma \mapsto \Delta \text{ iff } \mathbf{E}v(\mathbf{A}A \in \Gamma(v(A) = 1) \& \mathbf{A}B \in \Delta(v(B) = 0)) \\ \mathbf{G}_{\mathtt{f}} : \models_{\mathtt{f}} \Gamma \mapsto \Delta \text{ iff } \mathbf{E}v(\mathbf{A}A \in \Gamma(v(A) = 0) \& \mathbf{A}B \in \Delta(v(B) = 1)). \end{array}$$

Then, $\mathbf{G} = \mathbf{G}^{\mathtt{t}}$ and $\mathbf{G}' = \mathbf{G}_{\mathtt{t}}$.

G is taken as a combination of $\mathbf{T}^{\mathtt{f}}$ and $\mathbf{T}^{\mathtt{t}}$, and $\mathbf{G}'$ a combination of $\mathbf{T}_{\mathtt{t}}$ and $\mathbf{T}_{\mathtt{f}}$, denoted by

$$\begin{array}{l} \mathbf{G} = \mathbf{T}^{\mathtt{f}} \boxplus \mathbf{T}^{\mathtt{t}},\ \mathbf{G}^{\mathtt{f}} = \mathbf{T}^{\mathtt{t}} \boxplus \mathbf{T}^{\mathtt{f}}, \\ \mathbf{G}' = \mathbf{T}_{\mathtt{t}} \boxplus \mathbf{T}_{\mathtt{f}},\ \mathbf{G}_{\mathtt{f}} = \mathbf{T}_{\mathtt{f}} \boxplus \mathbf{T}_{\mathtt{t}}. \end{array}$$

For each tableau proof system, there is a corresponding R-calculus. The four R-calculi are classified into two classes: preserving $\mathtt{t}/\mathtt{f}$-validity or preserving $\mathtt{t}/\mathtt{f}$-satisfiability. Let $* \in \{\mathtt{t}, \mathtt{f}\}$.

- For $\mathbf{T}^*$-validity, there is an R-calculus $\mathbf{S}^*$ to preserve the $*$-validity, that is, for any theory Γ and formulas $A \in \Gamma$,

(i) $\Gamma|A \rightrightarrows \Gamma$ is provable in $\mathbf{S}^*$ if and only if $\Gamma - \{A\}$ is not $*$-valid.
(ii) $\Gamma|A \rightrightarrows \Gamma - \{A\}$ is provable in $\mathbf{S}^*$ if and only if $\Gamma - \{A\}$ is $*$-valid.

- For $\mathbf{T}_*$-validity, there is an R-calculus $\mathbf{S}_*$ to preserve the $*$-satisfiability, that is, for any theory Δ and formula B,

(i) $\Delta|B \rightrightarrows \Delta$ is provable in $\mathbf{S}_*$ if and only if $\Delta \cup \{B\}$ is not $*$-consistent.
(ii) $\Delta|B \rightrightarrows \Delta, B$ is provable in $\mathbf{S}_*$ if and only if $\Delta \cup \{B\}$ is $*$-consistent.

These R-calculi are reduced to the following rules:

- for $\mathbf{S}^{\mathtt{f}}$: for $A \in \Gamma$,

$$(0)\ \frac{\mathtt{f}\mathrm{inval}(\Gamma - \{A\})}{\Gamma|A \rightrightarrows \Gamma}\quad (-)\ \frac{\mathtt{f}\mathrm{val}(\Gamma - \{A\})}{\Gamma|A \rightrightarrows \Gamma - \{A\}}$$

- for $\mathbf{S}^{\mathtt{t}}$: for $B \in \Delta$,

$$(0)\ \frac{\mathtt{tinval}(\Delta-\{B\})}{\Delta|B\rightrightarrows\Delta}\quad(-)\ \frac{\mathtt{tval}(\Delta-\{B\})}{\Delta|B\rightrightarrows\Delta-\{B\}}$$

- for $\mathbf{S}_{\mathtt{t}}$:

$$(+)\ \frac{\mathtt{tcon}(\Gamma,A)}{\Gamma|A\rightrightarrows\Gamma,A}\quad(0)\ \frac{\mathtt{tincon}(\Gamma,A)}{\Gamma|A\rightrightarrows\Gamma}$$

- for $\mathbf{S}_{\mathtt{f}}$:

$$(+)\ \frac{\mathtt{fcon}(\Delta,B)}{\Delta|B\rightrightarrows\Delta,B}\quad(0)\ \frac{\mathtt{fincon}(\Delta)}{\Delta|B\rightrightarrows\Delta}$$

Let $*\in\{\mathtt{t},\mathtt{f}\}$. For sequents and co-sequents, there are R-calculi $\mathbf{R}^*$ and $\mathbf{R}_*$, respectively, such that $\mathbf{R}^*$ is a combination of $\mathbf{S}^{\mathtt{f}}$ and $\mathbf{S}^{\mathtt{t}}$, and $\mathbf{R}_*$ a combination of $\mathbf{S}_{\mathtt{f}}$ and $\mathbf{S}_{\mathtt{t}}$, that is,

$$\mathbf{R}^{\mathtt{t}}=\mathbf{S}^{\mathtt{f}}\boxplus\mathbf{S}^{\mathtt{t}}\ \mathbf{R}^{\mathtt{f}}=\mathbf{S}^{\mathtt{t}}\boxplus\mathbf{S}^{\mathtt{f}}$$
$$\mathbf{R}_{\mathtt{t}}=\mathbf{S}_{\mathtt{t}}\boxplus\mathbf{S}_{\mathtt{f}}\ \mathbf{R}_{\mathtt{f}}=\mathbf{S}_{\mathtt{f}}\boxplus\mathbf{S}_{\mathtt{t}}.$$

3.1 Tableau Proof Systems

Let a logical language of propositional logic contains the following symbols:

- variables: $p_0,p_1,\ldots$; and
- logical connectives: $\neg,\wedge,\vee$.

A formula A is a string of the following forms:

$$A::=p|\neg A_1|A_1\wedge A_2|A_1\vee A_2,$$

where $l::=p|\neg p$ is called a literal.

An assignment v is a function from variables to $\mathbf{B}_2=\{0,1\}$.

The truth-value $v(A)$ of formula A under assignment v is

$$v(A)=\begin{cases}v(p) & \text{if } A=p\\ 1-v(A_1) & \text{if } A=\neg A_1\\ \min\{v(A_1),v(A_2)\} & \text{if } A=A_1\wedge A_2\\ \max\{v(A_1),v(A_2)\} & \text{if } A=A_1\vee A_2.\end{cases}$$

A formula A is satisfied in v, denoted by $v\models A$, if $v(A)=1$; and A is valid, denoted by $\models A$, if A is satisfied in any assignment v.

Lemma 3.1.1 *The following equivalences hold for* $v(A)=0$:

$$\begin{aligned}A_1 \wedge A_2 &\equiv A_1 \underline{\vee} A_2\\ A_1 \vee A_2 &\equiv A_1 \underline{\wedge} A_2\\ \neg(A_1 \wedge A_2) &\equiv \neg A_1 \underline{\wedge} \neg A_2\\ \neg(A_1 \vee A_2) &\equiv \neg A_1 \underline{\vee} \neg A_2\end{aligned}$$

Proof For any assignment v,

(i) $v(A_1 \wedge A_2) = 0$ iff either $v(A_1) = 0$ or $v(A_2) = 0$;
(ii) $v(A_1 \vee A_2) = 0$ iff $v(A_1) = 0$ and $v(A_2) = 0$;
(iii) $v(\neg(A_1 \wedge A_2)) = 0$ iff $v(A_1 \wedge A_2) = 1$, iff $v(A_1) = 1$ and $v(A_2) = 1$, iff $v(\neg A_1) = 0$ and $v(\neg A_2) = 0$;
(iv) $v(\neg(A_1 \vee A_2)) = 0$ iff $v(A_1 \vee A_2) = 1$, iff either $v(A_1) = 1$ or $v(A_2) = 1$, iff either $v(\neg A_1) = 0$ or $v(\neg A_2) = 0$. □

Lemma 3.1.2 *The following equivalences hold for* $v(B) = 1$*:*

$$\begin{aligned}B_1 \wedge B_2 &\equiv B_1 \underline{\wedge} B_2\\ B_1 \vee B_2 &\equiv B_1 \underline{\vee} B_2\\ \neg(B_1 \wedge B_2) &\equiv \neg B_1 \underline{\vee} \neg B_2\\ \neg(B_1 \vee B_2) &\equiv \neg B_1 \underline{\wedge} \neg B_2\end{aligned}$$

where $\underline{\wedge}, \underline{\vee}$ *are* $\wedge, \vee$ *in semantics.*

Proof For any assignment v,

(i) $v(B_1 \wedge B_2) = 1$ iff $v(B_1) = 1$ and $v(B_2) = 1$;
(ii) $v(B_1 \vee B_2) = 1$ iff either $v(B_1) = 1$ or $v(B_2) = 1$;
(iii) $v(\neg(B_1 \wedge B_2)) = 1$ iff $v(B_1 \wedge B_2) = 0$, iff either $v(B_1) = 0$ or $v(B_2) = 0$, iff either $v(\neg B_1) = 1$ or $v(\neg B_2) = 1$;
(iv) $v(\neg(B_1 \vee B_2)) = 1$ iff $v(B_1 \vee B_2) = 0$, iff $v(B_1) = 0$ and $v(B_2) = 0$, iff $v(\neg B_1) = 1$ and $v(\neg B_2) = 1$. □

$v(A) \neq 1$ iff $v(A) = 0$, and $v(A) \neq 0$ iff $v(A) = 1$. Hence, we have the following equivalences

(i) for $v(A) \neq 1$:

$$\begin{aligned}A_1 \wedge A_2 &\cong A_1 \underline{\vee} A_2\\ A_1 \vee A_2 &\cong A_1 \underline{\wedge} A_2\\ \neg(A_1 \wedge A_2) &\cong \neg A_1 \underline{\wedge} \neg A_2\\ \neg(A_1 \vee A_2) &\cong \neg A_1 \underline{\vee} \neg A_2\end{aligned}$$

(ii) for $v(B) \neq 0$:

$$\begin{array}{rl} B_1 \wedge B_2 & \cong B_1 \underline{\wedge} B_2 \\ B_1 \vee B_2 & \cong B_1 \underline{\vee} B_2 \\ \neg(B_1 \wedge B_2) & \cong \neg B_1 \underline{\vee} \neg B_2 \\ \neg(B_1 \vee B_2) & \cong \neg B_1 \underline{\wedge} \neg B_2 \end{array}$$

A theory Γ/Δ is

- $\mathbf{T}^{\mathtt{f}}$-valid, denoted by $\models^{\mathtt{f}} \Gamma \Rightarrow$, if for any assignment v, there is a formula $A \in \Gamma$ such that $v(A) = 0$;
- $\mathbf{T}^{\mathtt{t}}$-valid, denoted by $\models^{\mathtt{t}} \Rightarrow \Delta$, if for any assignment v, there is a formula $B \in \Delta$ such that $v(B) = 1$.
- $\mathbf{T}_{\mathtt{t}}$-valid, denoted by $\models_{\mathtt{t}} \Gamma \Rightarrow$, if there is an assignment v such that for every formula $A \in \Gamma$, $v(A) = 1$;
- $\mathbf{T}_{\mathtt{f}}$-valid, denoted by $\models_{\mathtt{f}} \Rightarrow \Delta$, if there is an assignment v such that for every formula $B \in \Delta$, $v(B) = 0$.

$$\begin{array}{l} \models^{\mathtt{f}} \Gamma \Rightarrow \text{ iff } \mathbf{A}v\mathbf{E}A \in \Gamma(v(A) = 0) \\ \models^{\mathtt{t}} \Rightarrow \Delta \text{ iff } \mathbf{A}v\mathbf{E}B \in \Delta(v(B) = 1) \\ \models_{\mathtt{t}} \Gamma \Rightarrow \text{ iff } \mathbf{E}v\mathbf{A}C \in \Gamma(v(C) = 1) \\ \models_{\mathtt{f}} \Rightarrow \Delta \text{ iff } \mathbf{E}v\mathbf{A}D \in \Delta(v(D) = 0). \end{array}$$

3.1.1 Tableau Proof System $\mathbf{T}^{\mathtt{f}}$

A sequent $\Gamma \Rightarrow$ is $\mathbf{T}^{\mathtt{f}}$-valid, denoted by $\models^{\mathtt{f}} \Gamma \Rightarrow$, if for any assignment v, there is a formula $A \in \Gamma$ such that $v(A) = 0$. That is, $\Gamma \Rightarrow$ is valid if and only if Γ is unsatisfiable; that is, there is no assignment v such that $v(A) = 1$ for each formula $A \in \Gamma$.

Tableau proof system $\mathbf{T}^{\mathtt{f}}$ consists of the following axiom and deductions:

- **Axiom**:

$$(\mathtt{A}^{\mathtt{f}}) \frac{\mathbf{E}p(p, \neg p \in \Gamma)}{\Gamma \Rightarrow},$$

where Γ is a set of literals.

- **Deduction rules**:

$$\begin{array}{ll} (\neg\neg^L) \dfrac{\Gamma, A \Rightarrow}{\Gamma, \neg\neg A \Rightarrow} & \\ (\wedge^L) \dfrac{\begin{cases} \Gamma, A_1 \Rightarrow \\ \Gamma, A_2 \Rightarrow \end{cases}}{\Gamma, A_1 \wedge A_2 \Rightarrow} & (\neg\wedge^L) \dfrac{\left[\begin{array}{l} \Gamma, \neg A_1 \Rightarrow \\ \Gamma, \neg A_2 \Rightarrow \end{array} \right.}{\Gamma, \neg(A_1 \wedge A_2) \Rightarrow} \\ (\vee^L) \dfrac{\left[\begin{array}{l} \Gamma, A_1 \Rightarrow \\ \Gamma, A_2 \Rightarrow \end{array} \right.}{\Gamma, A_1 \vee A_2 \Rightarrow} & (\neg\vee^L) \dfrac{\begin{cases} \Gamma, \neg A_1 \Rightarrow \\ \Gamma, \neg A_2 \Rightarrow \end{cases}}{\Gamma, \neg(A_1 \vee A_2) \Rightarrow} \end{array}$$

Definition 3.1.3 A sequent $\Gamma \Rightarrow$ is provable in $\mathbf{T}^{\mathrm{f}}$, denoted by $\vdash^{\mathrm{f}} \Gamma \Rightarrow$, if there is a sequence $\Gamma_1 \Rightarrow, \ldots, \Gamma_n \Rightarrow\}$ of sequents such that $\Gamma_n = \Gamma$, and for each $1 \leq i \leq n$, $\Gamma_i \Rightarrow$ is an axiom or is deduced from the previous sequents by one of the deduction rules in $\mathbf{T}^{\mathrm{f}}$.

Theorem 3.1.4 (Soundness theorem) *For any sequent* $\Gamma \Rightarrow$, *if* $\vdash^{\mathrm{f}} \Gamma \Rightarrow$ *then* $\models^{\mathrm{f}} \Gamma \Rightarrow$.

Proof We prove that each axiom is valid and each deduction rule preserves the validity.

($\mathtt{A}^{\mathrm{f}}$) Assume that there is a variable p such that $p, \neg p \in \Gamma$. There is no assignment v such that $v \models \Gamma$. Then, $\models \Gamma \Rightarrow$.

($\neg\neg$) Assume that there is no assignment v such that $v \models \Gamma, A_1 \Rightarrow$. Then, there is no assignment v such that $v \models \Gamma, \neg\neg A_1 \Rightarrow$.

($\wedge$) Assume that there is no assignment v such that $v \models \Gamma, A_1 \Rightarrow$. Then, there is no assignment v such that $v \models \Gamma, A_1 \wedge A_2 \Rightarrow$.

($\neg\wedge$) Assume that there is no assignment v such that $v \models \Gamma, \neg A_1 \Rightarrow$; and there is no assignment v such that $v \models \Gamma, \neg A_2 \Rightarrow$. Then, there is no assignment v such that $v \models \Gamma, \neg(A_1 \wedge A_2) \Rightarrow$.

($\vee$) Assume that there is no assignment v such that $v \models \Gamma, A_1 \Rightarrow$; and there is no assignment v such that $v \models \Gamma, A_2 \Rightarrow$. Then, there is no assignment v such that $v \models \Gamma, A_1 \vee A_2 \Rightarrow$.

($\neg\vee$) Assume that there is no assignment v such that $v \models \Gamma, \neg A_1 \Rightarrow$. Then, there is no assignment v such that $v \models \Gamma, \neg(A_1 \vee A_2) \Rightarrow$. □

Theorem 3.1.5 (Completeness theorem) *For any sequent* $\Gamma \Rightarrow$, *if* $\models^{\mathrm{f}} \Gamma \Rightarrow$ *then* $\vdash^{\mathrm{f}} \Gamma \Rightarrow$.

Proof Given a sequent $\Gamma \Rightarrow$, we construct a tree T such that either

(i) for each branch ξ of T, there is a sequent $\Gamma' \Rightarrow$ at the leaf of ξ such that $\Gamma' \Rightarrow$ is an axiom, or
(ii) there is an assignment v such that $v \models \Gamma$.

T is constructed as follows:

- the root of T is $\Gamma \Rightarrow$;
- for a node ξ, if the sequent $\Gamma' \Rightarrow$ at ξ is literal then the node is a leaf;
- otherwise, ξ has the direct children nodes containing the following sequents:

$$\begin{cases} \Gamma_1, A \Rightarrow & \text{if } \Gamma_1, \neg\neg A \Rightarrow \in \xi \\ \left[\begin{array}{l} \Gamma_1, A_1 \Rightarrow \\ \Gamma_1, A_2 \Rightarrow \end{array}\right. & \text{if } \Gamma_1, A_1 \wedge A_2 \Rightarrow \in \xi \\ \left\{\begin{array}{l} \Gamma_1, A_1 \Rightarrow \\ \Gamma_1, A_2 \Rightarrow \end{array}\right. & \text{if } \Gamma_1, A_1 \vee A_2 \Rightarrow \in \xi \\ \left\{\begin{array}{l} \Gamma_1, \neg A_1 \Rightarrow \\ \Gamma_1, \neg A_2 \Rightarrow \end{array}\right. & \text{if } \Gamma_1, \neg(A_1 \wedge A_2) \Rightarrow \in \xi \\ \left[\begin{array}{l} \Gamma_1, \neg A_1 \Rightarrow \\ \Gamma_1, \neg A_2 \Rightarrow \end{array}\right. & \text{if } \Gamma_1, \neg(A_1 \vee A_2) \Rightarrow \in \xi \end{cases}$$

where $\left[\begin{array}{l}\delta_1\\ \delta_2\end{array}\right.$ represents that δ_1, δ_2 are at a same child node; and $\left\{\begin{array}{l}\delta_1\\ \delta_2\end{array}\right.$ represents that δ_1, δ_2 are at different direct child nodes.

Notice that $\left[\begin{array}{l}\Gamma_1, A_1 \Rightarrow\\ \Gamma_1, A_2 \Rightarrow\end{array}\right.$ is equivalent to $\Gamma_1, A_1, A_2 \Rightarrow$. Hence, there is one and only one sequent at each node of the tree.

Lemma 3.1.6 *If for each branch $\xi \subseteq T$, there is a sequent $\Gamma' \Rightarrow \in \xi$ which is an axiom in $\mathbf{T}^{\mathtt{f}}$ then T is a proof tree of $\Gamma \Rightarrow$.*

Proof By the definition of T, T is a proof tree of $\Gamma \Rightarrow$. □

Lemma 3.1.7 *If there is a branch $\xi \subseteq T$ such that each sequent $\Gamma' \Rightarrow \in \xi$ is not an axiom in $\mathbf{T}^{\mathtt{f}}$ then there is an assignment v such that $v \models \Gamma$.*

Proof Let ξ be a branch of T such that each sequent $\Gamma' \Rightarrow \in \xi$ is not an axiom in $\mathbf{T}^{\mathtt{f}}$.
Let

$$\Phi = \{A : A \in \Gamma' \Rightarrow \in \xi\}.$$

Define an assignment v as follows:

$$v(p) = 1 \text{ iff } p \in \Phi.$$

Then, v is well-defined, and for the sequent $\Gamma' \Rightarrow$ at the leaf node of ξ, $v \models \Gamma'$.

We prove by induction on nodes η of ξ that for each sequent $\Gamma' \Rightarrow$ at η, $v \models \Gamma'$.

Case $\Gamma' \Rightarrow = \Gamma_2, \neg\neg A \Rightarrow \in \eta$. Then, $\Gamma' \Rightarrow$ has a child node $\in \xi$ containing $\Gamma_2, A \Rightarrow$. By induction assumption, $v \models \Gamma_2, A$, which implies that $v \models \Gamma_2, \neg\neg A$.

Case $\Gamma' \Rightarrow = \Gamma_2, A_1 \wedge A_2 \Rightarrow \in \eta$. Then, $\Gamma' \Rightarrow$ has a child node $\in \xi$ containing $\Gamma_2, A_1 \Rightarrow$ and $\Gamma_2, A_2 \Rightarrow$. By induction assumption, $v \models \Gamma_2, A_1$ and $v \models \Gamma_2, A_2$, i.e., $v \models \Gamma_2, A_1 \wedge A_2$.

Case $\Gamma' \Rightarrow = \Gamma_2, A_1 \vee A_2 \Rightarrow \in \eta$. Then, $\Gamma' \Rightarrow$ has a child node $\in \xi$ containing $\Gamma_2, A_i \Rightarrow$. By induction assumption, $v \models \Gamma_2, A_i$. Hence, $v \models \Gamma_2, A_1 \vee A_2$.

Similar for other cases. □

3.1.2 Tableau Proof System $\mathbf{T}^{\mathtt{t}}$

A sequent $\Rightarrow \Delta$ is $\mathbf{T}^{\mathtt{t}}$-valid, denoted by $\models^{\mathtt{t}} \Rightarrow \Delta$, if for each assignment v, there is a $B \in \Delta$ such that $v(B) = 1$. That is, $\Rightarrow \Delta$ is valid if and only if $\neg\Delta$ is unsatisfiable.

Tableau proof system $\mathbf{T}^{\mathtt{t}}$ consists of the following axiom and deductions:

- **Axiom**:

$$(\mathtt{A}^{\mathtt{t}})\ \frac{\mathbf{E}q(q, \neg q \in \Delta)}{\Rightarrow \Delta},$$

where Δ is a set of literals.

- **Deduction rules**:

$$(\neg\neg^R)\ \frac{\Rightarrow B, \Delta}{\Rightarrow \neg\neg B, \Delta}$$

$$(\wedge^R)\ \frac{\left[\begin{array}{l}\Rightarrow B_1, \Delta\\ \Rightarrow B_2, \Delta\end{array}\right.}{\Rightarrow B_1 \wedge B_2, \Delta} \qquad (\neg\wedge^R)\ \frac{\left\{\begin{array}{l}\Rightarrow \neg B_1, \Delta\\ \Rightarrow \neg B_2, \Delta\end{array}\right.}{\Rightarrow \neg(B_1 \wedge B_2), \Delta}$$

$$(\vee^R)\ \frac{\left\{\begin{array}{l}\Rightarrow B_1, \Delta\\ \Rightarrow B_2, \Delta\end{array}\right.}{\Rightarrow B_1 \vee B_2, \Delta} \qquad (\neg\vee^R)\ \frac{\left[\begin{array}{l}\Rightarrow \neg B_1, \Delta\\ \Rightarrow \neg B_2, \Delta\end{array}\right.}{\Rightarrow \neg(B_1 \vee B_2), \Delta}$$

Definition 3.1.8 A sequent $\Rightarrow \Delta$ is provable in $\mathbf{T}^{\mathtt{t}}$, denoted by $\vdash^{\mathtt{t}} \Rightarrow \Delta$ if there is a sequence $\{\Rightarrow \Delta_1, \ldots, \Rightarrow \Delta_n\}$ of sequents such that $\Delta_n = \Delta$, and for each $1 \leq i \leq n$, $\Rightarrow \Delta_i$ is an axiom or is deduced from the previous sequents by one of the deduction rules in $\mathbf{T}^{\mathtt{t}}$.

Theorem 3.1.9 (Soundness theorem) *For any sequent* $\Rightarrow \Delta$, *if* $\vdash^{\mathtt{t}} \Rightarrow \Delta$ *then* $\models^{\mathtt{t}} \Rightarrow \Delta$.

Proof We prove that each axiom is valid and each deduction rule preserves the validity.

($\mathtt{A}^{\mathtt{t}}$) Assume that there is a variable q such that $q, \neg q \in \Delta$. Then, for any assignment v, $v \models \Delta$. Then, $\models \Rightarrow \Delta$.

($\wedge$) Assume that for any assignment v, $v \models \Rightarrow B_1, \Delta$ and $v \models \Rightarrow B_2, \Delta$. Then, for any assignment v, $v \models \Rightarrow B_1 \wedge B_2, \Delta$.

($\vee$) Assume that for any assignment v, $v \models \Rightarrow B_i, \Delta$. Then, for any assignment v, $v \models \Rightarrow B_1 \vee B_2, \Delta$.

($\neg\wedge$) Assume that for any assignment v, $v \models \Rightarrow \neg B_i, \Delta$. Then, for any assignment v, $v \models \Rightarrow \neg(B_1 \wedge B_2), \Delta$.

($\neg\vee$) Assume that for any assignment v, $v \models \Rightarrow \neg B_1, \Delta$ and $v \models \Rightarrow \neg B_2, \Delta$. Then, for any assignment v, $v \models \Rightarrow \neg(B_1 \vee B_2), \Delta$. □

Theorem 3.1.10 (Completeness theorem) *For any sequent* $\Rightarrow \Delta$, *if* $\models^{\mathtt{t}} \Rightarrow \Delta$ *then* $\vdash^{\mathtt{t}} \Rightarrow \Delta$.

Proof Given a sequent $\Rightarrow \Delta$, we construct a tree T such that either

(i) for each branch ξ of T, the sequent $\Rightarrow \Delta'$ at the leaf of ξ is an axiom, or
(ii) there is an assignment v such that $v \not\models \Delta$.

T is constructed as follows:

- the root of T is $\Rightarrow \Delta$;
- for a node ξ, if the sequent $\Rightarrow \Delta'$ at ξ is literal then the node is a leaf;
- otherwise, ξ has the direct children nodes containing the following sequents:

$$\begin{cases} \Rightarrow B, \Delta_1 & \text{if} \Rightarrow \neg\neg B, \Delta_1 \in \xi \\ \left\{\begin{array}{l} \Rightarrow B_1, \Delta_1 \\ \Rightarrow B_2, \Delta_1 \end{array}\right. & \text{if} \Rightarrow B_1 \wedge B_2, \Delta_1 \in \xi \\ \left[\begin{array}{l} \Rightarrow B_1, \Delta_1 \\ \Rightarrow B_2, \Delta_1 \end{array}\right. & \text{if} \Rightarrow B_1 \vee B_2, \Delta_1 \in \xi \\ \left[\begin{array}{l} \Rightarrow \neg B_1, \Delta_1 \\ \Rightarrow \neg B_2, \Delta_1 \end{array}\right. & \text{if} \Rightarrow \neg(B_1 \wedge B_2), \Delta_1 \in \xi \\ \left\{\begin{array}{l} \Rightarrow \neg B_1, \Delta_1 \\ \Rightarrow \neg B_2, \Delta_1 \end{array}\right. & \text{if} \Rightarrow \neg(B_1 \vee B_2), \Delta_1 \in \xi \end{cases}$$

Notice that two sequents $\left[\begin{array}{l} \Rightarrow B_1, \Delta_1 \\ \Rightarrow B_2, \Delta_1 \end{array}\right.$ are taken as one sequent $\Rightarrow B_1, B_2, \Delta_1$. Hence, there is one and only one sequent at each node of tree T.

Lemma 3.1.11 *If for each branch $\xi \subseteq T$, there is a sequent $\Rightarrow \Delta' \in \xi$ which is an axiom in $\mathbf{T}^{\mathrm{t}}$ then T is a proof tree of $\Rightarrow \Delta$.*

Proof By the definition of T, T is a proof tree of $\Rightarrow \Delta$. □

Lemma 3.1.12 *If there is a branch $\xi \subseteq T$ such that each sequent $\Rightarrow \Delta' \in \xi$ is not an axiom in $\mathbf{T}^{\mathrm{t}}$ then there is an assignment v such that $v \not\models \Delta$.*

Proof Let ξ be a branch of T such that each sequent $\Rightarrow \Delta' \in \xi$ is not an axiom in $\mathbf{T}^{\mathrm{t}}$.

Let

$$\Psi = \{B : B \in \Rightarrow \Delta' \in \xi\}.$$

Define an assignment v as follows:

$$v(q) = 0 \text{ iff } q \in \Psi.$$

Then, v is well-defined, and for the sequent $\Rightarrow \Delta'$ at the leaf node of ξ, $v \not\models \Delta'$.

We prove by induction on nodes η of ξ that for each sequent $\Rightarrow \Delta'$ at η, $v \not\models \Delta'$.

Case $\Rightarrow \Delta' = \Rightarrow \neg\neg B, \Delta_2 \in \eta$. Then, $\Rightarrow \Delta'$ has a child node $\in \xi$ containing $\Rightarrow B, \Delta_2$. By induction assumption, $v \not\models B, \Delta_2$, which implies that $v \not\models \neg\neg B, \Delta_2$.

Case $\Rightarrow \Delta' = \Rightarrow B_1 \wedge B_2, \Delta_2 \in \eta$. Then, $\Rightarrow \Delta'$ has a child node $\in \xi$ containing $\Rightarrow B_i, \Delta_2$. By induction assumption, $v \not\models B_i, \Delta_2$, i.e., $v \not\models B_1 \wedge B_2, \Delta_2$.

Case $\Rightarrow \Delta' = \Rightarrow B_1 \vee B_2, \Delta_2 \in \eta$. Then, $\Rightarrow \Delta'$ has a child node $\in \xi$ containing $\Rightarrow B_1, \Delta_2$ and $\Rightarrow B_2, \Delta_2$. By induction assumption, $v \not\models B_1, \Delta_2$ and $v \not\models B_2, \Delta_2$, i.e., $v \not\models B_1 \vee B_2, \Delta_2$.

Case $\Rightarrow \Delta' = \Rightarrow \neg(B_1 \wedge B_2), \Delta_2 \in \eta$. Then, $\Rightarrow \Delta'$ has a child node $\in \xi$ containing $\Rightarrow \neg B_1, \Delta_2$ and $\Rightarrow \neg B_2, \Delta_2$. By induction assumption, $v \not\models \neg B_1, \Delta_2$ and $v \not\models \neg B_2, \Delta_2$, i.e., $v \not\models \neg(B_1 \wedge B_2), \Delta_2$.

Case $\Rightarrow \Delta' = \Rightarrow \neg(B_1 \vee B_2), \Delta_2 \in \eta$. Then, $\Rightarrow \Delta'$ has a child node $\in \xi$ containing $\Rightarrow \neg B_i$. By induction assumption, $v \not\models \neg B_i, \Delta_2$, i.e., $v \not\models \neg(B_1 \vee B_2), \Delta_2$. □

3.1.3 Tableau Proof System $\mathbf{T}_{\mathtt{t}}$

A co-sequent $\Gamma \mapsto$ is $\mathbf{T}_{\mathtt{t}}$-valid, denoted by $\models \Gamma \mapsto$, if *there is an assignment* v such that for each formula $A \in \Gamma$, $v \models A$.

Tableau proof system $\mathbf{T}_{\mathtt{t}}$ consists of the following axiom and deductions:

- **Axiom**:

$$(\mathtt{A_t})\ \frac{\sim \mathbf{E}p(p, \neg p \in \Gamma)}{\Gamma \mapsto},$$

  where Γ is a set of literals.
- **Deduction rules**:

$$(\neg\neg^L)\ \frac{\Gamma, A \mapsto}{\Gamma, \neg\neg A \mapsto}$$

$$(\wedge^L)\ \frac{\begin{bmatrix}\Gamma, A_1 \mapsto \\ \Gamma, A_2 \mapsto\end{bmatrix}}{\Gamma, A_1 \wedge A_2 \mapsto} \quad (\neg\wedge^L)\ \frac{\begin{cases}\Gamma, \neg A_1 \mapsto \\ \Gamma, \neg A_2 \mapsto\end{cases}}{\Gamma, \neg(A_1 \wedge A_2) \mapsto}$$

$$(\vee^L)\ \frac{\begin{cases}\Gamma, A_1 \mapsto \\ \Gamma, A_2 \mapsto\end{cases}}{\Gamma, A_1 \vee A_2 \mapsto} \quad (\neg\vee^L)\ \frac{\begin{bmatrix}\Gamma, \neg A_1 \mapsto \\ \Gamma, \neg A_2 \mapsto\end{bmatrix}}{\Gamma, \neg(A_1 \vee A_2) \mapsto}$$

Definition 3.1.13 A co-sequent $\Gamma \mapsto$ is $\mathbf{T}_{\mathtt{t}}$-provable, denoted by $\vdash_{\mathtt{t}} \Gamma \mapsto$, if there is a sequence $\{\Gamma_1 \mapsto, \ldots, \Gamma_n \mapsto\}$ of co-sequents such that $\Gamma_n \mapsto = \Gamma \mapsto$, and for each $1 \leq i \leq n$, $\Gamma_i \mapsto$ is an axiom or is deduced from the previous co-sequents by one of the deduction rules.

Theorem 3.1.14 (Soundness theorem) *For any co-sequent* $\Gamma \mapsto$, *if* $\vdash_{\mathtt{t}} \Gamma \mapsto$ *then* $\models_{\mathtt{t}} \Gamma \mapsto$.

Proof We prove that each axiom is valid and each deduction rule preserves the validity.

$(\mathtt{A_t})$ Assume that $\sim \mathbf{E}p(p, \neg p \in \Gamma)$. Then, there is an assignment v such that $v \models \Gamma$, and defined as follows: for any variable p,

$$v(p) = 1 \text{ iff } p \in \Gamma.$$

Then, v is well-defined and $v \models \Gamma \Rightarrow$.

$(\wedge)$ Assume that there is an assignment v such that $v \models \Gamma, A_1 \Rightarrow$ and $v \models \Gamma, A_2 \Rightarrow$. Then, for this assignment v, $v \models \Gamma, A_1 \wedge A_2 \Rightarrow$.

$(\vee)$ Assume that there is an assignment v such that $v \models \Gamma, A_i \Rightarrow$. Then, for this assignment v, $v \models \Gamma, A_1 \vee A_2 \Rightarrow$.

$(\neg\wedge)$ Assume that there is an assignment v such that $v \models \Gamma, \neg A_i \Rightarrow$. Then, for this assignment v, $v \models \Gamma, \neg(A_1 \wedge A_2) \Rightarrow$.

$(\neg\vee)$ Assume that there is an assignment v such that $v \models \Gamma, \neg A_1 \Rightarrow$ and $v \models \Gamma, \neg A_2 \Rightarrow$. Then, for this assignment v, $v \models \Gamma, \neg(A_1 \vee A_2) \Rightarrow$. □

Theorem 3.1.15 (Completeness theorem) *For any co-sequent* $\Gamma \mapsto$, *if* $\models_{\mathtt{t}} \Gamma \mapsto$ *then* $\vdash_{\mathtt{t}} \Gamma \mapsto$.

Proof Given a co-sequent $\Gamma \mapsto$, we construct a tree T such that either

(i) for each branch ξ of T, the co-sequent $\Gamma' \mapsto$ at the leaf of ξ is an axiom in $\mathbf{T}^{\mathtt{f}}$, or

(ii) there is no assignment v such that $v \models \Gamma$.

T is constructed as follows:

- the root of T is $\Gamma \mapsto$;
- for a node ξ, if the co-sequent $\Gamma' \mapsto'$ at ξ is literal then the node is a leaf;
- otherwise, ξ has the direct children nodes containing the following co-sequents:

$$\begin{cases} \Gamma_1, A \mapsto_1 & \text{if } \Gamma_1, \neg\neg A \mapsto_1 \in \xi \\ \left[\begin{array}{l} \Gamma_1, A_1 \mapsto_1 \\ \Gamma_1, A_2 \mapsto_1 \end{array}\right. & \text{if } \Gamma_1, A_1 \wedge A_2 \mapsto_1 \in \xi \\ \left\{\begin{array}{l} \Gamma_1, A_1 \mapsto_1 \\ \Gamma_1, A_2 \mapsto_1 \end{array}\right. & \text{if } \Gamma_1, A_1 \vee A_2 \mapsto_1 \in \xi \\ \left\{\begin{array}{l} \Gamma_1, \neg A_1 \mapsto_1 \\ \Gamma_1, \neg A_2 \mapsto_1 \end{array}\right. & \text{if } \Gamma_1, \neg(A_1 \wedge A_2) \mapsto_1 \in \xi \\ \left[\begin{array}{l} \Gamma_1, \neg A_1 \mapsto_1 \\ \Gamma_1, \neg A_2 \mapsto_1 \end{array}\right. & \text{if } \Gamma_1, \neg(A_1 \vee A_2) \mapsto_1 \in \xi \end{cases}$$

Lemma 3.1.16 *If there is a branch* $\xi \subseteq T$ *such that* $\Gamma' \mapsto \in \xi$ *is an axiom in* $\mathbf{T}_{\mathtt{t}}$ *then* ξ *is a proof of* $\Gamma \mapsto$.

Proof By the definition of T, ξ is a proof of $\Gamma \mapsto$. □

Lemma 3.1.17 *If for each branch* $\xi \subseteq T$ *the co-sequent* $\Gamma' \mapsto \in \xi$ *is not an axiom in* $\mathbf{T}_{\mathtt{t}}$ *then* T *is a proof of sequent* $\Gamma \Rightarrow$ *in* $\mathbf{T}^{\mathtt{f}}$.

Proof We prove by induction on nodes η of T that for any co-sequent $\Gamma' \mapsto$ at $\eta, \vdash^{\mathtt{f}} \Gamma' \Rightarrow$.

Case $\Gamma' \mapsto$ being literal. By assumption, $\Gamma' \mapsto$ is not an axiom in $\mathbf{T}_{\mathtt{t}}$, and then $\Gamma' \Rightarrow$ is an axiom in $\mathbf{T}^{\mathtt{f}}$.

Case $\Gamma' \mapsto = \Gamma_2, \neg\neg A \mapsto \in \eta$. Then, $\Gamma' \mapsto$ has a child node containing $\Gamma_2, A \mapsto$. By induction assumption, $\vdash^{\mathtt{f}} \Gamma_2, A \Rightarrow$, and by $(\neg\neg^L) \in \mathbf{T}^{\mathtt{f}}, \vdash^{\mathtt{f}} \Gamma_2, \neg\neg A \Rightarrow$.

Case $\Gamma' \mapsto = \Gamma_2, A_1 \wedge A_2 \mapsto \in \eta$. Then, $\Gamma' \mapsto$ has a child node containing $\Gamma_2, A_1, A_2 \mapsto$. By induction assumption, $\vdash^{\mathtt{f}} \Gamma_2, A_1, A_2 \Rightarrow$, and by $(\wedge^L) \in \mathbf{T}^{\mathtt{f}}$, i.e., $\vdash^{\mathtt{f}} \Gamma_2, A_1 \wedge A_2 \Rightarrow$.

Case $\Gamma' \mapsto = \Gamma_2, \neg(A_1 \wedge A_2) \mapsto \in \eta$. Then, $\Gamma' \mapsto'$ has child node containing $\Gamma_2, \neg A_1 \mapsto$ or $\Gamma_2, \neg A_2 \mapsto$. By induction assumption, $\vdash^{\mathtt{f}} \Gamma_2, \neg A_1 \Rightarrow$, $\vdash^{\mathtt{f}} \Gamma_2, \neg A_2 \Rightarrow$, and by $(\neg\wedge^L) \in \mathbf{T}^{\mathtt{f}}$, i.e., $\vdash^{\mathtt{f}} \Gamma_2, \neg(A_1 \wedge A_2) \Rightarrow$.

Similar for other cases. □

3.1.4 Tableau Proof System $\mathbf{T}_{\pounds}$

A co-sequent $\mapsto \Delta$ is $\mathbf{T}_{\pounds}$-valid, denoted by $\models_{\pounds} \mapsto \Delta$, if *there is an assignment* v such that for each formula $B \in \Delta$, $v \not\models B$.

Tableau proof system $\mathbf{T}_{\pounds}$ consists of the following axiom and deductions:

- **Axiom**:

$$(\mathtt{A}_{\pounds})\ \frac{\sim \mathbf{E}p(p, \neg p \in \Delta)}{\mapsto \Delta},$$

where Δ is a set of literals.
- **Deduction rules**:

$$(\neg\neg^R)\ \frac{\mapsto B, \Delta}{\mapsto \neg\neg B, \Delta}$$

$$(\wedge^R)\ \frac{\begin{cases}\mapsto B_1, \Delta \\ \mapsto B_2, \Delta\end{cases}}{\mapsto B_1 \wedge B_2, \Delta} \quad (\neg\wedge^R)\ \frac{\left[\begin{array}{l}\mapsto \neg B_1, \Delta \\ \mapsto \neg B_2, \Delta\end{array}\right.}{\mapsto \neg(B_1 \wedge B_2), \Delta}$$

$$(\vee^R)\ \frac{\left[\begin{array}{l}\mapsto B_1, \Delta \\ \mapsto B_2, \Delta\end{array}\right.}{\mapsto B_1 \vee B_2, \Delta} \quad (\neg\vee^R)\ \frac{\begin{cases}\mapsto \neg B_1, \Delta \\ \mapsto \neg B_2, \Delta\end{cases}}{\mapsto \neg(B_1 \vee B_2), \Delta}$$

Definition 3.1.18 A co-sequent $\mapsto \Delta$ is provable in $\mathbf{T}_{\pounds}$, denoted by $\vdash_{\pounds} \mapsto \Delta$ if there is a sequence $\{\mapsto \Delta_1, \ldots, \mapsto \Delta_n\}$ of co-sequents such that $\Delta_n = \Delta$, and for each $1 \leq i \leq n$, $\mapsto \Delta_i$ is an axiom or is deduced from the previous co-sequents by one of the deduction rules.

Theorem 3.1.19 (Soundness theorem) *For any co-sequent* $\mapsto \Delta$, *if* $\vdash_{\pounds} \mapsto \Delta$ *then* $\models_{\pounds} \mapsto \Delta$. □

Theorem 3.1.20 (Completeness theorem) *For any co-sequent* $\mapsto \Delta$, *if* $\models_{\pounds} \mapsto \Delta$ *then* $\vdash_{\pounds} \mapsto \Delta$.

Proof Given a co-sequent $\mapsto \Delta$, we construct a tree T such that either

(i) for each branch ξ of T, the co-sequent $\mapsto \Delta'$ at the leaf of ξ is an axiom in $\mathbf{T}^{\mathtt{t}}$, or
(ii) there is a branch ξ which is a proof of $\Rightarrow \Delta$.

T is constructed as follows:

- the root of T is $\mapsto \Delta$;
- for a node ξ, if the co-sequent $\mapsto \Delta'$ at ξ is literal then the node is a leaf;
- otherwise, ξ has the direct children nodes containing the following co-sequents:

$$\begin{cases} \mapsto B, \Delta_1 & \text{if} \mapsto \neg\neg B, \Delta_1 \in \xi \\ \begin{cases} \mapsto B_1, \Delta_1 \\ \mapsto B_2, \Delta_1 \end{cases} & \text{if} \mapsto B_1 \wedge B_2, \Delta_1 \in \xi \\ \begin{bmatrix} \mapsto B_1, \Delta_1 \\ \mapsto B_2, \Delta_1 \end{bmatrix} & \text{if} \mapsto B_1 \vee B_2, \Delta_1 \in \xi \\ \begin{bmatrix} \mapsto \neg B_1, \Delta_1 \\ \mapsto \neg B_2, \Delta_1 \end{bmatrix} & \text{if} \mapsto \neg(B_1 \wedge B_2), \Delta_1 \in \xi \\ \begin{cases} \mapsto \neg B_1, \Delta_1 \\ \mapsto \neg B_2, \Delta_1 \end{cases} & \text{if} \mapsto \neg(B_1 \vee B_2), \Delta_1 \in \xi. \end{cases}$$

Lemma 3.1.21 *If there is a branch $\xi \subseteq T$ such that the co-sequent $\mapsto \Delta' \in \xi$ is an axiom in $\mathbf{T}_{\mathtt{f}}$ then ξ is a proof of $\mapsto \Delta$.*

Proof By the definition of T, ξ is a proof of $\mapsto \Delta$. □

Lemma 3.1.22 *If for each branch $\xi \subseteq T$, the co-sequent $\mapsto \Delta' \in \xi$ is not an axiom in $\mathbf{T}_{\mathtt{f}}$ then T is a proof of sequent $\Rightarrow \Delta$ in $\mathbf{T}^{\mathtt{t}}$.*

Proof We prove by induction on nodes η of T that for each co-sequent $\mapsto \Delta'$ at $\eta, \vdash^{\mathtt{t}} \Rightarrow \Delta'$.

Case $\mapsto \Delta'$ being literal. By assumption, $\mapsto \Delta'$ is not an axiom in $\mathbf{T}_{\mathtt{f}}$, and $\Rightarrow \Delta'$ is an axiom in $\mathbf{T}^{\mathtt{t}}$.

Case $\mapsto \Delta' = \mapsto \neg\neg B, \Delta_2 \in \eta$. Then, $\mapsto \Delta'$ has a child node containing $\mapsto A, \Delta_2$. By induction assumption, $\vdash^{\mathtt{t}} \Rightarrow A, \Delta_2$, and by $(\neg\neg^R) \in \mathbf{T}^{\mathtt{t}}, \vdash^{\mathtt{t}} \Rightarrow \neg\neg B, \Delta_2$.

Case $\mapsto \Delta' = \mapsto B_1 \wedge B_2, \Delta_2 \in \eta$. Then, $\mapsto \Delta'$ has children nodes containing $\mapsto B_1, \Delta_2$ and $\mapsto B_2, \Delta_2$, respectively. By induction assumption, $\vdash^{\mathtt{t}} \Rightarrow B_1, \Delta_2$ and $\vdash^{\mathtt{t}} \Rightarrow B_2, \delta_2$, and by $(\wedge^R) \in \mathbf{T}^{\mathtt{t}}$, i.e., $\vdash^{\mathtt{t}} \Rightarrow B_1 \wedge B_2, \Delta_2$.

Case $\mapsto \Delta' = \mapsto \neg(B_1 \wedge B_2), \Delta_2 \in \eta$. Then, $\mapsto \Delta'$ has a direct child node containing $\mapsto \neg B_1, \neg B_2, \Delta_2$. By induction assumption, $\vdash^{\mathtt{t}} \Rightarrow \neg B_1, \neg B_2, \Delta_2$, and by $(\neg\wedge^R) \in \mathbf{T}^{\mathtt{t}}$, i.e., $\vdash^{\mathtt{t}} \Rightarrow \neg(B_1 \wedge B_2), \Delta_2$.

Similar for other cases. □

3.2 R-Calculi for Theories

Let $* \in \{\mathtt{t}, \mathtt{f}\}$.

Given a theory $\mathbf{X}$ and formula $X \in \mathbf{X}$, a reduction $\mathbf{X}|X \rightrightarrows \mathbf{X} - \{X'\}$ is $\mathbf{G}^*$-valid, denoted by $\models^* \mathbf{X}|X \rightrightarrows \mathbf{X} - \{X'\}$, if

$$X' = \begin{cases} X \text{ if } \mathbf{X} - \{X\} \text{ is } \mathbf{G}^*\text{-valid} \\ \lambda \text{ otherwise.} \end{cases}$$

Given a theory $\mathbf{X}$ and formula X, a reduction $\mathbf{X}|X \rightrightarrows \mathbf{X}, X'$ is $\mathbf{G}_*$-valid, denoted by $\models_* \mathbf{X}|X \rightrightarrows \mathbf{X}, X'$, if

$$X' = \begin{cases} X \text{ if } \mathbf{X} \cup \{X\} \text{ is } \mathbf{G}_*\text{-valid} \\ \lambda \text{ otherwise.} \end{cases}$$

3.2.1 R-Calculus $\mathbf{S}^{\mathtt{f}}$

Let $A \in \Gamma$. A reduction $\Gamma|A \rightrightarrows \Gamma - \{A'\}$ is $\mathbf{S}^{\mathrm{f}}$-valid, denoted by $\models^{\mathrm{f}} \Gamma|A \rightrightarrows \Gamma - \{A'\}$, if

$$A' = \begin{cases} A \text{ if} \models^{\mathrm{f}} \Gamma - \{A\} \\ \lambda \text{ otherwie.} \end{cases}$$

Intuitively, a formula $A_1 \wedge A_2$ is extractable from Δ, if either A_1 or A_2 is extractable from Δ; and $A_1 \vee A_2$ is extractable from Δ if A_1 is extractable from Δ and A_2 is extractable from $\Delta - \{A_1\}$.

A formula $A_1 \wedge A_2$ is not extractable from Δ if both A_1 and A_2 are not extractable from Δ; and $A_1 \vee A_2$ is not extractable from Δ if either A_1 is not extractable from Δ or A_2 is not extractable from $\Delta - \{A_1\}$.

R-calculus $\mathbf{S}^{\mathrm{f}}$ consists of the following axioms and deductions:

• **Axioms**:

$$(\mathtt{A}^-)\ \frac{\mathbf{E}l' \neq l(l', \neg l' \in \Gamma)}{\Gamma|l \rightrightarrows \Gamma - \{l\}}, \qquad (\mathtt{A}^0)\ \frac{\sim \mathbf{E}l' \neq l(l', \neg l' \in \Gamma)}{\Gamma|l \rightrightarrows \Gamma},$$

where Γ is a set of literals and l is a literal.

• **Deduction rules**:

$$(\neg\neg^-)\ \frac{\Gamma|A \rightrightarrows \Gamma - \{A\}}{\Gamma|\neg\neg A \rightrightarrows \Gamma - \{\neg\neg A\}} \qquad (\neg\neg^0)\ \frac{\Gamma|A \rightrightarrows \Gamma}{\Gamma|\neg\neg A \rightrightarrows \Gamma}$$

$$(\wedge^-)\ \frac{\begin{cases}\Gamma|A_1 \rightrightarrows \Gamma - \{A_1\} \\ \Gamma|A_2 \rightrightarrows \Gamma - \{A_2\}\end{cases}}{\Gamma|A_1 \wedge A_2 \rightrightarrows \Gamma - \{A_1 \wedge A_2\}} \qquad (\wedge^0)\ \frac{\begin{bmatrix}\Gamma|A_1 \rightrightarrows \Gamma \\ \Gamma|A_2 \rightrightarrows \Gamma\end{bmatrix}}{\Gamma|A_1 \wedge A_2 \rightrightarrows \Gamma}$$

$$(\vee^-)\ \frac{\begin{bmatrix}\Gamma|A_1 \rightrightarrows \Gamma - \{A_1\} \\ \Gamma - \{A_1\}|A_2 \rightrightarrows \Gamma - \{A_1, A_2\}\end{bmatrix}}{\Gamma|A_1 \vee A_2 \rightrightarrows \Gamma - \{A_1 \vee A_2\}} \qquad (\vee^0)\ \frac{\begin{cases}\Gamma|A_1 \rightrightarrows \Gamma \\ \Gamma - \{A_1\}|A_2 \rightrightarrows \Gamma - \{A_1\}\end{cases}}{\Gamma|A_1 \vee A_2 \rightrightarrows \Gamma}$$

and

$$(\neg\wedge^-)\ \frac{\begin{bmatrix}\Gamma|\neg A_1 \rightrightarrows \Gamma - \{\neg A_1\} \\ \Gamma - \{\neg A_1\}|\neg A_2 \rightrightarrows \Gamma - \{\neg A_1, \neg A_2\}\end{bmatrix}}{\Gamma|\neg(A_1 \wedge A_2) \rightrightarrows \Gamma - \{\neg(A_1 \wedge A_2)\}}$$

$$(\neg\wedge^0)\ \frac{\begin{cases}\Gamma|\neg A_1 \rightrightarrows \Gamma \\ \Gamma - \{\neg A_1\}|\neg A_2 \rightrightarrows \Gamma - \{\neg A_1\}\end{cases}}{\Gamma|\neg(A_1 \wedge A_2) \rightrightarrows \Gamma}$$

$$(\neg\vee^+)\ \frac{\begin{cases}\Gamma|\neg A_1 \rightrightarrows \Gamma - \{\neg A_1\} \\ \Gamma|\neg A_2 \rightrightarrows \Gamma - \{\neg A_2\}\end{cases}}{\Gamma|\neg(A_1 \vee A_2) \rightrightarrows \Gamma - \{\neg(A_1 \vee A_2)\}}$$

$$(\neg\vee^-)\ \frac{\begin{bmatrix}\Gamma|\neg A_1 \rightrightarrows \Gamma \\ \Gamma|\neg A_2 \rightrightarrows \Gamma\end{bmatrix}}{\Gamma|\neg(A_1 \vee A_2) \rightrightarrows \Gamma}$$

Definition 3.2.1 A reduction $\Gamma|A \rightrightarrows \Gamma, A'$ is provable in $\mathbf{S}^{\mathtt{f}}$, denoted by $\vdash^{\mathtt{f}} \Gamma|A \rightrightarrows \Gamma - \{A'\}$, if there is a sequence $\{\delta_1, \ldots, \delta_n\}$ of reductions such that $\delta_n = \Gamma|A \rightrightarrows \Gamma - \{A'\}$, and for each $1 \leq i \leq n$, δ_i is an axiom or is deduced from the previous reductions by one of the deduction rules in $\mathbf{S}^{\mathtt{f}}$.

Theorem 3.2.2 (Soundness theorems) *For any reduction* $\Gamma|A \rightrightarrows \Gamma - \{A'\}$, *if* $\vdash^{\mathtt{f}} \Gamma|A \rightrightarrows \Gamma$ *then* $\models \Gamma|A \rightrightarrows \Gamma$; *and if* $\vdash^{\mathtt{f}} \Gamma|A \rightrightarrows \Gamma - \{A\}$ *then* $\models \Gamma|A \rightrightarrows \Gamma - \{A\}$.

Proof We prove that each axiom is valid and each deduction rule preserves the validity.

($\mathtt{A}^-$) Assume that Γ is a set of literals and there is a literal $l' \neq l$ such that $l', \neg l' \in \Gamma$. Then, $\vdash^{\mathtt{f}} \Gamma|l \rightrightarrows \Gamma - \{l\}$, and $\models^{\mathtt{f}} \Gamma|l \rightrightarrows \Gamma - \{l\}$, since Γ being $\mathbf{T}^{\mathtt{f}}$-valid implies $\Gamma - \{l\}$ being $\mathbf{T}^{\mathtt{f}}$-valid. Assume that there is no such a literal $l' \neq l$. Then, $\vdash^{\mathtt{f}} \Gamma|l \rightrightarrows \Gamma$, and $\models^{\mathtt{f}} \Gamma|l \rightrightarrows \Gamma$.

($\wedge^0$) Assume that $\Gamma - \{A_1\}$ being $\mathtt{f}$-valid implies Γ being $\mathtt{f}$-valid and $\Gamma - \{A_2\}$ being $\mathtt{f}$-valid implies Γ being $\mathtt{f}$-valid. Then, $\Gamma - \{A_1 \wedge A_2\}$ being $\mathtt{f}$-valid implies Γ being $\mathtt{f}$-valid.

($\wedge^-$) Assume that Γ being $\mathtt{f}$-valid implies $\Gamma - \{A_i\}$ being $\mathtt{f}$-valid. Then, Γ being $\mathtt{f}$-valid implies $\Gamma - \{A_1 \wedge A_2\}$ being $\mathtt{f}$-valid.

($\neg\wedge^-$) Assume that $\models^{\mathtt{f}} \Gamma$ implies $\models^{\mathtt{f}} \Gamma - \{\neg A_1\}$; and $\models^{\mathtt{f}} \Gamma - \{\neg A_1\}$ implies $\models^{\mathtt{f}} \Gamma - \{\neg A_1, \neg A_2\}$. Then, $\models^{\mathtt{f}} \Gamma$ implies $\models^{\mathtt{f}} \Gamma - \{\neg(A_1 \wedge A_2)\}$.

($\neg\wedge^0$) Assume that $\models^{\mathtt{f}} \Gamma$ implies $\models^{\mathtt{f}} \Gamma$, and $\models^{\mathtt{f}} \Gamma - \{\neg A_1\}$ implies $\models^{\mathtt{f}} \Gamma - \{\neg A_1\}$. Then, $\models^{\mathtt{f}} \Gamma$ implies $\models^{\mathtt{f}} \Gamma$.

Similar for other cases. □

Theorem 3.2.3 *For any reduction* $\Gamma|A \rightrightarrows \Gamma, A'$,

$$\textit{if} \models \Gamma|A \rightrightarrows \Gamma, A' \textit{ then} \vdash \Gamma|A \rightrightarrows \Gamma, A'.$$

Proof Given a reduction $\Gamma|A \rightrightarrows \Gamma - \{A\}$, we construct a tree T such that either

(i) T is a proof tree of $\Gamma|A \rightrightarrows \Gamma - \{A\}$; i.e., for each branch ξ of T, the reduction γ' at the leaf of ξ is an axiom of form ($\mathtt{A}^-$), or
(ii) there is a branch ξ of T such that ξ is a proof of $\Gamma|A \rightrightarrows \Gamma$.

T is constructed as follows:

- the root of T is $\Gamma|A \rightrightarrows \Gamma - \{A\}$;
- for a node ξ, if each reduction at ξ is of form $\Gamma'|l' \rightrightarrows \Gamma - \{l'\}$ then the node is a leaf;
- otherwise, ξ has the direct child containing the following reductions:

$$\begin{cases} \Gamma_1|A \rightrightarrows \Gamma_1 - \{A\} & \text{if } \Gamma_1(\neg\neg A) \in \xi \\ \begin{bmatrix} \Gamma_1|A_1 \rightrightarrows \Gamma_1 - \{A_1\} \\ \Gamma_1|A_2 \rightrightarrows \Gamma_1 - \{A_2\} \end{bmatrix} & \text{if } \Gamma_1(A_1 \wedge A_2) \in \xi \\ \begin{cases} \Gamma_1|\neg A_1 \rightrightarrows \Gamma_1 - \{\neg A_1\} \\ \Gamma_1 - \{\neg A_1\}|\neg A_2 \rightrightarrows \Gamma_1 - \{\neg A_1, \neg A_2\} \end{cases} & \text{if } \Gamma_1(\neg(A_1 \wedge A_2)) \in \xi \\ \begin{cases} \Gamma_1|A_1 \rightrightarrows \Gamma_1 - \{A_1\} \\ \Gamma_1 - \{A_1\}|A_2 \rightrightarrows \Gamma_1 - \{A_1, A_2\} \end{cases} & \text{if } \Gamma_1(A_1 \vee A_2) \in \xi \\ \begin{bmatrix} \Gamma_1|\neg A_1 \rightrightarrows \Gamma_1 - \{\neg A_1\} \\ \Gamma_1|\neg A_2 \rightrightarrows \Gamma_1 - \{\neg A_2\} \end{bmatrix} & \text{if } \Gamma_1(\neg(A_1 \vee A_2)) \in \xi. \end{cases}$$

where $\Gamma_1(A) = \Gamma_1|A \rightrightarrows \Gamma_1 - \{A\}$.

Lemma 3.2.4 *If for each branch $\xi \subseteq T$, there is a reduction $\Gamma'|A' \rightrightarrows \Gamma'|A' \in \xi$ which is an axiom in $\mathbf{S}^{\text{f}}$ then T is a proof tree of $\Gamma|A \rightrightarrows \Gamma, A$.*

Proof By the definition of T, T is a proof tree of $\Gamma|A \rightrightarrows \Gamma, A$. □

Lemma 3.2.5 *If there is a branch $\xi \subseteq T$ such that each reduction $\Gamma'|A' \rightrightarrows \Gamma', A' \in \xi$ is not an axiom in $\mathbf{S}^{\text{f}}$, then ξ is a proof of $\Gamma|A \rightrightarrows \Gamma$ in $\mathbf{S}^{\text{f}}$.*

Proof Let ξ be a branch of T such that each reduction $\Gamma'|A' \rightrightarrows \Gamma', A' \in \xi$ is not an axiom in $\mathbf{S}^{\text{f}}$.

We proved by induction on node $\eta \in \xi$ that for each reduction $\Gamma'|A' \rightrightarrows \Gamma', A' \in \eta$,

$$\vdash^{\text{f}} \Gamma'|A' \rightrightarrows \Gamma'.$$

Case $\Gamma'|l \rightrightarrows \Gamma', l \in \eta$ is a leaf node of ξ. Then, there is no literal $l' \neq l$ such that $l', \neg l' \in \Gamma'$. By $(\mathtt{A}_0^{\text{f}})$, $\vdash^{\text{f}} \Gamma'|l \rightrightarrows \Gamma'$.

Case $\delta = \Gamma_2|\neg\neg A \rightrightarrows \Gamma_2, \neg\neg A \in \eta$. Then, η has a child node $\in \xi$ containing reduction $\Gamma_2|A \rightrightarrows \Gamma_2, A$. By induction assumption, $\vdash^{\text{f}} \Gamma_2|A \rightrightarrows \Gamma_2$, and by $(\neg\neg^0)$, $\vdash^{\text{f}} \Gamma_2|\neg\neg A \rightrightarrows \Gamma_2$.

Case $\delta = \Gamma_2|A_1 \wedge A_2 \rightrightarrows \Gamma_2, A_1 \wedge A_2 \in \eta$. Then, η has a child node $\in \xi$ containing reductions $\Gamma_2|A_1 \rightrightarrows \Gamma_2, A_1$ and $\Gamma_2|A_2 \rightrightarrows \Gamma_2, A_2$. By induction assumption, $\vdash^{\text{f}} \Gamma_2|A_1 \rightrightarrows \Gamma_2$, and $\vdash^{\text{f}} \Gamma_2|A_2 \rightrightarrows \Gamma_2$. By $(\wedge^0)$, we have $\vdash^{\text{f}} \Gamma_2, A_1 \wedge A_2 \rightrightarrows \Gamma_2$.

Case $\delta = \Gamma_2|A_1 \vee A_2 \rightrightarrows \Gamma_2, A_1 \vee A_2 \in \eta$. Then, η has a child node $\in \xi$ containing reduction either $\Gamma_2|A_1 \rightrightarrows \Gamma_2, A_1$ or $\Gamma_2, A_1|A_2 \rightrightarrows \Gamma_2, A_1, A_2$. By induction assumption, either $\vdash^{\text{f}} \Gamma_2|A_1 \rightrightarrows \Gamma_2$ or $\vdash^{\text{f}} \Gamma_2, A_1|A_2 \rightrightarrows \Gamma_2, A_1$. By $(\vee^0)$, we have $\vdash^{\text{f}} \Gamma_2, A_1 \vee A_2 \rightrightarrows \Gamma$.

Similar for other cases. □

3.2.2 R-Calculus $\mathbf{S}^{\text{t}}$

Let $B \in \Delta$. A reduction $\Delta|B \rightrightarrows \Delta - \{B'\}$ is $\mathbf{S}^{\text{t}}$-valid, denoted by $\models^{\text{t}} \Delta|B \rightrightarrows \Delta - \{B'\}$, if

$$B' = \begin{cases} B \text{ if } \models^{\text{t}} \Delta - \{B\} \\ \lambda \text{ otherwise.} \end{cases}$$

Intuitively, a formula $B_1 \wedge B_2$ is extractable from Δ, if B_1 is extractable from Δ and B_2 is extractable from $\Delta - \{B_1\}$; and $B_1 \vee B_2$ is extractable from Δ if either B_1 or B_2 is extractable from Δ.

A formula $B_1 \wedge B_2$ is not extractable from Δ if either B_1 is not extractable from Δ or B_2 is not extractable from $\Delta - \{B_1\}$; and $B_1 \vee B_2$ is not extractable from Δ if both B_1 and B_2 are not extractable from Δ.

R-calculus $\mathbf{S}^{\mathtt{t}}$ consists of the following axioms and deductions:

- **Axioms**:

$$(\mathtt{A}^{\mathtt{t}}_{-})\ \frac{\mathbf{E}m' \neq m(m', \neg m' \notin \Delta)}{\Delta|m \rightrightarrows \Delta - \{m\}}, \quad (\mathtt{A}^{\mathtt{t}}_{0})\ \frac{\sim \mathbf{E}m' \neq m(m', \neg m' \notin \Delta)}{\Delta|m \rightrightarrows \Delta},$$

where Δ, m is literal.

- **Deduction rules**:

$$(\neg\neg^{-})\ \frac{\Delta|B \rightrightarrows \Delta - \{B\}}{\Delta|\neg\neg B \rightrightarrows \Delta - \{\neg\neg B\}} \qquad (\neg\neg^{0})\ \frac{\Delta|B \rightrightarrows \Delta}{\Delta|\neg\neg B \rightrightarrows \Delta}$$

$$(\wedge^{-})\ \frac{\left[\begin{array}{l}\Delta|B_1 \rightrightarrows \Delta - \{B_1\} \\ \Delta - \{B_1\}|B_2 \rightrightarrows \Delta - \{B_1, B_2\}\end{array}\right.}{\Delta|B_1 \wedge B_2 \rightrightarrows \Delta - \{B_1 \wedge B_2\}} \qquad (\wedge^{0})\ \frac{\left\{\begin{array}{l}\Delta|B_1 \rightrightarrows \Delta \\ \Delta - \{B_1\}|B_2 \rightrightarrows \Delta - \{B_1\}\end{array}\right.}{\Delta|B_1 \wedge B_2 \rightrightarrows \Delta}$$

$$(\vee^{-})\ \frac{\left\{\begin{array}{l}\Delta|B_1 \rightrightarrows \Delta - \{B_1\} \\ \Delta|B_2 \rightrightarrows \Delta - \{B_2\}\end{array}\right.}{\Delta|B_1 \vee B_2 \rightrightarrows \Delta - \{B_1 \vee B_2\}} \qquad (\vee^{0})\ \frac{\left[\begin{array}{l}\Delta|B_1 \rightrightarrows \Delta \\ \Delta|B_2 \rightrightarrows \Delta\end{array}\right.}{\Delta|B_1 \vee B_2 \rightrightarrows \Delta}$$

and

$$(\neg\wedge^{-})\ \frac{\left\{\begin{array}{l}\Delta|\neg B_1 \rightrightarrows \Delta - \{\neg B_1\} \\ \Delta|\neg B_2 \rightrightarrows \Delta - \{\neg B_2\}\end{array}\right.}{\Delta|\neg(B_1 \wedge B_2) \rightrightarrows \Delta - \{\neg(B_1 \wedge B_2)\}}$$

$$(\neg\wedge^{0})\ \frac{\left[\begin{array}{l}\Delta|\neg B_1 \rightrightarrows \Delta \\ \Delta|\neg B_2 \rightrightarrows \Delta\end{array}\right.}{\Delta|\neg(B_1 \wedge B_2) \rightrightarrows \Delta}$$

$$(\neg\vee^{-})\ \frac{\left[\begin{array}{l}\Delta|\neg B_1 \rightrightarrows \Delta - \{\neg B_1\} \\ \Delta - \{\neg B_1\}|\neg B_2 \rightrightarrows \Delta - \{\neg B_1, \neg B_2\}\end{array}\right.}{\Delta|\neg(B_1 \vee B_2) \rightrightarrows \Delta - \{\neg(B_1 \vee B_2)\}}$$

$$(\neg\vee^{-})\ \frac{\left\{\begin{array}{l}\Delta|\neg B_1 \rightrightarrows \Delta \\ \Delta - \{\neg B_1\}|\neg B_2 \rightrightarrows \Delta - \{\neg B_1\}\end{array}\right.}{\Delta|\neg(B_1 \vee B_2) \rightrightarrows \Delta}$$

Definition 3.2.6 A reduction $\Delta|B \rightrightarrows \Delta - \{B'\}$ is provable in $\mathbf{S}^{\mathtt{t}}$, denoted by $\vdash^{\mathtt{t}} \Delta|B \rightrightarrows \Delta - \{B'\}$, if there is a sequence $\{\delta_1, \ldots, \delta_n\}$ of reductions such that $\delta_n = \Delta|B \rightrightarrows \Delta - \{B'\}$, and for each $1 \leq i \leq n$, δ_i is an axiom or is deduced from the previous reductions by one of the deduction rules in $\mathbf{S}^{\mathtt{t}}$.

Theorem 3.2.7 (Soundness and completeness theorem) *For any reduction $\Delta|B \rightrightarrows \Delta - \{B'\}$, $\Delta|B \rightrightarrows \Delta - \{B'\}$ is provable in $\mathbf{S}^{\mathtt{t}}$ if and only if $\Delta|B \rightrightarrows \Delta - \{B'\}$ is valid, that is,*

$$\vdash^{\mathtt{t}} \Delta|B \rightrightarrows \Delta - \{B'\} \text{ iff } \models^{\mathtt{t}} \Delta|B \rightrightarrows \Delta - \{B'\}.$$

□

3.2.3 *R-Calculus* $\mathbf{S}_{\mathtt{t}}$

A reduction $\Gamma|C \rightrightarrows \Gamma, C'$ is $\mathbf{S}_{\mathtt{t}}$-valid, denoted by $\models_{\mathtt{t}} \Gamma|C \rightrightarrows \Gamma, C'$, if

$$C' = \begin{cases} C \text{ if } \models_{\mathtt{t}} \Gamma, C \\ \lambda \text{ otherwise.} \end{cases}$$

Intuitively, a formula $C_1 \wedge C_2$ is enumerable into Γ, if C_1 is enumerable into Γ and C_2 is enumerable into $\Gamma \cup \{C_1\}$; and $C_1 \vee C_2$ is enumerable into Γ if either C_1 or C_2 is enumerable into Γ.

$C_1 \wedge C_2$ is not enumerable into Γ, if either C_1 is not enumerable into Γ or C_2 is not enumerable into $\Gamma \cup \{C_1\}$; and $C_1 \vee C_2$ is not enumerable into Γ if both C_1 and C_2 are not enumerable into Γ.

R-calculus $\mathbf{S}_{\mathtt{t}}$ consists of the following axiom and deductions:

- **Axioms**:

$$(\mathtt{A}^0_{\mathtt{t}})\ \frac{\neg l \in \Gamma}{\Gamma|l \rightrightarrows \Gamma}, \quad (\mathtt{A}^+_{\mathtt{t}})\ \frac{\neg l \notin \Gamma}{\Gamma|l \rightrightarrows \Gamma, l},$$

  where Γ is a set of literals and l is a literal.
- **Deduction rules**:

$$(\neg\neg^+)\ \frac{\Gamma|C \rightrightarrows \Gamma, C}{\Gamma|\neg\neg C \rightrightarrows \Gamma, \neg\neg C} \qquad (\neg\neg^0)\ \frac{\Gamma|C \rightrightarrows \Gamma}{\Gamma|\neg\neg C \rightrightarrows \Gamma}$$

$$(\wedge^+)\ \frac{\left[\begin{array}{l}\Gamma|C_1 \rightrightarrows \Gamma, C_1 \\ \Gamma, C_1|C_2 \rightrightarrows \Gamma, C_1, C_2\end{array}\right.}{\Gamma|C_1 \wedge C_2 \rightrightarrows \Gamma, C_1 \wedge C_2} \qquad (\wedge^0)\ \frac{\left\{\begin{array}{l}\Gamma|C_1 \rightrightarrows \Gamma \\ \Gamma, C_1|C_2 \rightrightarrows \Gamma, C_1\end{array}\right.}{\Gamma|C_1 \wedge C_2 \rightrightarrows \Gamma}$$

$$(\neg\wedge^+)\ \frac{\left\{\begin{array}{l}\Gamma|\neg C_1 \rightrightarrows \Gamma, \neg C_1 \\ \Gamma|\neg C_2 \rightrightarrows \Gamma, \neg C_2\end{array}\right.}{\Gamma|\neg(C_1 \wedge C_2) \rightrightarrows \Gamma, \neg(C_1 \wedge C_2)} \qquad (\neg\wedge^0)\ \frac{\left[\begin{array}{l}\Gamma|\neg C_1 \rightrightarrows \Gamma \\ \Gamma|\neg C_2 \rightrightarrows \Gamma\end{array}\right.}{\Gamma|\neg(C_1 \wedge C_2) \rightrightarrows \Gamma}$$

$$(\vee^+)\ \frac{\left\{\begin{array}{l}\Gamma|C_1 \rightrightarrows \Gamma, C_1 \\ \Gamma|C_2 \rightrightarrows \Gamma, C_2\end{array}\right.}{\Gamma|C_1 \vee C_2 \rightrightarrows \Gamma, C_1 \vee C_2} \qquad (\vee^0)\ \frac{\left[\begin{array}{l}\Gamma|C_1 \rightrightarrows \Gamma \\ \Gamma|C_2 \rightrightarrows \Gamma\end{array}\right.}{\Gamma|C_1 \vee C_2 \rightrightarrows \Gamma}$$

$$(\neg\vee^+)\ \frac{\left[\begin{array}{l}\Gamma|\neg C_1 \rightrightarrows \Gamma, \neg C_1 \\ \Gamma, \neg C_1|\neg C_2 \rightrightarrows \Gamma, \neg C_1, \neg C_2\end{array}\right.}{\Gamma|\neg(C_1 \vee C_2) \rightrightarrows \Gamma, \neg(C_1 \vee C_2)} \qquad (\neg\vee^0)\ \frac{\left\{\begin{array}{l}\Gamma|\neg C_1 \rightrightarrows \Gamma \\ \Gamma|\neg C_2 \rightrightarrows \Gamma\end{array}\right.}{\Gamma|\neg(C_1 \vee C_2) \rightrightarrows \Gamma}$$

Definition 3.2.8 A reduction $\Gamma|C \rightrightarrows \Gamma, C'$ is provable in $\mathbf{S}_{\mathtt{t}}$, denoted by $\vdash_{\mathtt{t}} \Gamma|C \rightrightarrows \Gamma, C'$, if there is a sequence $\{\delta_1, \ldots, \delta_n\}$ of reductions such that $\delta_n = \Gamma|C \rightrightarrows \Gamma, C'$, and for each $1 \leq i \leq n$, δ_i is an axiom or is deduced from the previous reductions by one of the deduction rules in $\mathbf{S}_{\mathtt{t}}$.

Theorem 3.2.9 (Soundness and completeness theorem) *For any reduction* $\Gamma|C \rightrightarrows \Gamma, C'$,

$$\Gamma|C \rightrightarrows \Gamma, C' \text{ is provable in } \mathbf{S}_{\mathtt{t}} \text{ iff } \Gamma|C \rightrightarrows \Gamma, C' \text{ is valid.}$$

That is,

$$\vdash_{\mathtt{t}} \Gamma|C \rightrightarrows \Gamma, C' \text{ iff } \models_{\mathtt{t}} \Gamma|C \rightrightarrows \Gamma, C'.$$

Proof Here, we give the proof of completeness part. Given a reduction $\Gamma|C \rightrightarrows \Gamma|C'$, we construct a trees T such that either

(i) T is a proof tree of $\Gamma|C \rightrightarrows \Gamma, C$; i.e., for each branch ξ of T, there is a reduction γ at the leaf of ξ which is an axiom of form $(\mathtt{A}^+_{\mathtt{t}})$, or
(ii) there is a branch $\xi \in T$ such that ξ is a proof of $\Gamma|C \rightrightarrows \Gamma$.

T is constructed as follows:

- the root of T is $\Gamma|C \rightrightarrows \Gamma, C$;
- for a node ξ, if each reduction at ξ is of form $\Gamma'|l' \rightrightarrows \Gamma, l'$ then the node is a leaf;
- otherwise, ξ has the direct child containing the following reductions:

$$\begin{cases} \Gamma_1|C \rightrightarrows \Gamma_1, C & \text{if } \Gamma_1|\neg\neg C \rightrightarrows \Gamma_1, \neg\neg C \in \xi \\ \begin{cases} \Gamma_1|C_1 \rightrightarrows \Gamma_1, C_1 \\ \Gamma_1, C_1|C_2 \rightrightarrows \Gamma_1, C_1, C_2 \end{cases} & \text{if } \Gamma_1|C_1 \wedge C_2 \rightrightarrows \Gamma_1, C_1 \wedge C_2 \in \xi \\ \begin{bmatrix} \Gamma_1|\neg C_1 \rightrightarrows \Gamma_1, \neg C_1 \\ \Gamma_1|\neg C_2 \rightrightarrows \Gamma_1, \neg C_2 \end{bmatrix} & \text{if } \Gamma_1|\neg(C_1 \wedge C_2) \rightrightarrows \Gamma_1, \neg(C_1 \wedge C_2) \in \xi \\ \begin{bmatrix} \Gamma_1|C_1 \rightrightarrows \Gamma_1, C_1 \\ \Gamma_1|C_2 \rightrightarrows \Gamma_1, C_2 \end{bmatrix} & \text{if } \Gamma_1|C_1 \vee C_2 \rightrightarrows \Gamma_1, C_1 \vee C_2 \in \xi \\ \begin{cases} \Gamma_1|\neg C_1 \rightrightarrows \Gamma_1, \neg C_1 \\ \Gamma_1, \neg C_1|\neg C_2 \rightrightarrows \Gamma_1, \neg C_1, \neg C_2 \end{cases} & \text{if } \Gamma_1|\neg(C_1 \vee C_2) \rightrightarrows \Gamma_1, \neg(C_1 \vee C_2) \in \xi \end{cases}$$

Lemma 3.2.10 *If for each branch* $\xi \subseteq T$, *there is a reduction* $\Gamma'|C' \rightrightarrows \Gamma'|C' \in \xi$ *which is an axiom in* $\mathbf{S}_{\mathtt{t}}$ *then* T *is a proof tree of* $\Gamma|C \rightrightarrows \Gamma, C$ *in* $\mathbf{S}_{\mathtt{t}}$.

Proof By the definition of T, T is a proof tree of $\Gamma|C \rightrightarrows \Gamma, C$. □

Lemma 3.2.11 *If there is a branch* $\xi \subseteq T$ *such that each reduction* $\Gamma'|C' \rightrightarrows \Gamma', C' \in \xi$ *is not an axiom in* $\mathbf{S}_{\mathtt{t}}$, *then* ξ *is a proof of* $\Gamma|C \rightrightarrows \Gamma$.

Proof Let ξ be a branch of T such that any reduction $\Gamma'|C' \rightrightarrows \Gamma', C' \in \xi$ is not an axiom in $\mathbf{S}_{\mathtt{t}}$.

We proved by induction on node $\eta \in \xi$ that for each reduction $\Gamma'|C' \rightrightarrows \Gamma', C' \in \eta$, $\vdash^{\mathtt{f}} \Gamma'|C' \rightrightarrows \Gamma'$.

Case $\Gamma'|l \rightrightarrows \Gamma', l \in \eta$ is a leaf node of ξ. Then, there is no literal $l' \neq l$ such that $l', \neg l' \in \Gamma'$. Hence, $\vdash_{\mathtt{t}} \Gamma'|l \rightrightarrows \Gamma'$.

Case $\delta = \Gamma_2|\neg\neg C \rightrightarrows \Gamma_2, \neg\neg C \in \eta$. Then, η has a child node $\in \xi$ containing reduction $\Gamma_2|C \rightrightarrows \Gamma_2, C$. By induction assumption, $\vdash_{\mathtt{t}} \Gamma_2|C \rightrightarrows \Gamma_2$, and by $(\neg\neg^0)$, $\vdash_{\mathtt{t}} \Gamma_2|\neg\neg C \rightrightarrows \Gamma_2$.

Case $\delta = \Gamma_2|C_1 \wedge C_2 \rightrightarrows \Gamma_2, C_1 \wedge C_2 \in \eta$. Then, η has a child node $\in \xi$ containing reduction either $\Gamma_2|C_1 \rightrightarrows \Gamma_2, C_1$ or $\Gamma_2, C_1|C_2 \rightrightarrows \Gamma_2, C_1, C_2$. By induction assumption, either $\vdash_{\mathtt{t}} \Gamma_2|C_1 \rightrightarrows \Gamma_2$, or $\vdash_{\mathtt{t}} \Gamma_2, C_1|C_2 \rightrightarrows \Gamma_2, C_1$. By $(\wedge^0)$, we have $\vdash_{\mathtt{t}} \Gamma_2, C_1 \wedge C_2 \rightrightarrows \Gamma_2$.

Case $\delta = \Gamma_2|C_1 \vee C_2 \rightrightarrows \Gamma_2, C_1 \vee C_2 \in \eta$. Then, η has a child node $\in \xi$ containing reductions $\Gamma_2|C_1 \rightrightarrows \Gamma_2, C_1$ and $\Gamma_2|C_2 \rightrightarrows \Gamma_2, C_2$. By induction assumption, $\vdash_{\mathtt{t}} \Gamma_2|C_1 \rightrightarrows \Gamma_2$ and $\vdash_{\mathtt{t}} \Gamma_2|C_2 \rightrightarrows \Gamma_2$. By $(\vee^0)$, we have $\vdash_{\mathtt{t}} \Gamma_2, C_1 \vee C_2 \rightrightarrows \Gamma$.

Similar for other cases. □

3.2.4 R-Calculus $\mathbf{S}_{\mathtt{f}}$

A reduction $\Delta|D \rightrightarrows \Delta, D'$ is $\mathbf{S}_{\mathtt{f}}$-valid, denoted by $\models_{\mathtt{f}} \Delta|D \rightrightarrows \Delta, D'$, if

$$D' = \begin{cases} D \text{ if } \models_{\mathtt{f}} \Delta, D \\ \lambda \text{ otherwise.} \end{cases}$$

Intuitively, a formula $D_1 \wedge D_2$ is enumerable into Δ, if either D_1 or D_2 is enumerable into Δ; and $D_1 \vee D_2$ is enumerable into Δ if D_1 is enumerable into Δ and D_2 is enumerable into $\Delta \cup \{D_1\}$.

A formula $D_1 \wedge D_2$ is not enumerable into Δ, if both D_1 and D_2 are not enumerable into Δ; and $D_1 \vee D_2$ is not enumerable into Δ if either D_1 is not enumerable into Δ or D_2 is not enumerable into $\Delta \cup \{D_1\}$.

R-calculus $\mathbf{S}_{\mathtt{f}}$ consists of the following axiom and deductions:

- **Axioms**:

$$(\mathtt{A}_{\mathtt{f}}^0)\ \frac{\neg l \in \Delta}{\Delta|l \rightrightarrows \Delta}, \qquad (\mathtt{A}_{\mathtt{f}}^+)\ \frac{\neg l \notin \Delta}{\Delta|l \rightrightarrows \Delta, l}.$$

  where Δ is a set of literals and l is a literal.
- **Deduction rules**:

$$(\neg\neg^+)\ \frac{\Delta|D \rightrightarrows \Delta, D}{\Delta|\neg\neg D \rightrightarrows \Delta, \neg\neg D}$$

$$(\neg\neg^0)\ \frac{\Delta|D \rightrightarrows \Delta}{\Delta|\neg\neg D \rightrightarrows \Delta}$$

$$(\wedge^+)\ \frac{\begin{cases}\Delta|D_1 \rightrightarrows \Delta, D_1\\ \Delta|D_2 \rightrightarrows \Delta, D_2\end{cases}}{\Delta|D_1 \wedge D_2 \rightrightarrows \Delta, D_1 \wedge D_2}$$

$$(\wedge^0)\ \frac{\left[\begin{array}{l}\Delta|D_1 \rightrightarrows \Delta\\ \Delta|D_2 \rightrightarrows \Delta\end{array}\right.}{\Delta|D_1 \wedge D_2 \rightrightarrows \Delta}$$

$$(\neg\wedge^+)\ \frac{\left[\begin{array}{l}\Delta|\neg D_1 \rightrightarrows \Delta, \neg D_1\\ \Delta, \neg D_1|\neg D_2 \rightrightarrows \Delta, \neg D_1, \neg D_2\end{array}\right.}{\Delta|\neg(D_1 \wedge D_2) \rightrightarrows \Delta, \neg(D_1 \wedge D_2)}$$

$$(\neg\wedge^0)\ \frac{\begin{cases}\Delta|\neg D_1 \rightrightarrows \Delta\\ \Delta, \neg D_1|\neg D_2 \rightrightarrows \Delta, \neg D_1\end{cases}}{\Delta|\neg(D_1 \wedge D_2) \rightrightarrows \Delta}$$

$$(\vee^+)\ \frac{\left[\begin{array}{l}\Delta|D_1 \rightrightarrows \Delta, D_1\\ \Delta, D_1|D_2 \rightrightarrows \Delta, D_1, D_2\end{array}\right.}{\Delta|D_1 \vee D_2 \rightrightarrows \Delta, D_1 \vee D_2}$$

$$(\vee^0)\ \frac{\begin{cases}\Delta|D_1 \rightrightarrows \Delta\\ \Delta, D_1|D_2 \rightrightarrows \Delta, D_1\end{cases}}{\Delta|D_1 \vee D_2 \rightrightarrows \Delta}$$

$$(\neg\vee^+)\ \frac{\begin{cases}\Delta|\neg D_1 \rightrightarrows \Delta, \neg D_1\\ \Delta|\neg D_2 \rightrightarrows \Delta, \neg D_2\end{cases}}{\Delta|\neg(D_1 \vee D_2) \rightrightarrows \Delta, \neg(D_1 \vee D_2)}$$

$$(\neg\vee^0)\ \frac{\left[\begin{array}{l}\Delta|\neg D_1 \rightrightarrows \Delta\\ \Delta|\neg D_2 \rightrightarrows \Delta\end{array}\right.}{\Delta|\neg(D_1 \vee D_2) \rightrightarrows \Delta}$$

Definition 3.2.12 A reduction $\Delta|D \rightrightarrows \Delta, D'$ is provable in $\mathbf{S}_{\pounds}$, denoted by $\vdash \Delta|D \rightrightarrows \Delta, D'$, if there is a sequence $\{\delta_1, \ldots, \delta_n\}$ of reductions such that $\delta_n = \Delta|D \rightrightarrows \Delta, D'$, and for each $1 \leq i \leq n$, δ_i is an axiom or is deduced from the previous reductions by one of the deduction rules in $\mathbf{S}_{\pounds}$.

Theorem 3.2.13 (Soundness and completeness theorem) *For any reduction* $\Delta|D \rightrightarrows \Delta, D'$, $\Delta|D \rightrightarrows \Delta, D'$ *is provable in* $\mathbf{S}_{\pounds}$ *if and only if* $\Delta|D \rightrightarrows \Delta, D'$ *is* $\mathbf{S}_{\pounds}$*-valid. that is,*

$$\vdash_{\pounds} \Delta|D \rightrightarrows \Delta, D' \text{ iff } \models_{\pounds} \Delta|D \rightrightarrows \Delta, D'.$$

□

3.3 Gentzen Deduction Systems

A sequent $\Gamma \Rightarrow \Delta$ is

- $\mathbf{G}^{\mathtt{t}}$-valid, denoted by $\models^{\mathtt{t}} \Gamma \Rightarrow \Delta$, if for any assignment v, either $v(A) = 0$ for some formula $A \in \Gamma$ or $v(B) = 1$ for some formula $B \in \Delta$;
- $\mathbf{G}^{\mathtt{f}}$-valid, denoted by $\models^{\mathtt{f}} \Gamma \Rightarrow \Delta$, if for any assignment v, either $v(A) = 1$ for some formula $A \in \Gamma$ or $v(B) = 0$ for some formula $B \in \Delta$.

$$\begin{array}{l}\mathbf{G}^{\mathtt{t}} : \mathbf{A}v(\mathbf{E}A \in \Gamma(v(A) = 0) \text{ or } \mathbf{E}B \in \Delta(v(B) = 1)) \\ \mathbf{G}^{\mathtt{f}} : \mathbf{A}v(\mathbf{E}A \in \Gamma(v(A) = 1) \text{ or } \mathbf{E}B \in \Delta(v(B) = 0)).\end{array}$$

A co-sequent $\Gamma \mapsto \Delta$ is

- $\mathbf{G}_{\mathtt{t}}$-valid, denoted by $\models_{\mathtt{t}} \Gamma \mapsto \Delta$, if there is an assignment v such that $v(A) = 1$ for every formula $A \in \Gamma$ and $v(B) = 0$ for every formula $B \in \Delta$;
- $\mathbf{G}_{\mathtt{f}}$-valid, denoted by $\models_{\mathtt{f}} \Gamma \mapsto \Delta$, if there is an assignment v such that $v(A) = 0$ for every formula $A \in \Gamma$ and $v(B) = 1$ for every formula $B \in \Delta$.

$$\begin{array}{l}\mathbf{G}_{\mathtt{t}} : \mathbf{E}v(\mathbf{A}A \in \Gamma(v(A) = 1)\&\mathbf{A}B \in \Delta(v(B) = 0)) \\ \mathbf{G}_{\mathtt{f}} : \mathbf{E}v(\mathbf{A}A \in \Gamma(v(A) = 0)\&\mathbf{A}B \in \Delta(v(B) = 1)).\end{array}$$

A sequent $\Gamma \Rightarrow \Delta$/co-sequent $\Gamma \mapsto \Delta$ is atomic/literal if each formula in Γ and Δ is atomic/literal.

3.3.1 Gentzen Deduction System $\mathbf{G}^{\mathtt{t}}$

Intuitively, if for any assignment v, either $v \models \Gamma, A_1 \Rightarrow \Delta$ or $v \models \Gamma, A_2 \Rightarrow \Delta$ then for any assignment v, $v \models \Gamma, A_1 \wedge A_2 \Rightarrow \Delta$; and if for any assignment v, $v \models \Gamma \Rightarrow B_1, \Delta$ and $v \models \Gamma \Rightarrow B_2, \Delta$ then for any assignment v, $v \models \Gamma \Rightarrow B_1 \wedge B_2, \Delta$.

If for any assignment v, $v \models \Gamma, A_1 \Rightarrow \Delta$ and $v \models \Gamma, A_2 \Rightarrow \Delta$ then for any assignment v, $v \models \Gamma, A_1 \vee A_2 \Rightarrow \Delta$; and if for any assignment v, either $v \models \Gamma \Rightarrow B_1, \Delta$ or $v \models \Gamma \Rightarrow B_2, \Delta$ then for any assignment v, $v \models \Gamma \Rightarrow B_1 \vee B_2, \Delta$.

Gentzen deduction system $\mathbf{G}^{\mathtt{t}}$ consists of the following axiom and deductions:

- **Axiom**:

$$(\mathtt{A}^{\mathtt{t}}) \ \frac{\mathbf{E}p(p, \neg p \in \Gamma) \text{ or } \mathbf{E}q(q, \neg q \in \Delta) \text{ or } \Gamma \cap \Delta \neq \emptyset}{\Gamma \Rightarrow \Delta},$$

where Γ, Δ are sets of literals.

- **Deduction rules**:

$$(\neg\neg^L)\ \frac{\Gamma, A \Rightarrow \Delta}{\Gamma, \neg\neg A \Rightarrow \Delta} \qquad (\neg\neg^R)\ \frac{\Gamma \Rightarrow B, \Delta}{\Gamma \Rightarrow \neg\neg B, \Delta}$$

$$(\wedge^L)\ \frac{\begin{cases}\Gamma, A_1 \Rightarrow \Delta \\ \Gamma, A_2 \Rightarrow \Delta\end{cases}}{\Gamma, A_1 \wedge A_2 \Rightarrow \Delta} \qquad (\wedge^R)\ \frac{\left[\begin{array}{l}\Gamma \Rightarrow B_1, \Delta \\ \Gamma \Rightarrow B_2, \Delta\end{array}\right.}{\Gamma \Rightarrow B_1 \wedge B_2, \Delta}$$

$$(\vee^L)\ \frac{\left[\begin{array}{l}\Gamma, A_1 \Rightarrow \Delta \\ \Gamma, A_2 \Rightarrow \Delta\end{array}\right.}{\Gamma, A_1 \vee A_2 \Rightarrow \Delta} \qquad (\vee^R)\ \frac{\begin{cases}\Gamma \Rightarrow B_1, \Delta \\ \Gamma \Rightarrow B_2, \Delta\end{cases}}{\Gamma \Rightarrow B_1 \vee B_2, \Delta}$$

and

$$(\neg\wedge^L)\ \frac{\left[\begin{array}{l}\Gamma, \neg A_1 \Rightarrow \Delta \\ \Gamma, \neg A_2 \Rightarrow \Delta\end{array}\right.}{\Gamma, \neg(A_1 \wedge A_2) \Rightarrow \Delta} \qquad (\neg\wedge^R)\ \frac{\begin{cases}\Gamma \Rightarrow \neg B_1, \Delta \\ \Gamma \Rightarrow \neg B_2, \Delta\end{cases}}{\Gamma \Rightarrow \neg(B_1 \wedge B_2), \Delta}$$

$$(\neg\vee^L)\ \frac{\begin{cases}\Gamma, \neg A_1 \Rightarrow \Delta \\ \Gamma, \neg A_2 \Rightarrow \Delta\end{cases}}{\Gamma, \neg(A_1 \vee A_2) \Rightarrow \Delta} \qquad (\neg\vee^R)\ \frac{\left[\begin{array}{l}\Gamma \Rightarrow \neg B_1, \Delta \\ \Gamma \Rightarrow \neg B_2, \Delta\end{array}\right.}{\Gamma \Rightarrow \neg(B_1 \vee B_2), \Delta}$$

Definition 3.3.1 A sequent $\Gamma \Rightarrow \Delta$ is provable in $\mathbf{G}^{\mathrm{t}}$, denoted by $\vdash^{\mathrm{t}} \Gamma \Rightarrow \Delta$, if there is a sequence $\{\Gamma_1 \Rightarrow \Delta_1, \ldots, \Gamma_n \Rightarrow \Delta_n\}$ of sequents such that $\Gamma_n \Rightarrow \Delta_n = \Gamma \Rightarrow \Delta$, and for each $1 \leq i \leq n$, $\Gamma_i \Rightarrow \Delta_i$ is an axiom or is deduced from the previous sequents by one of the deduction rules in $\mathbf{G}^{\mathrm{t}}$.

Theorem 3.3.2 (Soundness theorem) *For any sequent* $\Gamma \Rightarrow \Delta$, *if* $\vdash^{\mathrm{t}} \Gamma \Rightarrow \Delta$, *then* $\models^{\mathrm{t}} \Gamma \Rightarrow \Delta$.

Proof We prove that each axiom is valid and each deduction rule preserves the validity.

$(\mathtt{A}^{\mathrm{t}})$ For any assignment v, (i) assume that $p, \neg p \in \Gamma$, then either $v(p) = 0$ or $v(\neg p) = 0$; (ii) assume that $q, \neg q \in \Delta$, then either $v(q) = 1$ or $v(\neg q) = 1$; (iii) $l \in \Gamma \cap \Delta$, then either $v(l) = 0$ or $v(\neg l) = 1$. In each case, $v \models^{\mathrm{t}} \Gamma \Rightarrow \Delta$.

$(\neg\neg^L)$ Assume that for any assignment v, $v \models^{\mathrm{t}} \Gamma, A \Rightarrow \Delta$. Then, for any assignment v, if $v \models \Gamma, \neg\neg A$ then $v \models A$, and by induction assumption, $v \models \Delta$.

$(\neg\neg^R)$ Assume that for any assignment v, $v \models^{\mathrm{t}} \Gamma \Rightarrow B, \Delta$. Then, for any assignment v, if $v \models \Gamma$ then by induction assumption, $v \models B, \Delta$, and $v \models \neg\neg B, \Delta$, because $v(\neg\neg B) = v(B)$.

$(\wedge^L)$ Assume that for any assignment v, $v \models^{\mathrm{t}} \Gamma, A_i \Rightarrow \Delta$. Then, for any assignment v, if $v \models \Gamma, A_1 \wedge A_2$ then $v \models \Gamma, A_i$ and by induction assumption, $v \models \Delta$.

$(\wedge^R)$ Assume that for any assignment v, $v \models^{\mathrm{t}} \Gamma \Rightarrow B_1, \Delta$ and $v \models \Gamma \Rightarrow B_2, \Delta$. Then, for any assignment v, if $v \models \Gamma$ then there are two cases: (i) $v \not\models B_1$ or $v \not\models B_2$. By induction assumption, $v \models \Delta$; (ii) $v \models B_1$ and $v \models B_2$. Then, $v \models B_1 \wedge B_2$, and hence, $v \models B_1 \wedge B_2, \Delta$.

$(\vee^L)$ Assume that for any assignment v, $v \models^{\mathrm{t}} \Gamma, A_1 \Rightarrow \Delta$ and $v \models^{\mathrm{t}} \Gamma, A_2 \Rightarrow \Delta$. Then, for any assignment v, if $v \models \Gamma, A_1 \vee A_2$ then $v \models \Gamma$ and $v \models A_1 \vee A_2$. There are two cases: If $v \models A_1$, then by induction assumption, $v \models \Delta$; and if $v \models A_2$, then by induction assumption, $v \models \Delta$.

$(\vee^R)$ Assume that for any assignment v, $v \models^{\mathsf{t}} \Gamma \Rightarrow B_1, \Delta$. Then, for any assignment v, if $v \models \Gamma$ then by assumption, $v \models B_1, \Delta$. There are two cases: $v \models B_1$, then $v \models B_1 \vee B_2$, and hence, $v \models B_1 \vee B_2, \Delta$; and $v \models \Delta$, and hence, $v \models^{\mathsf{t}} B_1 \vee B_2, \Delta$.

Similar for other cases. □

Theorem 3.3.3 (Completeness theorem) *For any sequent* $\Gamma \Rightarrow \Delta$, *if* $\models^{\mathsf{t}} \Gamma \Rightarrow \Delta$ *then* $\vdash^{\mathsf{t}} \Gamma \Rightarrow \Delta$.

Proof Given a sequent $\Gamma \Rightarrow \Delta$, we construct a tree T such that either

(i) for each branch ξ of T, the sequent $\Gamma' \Rightarrow \Delta'$ at the leaf of ξ is an axiom, or
(ii) there is an assignment v such that $v \not\models \Gamma \Rightarrow \Delta$.

T is constructed as follows:

- the root of T is $\Gamma \Rightarrow \Delta$;
- for a node ξ, if the sequent $\Gamma' \Rightarrow \Delta'$ at ξ is atomic then the node is a leaf;
- otherwise, ξ has the direct children nodes containing the following sequents:

$$\begin{cases} \Gamma_1, A_1, A_2 \Rightarrow \Delta_1 & \text{if } \Gamma_1, A_1 \wedge A_2 \Rightarrow \Delta_1 \in \xi \\ \begin{cases} \Gamma_1 \Rightarrow B_1, \Delta_1 \\ \Gamma_1 \Rightarrow B_2, \Delta_1 \end{cases} & \text{if } \Gamma_1 \Rightarrow B_1 \wedge B_2, \Delta_1 \in \xi \\ \begin{cases} \Gamma_1, A_1 \Rightarrow \Delta_1 \\ \Gamma_1, A_2 \Rightarrow \Delta_1 \end{cases} & \text{if } \Gamma_1, A_1 \vee A_2 \Rightarrow \Delta_1 \in \xi \\ \Gamma_1 \Rightarrow B_1, B_2, \Delta_1 & \text{if } \Gamma_1 \Rightarrow B_1 \vee B_2, \Delta_1 \in \xi \end{cases}$$

and

$$\begin{cases} \Gamma_1, A \Rightarrow \Delta_1 & \text{if } \Gamma_1, \neg\neg A \Rightarrow \Delta_1 \in \xi \\ \Gamma_1 \Rightarrow B, \Delta_1 & \text{if } \Gamma_1 \Rightarrow \neg\neg B, \Delta_1 \in \xi \\ \begin{cases} \Gamma_1, \neg A_1 \Rightarrow \Delta_1 \\ \Gamma_1, \neg A_2 \Rightarrow \Delta_1 \end{cases} & \text{if } \Gamma_1, \neg(A_1 \wedge A_2) \Rightarrow \Delta_1 \in \xi \\ \Gamma_1 \Rightarrow \neg B_1, \neg B_2, \Delta_1 & \text{if } \Gamma_1 \Rightarrow \neg(B_1 \wedge B_2), \Delta_1 \in \xi \\ \Gamma_1, \neg A_1, \neg A_2 \Rightarrow \Delta_1 & \text{if } \Gamma_1, \neg(A_1 \vee A_2) \Rightarrow \Delta_1 \in \xi \\ \begin{cases} \Gamma_1 \Rightarrow \neg B_1, \Delta_1 \\ \Gamma_1 \Rightarrow \neg B_2, \Delta_1 \end{cases} & \text{if } \Gamma_1 \Rightarrow \neg(B_1 \vee B_2), \Delta_1 \in \xi \end{cases}$$

Lemma 3.3.4 *If for each branch* $\xi \subseteq T$, *there is a sequent* $\Gamma' \Rightarrow \Delta' \in \xi$ *which is an axiom in* $\mathbf{G}^{\mathsf{t}}$, *then* T *is a proof tree of* $\Gamma \Rightarrow \Delta$.

Proof By the definition of T, T is a proof tree of $\Gamma \Rightarrow \Delta$. □

Lemma 3.3.5 *If there is a branch* $\xi \subseteq T$ *such that each sequent* $\Gamma' \Rightarrow \Delta' \in \xi$ *is not an axiom in* $\mathbf{G}^{\mathsf{t}}$, *then there is an assignment* v *such that* $v \not\models^{\mathsf{t}} \Gamma \Rightarrow \Delta$.

Proof Let ξ be a branch of T such that each sequent $\Gamma' \Rightarrow \Delta' \in \xi$ is not an axiom in $\mathbf{G}^{\mathsf{t}}$.

Let

$$\Phi = \{A \in \Gamma' : \Gamma' \Rightarrow \Delta' \in \xi\}$$
$$\Psi = \{B \in \Delta' : \Gamma' \Rightarrow \Delta' \in \xi\}.$$

Define an assignment v as follows:

$$v(p) = \begin{cases} 1 \text{ if } p \in \Phi \text{ or } \neg p \in \Psi \\ 0 \text{ otherwise.} \end{cases}$$

Then, v is well-defined and $v \not\models^{\mathtt{t}} \Gamma_0 \Rightarrow \Delta_0$, where $\Gamma_0 \Rightarrow \Delta_0$ is a sequent at the leaf node of ξ.

We prove by induction on nodes η of ξ that for each sequent $\Gamma' \Rightarrow \Delta'$ at η, $v \not\models^{\mathtt{t}} \Gamma' \Rightarrow \Delta'$.

Case $\Gamma' \Rightarrow \Delta' = \Gamma_2, \neg\neg A \Rightarrow \Delta_2 \in \eta$. Then, $\Gamma' \Rightarrow \Delta'$ has a child node $\in \xi$ containing $\Gamma_2, A \Rightarrow \Delta_2 \in \eta$. By induction assumption, $v \models \Gamma_2, A$ and $v \not\models \Delta_2$, which implies that $v \models \Gamma_2, \neg\neg A$ and $v \not\models \Delta_2$.

Case $\Gamma' \Rightarrow \Delta' = \Gamma_2 \Rightarrow \neg\neg B, \Delta_2 \in \eta$. Then, $\Gamma' \Rightarrow \Delta'$ has a child node $\in \xi$ containing $\Gamma_2 \Rightarrow B, \Delta_2 \in \eta$. By induction assumption, $v \models \Gamma_2$ and $v \not\models B, \Delta_2$, which implies that $v \models \Gamma_2$ and $v \not\models \neg\neg B, \Delta_2$.

Case $\Gamma' \Rightarrow \Delta' = \Gamma_2, A_1 \wedge A_2 \Rightarrow \Delta_2 \in \eta$. Then, $\Gamma' \Rightarrow \Delta'$ has a child node $\in \xi$ containing $\Gamma_2, A_1 \Rightarrow \Delta_2 \in \eta$ and $\Gamma_2, A_2 \Rightarrow \Delta_2 \in \eta$. By induction assumption, $v \models \Gamma_2, A_1$; $v \models \Gamma_2, A_2$ and $v \not\models \Delta_2$, i.e., $v \models \Gamma_2, A_1 \wedge A_2$ and $v \not\models \Delta_2$.

Case $\Gamma' \Rightarrow \Delta' = \Gamma_2 \Rightarrow B_1 \wedge B_2, \Delta_2 \in \eta$. Then, $\Gamma' \Rightarrow \Delta'$ has a child node $\in \xi$ containing $\Gamma_2 \Rightarrow B_i, \Delta_2 \in \eta$. By induction assumption, $v \models \Gamma_2$, and $v \not\models B_i, \Delta_2$, i.e., $v \models \Gamma_2$ and $v \not\models B_1 \wedge B_2, \Delta_2$.

Case $\Gamma' \Rightarrow \Delta' = \Gamma_2, A_1 \vee A_2 \Rightarrow \Delta_2 \in \eta$. Then, $\Gamma' \Rightarrow \Delta'$ has a child node $\in \xi$ containing $\Gamma_2, A_i \Rightarrow \Delta_2 \in \eta$. By induction assumption, $v \models \Gamma_2, A_i$ and $v \not\models \Delta_2$. Hence, $v \models \Gamma_2, A_1 \vee A_2$ and $v \not\models \Delta_2$.

Case $\Gamma' \Rightarrow \Delta' = \Gamma_2 \Rightarrow B_1 \vee B_2, \Delta_2 \in \eta$. Then, $\Gamma' \Rightarrow \Delta'$ has a child node $\in \xi$ containing $\Gamma_2 \Rightarrow B_1, \Delta_2 \in \eta$ and $\Gamma_2 \Rightarrow B_2, \Delta_2 \in \eta$. By induction assumption, $v \models \Gamma_2$, $v \not\models B_1, \Delta_2$ and $v \not\models B_2, \Delta_2$. Hence, $v \models \Gamma_2$ and $v \not\models B_1 \vee B_2, \Delta_2$.

Similar for other cases. □

3.3.2 Gentzen Deduction System $\mathbf{G}^{\mathtt{f}}$

A sequent $\Gamma \Rightarrow \Delta$ is $\mathtt{f}$-valid, denoted by $\models^{\mathtt{f}} \Gamma \Rightarrow \Delta$, if for any assignment v, either $v(A) = 1$ for some $A \in \Gamma$, or $v(B) = 0$ for some $B \in \Delta$.

Gentzen deduction system $\mathbf{G}^{\mathtt{f}}$ consists of the following axiom and deductions:

- **Axiom**:

$$(\mathtt{A}^{\mathtt{f}}) \ \frac{\mathbf{E}p(p, \neg p \in \Gamma) \text{ or } \mathbf{E}q(q, \neg q \in \Delta) \text{ or } \Gamma \cap \Delta \neq \emptyset}{\Gamma \Rightarrow \Delta},$$

where Γ, Δ are sets of literals.

- **Deduction rules**:

$$(\wedge^L)\ \frac{\left[\begin{array}{l}\Gamma, A_1 \Rightarrow \Delta\\ \Gamma, A_2 \Rightarrow \Delta\end{array}\right.}{\Gamma, A_1 \wedge A_2 \Rightarrow \Delta} \quad (\wedge^R)\ \frac{\left\{\begin{array}{l}\Gamma \Rightarrow B_1, \Delta\\ \Gamma \Rightarrow B_2, \Delta\end{array}\right.}{\Gamma \Rightarrow B_1 \wedge B_2, \Delta}$$
$$(\vee^L)\ \frac{\left\{\begin{array}{l}\Gamma, A_1 \Rightarrow \Delta\\ \Gamma, A_2 \Rightarrow \Delta\end{array}\right.}{\Gamma, A_1 \vee A_2 \Rightarrow \Delta} \quad (\vee^R)\ \frac{\left[\begin{array}{l}\Gamma \Rightarrow B_1, \Delta\\ \Gamma \Rightarrow B_2, \Delta\end{array}\right.}{\Gamma \Rightarrow B_1 \vee B_2, \Delta}$$

and

$$(\neg\neg^L)\ \frac{\Gamma, A \Rightarrow \Delta}{\Gamma, \neg\neg A \Rightarrow \Delta} \quad (\neg\neg^R)\ \frac{\Gamma \Rightarrow B, \Delta}{\Gamma \Rightarrow \neg\neg B, \Delta}$$
$$(\neg\wedge^L)\ \frac{\left\{\begin{array}{l}\Gamma, \neg A_1 \Rightarrow \Delta\\ \Gamma, \neg A_2 \Rightarrow \Delta\end{array}\right.}{\Gamma, \neg(A_1 \wedge A_2) \Rightarrow \Delta} \quad (\neg\wedge^R)\ \frac{\left[\begin{array}{l}\Gamma \Rightarrow \neg B_1, \Delta\\ \Gamma \Rightarrow \neg B_2, \Delta\end{array}\right.}{\Gamma \Rightarrow \neg(B_1 \wedge B_2), \Delta}$$
$$(\neg\vee^L)\ \frac{\left[\begin{array}{l}\Gamma, \neg A_1 \Rightarrow \Delta\\ \Gamma, \neg A_2 \Rightarrow \Delta\end{array}\right.}{\Gamma, \neg(A_1 \vee A_2) \Rightarrow \Delta} \quad (\neg\vee^R)\ \frac{\left\{\begin{array}{l}\Gamma \Rightarrow \neg B_1, \Delta\\ \Gamma \Rightarrow \neg B_2, \Delta\end{array}\right.}{\Gamma \Rightarrow \neg(B_1 \vee B_2), \Delta}$$

Definition 3.3.6 A sequent $\Gamma \Rightarrow \Delta$ is provable in $\mathbf{G}^{\mathrm{f}}$, denoted by $\vdash^{\mathrm{f}} \Gamma \Rightarrow \Delta$, if there is a sequence $\{\Gamma_1 \Rightarrow \Delta_1, \ldots, \Gamma_n \Rightarrow \Delta_n\}$ of sequents such that $\Gamma_n \Rightarrow \Delta_n = \Gamma \Rightarrow \Delta$, and for each $1 \leq i \leq n$, $\Gamma_i \Rightarrow \Delta_i$ is an axiom or is deduced from the previous sequents by one of the deduction rules in $\mathbf{G}^{\mathrm{f}}$.

Theorem 3.3.7 (Soundness and completeness theorem) *For any sequent* $\Gamma \Rightarrow \Delta$,

$$\vdash^{\mathrm{f}} \Gamma \Rightarrow \Delta \textit{ iff } \models^{\mathrm{f}} \Gamma \Rightarrow \Delta.$$

□

3.3.3 Gentzen Deduction System $\mathbf{G}_{\mathrm{t}}$

A co-sequent $\Gamma \mapsto \Delta$ is $\mathbf{G}_{\mathrm{t}}$-valid, denoted by $\models_{\mathrm{t}} \Gamma \mapsto \Delta$, if there is an assignment v such that $v \models \Gamma$ and $v \models \Delta$, where $v \models \Gamma$ means that for each formula $A \in \Delta, v(A) = 1$; and $v \models \Delta$ means that for each formula $B \in \Delta, v(B) = 0$.

Intuitively, if there is an assignment v such that $v \models \Gamma, A_1 \mapsto \Delta$ and $v \models \Gamma, A_2 \mapsto \Delta$ then for this assignment v, $v \models \Gamma, A_1 \wedge A_2 \mapsto \Delta$; and if there is an assignment v such that either $v \models \Gamma \mapsto B_1, \Delta$ or $v \models \Gamma \mapsto B_2, \Delta$ then for this assignment v, $v \models \Gamma \mapsto B_1 \wedge B_2, \Delta$.

If there is an assignment v such that either $v \models \Gamma, A_1 \mapsto \Delta$ or $v \models \Gamma, A_2 \mapsto \Delta$ then for this assignment v, $v \models \Gamma, A_1 \vee A_2 \mapsto \Delta$; and if there is an assignment v such that $v \models \Gamma \mapsto B_1, \Delta$ and $v \models \Gamma \mapsto B_2, \Delta$ then for this assignment v, $v \models \Gamma \mapsto B_1 \vee B_2, \Delta$.

Gentzen deduction system $\mathbf{G}_{\mathrm{t}}$ consists of the following axiom and deductions:

- **Axiom**:

$$\frac{\mathrm{con}(\Gamma)\&\mathrm{inval}(\Delta)\&\Gamma\cap\Delta=\emptyset}{\Gamma\mapsto\Delta},$$

where Γ, Δ are sets of literals.

- **Deduction rules**:

$$\begin{array}{llll}
(\neg\neg^L) & \dfrac{\Gamma, A\mapsto\Delta}{\Gamma,\neg\neg A\mapsto\Delta} & (\neg\neg^R) & \dfrac{\Gamma\mapsto B,\Delta}{\Gamma\mapsto\neg\neg B,\Delta}\\
(\wedge^L) & \dfrac{\left[\begin{array}{l}\Gamma, A_1\mapsto\Delta\\ \Gamma, A_2\mapsto\Delta\end{array}\right.}{\Gamma, A_1\wedge A_2\mapsto\Delta} & (\wedge^R) & \dfrac{\left\{\begin{array}{l}\Gamma\mapsto B_1,\Delta\\ \Gamma\mapsto B_2,\Delta\end{array}\right.}{\Gamma\mapsto B_1\wedge B_2,\Delta}\\
(\vee^L) & \dfrac{\left\{\begin{array}{l}\Gamma, A_1\mapsto\Delta\\ \Gamma, A_2\mapsto\Delta\end{array}\right.}{\Gamma, A_1\vee A_2\mapsto\Delta} & (\vee^R) & \dfrac{\left[\begin{array}{l}\Gamma\mapsto B_1,\Delta\\ \Gamma\mapsto B_2,\Delta\end{array}\right.}{\Gamma\mapsto B_1\vee B_2,\Delta}
\end{array}$$

and

$$\begin{array}{llll}
(\neg\wedge^L) & \dfrac{\left\{\begin{array}{l}\Gamma,\neg A_1\mapsto\Delta\\ \Gamma,\neg A_2\mapsto\Delta\end{array}\right.}{\Gamma,\neg(A_1\wedge A_2)\mapsto\Delta} & (\neg\wedge^R) & \dfrac{\left[\begin{array}{l}\Gamma\mapsto\neg B_1,\Delta\\ \Gamma\mapsto\neg B_2,\Delta\end{array}\right.}{\Gamma\mapsto\neg(B_1\wedge B_2),\Delta}\\
(\neg\vee^L) & \dfrac{\left[\begin{array}{l}\Gamma,\neg A_1\mapsto\Delta\\ \Gamma,\neg A_2\mapsto\Delta\end{array}\right.}{\Gamma,\neg(A_1\vee A_2)\mapsto\Delta} & (\neg\vee^R) & \dfrac{\left\{\begin{array}{l}\Gamma\mapsto\neg B_1,\Delta\\ \Gamma\mapsto\neg B_2,\Delta\end{array}\right.}{\Gamma\mapsto\neg(B_1\vee B_2),\Delta}
\end{array}$$

Definition 3.3.8 A co-sequent $\Gamma\mapsto\Delta$ is provable in $\mathbf{G}_{\mathrm{t}}$, denoted by $\vdash_{\mathrm{t}}\Gamma\mapsto\Delta$, if there is a sequence $\{\Gamma_1\mapsto\Delta_1,\ldots,\Gamma_n\mapsto\Delta_n\}$ of co-sequents such that $\Gamma_n\mapsto\Delta_n=\Gamma\mapsto\Delta$, and for each $1\leq i\leq n$, $\Gamma_i\mapsto\Delta_i$ is an axiom or is deduced from the previous co-sequents by one of the deduction rules in $\mathbf{G}_{\mathrm{t}}$.

Theorem 3.3.9 (Soundness theorem) *For any co-sequent* $\Gamma\mapsto\Delta$, *if* $\vdash_{\mathrm{t}}\Gamma\mapsto\Delta$ *then* $\models_{\mathrm{t}}\Gamma\mapsto\Delta$.

Proof We prove that each axiom is valid and each deduction rule preserves the validity.

Assume that $\mathrm{con}(\Gamma)$, $\mathrm{inval}(\Delta)$ and $\Gamma\cap\Delta=\emptyset$. Then, we define an assignment v such that for any variable p,

$$v(p)=1 \text{ iff } p\in\Gamma \text{ or } \neg p\in\Delta.$$

Then, $v\models\Gamma\mapsto\Delta$.

$(\neg\neg^L)$ Assume that there is an assignment v such that $v\models\Gamma, A\mapsto\Delta$. Then, for this very assignment v, $v\models\Gamma,\neg\neg A\mapsto\Delta$.

$(\neg\neg^R)$ Assume that there is an assignment v such that $v\models\Gamma\mapsto B,\Delta$. Then, for this very assignment v, $v\models\Gamma\mapsto\neg\neg B,\Delta$.

($\wedge^L$) Assume that there is an assignment v such that $v \models \Gamma, A_1 \mapsto \Delta$ and $v \models \Gamma, A_2 \mapsto \Delta$. Then, for this very assignment v, $v \models \Gamma, A_1 \wedge A_2 \mapsto \Delta$.

($\wedge^R$) Assume that there is an assignment v such that $v \models \Gamma \mapsto B_1, \Delta$. Then, for this very assignment v, $v \models \Gamma \mapsto B_1 \wedge B_2, \Delta$.

($\neg\wedge^L$) Assume that there is an assignment v such that $v \models \Gamma, \neg A_1 \mapsto \Delta$. Then, for this very assignment v, $v \models \Gamma, \neg(A_1 \wedge A_2) \mapsto \Delta$.

($\neg\wedge^R$) Assume that there is an assignment v such that $v \models \Gamma \mapsto \neg B_1, \Delta$ and $v \models \Gamma \mapsto \neg B_2, \Delta$. Then, for this very assignment v, $v \models \Gamma \mapsto \neg(B_1 \wedge B_2), \Delta$.

Similar for other cases. □

Theorem 3.3.10 (Completeness theorem) *For any co-sequent* $\Gamma \mapsto \Delta$, *if* $\models_{\mathrm{t}} \Gamma \mapsto \Delta$ *then* $\vdash_{\mathrm{t}} \Gamma \mapsto \Delta$.

Proof Given a co-sequent $\Gamma \mapsto \Delta$, we construct a tree T such that either

(i) there is a branch ξ of T which is a proof of $\Gamma \mapsto \Delta$, or
(ii) $\vdash^{\mathrm{t}} \Gamma \Rightarrow \Delta$.

T is constructed as follows:

- the root of T is $\Gamma \mapsto \Delta$;
- for a node ξ, if co-sequent $\Gamma' \mapsto \Delta'$ at ξ is literal then the node is a leaf;
- otherwise, ξ has direct children nodes containing the following co-sequents:

$$\begin{cases} \left[\begin{array}{l} \Gamma_1, A_1 \mapsto \Delta_1 \\ \Gamma_1, A_2 \mapsto \Delta_1 \end{array}\right. & \text{if } \Gamma_1, A_1 \wedge A_2 \mapsto \Delta_1 \in \xi \\ \left\{\begin{array}{l} \Gamma_1 \mapsto B_1, \Delta_1 \\ \Gamma_1 \mapsto B_2, \Delta_1 \end{array}\right. & \text{if } \Gamma_1 \mapsto B_1 \wedge B_2, \Delta_1 \in \xi \\ \left\{\begin{array}{l} \Gamma_1, A_1 \mapsto \Delta_1 \\ \Gamma_1, A_2 \mapsto \Delta_1 \end{array}\right. & \text{if } \Gamma_1, A_1 \vee A_2 \mapsto \Delta_1 \in \xi \\ \left[\begin{array}{l} \Gamma_1 \mapsto B_1, \Delta_1 \\ \Gamma_1 \mapsto B_2, \Delta_1 \end{array}\right. & \text{if } \Gamma_1 \mapsto B_1 \vee B_2, \Delta_1 \in \xi \end{cases}$$

and

$$\begin{cases} \Gamma_1, A \mapsto \Delta_1 & \text{if } \Gamma_1, \neg\neg A \mapsto \Delta_1 \in \xi \\ \Gamma_1 \mapsto B, \Delta_1 & \text{if } \Gamma_1 \mapsto \neg\neg B, \Delta_1 \in \xi \\ \left\{\begin{array}{l} \Gamma_1, \neg A_1 \mapsto \Delta_1 \\ \Gamma_1, \neg A_2 \mapsto \Delta_1 \end{array}\right. & \text{if } \Gamma_1, \neg(A_1 \wedge A_2) \mapsto \Delta_1 \in \xi \\ \left[\begin{array}{l} \Gamma_1 \mapsto \neg B_1, \Delta_1 \\ \Gamma_1 \mapsto \neg B_2, \Delta_1 \end{array}\right. & \text{if } \Gamma_1 \mapsto \neg(B_1 \wedge B_2), \Delta_1 \in \xi \\ \left[\begin{array}{l} \Gamma_1, \neg A_1 \mapsto \Delta_1 \\ \Gamma_1, \neg A_2 \mapsto \Delta_1 \end{array}\right. & \text{if } \Gamma_1, \neg(A_1 \vee A_2) \mapsto \Delta_1 \in \xi \\ \left\{\begin{array}{l} \Gamma_1 \mapsto \neg B_1, \Delta_1 \\ \Gamma_1 \mapsto \neg B_2, \Delta_1 \end{array}\right. & \text{if } \Gamma_1 \mapsto \neg(B_1 \vee B_2), \Delta_1 \in \xi. \end{cases}$$

Lemma 3.3.11 *If there is a branch* $\xi \subseteq T$ *such that each co-sequent* $\Gamma' \mapsto \Delta'$ *at the leaf node of* ξ *is an axiom in* $\mathbf{G}_{\mathrm{t}}$ *then* ξ *is a proof of* $\Gamma \mapsto \Delta$.

Proof By the definition of T, ξ is a proof of $\Gamma \mapsto \Delta$. □

Lemma 3.3.12 *If for each branch $\xi \subseteq T$, there is a co-sequent $\Gamma' \mapsto \Delta'$ at the leaf node of ξ which is not an axiom in $\mathbf{G}_{\mathtt{t}}$ then T is a proof tree of sequent $\Gamma \Rightarrow \Delta$ in $\mathbf{G}^{\mathtt{t}}$.*

Proof We prove by induction on nodes ξ of T that for some co-sequent $\Gamma' \mapsto \Delta'$ at $\xi, \vdash^{\mathtt{t}} \Gamma' \Rightarrow \Delta'$.

Case ξ being a leaf node. By assumption, a co-sequent $\Gamma' \mapsto \Delta' \in \xi$ is not an axiom in $\mathbf{G}_{\mathtt{t}}$. Then, $\Gamma' \Rightarrow \Delta'$ is an axiom in $\mathbf{G}^{\mathtt{t}}$.

Case $\Gamma' \mapsto \Delta' = \Gamma_2, \neg\neg A \mapsto \Delta_2 \in \xi$. Then, $\Gamma' \mapsto \Delta'$ has a child node containing $\Gamma_2, A \mapsto \Delta_2$. By induction assumption, $\vdash^{\mathtt{t}} \Gamma_2, A \Rightarrow \Delta_2$, and by $(\neg\neg^L), \vdash^{\mathtt{t}} \Gamma_2, \neg\neg A \Rightarrow \Delta_2$.

Case $\Gamma' \mapsto \Delta' = \Gamma_2, A_1 \wedge A_2 \mapsto \Delta_2 \in \xi$. Then, $\Gamma' \mapsto \Delta'$ has a child node containing $\Gamma_2, A_1, A_2 \mapsto \Delta_2$. By induction assumption, $\vdash^{\mathtt{t}} \Gamma_2, A_1, A_2 \Rightarrow \Delta_2$, and by $(\wedge^L), \vdash^{\mathtt{t}} \Gamma_2, A_1 \wedge A_2 \Rightarrow \Delta_2$.

Case $\Gamma' \mapsto \Delta' = \Gamma_2, \neg(A_1 \wedge A_2) \mapsto \Delta_2 \in \xi$. Then, $\Gamma' \mapsto \Delta'$ has children nodes containing $\Gamma_2, \neg A_1 \mapsto \Delta_2$ and $\Gamma_2, \neg A_2 \mapsto \Delta_2$, respectively. By induction assumption, $\vdash^{\mathtt{t}} \Gamma_2, \neg A_1 \Rightarrow \Delta_2, \vdash^{\mathtt{t}} \Gamma_2, \neg A_2 \Rightarrow \Delta_2$, and by $(\neg\wedge^L)$, $\vdash^{\mathtt{t}} \Gamma_2, \neg(A_1 \wedge A_2) \Rightarrow \Delta_2$.

Similar for other cases. □

We rewrite the proof of the completeness theorem for $\mathbf{G}^{\mathtt{t}}$ as follows:

Lemma 3.3.13 *If for each branch $\xi \subseteq T$, there is a sequent $\Gamma' \Rightarrow \Delta' \in \xi$ which is an axiom in $\mathbf{G}^{\mathtt{t}}$ then T is a proof tree of $\Gamma \Rightarrow \Delta$.*

Proof By the definition of T, T is a proof tree of $\Gamma \Rightarrow \Delta$. □

Lemma 3.3.14 *If there is a branch $\xi \subseteq T$ such that each sequent $\Gamma' \Rightarrow \Delta' \in \xi$ is not an axiom in $\mathbf{G}^{\mathtt{t}}$ then $\vdash_{\mathtt{t}} \Gamma \mapsto \Delta$.*

Proof Let ξ be a branch of T such that each sequent $\Gamma' \Rightarrow \Delta' \in \xi$ is not an axiom in $\mathbf{G}^{\mathtt{t}}$. Let

$$\Phi = \{A \in \Gamma' : \Gamma' \Rightarrow \Delta' \in \xi\}$$
$$\Psi = \{B \in \Delta' : \Gamma' \Rightarrow \Delta' \in \xi\}.$$

Define an assignment v as follows:

$$v(p) = \begin{cases} 1 \text{ if } p \in \Phi \text{ or } \neg p \in \Psi \\ 0 \text{ otherwise.} \end{cases}$$

Then, v is well-defined and $v \models \Gamma_0 \mapsto \Delta_0$, where $\Gamma_0 \Rightarrow \Delta_0$ is a sequent at the leaf node of ξ. That is, $\vdash_{\mathtt{t}} \Gamma_0 \mapsto \Delta_0$.

We prove by induction on nodes η of ξ that for each sequent $\Gamma' \Rightarrow \Delta'$ at $\eta, \vdash_{\mathtt{t}} \Gamma' \mapsto \Delta'$.

Case $\Gamma' \Rightarrow \Delta' = \Gamma_2, \neg\neg A \Rightarrow \Delta_2 \in \eta$. Then, $\Gamma' \Rightarrow \Delta'$ has a direct child node $\in \xi$ containing $\Gamma_2, A \Rightarrow \Delta_2 \in \eta$. By induction assumption, $\vdash_{\mathtt{t}} \Gamma_2, A \mapsto \Delta_2$, which implies that $\vdash_{\mathtt{t}} \Gamma_2, \neg\neg A \mapsto \Delta_2$.

Case $\Gamma' \Rightarrow \Delta' = \Gamma_2, A_1 \wedge A_2 \Rightarrow \Delta_2 \in \eta$. Then, $\Gamma' \Rightarrow \Delta'$ has a direct child node $\in \xi$ containing $\Gamma_2, A_1 \Rightarrow \Delta_2 \in \eta$ and $\Gamma_2, A_2 \Rightarrow \Delta_2 \in \eta$. By induction assumption, $\vdash_{\mathtt{t}} \Gamma_2, A_1 \mapsto \Delta_2$; and $\vdash_{\mathtt{t}} \Gamma_2, A_2 \mapsto \Delta_2$, i.e., $\vdash_{\mathtt{t}} \Gamma_2, A_1 \wedge A_2 \mapsto \Delta_2$.

Case $\Gamma' \Rightarrow \Delta' = \Gamma_2, A_1 \vee A_2 \Rightarrow \Delta_2 \in \eta$. Then, $\Gamma' \Rightarrow \Delta'$ has a direct child node $\in \xi$ containing $\Gamma_2, A_i \Rightarrow \Delta_2 \in \eta$. By induction assumption, $\vdash_{\mathtt{t}} \Gamma_2, A_i \mapsto \Delta_2$. By $(\vee^L)$ in $\mathbf{G}_{\mathtt{t}}$, $\vdash_{\mathtt{t}} \Gamma_2, A_1 \vee A_2 \mapsto \Delta_2$.

Similar for other cases. □

There is a question: whether we can use the following tree to prove the completeness theorem of $\mathbf{G}_{\mathtt{t}}$:

$$\begin{cases} \begin{bmatrix} \Gamma_1, A_1 \mapsto \Delta_1 \\ \Gamma_1, A_2 \mapsto \Delta_1 \end{bmatrix} & \text{if } \Gamma_1, A_1 \wedge A_2 \mapsto \Delta_1 \in \xi \\ \begin{cases} \Gamma_1 \mapsto B_1, \Delta_1 \\ \Gamma_1 \mapsto B_2, \Delta_1 \end{cases} & \text{if } \Gamma_1 \mapsto B_1 \wedge B_2, \Delta_1 \in \xi \\ \begin{cases} \Gamma_1, A_1 \mapsto \Delta_1 \\ \Gamma_1, A_2 \mapsto \Delta_1 \end{cases} & \text{if } \Gamma_1, A_1 \vee A_2 \mapsto \Delta_1 \in \xi \\ \begin{bmatrix} \Gamma_1 \mapsto B_1, \Delta_1 \\ \Gamma_1 \mapsto B_2, \Delta_1 \end{bmatrix} & \text{if } \Gamma_1 \mapsto B_1 \vee B_2, \Delta_1 \in \xi \end{cases}$$

and

$$\begin{cases} \Gamma_1, A \mapsto \Delta_1 & \text{if } \Gamma_1, \neg\neg A \mapsto \Delta_1 \in \xi \\ \Gamma_1 \mapsto B, \Delta_1 & \text{if } \Gamma_1 \mapsto \neg\neg B, \Delta_1 \in \xi \\ \begin{cases} \Gamma_1, \neg A_1 \mapsto \Delta_1 \\ \Gamma_1, \neg A_2 \mapsto \Delta_1 \end{cases} & \text{if } \Gamma_1, \neg(A_1 \wedge A_2) \mapsto \Delta_1 \in \xi \\ \begin{bmatrix} \Gamma_1 \mapsto \neg B_1, \Delta_1 \\ \Gamma_1 \mapsto \neg B_2, \Delta_1 \end{bmatrix} & \text{if } \Gamma_1 \mapsto \neg(B_1 \wedge B_2), \Delta_1 \in \xi \\ \begin{bmatrix} \Gamma_1, \neg A_1 \mapsto \Delta_1 \\ \Gamma_1, \neg A_2 \mapsto \Delta_1 \end{bmatrix} & \text{if } \Gamma_1, \neg(A_1 \vee A_2) \mapsto \Delta_1 \in \xi \\ \begin{cases} \Gamma_1 \mapsto \neg B_1, \Delta_1 \\ \Gamma_1 \mapsto \neg B_2, \Delta_1 \end{cases} & \text{if } \Gamma_1 \mapsto \neg(B_1 \vee B_2), \Delta_1 \in \xi. \end{cases}$$

3.3.4 Gentzen Deduction System $\mathbf{G}_{\mathtt{f}}$

A co-sequent $\Gamma \mapsto \Delta$ is $\mathbf{G}_{\mathtt{f}}$-valid, denoted by $\models_{\mathtt{f}} \Gamma \mapsto \Delta$, if there is an assignment v such that for each formula $A \in \Delta$, $v(A) = 0$; and for each formula $B \in \Delta$, $v(B)=1$.

Gentzen deduction system $\mathbf{G}_{\mathtt{f}}$ consists of the following axiom and deductions:

- **Axiom**:

$$(\mathbb{A}_{\mathtt{f}})\ \frac{\mathrm{con}(\Gamma)\&\mathrm{inval}(\Delta)\&\Gamma\cap\Delta=\emptyset}{\Gamma\mapsto\Delta},$$

where Γ, Δ are sets of literals.

- **Deduction rules**:

$$
\begin{array}{ll}
(\wedge^L)\ \dfrac{\left\{\begin{array}{l}\Gamma, A_1\mapsto\Delta\\ \Gamma, A_2\mapsto\Delta\end{array}\right.}{\Gamma, A_1\wedge A_2\mapsto\Delta} & (\wedge^R)\ \dfrac{\left[\begin{array}{l}\Gamma\mapsto B_1,\Delta\\ \Gamma\mapsto B_2,\Delta\end{array}\right.}{\Gamma\mapsto B_1\wedge B_2,\Delta}\\
(\vee^L)\ \dfrac{\left[\begin{array}{l}\Gamma, A_1\mapsto\Delta\\ \Gamma, A_2\mapsto\Delta\end{array}\right.}{\Gamma, A_1\vee A_2\mapsto\Delta} & (\vee^R)\ \dfrac{\left\{\begin{array}{l}\Gamma\mapsto B_1,\Delta\\ \Gamma\mapsto B_2,\Delta\end{array}\right.}{\Gamma\mapsto B_1\vee B_2,\Delta}
\end{array}
$$

and

$$
\begin{array}{ll}
(\neg\neg^L)\ \dfrac{\Gamma, A\mapsto\Delta}{\Gamma,\neg\neg A\mapsto\Delta} & (\neg\neg^R)\ \dfrac{\Gamma\mapsto B,\Delta}{\Gamma\mapsto\neg\neg B,\Delta}\\
(\neg\wedge^L)\ \dfrac{\left[\begin{array}{l}\Gamma,\neg A_1\mapsto\Delta\\ \Gamma,\neg A_1\mapsto\Delta\end{array}\right.}{\Gamma,\neg(A_1\wedge A_2)\mapsto\Delta} & (\neg\wedge^R)\ \dfrac{\left\{\begin{array}{l}\Gamma\mapsto\neg B_1,\Delta\\ \Gamma\mapsto\neg B_2,\Delta\end{array}\right.}{\Gamma\mapsto\neg(B_1\wedge B_2),\Delta}\\
(\neg\vee^L)\ \dfrac{\left\{\begin{array}{l}\Gamma,\neg A_1\mapsto\Delta\\ \Gamma,\neg A_2\mapsto\Delta\end{array}\right.}{\Gamma,\neg(A_1\vee A_2)\mapsto\Delta} & (\neg\vee^R)\ \dfrac{\left[\begin{array}{l}\Gamma\mapsto\neg B_1,\Delta\\ \Gamma\mapsto\neg B_2,\Delta\end{array}\right.}{\Gamma\mapsto\neg(B_1\vee B_2),\Delta}
\end{array}
$$

Definition 3.3.15 A co-sequent $\Gamma\mapsto\Delta$ is provable, denoted by $\vdash_{\mathtt{f}}\Gamma\mapsto\Delta$, if there is a sequence $\{\Gamma_1\mapsto\Delta_1,\ldots,\Gamma_n\mapsto\Delta_n\}$ of co-sequents such that $\Gamma_n\mapsto\Delta_n=\Gamma\mapsto\Delta$, and for each $1\le i\le n$, $\Gamma_i\mapsto\Delta_i$ is an axiom or is deduced from the previous co-sequents by one of the deduction rules in $\mathbf{G}_{\mathtt{f}}$.

Theorem 3.3.16 (Soundness and completeness theorem) *For any co-sequent* $\Gamma\mapsto\Delta$,

$$\vdash_{\mathtt{f}}\Gamma\mapsto\Delta \textit{ iff } \models_{\mathtt{f}}\Gamma\mapsto\Delta.$$

□

3.4 R-Calculi for Sequents

Let $*\in\{\mathtt{t},\mathtt{f}\}$.

Given two theories Γ,Δ and formulas $A\in\Gamma, B\in\Delta$, a reduction $\Gamma\Rightarrow\Delta|(A,B)\rightrightarrows\Gamma'\Rightarrow\Delta'$ is $\mathbf{R}^*$-valid, denoted by $\models^*\Gamma\Rightarrow\Delta|(A,B)\rightrightarrows\Gamma'\Rightarrow\Delta'$, if

$$
\begin{array}{l}
\Gamma'=\left\{\begin{array}{ll}\Gamma-\{A\} & \text{if}\models^*\Gamma-\{A\}\Rightarrow\Delta\\ \Gamma & \text{otherwise;}\end{array}\right.\\
\Delta'=\left\{\begin{array}{ll}\Delta-\{B\} & \text{if}\models^*\Gamma'\Rightarrow\Delta-\{B\}\\ \Delta & \text{otherwise;}\end{array}\right.
\end{array}
$$

Given two theories Γ, Δ and formulas A, B, a reduction $\Gamma \Rightarrow \Delta|(A, B) \rightrightarrows \Gamma' \Rightarrow \Delta'$ is $\mathbf{R}_*$-valid, denoted by $\models_* \Gamma \Rightarrow \Delta|(A, B) \rightrightarrows \Gamma' \Rightarrow \Delta'$, if

$$\Gamma' = \begin{cases} \Gamma \cup \{A\} & \text{if} \models_* \Gamma, A \Rightarrow \Delta \\ \Gamma & \text{otherwise;} \end{cases}$$
$$\Delta' = \begin{cases} \Delta \cup \{B\} & \text{if} \models_* \Gamma' \Rightarrow \Delta, B \\ \Delta & \text{otherwise;} \end{cases}$$

3.4.1 R-Calculus $\mathbf{R}^{\mathtt{t}}$

Given two theories Γ, Δ and formulas $A \in \Gamma, B \in \Delta$, a reduction $\Gamma \Rightarrow \Delta|(A, B) \rightrightarrows \Gamma' \Rightarrow \Delta'$ is $\mathbf{R}^{\mathtt{t}}$-valid, denoted by $\models^{\mathtt{t}} \Gamma \Rightarrow \Delta|(A, B) \rightrightarrows \Gamma' \Rightarrow \Delta'$, if

$$\Gamma' = \begin{cases} \Gamma - \{A\} & \text{if} \models^{\mathtt{t}} \Gamma - \{A\} \Rightarrow \Delta \\ \Gamma & \text{otherwise;} \end{cases}$$
$$\Delta' = \begin{cases} \Delta - \{B\} & \text{if} \models^{\mathtt{t}} \Gamma' \Rightarrow \Delta - \{B\} \\ \Delta & \text{otherwise;} \end{cases}$$

R-calculus $\mathbf{R}^{\mathtt{t}}$ consists of the following axioms and deductions:

- **Axioms**:

$$(\mathtt{A}^{\mathtt{t}}_{-})\ \frac{\mathbf{E}l' \neq l(l', \neg l' \in \Gamma \text{ or } l' \in \Gamma \cap \Delta)}{\Gamma \Rightarrow \Delta|(l, m) \Rightarrow \Gamma - \{l\} \rightrightarrows \Delta|m},$$
$$(\mathtt{A}^{\mathtt{t}}_{0})\ \frac{\sim \mathbf{E}l' \neq l(l', \neg l' \in \Gamma \text{ or } l' \in \Gamma \cap \Delta)}{\Gamma \Rightarrow \Delta|(l, m) \Rightarrow \Gamma \rightrightarrows \Delta|m},$$
$$(\mathtt{A}^{\mathtt{t}}_{-})\ \frac{\mathbf{E}m' \neq m(m', \neg m' \in \Delta \text{ or } m' \in \Gamma' \cap \Delta)}{\Gamma' \Rightarrow \Delta|m \Rightarrow \Gamma' \rightrightarrows \Delta - \{m\}},$$
$$(\mathtt{A}^{\mathtt{t}}_{0})\ \frac{\sim \mathbf{E}m' \neq m(m', \neg m' \in \Delta \text{ or } m' \in \Gamma' \cap \Delta)}{\Gamma' \Rightarrow \Delta|m \Rightarrow \Gamma' \rightrightarrows \Delta},$$

  where l, m are literals and Γ, Δ are sets of literals.
- **Deduction rules**:

$$(\wedge^{L-})\ \frac{\begin{cases} \Gamma \Rightarrow \Delta|(A_1, B) \rightrightarrows \Gamma - \{A_1\} \Rightarrow \Delta|B \\ \Gamma \Rightarrow \Delta|(A_2, B) \rightrightarrows \Gamma - \{A_2\} \Rightarrow \Delta|B \end{cases}}{\Gamma \Rightarrow \Delta|(A_1 \wedge A_2, B) \rightrightarrows \Gamma - \{A_1 \wedge A_2\} \Rightarrow \Delta|B}$$
$$(\wedge^{L0})\ \frac{\begin{bmatrix} \Gamma \Rightarrow \Delta|(A_1, B) \rightrightarrows \Gamma \Rightarrow \Delta|B \\ \Gamma \Rightarrow \Delta|(A_2, B) \rightrightarrows \Gamma \Rightarrow \Delta|B \end{bmatrix}}{\Gamma \Rightarrow \Delta|(A_1 \wedge A_2, B) \rightrightarrows \Gamma \Rightarrow \Delta|B}$$
$$(\wedge^{R-})\ \frac{\begin{bmatrix} \Gamma' \Rightarrow \Delta|B_1 \rightrightarrows \Gamma' \Rightarrow \Delta - \{B_1\} \\ \Gamma' \Rightarrow \Delta - \{B_1\}|B_2 \rightrightarrows \Gamma' \Rightarrow \Delta - \{B_1, B_2\} \end{bmatrix}}{\Gamma' \Rightarrow \Delta|B_1 \wedge B_2 \rightrightarrows \Gamma' \Rightarrow \Delta - \{B_1 \wedge B_2\}}$$
$$(\wedge^{R0})\ \frac{\begin{cases} \Gamma' \Rightarrow \Delta|B_1 \rightrightarrows \Gamma' \Rightarrow \Delta \\ \Gamma' \Rightarrow \Delta - \{B_1\}|B_2 \rightrightarrows \Gamma' \Rightarrow \Delta - \{B_1\} \end{cases}}{\Gamma' \Rightarrow \Delta|B_1 \wedge B_2 \rightrightarrows \Gamma' \Rightarrow \Delta}$$

and

$$(\vee^{L-})\ \frac{\left[\begin{array}{l}\Gamma \Rightarrow \Delta|(A_1, B) \rightrightarrows \Gamma - \{A_1\} \Rightarrow \Delta|B \\ \Gamma - \{A_1\} \Rightarrow \Delta|(A_2, B) \rightrightarrows \Gamma - \{A_1, A_2\} \Rightarrow \Delta|B\end{array}\right.}{\Gamma \Rightarrow \Delta|(A_1 \vee A_2, B) \rightrightarrows \Gamma - \{A_1 \vee A_2\} \Rightarrow \Delta|B}$$

$$(\vee^{L0})\ \frac{\left\{\begin{array}{l}\Gamma \Rightarrow \Delta|(A_1, B) \rightrightarrows \Gamma \Rightarrow \Delta|B \\ \Gamma - \{A_1\} \Rightarrow \Delta|(A_2, B) \rightrightarrows \Gamma - \{A_1\} \Rightarrow \Delta|B\end{array}\right.}{\Gamma \Rightarrow \Delta|(A_1 \vee A_2, B) \rightrightarrows \Gamma \Rightarrow \Delta|B}$$

$$(\vee^{R-})\ \frac{\left\{\begin{array}{l}\Gamma' \Rightarrow \Delta|B_1 \rightrightarrows \Gamma' \Rightarrow \Delta - \{B_1\} \\ \Gamma' \Rightarrow \Delta|B_2 \rightrightarrows \Gamma' \Rightarrow \Delta - \{B_2\}\end{array}\right.}{\Gamma' \Rightarrow \Delta|B_1 \vee B_2 \rightrightarrows \Gamma' \Rightarrow \Delta - \{B_1 \vee B_2\}}$$

$$(\vee^{R0})\ \frac{\left[\begin{array}{l}\Gamma' \Rightarrow \Delta|B_1 \rightrightarrows \Gamma' \Rightarrow \Delta \\ \Gamma' \Rightarrow \Delta|B_2 \rightrightarrows \Gamma' \Rightarrow \Delta\end{array}\right.}{\Gamma' \Rightarrow \Delta|B_1 \vee B_2 \rightrightarrows \Gamma' \Rightarrow \Delta}$$

and

$$(\neg\neg^{L-})\ \frac{\Gamma \Rightarrow \Delta|(A, B) \Rightarrow \Gamma - \{A\} \Rightarrow \Delta|B}{\Gamma \Rightarrow \Delta|(\neg\neg A, B) \rightrightarrows \Gamma - \{\neg\neg A\} \Rightarrow \Delta|B}$$

$$(\neg\neg^{L0})\ \frac{\Gamma \Rightarrow \Delta|(A, B) \Rightarrow \Gamma \Rightarrow \Delta|B}{\Gamma \Rightarrow \Delta|(\neg\neg A, B) \rightrightarrows \Gamma \Rightarrow \Delta|B}$$

$$(\neg\neg^{R-})\ \frac{\Gamma' \Rightarrow \Delta|B \Rightarrow \Gamma' \Rightarrow \Delta - \{B\}}{\Gamma' \Rightarrow \Delta|\neg\neg B \rightrightarrows \Gamma' \Rightarrow \Delta - \{\neg\neg B\}}$$

$$(\neg\neg^{R0})\ \frac{\Gamma' \Rightarrow \Delta|B \Rightarrow \Gamma' \Rightarrow \Delta}{\Gamma' \Rightarrow \Delta|\neg\neg B \rightrightarrows \Gamma' \Rightarrow \Delta}$$

and

$$(\neg\wedge^{L-})\ \frac{\left[\begin{array}{l}\Gamma \Rightarrow \Delta|(\neg A_1, B) \rightrightarrows \Gamma - \{\neg A_1\} \Rightarrow \Delta|B \\ \Gamma - \{\neg A_1\} \Rightarrow \Delta|(\neg A_2, B) \rightrightarrows \Gamma - \{\neg A_1, \neg A_2\} \Rightarrow \Delta|B\end{array}\right.}{\Gamma \Rightarrow \Delta|(\neg(A_1 \wedge A_2), B) \rightrightarrows \Gamma - \{\neg(A_1 \wedge A_2)\} \Rightarrow \Delta|B}$$

$$(\neg\wedge^{L0})\ \frac{\left\{\begin{array}{l}\Gamma \Rightarrow \Delta|(\neg A_1, B) \rightrightarrows \Gamma \Rightarrow \Delta|B \\ \Gamma - \{\neg A_1\} \Rightarrow \Delta|(\neg A_2, B) \rightrightarrows \Gamma - \{\neg A_1\} \Rightarrow \Delta|B\end{array}\right.}{\Gamma \Rightarrow \Delta|(\neg(A_1 \wedge A_2), B) \rightrightarrows \Gamma \Rightarrow \Delta|B}$$

$$(\neg\wedge^{R-})\ \frac{\left\{\begin{array}{l}\Gamma' \Rightarrow \Delta|\neg B_1 \rightrightarrows \Gamma' \Rightarrow \Delta - \{\neg B_1\} \\ \Gamma' \Rightarrow \Delta|\neg B_2 \rightrightarrows \Gamma' \Rightarrow \Delta - \{\neg B_2\}\end{array}\right.}{\Gamma' \Rightarrow \Delta|\neg(B_1 \wedge B_2) \rightrightarrows \Gamma' \Rightarrow \Delta - \{\neg B_1 \wedge \neg B_2\}}$$

$$(\neg\wedge^{R0})\ \frac{\left[\begin{array}{l}\Gamma' \Rightarrow \Delta|\neg B_1 \rightrightarrows \Gamma' \Rightarrow \Delta \\ \Gamma' \Rightarrow \Delta|\neg B_2 \rightrightarrows \Gamma' \Rightarrow \Delta\end{array}\right.}{\Gamma' \Rightarrow \Delta|\neg(B_1 \wedge B_2) \rightrightarrows \Gamma' \Rightarrow \Delta}$$

and

$$(\neg\vee^{L-})\ \frac{\begin{cases}\Gamma \Rightarrow \Delta|(\neg A_1, B) \rightrightarrows \Gamma - \{\neg A_1\} \Rightarrow \Delta|B \\ \Gamma \Rightarrow \Delta|(\neg A_2, B) \rightrightarrows \Gamma - \{\neg A_2\} \Rightarrow \Delta|B\end{cases}}{\Gamma \Rightarrow \Delta|(\neg(A_1 \vee A_2), B) \rightrightarrows \Gamma - \{\neg(A_1 \vee A_2)\} \Rightarrow \Delta|B}$$

$$(\neg\vee^{L0})\ \frac{\begin{bmatrix}\Gamma \Rightarrow \Delta|(\neg A_1, B) \rightrightarrows \Gamma \Rightarrow \Delta|B \\ \Gamma \Rightarrow \Delta|(\neg A_2, B) \rightrightarrows \Gamma \Rightarrow \Delta|B\end{bmatrix}}{\Gamma \Rightarrow \Delta|(\neg(A_1 \vee A_2), B) \rightrightarrows \Gamma \Rightarrow \Delta|B}$$

$$(\neg\vee^{R-})\ \frac{\begin{bmatrix}\Gamma' \Rightarrow \Delta|\neg B_1 \rightrightarrows \Gamma' \Rightarrow \Delta - \{\neg B_1\} \\ \Gamma' \Rightarrow \Delta - \{\neg B_1\}|\neg B_2 \rightrightarrows \Gamma' \Rightarrow \Delta - \{\neg B_1, \neg B_2\}\end{bmatrix}}{\Gamma' \Rightarrow \Delta|\neg(B_1 \vee B_2) \rightrightarrows \Gamma' \Rightarrow \Delta - \{\neg(B_1 \vee B_2)\}}$$

$$(\neg\vee^{R0})\ \frac{\begin{cases}\Gamma' \Rightarrow \Delta|\neg B_1 \rightrightarrows \Gamma' \Rightarrow \Delta \\ \Gamma' \Rightarrow \Delta - \{\neg B_1\}|\neg B_2 \rightrightarrows \Gamma' \Rightarrow \Delta - \{\neg B_1\}\end{cases}}{\Gamma \Rightarrow \Delta|\neg(B_1 \vee B_2) \rightrightarrows \Gamma' \Rightarrow \Delta}$$

Definition 3.4.1 A reduction $\delta = \Gamma \Rightarrow \Delta|(A, B) \rightrightarrows \Gamma' \Rightarrow \Delta'$ is provable in $\mathbf{R}^{\mathtt{t}}$, denoted by $\vdash^{\mathtt{t}} \delta$, if there is a sequence $\{\delta_1, \ldots, \delta_n\}$ of reductions such that $\delta_n = \delta$, and for each $1 \leq i \leq n$, δ_i is an axiom or is deduced from the previous reductions by one of the deduction rules in $\mathbf{R}^{\mathtt{t}}$.

Theorem 3.4.2 (Soundness and completeness theorem) *For any reduction* $\Gamma \Rightarrow \Delta|(A, B) \rightrightarrows \Gamma' \Rightarrow \Delta'$, *where* $A \in \Gamma, B \in \Delta$,

$$\vdash^{\mathtt{t}} \Gamma \Rightarrow \Delta|(A, B) \rightrightarrows \Gamma' \Rightarrow \Delta'$$

if and only if

$$\models^{\mathtt{t}} \Gamma \Rightarrow \Delta|(A, B) \rightrightarrows \Gamma' \Rightarrow \Delta'.$$

□

3.4.2 R-Calculus $\mathbf{R}^{\mathtt{f}}$

Given two theories Γ, Δ and formulas $A \in \Gamma, B \in \Delta$, a reduction $\Gamma \Rightarrow \Delta|(A, B) \rightrightarrows \Gamma' \Rightarrow \Delta'$ is $\mathbf{R}^{\mathtt{f}}$-valid, denoted by $\models^{\mathtt{f}} \Gamma \Rightarrow \Delta|(A, B) \rightrightarrows \Gamma' \Rightarrow \Delta'$, if

$$\Gamma' = \begin{cases}\Gamma - \{A\} & \text{if} \models^{\mathtt{f}} \Gamma - \{A\} \Rightarrow \Delta \\ \Gamma & \text{otherwise;}\end{cases}$$
$$\Delta' = \begin{cases}\Delta - \{B\} & \text{if} \models^{\mathtt{f}} \Gamma' \Rightarrow \Delta - \{B\} \\ \Delta & \text{otherwise;}\end{cases}$$

R-calculus $\mathbf{R}^{\mathtt{f}}$ consists of the following axioms and deductions:

- **Axioms**:

$$(\mathtt{A}_{+}^{\mathtt{f}})\ \frac{\mathbf{E}l' \neq l(l', \neg l' \in \Gamma \text{ or } l' \in \Gamma \cap \Delta)}{\Gamma \Rightarrow \Delta|(l, m) \Rightarrow \Gamma - \{l\} \rightrightarrows \Delta|m},$$

$$(\mathtt{A}_{0}^{\mathtt{f}})\ \frac{\sim \mathbf{E}l' \neq l(l', \neg l' \in \Gamma \text{ or } l' \in \Gamma \cap \Delta)}{\Gamma \Rightarrow \Delta|(l, m) \Rightarrow \Gamma \rightrightarrows \Delta|m},$$

$$(\mathtt{A}_{+}^{\mathtt{f}})\ \frac{\mathbf{E}m' \neq m(m', \neg m' \in \Delta \text{ or } m' \in \Gamma' \cap \Delta)}{\Gamma' \Rightarrow \Delta|m \Rightarrow \Gamma' \rightrightarrows \Delta - \{m\}},$$

$$(\mathtt{A}_{0}^{\mathtt{f}})\ \frac{\sim \mathbf{E}m' \neq m(m', \neg m' \in \Delta \text{ or } m' \in \Gamma' \cap \Delta)}{\Gamma' \Rightarrow \Delta|m \Rightarrow \Gamma' \rightrightarrows \Delta},$$

where l, m are literals, Γ, Δ are sets of literals.

- **Deduction rules**:

$$(\wedge^{L-})\ \frac{\left[\begin{array}{l} \Gamma \Rightarrow \Delta|(A_1, B) \rightrightarrows \Gamma - \{A_1\} \Rightarrow \Delta|B \\ \Gamma - \{A_1\} \Rightarrow \Delta|(A_2, B) \rightrightarrows \Gamma - \{A_1, A_2\} \Rightarrow \Delta|B \end{array}\right.}{\Gamma \Rightarrow \Delta|(A_1 \wedge A_2, B) \rightrightarrows \Gamma - \{A_1 \wedge A_2\} \Rightarrow \Delta|B}$$

$$(\wedge^{L0})\ \frac{\left\{\begin{array}{l} \Gamma \Rightarrow \Delta|(A_1, B) \rightrightarrows \Gamma \Rightarrow \Delta|B \\ \Gamma - \{A_1\} \Rightarrow \Delta'|(A_2, B) \rightrightarrows \Gamma - \{A_1\} \Rightarrow \Delta|B \end{array}\right.}{\Gamma \Rightarrow \Delta|(A_1 \wedge A_2, B) \rightrightarrows \Gamma \Rightarrow \Delta|B}$$

$$(\wedge^{R-})\ \frac{\left[\begin{array}{l} \Gamma' \Rightarrow \Delta|B_1 \rightrightarrows \Gamma' \Rightarrow \Delta - \{B_1\} \\ \Gamma' \Rightarrow \Delta|B_2 \rightrightarrows \Gamma' \Rightarrow \Delta - \{B_2\} \end{array}\right.}{\Gamma' \Rightarrow \Delta|B_1 \wedge B_2 \rightrightarrows \Gamma' \Rightarrow \Delta - \{B_1 \wedge B_2\}}$$

$$(\wedge^{R0})\ \frac{\left\{\begin{array}{l} \Gamma' \Rightarrow \Delta|B_1 \rightrightarrows \Gamma' \Rightarrow \Delta \\ \Gamma' \Rightarrow \Delta|B_2 \rightrightarrows \Gamma' \Rightarrow \Delta \end{array}\right.}{\Gamma' \Rightarrow \Delta|B_1 \wedge B_2 \rightrightarrows \Gamma' \Rightarrow \Delta}$$

and

$$(\vee^{L-})\ \frac{\left\{\begin{array}{l} \Gamma \Rightarrow \Delta|(A_1, B) \rightrightarrows \Gamma - \{A_1\} \Rightarrow \Delta|B \\ \Gamma \Rightarrow \Delta|(A_2, B) \rightrightarrows \Gamma - \{A_2\} \Rightarrow \Delta|B \end{array}\right.}{\Gamma \Rightarrow \Delta|(A_1 \vee A_2, B) \rightrightarrows \Gamma - \{A_1 \vee A_2\} \Rightarrow \Delta|B}$$

$$(\vee^{L0})\ \frac{\left[\begin{array}{l} \Gamma \Rightarrow \Delta|(A_1, B) \rightrightarrows \Gamma \Rightarrow \Delta|B \\ \Gamma \Rightarrow \Delta'|(A_2, B) \rightrightarrows \Gamma \Rightarrow \Delta|B \end{array}\right.}{\Gamma \Rightarrow \Delta|(A_1 \vee A_2, B) \rightrightarrows \Gamma \Rightarrow \Delta|B}$$

$$(\vee^{R-})\ \frac{\left\{\begin{array}{l} \Gamma' \Rightarrow \Delta|B_1 \rightrightarrows \Gamma' \Rightarrow \Delta - \{B_1\} \\ \Gamma' \Rightarrow \Delta - \{B_1\}|B_2 \rightrightarrows \Gamma' \Rightarrow \Delta - \{B_1, B_2\} \end{array}\right.}{\Gamma' \Rightarrow \Delta|B_1 \vee B_2 \rightrightarrows \Gamma' \Rightarrow \Delta - \{B_1 \vee B_2\}}$$

$$(\vee^{R0})\ \frac{\left[\begin{array}{l} \Gamma' \Rightarrow \Delta|B_1 \rightrightarrows \Gamma' \Rightarrow \Delta \\ \Gamma' \Rightarrow \Delta - \{B_1\}|B_2 \rightrightarrows \Gamma' \Rightarrow \Delta - \{B_1\} \end{array}\right.}{\Gamma' \Rightarrow \Delta|B_1 \vee B_2 \rightrightarrows \Gamma' \Rightarrow \Delta}$$

and

$$(\neg\neg^{L-})\ \frac{\Gamma \Rightarrow \Delta|(A, B) \Rightarrow \Gamma - \{A\} \Rightarrow \Delta|B}{\Gamma \Rightarrow \Delta|(\neg\neg A, B) \rightrightarrows \Gamma - \{\neg\neg A\} \Rightarrow \Delta|B}$$

$$(\neg\neg^{L0})\ \frac{\Gamma \Rightarrow \Delta|(A, B) \Rightarrow \Gamma \Rightarrow \Delta|B}{\Gamma \Rightarrow \Delta|(\neg\neg A, B) \rightrightarrows \Gamma \Rightarrow \Delta|B}$$

$$(\neg\neg^{R-})\ \frac{\Gamma' \Rightarrow \Delta|B \Rightarrow \Gamma' \Rightarrow \Delta - \{B\}}{\Gamma' \Rightarrow \Delta|\neg\neg B \rightrightarrows \Gamma' \Rightarrow \Delta - \{\neg\neg B\}}$$

$$(\neg\neg^{R0})\ \frac{\Gamma' \Rightarrow \Delta|B \Rightarrow \Gamma' \Rightarrow \Delta}{\Gamma' \Rightarrow \Delta|\neg\neg B \rightrightarrows \Gamma' \Rightarrow \Delta}$$

and

$$(\neg\wedge^{L-})\ \frac{\begin{cases}\Gamma \Rightarrow \Delta|(\neg A_1, B) \rightrightarrows \Gamma - \{\neg A_1\} \Rightarrow \Delta|B\\ \Gamma \Rightarrow \Delta|(\neg A_2, B) \rightrightarrows \Gamma - \{\neg A_2\} \Rightarrow \Delta|B\end{cases}}{\Gamma \Rightarrow \Delta|(\neg(A_1 \wedge A_2), B) \rightrightarrows \Gamma - \{\neg(A_1 \wedge A_2)\} \Rightarrow \Delta|B}$$

$$(\neg\wedge^{L0})\ \frac{\left[\begin{array}{l}\Gamma \Rightarrow \Delta|(\neg A_1, B) \rightrightarrows \Gamma \Rightarrow \Delta|B\\ \Gamma \Rightarrow \Delta'|(\neg A_2, B) \rightrightarrows \Gamma \Rightarrow \Delta|B\end{array}\right.}{\Gamma \Rightarrow \Delta|(\neg(A_1 \wedge A_2), B) \rightrightarrows \Gamma \Rightarrow \Delta|B}$$

$$(\neg\wedge^{R-})\ \frac{\left[\begin{array}{l}\Gamma' \Rightarrow \Delta|\neg B_1 \rightrightarrows \Gamma' \Rightarrow \Delta - \{\neg B_1\}\\ \Gamma' \Rightarrow \Delta - \{\neg B_1\}|\neg B_2 \rightrightarrows \Gamma' \Rightarrow \Delta - \{\neg B_1, \neg B_2\}\end{array}\right.}{\Gamma' \Rightarrow \Delta|\neg(B_1 \wedge B_2) \rightrightarrows \Gamma' \Rightarrow \Delta - \{\neg(B_1 \wedge B_2)\}}$$

$$(\neg\wedge^{R0})\ \frac{\begin{cases}\Gamma' \Rightarrow \Delta|\neg B_1 \rightrightarrows \Gamma' \Rightarrow \Delta\\ \Gamma' \Rightarrow \Delta - \{\neg B_1\}|\neg B_2 \rightrightarrows \Gamma' \Rightarrow \Delta - \{\neg B_1\}\end{cases}}{\Gamma' \Rightarrow \Delta|\neg(B_1 \wedge B_2) \rightrightarrows \Gamma' \Rightarrow \Delta}$$

and

$$(\neg\vee^{L-})\ \frac{\left[\begin{array}{l}\Gamma \Rightarrow \Delta|(\neg A_1, B) \rightrightarrows \Gamma - \{\neg A_1\} \Rightarrow \Delta|B\\ \Gamma - \{\neg A_1\} \Rightarrow \Delta|(\neg A_2, B) \rightrightarrows \Gamma - \{\neg A_1, \neg A_2\} \Rightarrow \Delta|B\end{array}\right.}{\Gamma \Rightarrow \Delta|(\neg(A_1 \vee A_2), B) \rightrightarrows \Gamma - \{\neg(A_1 \vee A_2)\} \Rightarrow \Delta|B}$$

$$(\neg\vee^{L0})\ \frac{\begin{cases}\Gamma \Rightarrow \Delta|(\neg A_1, B) \rightrightarrows \Gamma \Rightarrow \Delta|B\\ \Gamma - \{\neg A_1\} \Rightarrow \Delta'|(\neg A_2, B) \rightrightarrows \Gamma - \{\neg A_1\} \Rightarrow \Delta|B\end{cases}}{\Gamma \Rightarrow \Delta|(\neg(A_1 \vee A_2), B) \rightrightarrows \Gamma \Rightarrow \Delta|B}$$

$$(\neg\vee^{R-})\ \frac{\begin{cases}\Gamma' \Rightarrow \Delta|\neg B_1 \rightrightarrows \Gamma' \Rightarrow \Delta - \{\neg B_1\}\\ \Gamma' \Rightarrow \Delta|\neg B_2 \rightrightarrows \Gamma' \Rightarrow \Delta - \{\neg B_2\}\end{cases}}{\Gamma' \Rightarrow \Delta|\neg(B_1 \vee B_2) \rightrightarrows \Gamma' \Rightarrow \Delta - \{\neg(B_1 \vee B_2)\}}$$

$$(\neg\vee^{R0})\ \frac{\left[\begin{array}{l}\Gamma' \Rightarrow \Delta|\neg B_1 \rightrightarrows \Gamma' \Rightarrow \Delta\\ \Gamma' \Rightarrow \Delta|\neg B_2 \rightrightarrows \Gamma' \Rightarrow \Delta\end{array}\right.}{\Gamma' \Rightarrow \Delta|\neg(B_1 \vee B_2) \rightrightarrows \Gamma' \Rightarrow \Delta}$$

Definition 3.4.3 A reduction $\delta = \Gamma \Rightarrow \Delta|(A, B) \rightrightarrows \Gamma' \Rightarrow \Delta'$ is provable in $\mathbf{R}^{\text{f}}$, denoted by $\vdash^{\text{f}} \delta$, if there is a sequence $\{\delta_1, \ldots, \delta_n\}$ of reductions such that $\delta_n = \delta$, and for each $1 \leq i \leq n$, δ_i is an axiom or is deduced from the previous reductions by one of the deduction rules in $\mathbf{R}^{\text{f}}$.

Theorem 3.4.4 (Soundness and completeness theorem) *For any reduction* $\Gamma \Rightarrow \Delta|(A,B) \rightrightarrows \Gamma' \Rightarrow \Delta'$,

$$\vdash^{\mathtt{f}} \Gamma \Rightarrow \Delta|(A,B) \rightrightarrows \Gamma' \Rightarrow \Delta'$$

if and only if

$$\models^{\mathtt{f}} \Gamma \Rightarrow \Delta|(A,B) \rightrightarrows \Gamma' \Rightarrow \Delta'.$$

□

3.4.3 *R-Calculus* $\mathbf{R}_{\mathtt{t}}$

Given two theories Γ, Δ and formulas A, B, a reduction $\Gamma \Rightarrow \Delta|(A,B) \rightrightarrows \Gamma' \Rightarrow \Delta'$ is $\mathbf{R}_{\mathtt{t}}$-valid, denoted by $\models_{\mathtt{t}} \Gamma \Rightarrow \Delta|(A,B) \rightrightarrows \Gamma' \Rightarrow \Delta'$, if

$$\Gamma' = \begin{cases} \Gamma \cup \{A\} \text{ if } \models_{\mathtt{t}} \Gamma, A \Rightarrow \Delta \\ \Gamma \quad \text{otherwise;} \end{cases}$$
$$\Delta' = \begin{cases} \Delta \cup \{B\} \text{ if } \models_{\mathtt{t}} \Gamma' \Rightarrow \Delta, B \\ \Delta \quad \text{otherwise;} \end{cases}$$

R-calculus $\mathbf{R}_{\mathtt{t}}$ consists of the following axioms and deductions:

- **Axioms**:

$$(\mathtt{A}_{\mathtt{t}}^{+L})\ \frac{\neg l \notin \Gamma \& l \notin \Delta}{\Gamma \mapsto \Delta|(l,m) \rightrightarrows \Gamma, l \mapsto \Delta|m},\quad (\mathtt{A}_{\mathtt{t}}^{0L})\ \frac{\neg l \in \Gamma \text{ or } l \in \Delta}{\Gamma \mapsto \Delta|(l,m) \rightrightarrows \Gamma \mapsto \Delta|m},$$
$$(\mathtt{A}_{\mathtt{t}}^{+R})\ \frac{m \notin \Gamma \& \neg m \notin \Delta}{\Gamma' \mapsto \Delta|m \rightrightarrows \Gamma' \mapsto \Delta, m},\quad (\mathtt{A}_{\mathtt{t}}^{0R})\ \frac{m \in \Gamma \text{ or } \neg m \in \Delta}{\Gamma' \mapsto \Delta|m \rightrightarrows \Gamma' \mapsto \Delta},$$

where l, m are literals, and Δ, Γ are sets of literals.
- **Deduction rules**:

$$(\wedge^{L+})\ \frac{\left[\begin{array}{l}\Gamma \mapsto \Delta|(A_1,B) \rightrightarrows \Gamma, A_1 \mapsto \Delta|B \\ \Gamma, A_1 \mapsto \Delta|(A_2,B) \rightrightarrows \Gamma, A_1, A_2 \mapsto \Delta|B\end{array}\right.}{\Gamma \mapsto \Delta|(A_1 \wedge A_2, B) \rightrightarrows \Gamma, A_1 \wedge A_2 \mapsto \Delta|B}$$

$$(\wedge^{L0})\ \frac{\left\{\begin{array}{l}\Gamma \mapsto \Delta|(A_1,B) \rightrightarrows \Gamma \mapsto \Delta|B \\ \Gamma, A_1 \mapsto \Delta|(A_2,B) \rightrightarrows \Gamma, A_1 \mapsto \Delta|B\end{array}\right.}{\Gamma \mapsto \Delta|(A_1 \wedge A_2, B) \rightrightarrows \Gamma \mapsto \Delta|B}$$

$$(\wedge^{R+})\ \frac{\left\{\begin{array}{l}\Gamma' \mapsto \Delta|B_1 \rightrightarrows \Gamma' \mapsto \Delta, B_1 \\ \Gamma' \mapsto \Delta|B_2 \rightrightarrows \Gamma' \mapsto \Delta, B_2\end{array}\right.}{\Gamma' \mapsto \Delta|B_1 \wedge B_2 \rightrightarrows \Gamma' \mapsto \Delta, B_1 \wedge B_2}$$

$$(\wedge^{R0})\ \frac{\left[\begin{array}{l}\Gamma' \mapsto \Delta|B_1 \rightrightarrows \Gamma' \mapsto \Delta \\ \Gamma' \mapsto \Delta|B_2 \rightrightarrows \Gamma' \mapsto \Delta\end{array}\right.}{\Gamma' \mapsto \Delta|B_1 \wedge B_2 \rightrightarrows \Gamma' \mapsto \Delta}$$

and

$$(\vee^{L+})\ \frac{\begin{cases}\Gamma \mapsto \Delta|(A_1, B) \rightrightarrows \Gamma, A_1 \mapsto \Delta|B\\ \Gamma \mapsto \Delta|(A_2, B) \rightrightarrows \Gamma, A_2 \mapsto \Delta|B\end{cases}}{\Gamma \mapsto \Delta|(A_1 \vee A_2, B) \rightrightarrows \Gamma, A_1 \vee A_2 \mapsto \Delta|B}$$

$$(\vee^{L0})\ \frac{\begin{bmatrix}\Gamma \mapsto \Delta|(A_1, B) \rightrightarrows \Gamma \mapsto \Delta|B\\ \Gamma \mapsto \Delta|(A_2, B) \rightrightarrows \Gamma \mapsto \Delta|B\end{bmatrix}}{\Gamma \mapsto \Delta|(A_1 \vee A_2, B) \rightrightarrows \Gamma \mapsto \Delta|B}$$

$$(\vee^{R-})\ \frac{\begin{bmatrix}\Gamma' \mapsto \Delta|B_1 \rightrightarrows \Gamma' \mapsto \Delta, B_1\\ \Gamma' \mapsto \Delta, B_1|B_2 \rightrightarrows \Gamma' \mapsto \Delta, B_1, B_2\end{bmatrix}}{\Gamma' \mapsto \Delta|B_1 \vee B_2 \rightrightarrows \Gamma' \mapsto \Delta, B_1 \vee B_2}$$

$$(\vee^{R0})\ \frac{\begin{cases}\Gamma' \mapsto \Delta|B_1 \rightrightarrows \Gamma' \mapsto \Delta\\ \Gamma' \mapsto \Delta, B_1|B_2 \rightrightarrows \Gamma' \mapsto \Delta, B_1\end{cases}}{\Gamma' \mapsto \Delta|B_1 \vee B_2 \rightrightarrows \Gamma' \mapsto \Delta}$$

and

$$(\neg\neg^{L+})\ \frac{\Gamma \mapsto \Delta|(A, B) \rightrightarrows \Gamma, A \mapsto \Delta|B}{\Gamma \mapsto \Delta|(\neg\neg A, B) \rightrightarrows \Gamma, \neg\neg A \mapsto \Delta|B}$$

$$(\neg\neg^{L0})\ \frac{\Gamma \mapsto \Delta|(A, B) \rightrightarrows \Gamma \mapsto \Delta|B}{\Gamma \mapsto \Delta|(\neg\neg A, B) \rightrightarrows \Gamma \mapsto \Delta|B}$$

$$(\neg\neg^{R+})\ \frac{\Gamma' \mapsto \Delta|B \rightrightarrows \Gamma' \mapsto \Delta, B}{\Gamma' \mapsto \Delta|\neg\neg B \rightrightarrows \Gamma' \mapsto \Delta, \neg\neg B}$$

$$(\neg\neg^{R0})\ \frac{\Gamma' \mapsto \Delta|B \rightrightarrows \Gamma' \mapsto \Delta}{\Gamma' \mapsto \Delta|\neg\neg B \rightrightarrows \Gamma \mapsto \Delta}$$

and

$$(\neg\wedge^{L+})\ \frac{\begin{cases}\Gamma \mapsto \Delta|(\neg A_1, B) \rightrightarrows \Gamma, \neg A_1 \mapsto \Delta|B\\ \Gamma \mapsto \Delta|(\neg A_2, B) \rightrightarrows \Gamma, \neg A_2 \mapsto \Delta|B\end{cases}}{\Gamma \mapsto \Delta|(\neg(A_1 \wedge A_2), B) \rightrightarrows \Gamma, \neg(A_1 \wedge A_2) \mapsto \Delta|B}$$

$$(\neg\wedge^{L0})\ \frac{\begin{bmatrix}\Gamma \mapsto \Delta|(\neg A_1, B) \rightrightarrows \Gamma \mapsto \Delta|B\\ \Gamma \mapsto \Delta|(\neg A_2, B) \rightrightarrows \Gamma \mapsto \Delta|B\end{bmatrix}}{\Gamma \mapsto \Delta|(\neg(A_1 \wedge A_2), B) \rightrightarrows \Gamma \mapsto \Delta|B}$$

$$(\neg\wedge^{R+})\ \frac{\begin{bmatrix}\Gamma' \mapsto \Delta|\neg B_1 \rightrightarrows \Gamma' \mapsto \Delta, \neg B_1\\ \Gamma' \mapsto \Delta, \neg B_1|\neg B_2 \rightrightarrows \Gamma' \mapsto \Delta, \neg B_1, \neg B_2\end{bmatrix}}{\Gamma' \mapsto \Delta|\neg(B_1 \wedge B_2) \rightrightarrows \Gamma' \mapsto \Delta, \neg(B_1 \wedge B_2)}$$

$$(\neg\wedge^{R0})\ \frac{\begin{cases}\Gamma' \mapsto \Delta|\neg B_1 \rightrightarrows \Gamma' \mapsto \Delta\\ \Gamma' \mapsto \Delta, \neg B_1|\neg B_2 \rightrightarrows \Gamma' \mapsto \Delta, \neg B_1\end{cases}}{\Gamma' \mapsto \Delta|\neg(B_1 \wedge B_2) \rightrightarrows \Gamma' \mapsto \Delta}$$

and

$$(\neg\vee^{L+})\ \frac{\left[\begin{array}{l}\Gamma \mapsto \Delta|(\neg A_1, B) \rightrightarrows \Gamma, \neg A_1 \mapsto \Delta|B \\ \Gamma, \neg A_1 \mapsto \Delta|(\neg A_2, B) \rightrightarrows \Gamma, \neg A_1, \neg A_2 \mapsto \Delta|B\end{array}\right.}{\Gamma \mapsto \Delta|(\neg(A_1 \vee A_2), B) \rightrightarrows \Gamma, \neg(A_1 \vee A_2) \mapsto \Delta|B}$$

$$(\neg\vee^{L0})\ \frac{\left\{\begin{array}{l}\Gamma \mapsto \Delta|(\neg A_1, B) \rightrightarrows \Gamma \mapsto \Delta|B \\ \Gamma, \neg A_1 \mapsto \Delta|(\neg A_2, B) \rightrightarrows \Gamma, \neg A_1 \mapsto \Delta|B\end{array}\right.}{\Gamma \mapsto \Delta|(\neg(A_1 \vee A_2), B) \rightrightarrows \Gamma \mapsto \Delta|B}$$

$$(\neg\vee^{R+})\ \frac{\left\{\begin{array}{l}\Gamma' \mapsto \Delta|\neg B_1 \rightrightarrows \Gamma' \mapsto \Delta, \neg B_1 \\ \Gamma' \mapsto \Delta|\neg B_2 \rightrightarrows \Gamma' \mapsto \Delta, \neg B_2\end{array}\right.}{\Gamma' \mapsto \Delta|\neg(B_1 \vee B_2) \rightrightarrows \Gamma' \mapsto \Delta, \neg(B_1 \vee B_2)}$$

$$(\neg\vee^{R0})\ \frac{\left[\begin{array}{l}\Gamma' \mapsto \Delta|\neg B_1 \rightrightarrows \Gamma' \mapsto \Delta \\ \Gamma' \mapsto \Delta|\neg B_2 \rightrightarrows \Gamma' \mapsto \Delta\end{array}\right.}{\Gamma' \mapsto \Delta|\neg(B_1 \vee B_2) \rightrightarrows \Gamma' \mapsto \Delta}$$

Definition 3.4.5 A reduction $\delta = \Gamma \mapsto \Delta|(A, B) \rightrightarrows \Gamma' \mapsto \Delta'$ is provable in $\mathbf{R}_{\mathtt{t}}$, denoted by $\vdash_{\mathtt{t}} \delta$, if there is a sequence $\{\delta_1, \ldots, \delta_n\}$ of reductions such that $\delta_n = \delta$, and for each $1 \leq i \leq n$, δ_i is an axiom or is deduced from the previous reductions by one of the deduction rules in $\mathbf{R}_{\mathtt{t}}$.

Theorem 3.4.6 (Soundness and completeness theorem) *For any reduction* $\Gamma \mapsto \Delta|(A, B) \rightrightarrows \Gamma' \mapsto \Delta'$,

$$\vdash_{\mathtt{t}} \Gamma \mapsto \Delta|(A, B) \rightrightarrows \Gamma' \mapsto \Delta'$$

if and only if

$$\models_{\mathtt{t}} \Gamma \mapsto \Delta|(A, B) \rightrightarrows \Gamma' \mapsto \Delta'.$$

Proof Here we give the proof of completeness part. Given a reduction $\Gamma \Rightarrow \Delta|(A, B) \rightrightarrows \Gamma, A \Rightarrow \Delta|B$, we construct a trees T such that either

(i) T is a proof tree of $\Gamma \Rightarrow \Delta|(A, B) \rightrightarrows \Gamma, A \Rightarrow \Delta|B$; i.e., for each branch ξ of T, there is a reduction γ at the leaf of ξ which is an axiom of form $(\mathtt{A}_{\mathtt{t}}^{+L})$, or
(ii) there is a branch $\xi \in T$ such that ξ is a proof of $\Gamma \Rightarrow \Delta|(A, B) \rightrightarrows \Gamma \Rightarrow \Delta|B$.

T is constructed as follows:

- the root of T is $\Gamma \Rightarrow \Delta|(A, B) \rightrightarrows \Gamma, A \Rightarrow \Delta|B$;
- for a node ξ, if each reduction at ξ is of form $\Gamma' \Rightarrow \Delta'|(l', B) \rightrightarrows \Gamma, l' \Rightarrow \Delta|B$ then the node is a leaf;
- otherwise, ξ has the direct children nodes containing the following reductions:

$$\begin{cases}
\Gamma_1 \Rightarrow \Delta_1|(A,B) \rightrightarrows \Gamma_1, A \Rightarrow \Delta_1|B & \text{if } \Gamma_1(\neg\neg A) \in \xi \\
\left[\begin{array}{l} \Gamma_1 \Rightarrow \Delta_1|(A_1,B) \rightrightarrows \Gamma_1, A_1 \Rightarrow \Delta_1|B \\ \Gamma_1, A_1 \Rightarrow \Delta_1|(A_2,B) \rightrightarrows \Gamma_1, A_1, A_2 \Rightarrow \Delta_1|B \end{array}\right. & \text{if } \Gamma_1(A_1 \wedge A_2) \in \xi \\
\left\{\begin{array}{l} \Gamma_1 \Rightarrow \Delta_1|(\neg A_1,B) \rightrightarrows \Gamma_1, \neg A_1 \Rightarrow \Delta_1|B \\ \Gamma_1 \Rightarrow \Delta_1|(\neg A_2,B) \rightrightarrows \Gamma_1, \neg A_2 \Rightarrow \Delta_1|B \end{array}\right. & \text{if } \Gamma_1(\neg(A_1 \wedge A_2)) \in \xi \\
\left\{\begin{array}{l} \Gamma_1 \Rightarrow \Delta_1|(A_1,B) \rightrightarrows \Gamma_1, A_1 \Rightarrow \Delta_1|B \\ \Gamma_1 \Rightarrow \Delta_1|(A_2,B) \rightrightarrows \Gamma_1, A_2 \Rightarrow \Delta_1|B \end{array}\right. & \text{if } \Gamma_1(A_1 \vee A_2) \in \xi \\
\left[\begin{array}{l} \Gamma_1 \Rightarrow \Delta_1|(\neg A_1,B) \rightrightarrows \Gamma_1, \neg A_1 \Rightarrow \Delta_1|B \\ \Gamma_1, \neg A_1 \Rightarrow \Delta_1|(\neg A_2,B) \rightrightarrows \Gamma_1, \neg A_1, \neg A_2 \Rightarrow \Delta_1|B \end{array}\right. & \text{if } \Gamma_1(\neg(A_1 \vee A_2)) \in \xi
\end{cases}$$

where $\Gamma_1(A) = \Gamma_1 \Rightarrow \Delta_1|(A,B) \rightrightarrows \Gamma_1, A \Rightarrow \Delta_1|B$.

Lemma 3.4.7 *If for each branch $\xi \subseteq T$, there is a reduction $\Gamma_1 \Rightarrow \Delta_1|(A,B) \rightrightarrows \Gamma_1, A \Rightarrow \Delta|B \in \xi$ which is an axiom in $\mathbf{R}_{\mathtt{t}}$ then T is a proof tree of $\Gamma \Rightarrow \Delta|(A,B) \rightrightarrows \Gamma, A \Rightarrow \Delta|B$ in $\mathbf{R}_{\mathtt{t}}$.*

Proof By the definition of T, T is a proof tree of $\Gamma \Rightarrow \Delta|(A,B) \rightrightarrows \Gamma, A \Rightarrow \Delta|B$. □

Lemma 3.4.8 *If there is a branch $\xi \subseteq T$ such that each reduction $\Gamma_1 \Rightarrow \Delta_1|(A',B) \rightrightarrows \Gamma_1, A' \Rightarrow \Delta|B \in \xi$ is not an axiom in $\mathbf{R}_{\mathtt{t}}$ then ξ is a proof of $\Gamma \Rightarrow \Delta|(A,B) \rightrightarrows \Gamma \Rightarrow \Delta|B$ in $\mathbf{R}_{\mathtt{t}}$.*

Proof Let ξ be a branch of T such that any reduction $\Gamma_1 \Rightarrow \Delta_1|(A',B) \rightrightarrows \Gamma_1, A' \Rightarrow \Delta|B \in \xi$ is not an axiom in $\mathbf{R}_{\mathtt{t}}$.

We proved by induction on node $\eta \in \xi$ that for each reduction $\Gamma' \Rightarrow \Delta|(A',B) \rightrightarrows \Gamma' \Rightarrow \Delta|B \in \eta, \vdash_{\mathtt{t}} \Gamma' \Rightarrow \Delta|(A',B) \rightrightarrows \Gamma' \Rightarrow \Delta|B$.

Case $\Gamma' \Rightarrow \Delta'|(l,m) \rightrightarrows \Gamma', l \Rightarrow \Delta'|m \in \eta$ is a leaf node of ξ. Then, $\neg l \in \Gamma'$ or $l\Delta'$, and $\vdash_{\mathtt{t}} \Gamma' \Rightarrow \Delta'|(l,m) \rightrightarrows \Gamma' \Rightarrow \Delta'|m$.

Case $\delta = \Gamma_2 \Rightarrow \Delta_2|(\neg\neg A,B) \rightrightarrows \Gamma_2, \neg\neg A \Rightarrow \Delta_2|B \in \eta$. Then, η has a child node $\in \xi$ containing reduction $\Gamma_2 \Rightarrow \Delta_2|(A,B) \rightrightarrows \Gamma_2, A \Rightarrow \Delta_2|B$. By induction assumption, $\vdash_{\mathtt{t}} \Gamma_2 \Rightarrow \Delta_2|(A,B) \rightrightarrows \Gamma_2 \Rightarrow \Delta_2|B$, and by $(\neg\neg^0), \vdash_{\mathtt{t}} \Gamma_2 \Rightarrow \Delta_2|(\neg\neg A,B) \rightrightarrows \Gamma_2 \Rightarrow \Delta_2|B$.

Case $\delta = \Gamma_2 \Rightarrow \Delta_2|(A_1 \wedge A_2,B) \rightrightarrows \Gamma_2, A_1 \wedge A_2 \Rightarrow \Delta_2|B \in \eta$. Then, η has a child node $\in \xi$ containing reduction $\Gamma_2 \Rightarrow \Delta_2|(A_1,B) \rightrightarrows \Gamma_2, A_1 \Rightarrow \Delta_2|B$ or $\Gamma_2, A_1 \Rightarrow \Delta_2|(A_2,B) \rightrightarrows \Gamma_2, A_1, A_2 \Rightarrow \Delta_2|B$. By induction assumption, either $\vdash_{\mathtt{t}} \Gamma_2 \Rightarrow \Delta_2|(A_1,B) \rightrightarrows \Gamma_2 \Rightarrow \Delta_2|B$, or $\vdash_{\mathtt{t}} \Gamma_2, A_1 \Rightarrow \Delta_2|(A_2,B) \rightrightarrows \Gamma_2, A_1 \Rightarrow \Delta_2|B$. By $(\wedge^0)$, we have $\vdash_{\mathtt{t}} \Gamma_2 \Rightarrow \Delta_2|(A_1 \wedge A_2,B) \rightrightarrows \Gamma_2 \Rightarrow \Delta_2|B$.

Case $\delta = \Gamma_2 \Rightarrow \Delta_2|(A_1 \vee A_2,B) \rightrightarrows \Gamma_2 \Rightarrow \Delta_2|B \in \eta$. Then, η has a child node $\in \xi$ containing reductions $\Gamma_2 \Rightarrow \Delta_2|(A_1,B) \rightrightarrows \Gamma_2, A_1 \Rightarrow \Delta_2|B$ and $\Gamma_2 \Rightarrow \Delta_2|(A_2,B) \rightrightarrows \Gamma_2, A_2 \Rightarrow \Delta_2|B$. By induction assumption, $\vdash_{\mathtt{t}} \Gamma_2 \Rightarrow \Delta_2|(A_1,B) \rightrightarrows \Gamma_2 \Rightarrow \Delta_2|B$ and $\vdash_{\mathtt{t}} \Gamma_2 \Rightarrow \Delta_2|(A_2,B) \rightrightarrows \Gamma_2 \Rightarrow \Delta_2|B$. By $(\vee^0)$, we have $\vdash_{\mathtt{t}} \Gamma_2 \Rightarrow \Delta_2|(A_1 \vee A_2,B) \rightrightarrows \Gamma_2 \Rightarrow \Delta_2|B$.

Similar for other cases. □

Given a reduction $\Gamma \Rightarrow \Delta|B \rightrightarrows \Gamma \Rightarrow \Delta'$, we construct a trees T' such that either

(i) T' is a proof tree of $\Gamma \Rightarrow \Delta|B \rightrightarrows \Gamma \Rightarrow \Delta, B$; i.e., for each branch ξ of T', there is a reduction γ at the leaf of ξ which is an axiom of form $(\mathtt{A}_{\mathtt{t}}^{+R})$, or
(ii) there is a branch $\xi \in T'$ such that ξ is a proof of $\Gamma \Rightarrow \Delta|B \rightrightarrows \Gamma \Rightarrow \Delta$.

T' is constructed as follows:

- the root of T' is $\Gamma \Rightarrow \Delta|B \rightrightarrows \Gamma \Rightarrow \Delta, B$;
- for a node ξ, if each reduction at ξ is of form $\Gamma \Rightarrow \Delta|m \rightrightarrows \Gamma \Rightarrow \Delta, m$ then the node is a leaf;
- otherwise, ξ has the direct children nodes containing the following reductions:

$$\begin{cases} \Gamma_1 \Rightarrow \Delta_1|B \rightrightarrows \Gamma_1 \Rightarrow \Delta_1, B & \text{if } \Delta_1(\neg\neg B) \in \xi \\ \begin{cases} \Gamma_1 \Rightarrow \Delta_1|B_1 \rightrightarrows \Gamma_1 \Rightarrow \Delta_1, B_1 \\ \Gamma_1 \Rightarrow \Delta_1|B_2 \rightrightarrows \Gamma_1 \Rightarrow \Delta_1, B_2 \end{cases} & \text{if } \Delta_1(B_1 \wedge B_2) \in \xi \\ \begin{bmatrix} \Gamma_1 \Rightarrow \Delta_1|\neg B_1 \rightrightarrows \Gamma_1 \Rightarrow \Delta_1, \neg B_1 \\ \Gamma_1 \Rightarrow \Delta_1, \neg B_1|\neg B_2 \rightrightarrows \Gamma_1 \Rightarrow \Delta_1, \neg B_1, \neg B_2 \end{bmatrix} & \text{if } \Delta_1(\neg(B_1 \wedge B_2)) \in \xi \\ \begin{bmatrix} \Gamma_1 \Rightarrow \Delta_1|B_1 \rightrightarrows \Gamma_1 \Rightarrow \Delta_1, B_1 \\ \Gamma_1 \Rightarrow \Delta_1, B_1|B_2 \rightrightarrows \Gamma_1 \Rightarrow \Delta_1, B_1, B_2 \end{bmatrix} & \text{if } \Delta_1(B_1 \vee B_2) \in \xi \\ \begin{cases} \Gamma_1 \Rightarrow \Delta_1|\neg B_1 \rightrightarrows \Gamma_1 \Rightarrow \Delta_1, \neg B_1 \\ \Gamma_1 \Rightarrow \Delta_1|\neg B_2 \rightrightarrows \Gamma_1 \Rightarrow \Delta_1, \neg B_2 \end{cases} & \text{if } \Delta_1(\neg(B_1 \vee B_2)) \in \xi \end{cases}$$

where $\Delta_1(B) = \Gamma_1 \Rightarrow \Delta_1|B \rightrightarrows \Gamma_1 \Rightarrow \Delta_1, B$.

Lemma 3.4.9 *If for each branch $\xi \subseteq T'$, there is a reduction $\Gamma_1 \Rightarrow \Delta_1|B' \rightrightarrows \Gamma_1 \Rightarrow \Delta, B' \in \xi$ which is an axiom in $\mathbf{R}_{\mathtt{t}}$ then T' is a proof tree of $\Gamma \Rightarrow \Delta|B \rightrightarrows \Gamma \Rightarrow \Delta, B$ in $\mathbf{R}_{\mathtt{t}}$.*

Proof By the definition of T', T' is a proof tree of $\Gamma \Rightarrow \Delta|B \rightrightarrows \Gamma \Rightarrow \Delta, B$. □

Lemma 3.4.10 *If there is a branch $\xi \subseteq T'$ such that reductions $\Gamma_1 \Rightarrow \Delta_1|B' \rightrightarrows \Gamma_1 \Rightarrow \Delta, B' \in \xi$ are not axioms in $\mathbf{R}_{\mathtt{t}}$ then ξ is a proof of $\Gamma \Rightarrow \Delta|B \rightrightarrows \Gamma \Rightarrow \Delta$.*

Proof Let ξ be a branch of T' such that any reduction $\Gamma_1 \Rightarrow \Delta_1|B' \rightrightarrows \Gamma_1 \Rightarrow \Delta_1, B' \in \xi$ is not an axiom in $\mathbf{R}_{\mathtt{t}}$.

We proved by induction on node $\eta \in \xi$ that for each reduction $\Gamma' \Rightarrow \Delta|B' \rightrightarrows \Gamma' \Rightarrow \Delta, B' \in \eta$,

$$\vdash_{\mathtt{t}} \Gamma' \Rightarrow \Delta|B' \rightrightarrows \Gamma' \Rightarrow \Delta.$$

Case $\Gamma' \Rightarrow \Delta|m \rightrightarrows \Gamma' \Rightarrow \Delta, m \in \eta$ is a leaf node of ξ. Then,

$$\neg m \in \Delta \text{ or } m \in \Gamma'.$$

$\vdash_{\mathtt{t}} \Gamma' \Rightarrow \Delta|m \rightrightarrows \Gamma' \Rightarrow \Delta.$

Case $\delta = \Gamma_2 \Rightarrow \Delta_2|\neg\neg B \rightrightarrows \Gamma_2 \Rightarrow \Delta_2, \neg\neg B \in \eta$. Then, η has a child node $\in \xi$ containing reduction $\Gamma_2 \Rightarrow \Delta_2|B \rightrightarrows \Gamma_2 \Rightarrow \Delta_2, B$. By induction assumption, $\vdash_{\mathfrak{t}} \Gamma_2 \Rightarrow \Delta_2|B \rightrightarrows \Gamma_2 \Rightarrow \Delta_2$, and by $(\neg\neg^0)$, $\vdash_{\mathfrak{t}} \Gamma_2 \Rightarrow \Delta_2|\neg\neg B \rightrightarrows \Gamma_2 \Rightarrow \Delta_2$.

Case $\delta = \Gamma_2 \Rightarrow \Delta_2|B_1 \wedge B_2 \rightrightarrows \Gamma_2 \Rightarrow \Delta_2, B_1 \wedge B_2 \in \eta$. Then, η has a child node $\in \xi$ containing reductions $\Gamma_2 \Rightarrow \Delta_2|B_1 \rightrightarrows \Gamma_2 \Rightarrow \Delta_2, B_1$ and $\Gamma_2 \Rightarrow \Delta_2|B_2 \rightrightarrows \Gamma_2 \Rightarrow \Delta_2, B_2$. By induction assumption, $\vdash_{\mathfrak{t}} \Gamma_2 \Rightarrow \Delta_2|B_1 \rightrightarrows \Gamma_2 \Rightarrow \Delta_2$, and $\vdash_{\mathfrak{t}} \Gamma_2 \Rightarrow \Delta_2|B_2 \rightrightarrows \Gamma_2 \Rightarrow \Delta_2$. By $(\wedge^0)$, we have $\vdash_{\mathfrak{t}} \Gamma_2 \Rightarrow \Delta_2|B_1 \wedge B_2 \rightrightarrows \Gamma_2 \Rightarrow \Delta_2$.

Case $\delta = \Gamma_2 \Rightarrow \Delta_2|B_1 \vee B_2 \rightrightarrows \Gamma_2 \Rightarrow \Delta_2, B_1 \vee B_2 \in \eta$. Then, η has a child node $\in \xi$ containing reduction $\Gamma_2 \Rightarrow \Delta_2|B_1 \rightrightarrows \Gamma_2 \Rightarrow \Delta_2, B_1$ or $\Gamma_2 \Rightarrow \Delta_2, B_1|B_2 \rightrightarrows \Gamma_2 \Rightarrow \Delta_2, B_1, B_2$. By induction assumption, either $\vdash_{\mathfrak{t}} \Gamma_2 \Rightarrow \Delta_2|B_1 \rightrightarrows \Gamma_2 \Rightarrow \Delta_2$ or $\vdash_{\mathfrak{t}} \Gamma_2 \Rightarrow \Delta_2, B_1|B_2 \rightrightarrows \Gamma_2 \Rightarrow \Delta_2, B_1$. By $(\vee^0)$, we have $\vdash_{\mathfrak{t}} \Gamma_2 \Rightarrow \Delta_2|B_1 \vee B_2 \rightrightarrows \Gamma_2 \Rightarrow \Delta_2$.

Similar for other cases. □

3.4.4 R-Calculus $\mathbf{R}_{\mathfrak{f}}$

Given two theories Γ, Δ and formulas A, B, a reduction $\Gamma \Rightarrow \Delta|(A, B) \rightrightarrows \Gamma' \Rightarrow \Delta'$ is $\mathbf{R}_{\mathfrak{f}}$-valid, denoted by $\models_{\mathfrak{f}} \Gamma \Rightarrow \Delta|(A, B) \rightrightarrows \Gamma' \Rightarrow \Delta'$, if

$$\Gamma' = \begin{cases} \Gamma \cup \{A\} & \text{if } \models_{\mathfrak{f}} \Gamma, A \Rightarrow \Delta \\ \Gamma & \text{otherwise;} \end{cases}$$
$$\Delta' = \begin{cases} \Delta \cup \{B\} & \text{if } \models_{\mathfrak{f}} \Gamma' \Rightarrow \Delta, B \\ \Delta & \text{otherwise;} \end{cases}$$

R-calculus $\mathbf{R}_{\mathfrak{f}}$ consists of the following axioms and deductions:

- **Axioms**:

$$(\mathrm{A}_{\mathfrak{f}}^{+L}) \ \frac{\neg l \notin \Gamma \& l \notin \Delta}{\Gamma \mapsto \Delta|(l, m) \rightrightarrows \Gamma, l \mapsto \Delta|m},$$
$$(\mathrm{A}_{\mathfrak{f}}^{0L}) \ \frac{\neg l \in \Gamma \text{ or } l \in \Delta}{\Gamma \mapsto \Delta|(l, m) \rightrightarrows \Gamma \mapsto \Delta|m},$$
$$(\mathrm{A}_{\mathfrak{f}}^{+R}) \ \frac{m \notin \Gamma \& \neg m \notin \Delta}{\Gamma' \mapsto \Delta|m \rightrightarrows \Gamma' \mapsto \Delta, m},$$
$$(\mathrm{A}_{\mathfrak{f}}^{0R}) \ \frac{m \in \Gamma \text{ or } \neg m \in \Delta}{\Gamma' \mapsto \Delta|m \rightrightarrows \Gamma' \mapsto \Delta},$$

where l, m are literals, and Δ, Γ are sets of literals.

- **Deduction rules**:

$$(\wedge^{L+})\ \frac{\left\{\begin{array}{l}\Gamma \mapsto \Delta|(A_1, B) \rightrightarrows \Gamma, A_1 \mapsto \Delta|B \\ \Gamma \mapsto \Delta|(A_2, B) \rightrightarrows \Gamma, A_2 \mapsto \Delta|B\end{array}\right.}{\Gamma \mapsto \Delta|(A_1 \wedge A_2, B) \rightrightarrows \Gamma, A_1 \wedge A_2 \mapsto \Delta|B}$$

$$(\wedge^{L0})\ \frac{\left[\begin{array}{l}\Gamma \mapsto \Delta|(A_1, B) \rightrightarrows \Gamma \mapsto \Delta|B \\ \Gamma \mapsto \Delta|(A_2, B) \rightrightarrows \Gamma \mapsto \Delta|B\end{array}\right.}{\Gamma \mapsto \Delta|(A_1 \wedge A_2, B) \rightrightarrows \Gamma \mapsto \Delta|B}$$

$$(\wedge^{R+})\ \frac{\left[\begin{array}{l}\Gamma' \mapsto \Delta|B_1 \rightrightarrows \Gamma' \mapsto \Delta, B_1 \\ \Gamma' \mapsto \Delta, B_1|B_2 \rightrightarrows \Gamma' \mapsto \Delta, B_1, B_2\end{array}\right.}{\Gamma' \mapsto \Delta|B_1 \wedge B_2 \rightrightarrows \Gamma' \mapsto \Delta, B_1 \wedge B_2}$$

$$(\wedge^{R0})\ \frac{\left\{\begin{array}{l}\Gamma' \mapsto \Delta|B_1 \rightrightarrows \Gamma' \mapsto \Delta \\ \Gamma' \mapsto \Delta, B_1|B_2 \rightrightarrows \Gamma' \mapsto \Delta, B_1\end{array}\right.}{\Gamma' \mapsto \Delta|B_1 \wedge B_2 \rightrightarrows \Gamma' \mapsto \Delta}$$

and

$$(\vee^{L+})\ \frac{\left[\begin{array}{l}\Gamma \mapsto \Delta|(A_1, B) \rightrightarrows \Gamma, A_1 \mapsto \Delta|B \\ \Gamma, A_1 \mapsto \Delta|(A_2, B) \rightrightarrows \Gamma, A_1, A_2 \mapsto \Delta|B\end{array}\right.}{\Gamma \mapsto \Delta|(A_1 \vee A_2, B) \rightrightarrows \Gamma, A_1 \vee A_2 \mapsto \Delta|B}$$

$$(\vee^{L0})\ \frac{\left\{\begin{array}{l}\Gamma \mapsto \Delta|(A_1, B) \rightrightarrows \Gamma \mapsto \Delta|B \\ \Gamma, A_1 \mapsto \Delta|(A_2, B) \rightrightarrows \Gamma, A_1 \mapsto \Delta|B\end{array}\right.}{\Gamma \mapsto \Delta|(A_1 \vee A_2, B) \rightrightarrows \Gamma \mapsto \Delta|B}$$

$$(\vee^{R-})\ \frac{\left\{\begin{array}{l}\Gamma' \mapsto \Delta|B_1 \rightrightarrows \Gamma' \mapsto \Delta, B_1 \\ \Gamma' \mapsto \Delta|B_2 \rightrightarrows \Gamma' \mapsto \Delta, B_2\end{array}\right.}{\Gamma' \mapsto \Delta|B_1 \vee B_2 \rightrightarrows \Gamma' \mapsto \Delta, B_1 \vee B_2}$$

$$(\vee^{R0})\ \frac{\left[\begin{array}{l}\Gamma' \mapsto \Delta|B_1 \rightrightarrows \Gamma' \mapsto \Delta \\ \Gamma' \mapsto \Delta|B_2 \rightrightarrows \Gamma' \mapsto \Delta\end{array}\right.}{\Gamma' \mapsto \Delta|B_1 \vee B_2 \rightrightarrows \Gamma' \mapsto \Delta}$$

and

$$(\neg\neg^{L+})\ \frac{\Gamma \mapsto \Delta|(A, B) \rightrightarrows \Gamma, A \mapsto \Delta|B}{\Gamma \mapsto \Delta|(\neg\neg A, B) \rightrightarrows \Gamma, \neg\neg A \mapsto \Delta|B}$$

$$(\neg\neg^{L0})\ \frac{\Gamma \mapsto \Delta|(A, B) \rightrightarrows \Gamma \mapsto \Delta|B}{\Gamma \mapsto \Delta|(\neg\neg A, B) \rightrightarrows \Gamma \mapsto \Delta|B}$$

$$(\neg\neg^{R+})\ \frac{\Gamma' \mapsto \Delta|B \rightrightarrows \Gamma' \mapsto \Delta, B}{\Gamma' \mapsto \Delta|\neg\neg B \rightrightarrows \Gamma' \mapsto \Delta, \neg\neg B}$$

$$(\neg\neg^{R0})\ \frac{\Gamma' \mapsto \Delta|B \rightrightarrows \Gamma' \mapsto \Delta}{\Gamma' \mapsto \Delta|\neg\neg B \rightrightarrows \Gamma' \mapsto \Delta}$$

and

$$(\neg\wedge^{L+})\ \left[\begin{array}{l}\Gamma \mapsto \Delta|(\neg A_1, B) \rightrightarrows \Gamma, \neg A_1 \mapsto \Delta|B \\ \Gamma, \neg A_1 \mapsto \Delta|(\neg A_2, B) \rightrightarrows \Gamma, \neg A_1, \neg A_2 \mapsto \Delta|B\end{array}\right. \over \Gamma \mapsto \Delta|(\neg(A_1 \wedge A_2), B) \rightrightarrows \Gamma, \neg(A_1 \wedge A_2) \mapsto \Delta|B$$

$$(\neg\wedge^{L0})\ \left\{\begin{array}{l}\Gamma \mapsto \Delta|(\neg A_1, B) \rightrightarrows \Gamma \mapsto \Delta|B \\ \Gamma, \neg A_1 \mapsto \Delta|(\neg A_2, B) \rightrightarrows \Gamma, \neg A_1 \mapsto \Delta|B\end{array}\right. \over \Gamma \mapsto \Delta|(\neg(A_1 \wedge A_2), B) \rightrightarrows \Gamma \mapsto \Delta|B$$

$$(\neg\wedge^{R+})\ \left\{\begin{array}{l}\Gamma' \mapsto \Delta|\neg B_1 \rightrightarrows \Gamma' \mapsto \Delta, \neg B_1 \\ \Gamma' \mapsto \Delta|\neg B_2 \rightrightarrows \Gamma' \mapsto \Delta, \neg B_2\end{array}\right. \over \Gamma' \mapsto \Delta|\neg(B_1 \wedge B_2) \rightrightarrows \Gamma' \mapsto \Delta, \neg(B_1 \wedge B_2)$$

$$(\neg\wedge^{R0})\ \left[\begin{array}{l}\Gamma' \mapsto \Delta|\neg B_1 \rightrightarrows \Gamma' \mapsto \Delta \\ \Gamma' \mapsto \Delta|\neg B_2 \rightrightarrows \Gamma' \mapsto \Delta\end{array}\right. \over \Gamma' \mapsto \Delta|\neg(B_1 \wedge B_2) \rightrightarrows \Gamma' \mapsto \Delta$$

and

$$(\neg\vee^{L+})\ \left\{\begin{array}{l}\Gamma \mapsto \Delta|(\neg A_1, B) \rightrightarrows \Gamma, \neg A_1 \mapsto \Delta|B \\ \Gamma \mapsto \Delta|(\neg A_2, B) \rightrightarrows \Gamma, \neg A_2 \mapsto \Delta|B\end{array}\right. \over \Gamma \mapsto \Delta|(\neg(A_1 \vee A_2), B) \rightrightarrows \Gamma, \neg(A_1 \vee A_2) \mapsto \Delta|B$$

$$(\neg\vee^{L0})\ \left[\begin{array}{l}\Gamma \mapsto \Delta|(\neg A_1, B) \rightrightarrows \Gamma \mapsto \Delta|B \\ \Gamma \mapsto \Delta|(\neg A_2, B) \rightrightarrows \Gamma \mapsto \Delta|B\end{array}\right. \over \Gamma \mapsto \Delta|(\neg(A_1 \vee A_2), B) \rightrightarrows \Gamma \mapsto \Delta|B$$

$$(\neg\vee^{R+})\ \left[\begin{array}{l}\Gamma' \mapsto \Delta|\neg B_1 \rightrightarrows \Gamma' \mapsto \Delta, \neg B_1 \\ \Gamma' \mapsto \Delta, \neg B_1|\neg B_2 \rightrightarrows \Gamma' \mapsto \Delta, \neg B_1, \neg B_2\end{array}\right. \over \Gamma' \mapsto \Delta|\neg(B_1 \vee B_2) \rightrightarrows \Gamma' \mapsto \Delta, \neg(B_1 \vee B_2)$$

$$(\neg\vee^{R0})\ \left\{\begin{array}{l}\Gamma' \mapsto \Delta|\neg B_1 \rightrightarrows \Gamma' \mapsto \Delta \\ \Gamma' \mapsto \Delta, \neg B_1|\neg B_2 \rightrightarrows \Gamma' \mapsto \Delta, \neg B_1\end{array}\right. \over \Gamma' \mapsto \Delta|\neg(B_1 \vee B_2) \rightrightarrows \Gamma' \mapsto \Delta$$

Definition 3.4.11 A reduction $\delta = \Gamma \mapsto \Delta|(A, B) \rightrightarrows \Gamma' \mapsto \Delta'$ is provable in $\mathbf{R}_{\mathfrak{f}}$, denoted by $\vdash_{\mathfrak{f}} \delta$, if there is a sequence $\{\delta_1, \ldots, \delta_n\}$ of reductions such that $\delta_n = \delta$, and for each $1 \leq i \leq n$, δ_i is an axiom or is deduced from the previous reductions by one of the deduction rules in $\mathbf{R}_{\mathfrak{f}}$.

Theorem 3.4.12 (Soundness and completeness theorem) *For any reduction* $\Gamma \mapsto \Delta|(A, B) \rightrightarrows \Gamma' \mapsto \Delta'$,

$$\vdash_{\mathfrak{f}} \Gamma \mapsto \Delta|(A, B) \rightrightarrows \Gamma' \mapsto \Delta'$$

if and only if

$$\models_{\mathfrak{f}} \Gamma \mapsto \Delta|(A, B) \rightrightarrows \Gamma' \mapsto \Delta'.$$

□

3.5 Conclusions

Tableau proof systems:

validity	negation
$\mathbf{T}^{\mathtt{f}}: \mathbf{A}v\mathbf{E}A \in \Gamma(v(A)=0)$	$\mathbf{T}_{\mathtt{t}}: \mathbf{E}v\mathbf{A}A \in \Gamma(v(A)=1)$
$\mathbf{T}^{\mathtt{t}}: \mathbf{A}v\mathbf{E}B \in \Delta(v(B)=1)$	$\mathbf{T}_{\mathtt{f}}: \mathbf{E}v\mathbf{A}B \in \Delta(v(A)=0)$

R-calculi:

- for theories:

$$\begin{array}{l} \mathbf{S}^{\mathtt{t}}: \vdash^{\mathtt{t}} \Delta|B \rightrightarrows \Delta-\{B\} \text{ iff } \vdash^{\mathtt{t}} \Delta-\{B\} \\ \quad\ \vdash^{\mathtt{t}} \Delta|B \rightrightarrows \Delta \text{ iff } \nvdash^{\mathtt{t}} \Delta-\{B\} \\ \mathbf{S}^{\mathtt{f}}: \vdash^{\mathtt{f}} \Gamma|A \rightrightarrows \Gamma-\{A\} \text{ iff } \vdash^{\mathtt{f}} \Gamma-\{A\} \\ \quad\ \vdash^{\mathtt{f}} \Gamma|A \rightrightarrows \Gamma \text{ iff } \nvdash^{\mathtt{f}} \Gamma-\{\neg A\} \\ \hline \mathbf{S}_{\mathtt{t}}: \vdash_{\mathtt{t}} \Delta|B \Rightarrow \Delta, B \text{ iff } \Delta \nvdash \neg B \\ \quad\ \vdash_{\mathtt{t}} \Delta|B \Rightarrow \Delta \text{ iff } \Delta \vdash \neg B \\ \mathbf{S}_{\mathtt{f}}: \vdash_{\mathtt{f}} \Gamma|A \Rightarrow \Gamma, A \text{ iff } \Gamma \nvdash \neg A \\ \quad\ \vdash_{\mathtt{f}} \Gamma|A \Rightarrow \Gamma \text{ iff } \Gamma \vdash \neg A \end{array}$$

- for sequents:

validity	$\mathbf{G}^{\mathtt{t}} \vdash \Gamma \Rightarrow \Delta$ iff $\mathbf{A}v(\mathbf{E}A \in \Gamma(v(A)=0)$ or $\mathbf{E}B \in \Delta(v(B)=1))$
negation	$\mathbf{G}_{\mathtt{t}} \vdash \Gamma \mapsto \Delta$ iff $\mathbf{E}v(\mathbf{A}A \in \Gamma(v(A)=1)\&\mathbf{A}B \in \Delta(v(B)=0))$
validity	$\mathbf{G}^{\mathtt{f}} \vdash \Gamma \Rightarrow \Delta$ iff $\mathbf{A}v(\mathbf{E}A \in \Gamma(v(A)=1)$ or $\mathbf{E}B \in \Delta(v(B)=0))$
negation	$\mathbf{G}_{\mathtt{f}} \vdash \Gamma \mapsto \Delta$ iff $\mathbf{E}v(\mathbf{A}A \in \Gamma(v(A)=0)\&\mathbf{A}B \in \Delta(v(B)=1))$

and for $\mathbf{R}^{\mathtt{t}}$, we have the following equivalences:

$$\begin{array}{l} \vdash^{\mathtt{t}} \Gamma \Rightarrow \Delta|(A,B) \rightrightarrows \Gamma-\{A\} \Rightarrow \Delta-\{B\} \text{ iff } \models^{\mathtt{t}} \Gamma-\{A\} \Rightarrow \Delta-\{B\} \\ \vdash^{\mathtt{t}} \Gamma \Rightarrow \Delta|(A,B) \rightrightarrows \Gamma-\{A\} \Rightarrow \Delta \text{ iff } \models^{\mathtt{t}} \Gamma-\{A\} \Rightarrow \Delta \\ \vdash^{\mathtt{t}} \Gamma \Rightarrow \Delta|(A,B) \rightrightarrows \Gamma \Rightarrow \Delta-\{B\} \text{ iff } \models^{\mathtt{t}} \Gamma \Rightarrow \Delta-\{B\} \\ \vdash^{\mathtt{t}} \Gamma \Rightarrow \Delta|(A,B) \rightrightarrows \Gamma \Rightarrow \Delta \text{ iff } \left[\begin{array}{l} \not\models^{\mathtt{t}} \Gamma-\{A\} \Rightarrow \Delta-\{B\} \\ \not\models^{\mathtt{t}} \Gamma-\{A\} \Rightarrow \Delta \\ \not\models^{\mathtt{t}} \Gamma \Rightarrow \Delta-\{B\} \end{array}\right. \end{array}$$

and for $\mathbf{R}^{\mathtt{f}}$, we have the following equivalences:

$$\begin{array}{l} \vdash^{\mathtt{f}} \Gamma \Rightarrow \Delta|(A,B) \rightrightarrows \Gamma-\{A\} \Rightarrow \Delta-\{B\} \text{ iff } \models^{\mathtt{f}} \Gamma-\{A\} \Rightarrow \Delta-\{B\} \\ \vdash^{\mathtt{f}} \Gamma \Rightarrow \Delta|(A,B) \rightrightarrows \Gamma-\{A\} \Rightarrow \Delta \text{ iff } \models^{\mathtt{f}} \Gamma-\{A\} \Rightarrow \Delta \\ \vdash^{\mathtt{f}} \Gamma \Rightarrow \Delta|(A,B) \rightrightarrows \Gamma \Rightarrow \Delta-\{B\} \text{ iff } \models^{\mathtt{f}} \Gamma \Rightarrow \Delta-\{B\} \\ \vdash^{\mathtt{f}} \Gamma \Rightarrow \Delta|(A,B) \rightrightarrows \Gamma \Rightarrow \Delta \text{ iff } \left[\begin{array}{l} \not\models^{\mathtt{f}} \Gamma-\{A\} \Rightarrow \Delta-\{B\} \\ \not\models^{\mathtt{f}} \Gamma-\{A\} \Rightarrow \Delta \\ \not\models^{\mathtt{f}} \Gamma \Rightarrow \Delta-\{B\} \end{array}\right. \end{array}$$

- co-sequents: for $\mathbf{R}_{\mathtt{t}}$, we have the following equivalences:

$$\vdash_{\mathtt{t}} \Gamma \mapsto \Delta|(A,B) \rightrightarrows \Gamma, A \mapsto \Delta, B \text{ iff } \models_{\mathtt{t}} \Gamma, A \mapsto \Delta, B$$
$$\vdash_{\mathtt{t}} \Gamma \mapsto \Delta|(A,B) \rightrightarrows \Gamma, A \mapsto \Delta \text{ iff } \models_{\mathtt{t}} \Gamma, A \mapsto \Delta$$
$$\vdash_{\mathtt{t}} \Gamma \mapsto \Delta|(A,B) \rightrightarrows \Gamma \mapsto \Delta, B \text{ iff } \models_{\mathtt{t}} \Gamma \mapsto \Delta, B$$
$$\vdash_{\mathtt{t}} \Gamma \mapsto \Delta|(A,B) \rightrightarrows \Gamma \mapsto \Delta \text{ iff } \left[\begin{array}{l} \not\models_{\mathtt{t}} \Gamma - \{A\} \mapsto \Delta - \{B\} \\ \not\models_{\mathtt{t}} \Gamma - \{A\} \mapsto \Delta \\ \not\models_{\mathtt{t}} \Gamma \mapsto \Delta - \{B\} \end{array}\right.$$

and for $\mathbf{R}_{\mathtt{f}}$, we have the following equivalences:

$$\vdash_{\mathtt{f}} \Gamma \mapsto \Delta|(A,B) \rightrightarrows \Gamma, A \mapsto \Delta, B \text{ iff } \models_{\mathtt{f}} \Gamma, A \mapsto \Delta, B$$
$$\vdash_{\mathtt{f}} \Gamma \mapsto \Delta|(A,B) \rightrightarrows \Gamma, A \mapsto \Delta \text{ iff } \models_{\mathtt{f}} \Gamma, A \mapsto \Delta$$
$$\vdash_{\mathtt{f}} \Gamma \mapsto \Delta|(A,B) \rightrightarrows \Gamma \mapsto \Delta, B \text{ iff } \models_{\mathtt{f}} \Gamma \mapsto \Delta, B$$
$$\vdash_{\mathtt{f}} \Gamma \mapsto \Delta|(A,B) \rightrightarrows \Gamma \mapsto \Delta \text{ iff } \left[\begin{array}{l} \not\models_{\mathtt{f}} \Gamma - \{A\} \mapsto \Delta - \{B\} \\ \not\models_{\mathtt{f}} \Gamma - \{A\} \mapsto \Delta \\ \not\models_{\mathtt{f}} \Gamma \mapsto \Delta - \{B\} \end{array}\right.$$

References

Cao, C., Sui, Y., Wang, Y.: The nonmonotonic propositional logics. Artif. Intell. Res. **5**, 111–120 (2016)

Clark, K.: Negation as failure. In: Readings in Nonmonotonic Reasoning, pp. 311–325. Morgan Kaufmann Publishers (1987)

Ginsberg, M.L. (ed.): Readings in Nonmonotonic Reasoning. Morgan Kaufmann, San Francisco (1987)

Hähnle, R.: Advanced many-valued logics. In: Gabbay, D., Guenthner, F. (eds.) Handbook of Philosophical Logic, vol. 2, pp. 297–395. Kluwer, Dordrecht (2001)

Li, W.: R-calculus: an inference system for belief revision. Comput. J. **50**, 378–390 (2007)

Li, W.: Mathematical logic, foundations for information science. In: Progress in Computer Science and Applied Logic, vol. 25. Birkhäuser (2010)

Malinowski, G.: Many-valued logic and its philosophy. In: Gabbay, D.M., Woods, J. (eds.) Handbook of the History of Logic, vol. 8. Elsevier, The Many Valued and Nonmonotonic Turn in Logic (2009)

Reiter, R.: A logic for default reasoning. Artif. Intell. **13**, 81–132 (1980)

Takeuti, G.: Proof theory. In: Barwise, J. (ed.) Handbook of Mathematical Logic. Studies in Logic and the Foundations of Mathematics. North-Holland, Amsterdam, NL (1987)

Urquhart, A.: Basic many-valued logic. In: Gabbay, D., Guenthner, F. (eds.) Handbook of Philosophical Logic, 2nd edn, vol. 2, pp. 249–295. Kluwer, Dordrecht (2001)

Chapter 4
R-Calculi $\mathbf{R}^{Q_1Q_2}/\mathbf{R}_{Q_1Q_2}$

Let $Q_1, Q_2 \in \{\mathbf{A}, \mathbf{E}\}$. We consider $\mathbf{G}^{Q_1Q_2}$-valid sequents and $\mathbf{G}_{Q_1Q_2}$-valid co-sequents, Gentzen deduction systems $\mathbf{G}^{Q_1Q_2}$, $\mathbf{G}_{Q_1Q_2}$, and corresponding R-calculi $\mathbf{R}^{Q_1Q_2}$, $\mathbf{R}_{Q_1Q_2}$.

(1) Gentzen deduction systems (Li 2010; Takeuti and Barwise 1987):

- $\mathbf{G}^{Q_1Q_2}$-validity: A sequent $\Gamma \Rightarrow \Delta$ is $\mathbf{G}^{Q_1Q_2}$-valid, denoted by $\models^{Q_1Q_2} \Gamma \Rightarrow \Delta$, if for any assignment v, either $Q_1A \in \Gamma(v(A) = 0)$ or $Q_2B \in \Delta(v(B) = 1)$.
- $\mathbf{G}_{Q_1Q_2}$-validity: A co-sequent $\Gamma \mapsto \Delta$ is $\mathbf{G}_{Q_1Q_2}$-valid, denoted by $\models_{Q_1Q_2} \Gamma \mapsto \Delta$, if there is an assignment v such that $Q_1A \in \Gamma(v(A) = 1)$ and $Q_2B \in \Delta(v(B) = 0)$.
  There are eight kinds of sequents and co-sequents, and sound and complete Gentzen deduction systems:

sequents	$\mathbf{G}^{\mathbf{EE}}$	$\mathbf{G}^{\mathbf{EA}}$	$\mathbf{G}^{\mathbf{AE}}$	$\mathbf{G}^{\mathbf{AA}}$
co-sequents	$\mathbf{G}^{\mathbf{AA}}$	$\mathbf{G}^{\mathbf{AE}}$	$\mathbf{G}^{\mathbf{EA}}$	$\mathbf{G}^{\mathbf{EE}}$.

(2) R-calculi (Alchourrón et al. 1985; Cao et al. 2016; Darwiche and Pearl 1997; Fermé and Hansson 2011; Gärdenfors and Rott 1995; Ginsberg 1987; Li 2007; Reiter 1980):

- $\mathbf{R}^{Q_1Q_2}$-validity: Given a sequent $\Gamma \Rightarrow \Delta$ and pair (A, B) of formulas such that $A \in \Gamma$ and $B \in \Delta$, the result of $\Gamma \Rightarrow \Delta$ $\mathbf{G}^{Q_1Q_2}$-revising (A, B) is sequent $\Gamma' \Rightarrow \Delta'$, denoted by

$$\models^{Q_1Q_2} \Gamma \Rightarrow \Delta|(A, B) \rightrightarrows \Gamma' \Rightarrow \Delta',$$

W. Li and Y. Sui, *R-Calculus, IV: Propositional Logic*,
Perspectives in Formal Induction, Revision and Evolution,
https://doi.org/10.1007/978-981-19-8633-8_4

and we say that reduction $\Gamma \Rightarrow \Delta|(A, B) \rightrightarrows \Gamma' \Rightarrow \Delta'$ is $\mathbf{R}^{Q_1Q_2}$-valid, where

$$\Gamma' = \begin{cases} \Gamma, A \text{ if } \models^{Q_1Q_2} \Gamma, A \Rightarrow \Delta \\ \Gamma \quad \text{otherwise;} \end{cases}$$
$$\Delta' = \begin{cases} \Delta, B \text{ if } \models^{Q_1Q_2} \Gamma' \Rightarrow \Delta, B \\ \Delta \quad \text{otherwise.} \end{cases}$$

- $\mathbf{R}_{Q_1Q_2}$-validity: Given a sequent $\Gamma \mapsto \Delta$ and pair (A, B) of formulas, the result of $\Gamma \mapsto \Delta$ $\mathbf{G}_{Q_1Q_2}$-revising (A, B) is denoted by

$$\models_{Q_1Q_2} \Gamma \mapsto \Delta|(A, B) \rightrightarrows \Gamma' \mapsto \Delta',$$

and we say that reduction $\Gamma \mapsto \Delta|(A, B) \rightrightarrows \Gamma' \mapsto \Delta'$ is $\mathbf{R}_{Q_1Q_2}$-valid, if

$$\Gamma' = \begin{cases} \Gamma \cup \{A\} \text{ if } \vdash_{Q_1Q_2} \Gamma, A \mapsto \Delta \\ \Gamma \quad \text{otherwise;} \end{cases}$$
$$\Delta' = \begin{cases} \Delta \cup \{B\} \text{ if } \vdash_{Q_1Q_2} \Gamma' \mapsto \Delta, B \\ \Delta \quad \text{otherwise.} \end{cases}$$

Correspondingly, there are eight kinds of sound and complete R-calculi for sequents and co-sequents:

sequents	$\mathbf{R}^{\mathbf{EE}}$ $\mathbf{R}^{\mathbf{EA}}$ $\mathbf{R}^{\mathbf{AE}}$ $\mathbf{R}^{\mathbf{AA}}$
co-sequents	$\mathbf{R}^{\mathbf{AA}}$ $\mathbf{R}^{\mathbf{AE}}$ $\mathbf{R}^{\mathbf{EA}}$ $\mathbf{R}^{\mathbf{EE}}$.

There are eight basic forms of axioms:

$$(\mathbb{A}^{\mathbf{EE}}) \frac{\text{incon}(\Gamma) \text{ or val}(\Delta) \text{ or } \Gamma \cap \Delta \neq \emptyset}{\Gamma \Rightarrow \Delta}$$
$$(\mathbb{A}^{\mathbf{AE}}) \frac{\text{val}(\Delta) \text{ or } \Gamma \subseteq \Delta}{\Gamma \Rightarrow \Delta}$$
$$(\mathbb{A}^{\mathbf{EA}}) \frac{\text{incon}(\Gamma) \text{ or } \Delta \subseteq \Gamma}{\Gamma \Rightarrow \Delta}$$
$$(\mathbb{A}^{\mathbf{AA}}) \frac{\Gamma = \emptyset \text{ or } \Delta = \emptyset \text{ or } \Delta = \Gamma = \{l\}}{\Gamma \Rightarrow \Delta}$$

and

$$(\mathbb{A}_{\mathbf{AA}}) \frac{\text{con}(\Gamma)\&\text{inval}(\Delta)\&\Gamma \cap \Delta = \emptyset}{\Gamma \mapsto \Delta}$$
$$(\mathbb{A}_{\mathbf{EA}}) \frac{\text{inval}(\Delta)\&\Delta \nsubseteq \Gamma}{\Gamma \mapsto \Delta}$$
$$(\mathbb{A}_{\mathbf{AE}}) \frac{\text{con}(\Gamma)\&\Gamma \nsubseteq \Delta}{\Gamma \mapsto \Delta}$$
$$(\mathbb{A}_{\mathbf{EE}}) \frac{\Gamma \neq \emptyset \neq \Delta\&\Delta \neq \Gamma \neq \{l\}}{\Gamma \mapsto \Delta}$$

where $\Delta \neq \Gamma \neq \{l\}$ denotes that either $\Delta \neq \{l\}$ or $\Gamma \neq \{l\}$.

We have the following equivalences:

$$\mathbf{G}^{\mathtt{t}} = \mathbf{G}^{\mathbf{EE}},\ \mathbf{G}^{\mathtt{f}} = --;$$
$$\mathbf{G}_{\mathtt{t}} = \mathbf{G}_{\mathbf{AA}},\ \mathbf{G}_{\mathtt{f}} = --.$$

4.1 Gentzen Deduction Systems $\mathbf{G}^{Q_1Q_2}$

Let $Q_1, Q_2 \in \{\mathbf{A}, \mathbf{E}\}$.

Definition 4.1.1 A sequent $\Gamma \Rightarrow \Delta$ is $\mathbf{G}^{Q_1Q_2}$-valid, denoted by $\models^{Q_1Q_2} \Gamma \Rightarrow \Delta$, if for any assignment v, either $Q_1A \in \Gamma(v(A) = 0)$ or $Q_2B \in \Delta(v(B) = 1)$.

4.1.1 Axioms

Let Γ, Δ be sets of literals. We define

$$\begin{array}{rl} \text{incon}(\Gamma) & \text{iff } \mathbf{E}l(l, \neg l \in \Gamma) \\ \text{con}(\Gamma) & \text{iff } \sim \mathbf{E}l(l, \neg l \in \Gamma) \\ \text{val}(\Delta) & \text{iff } \mathbf{E}l(l, \neg l \in \Delta) \\ \text{inval}(\Delta) & \text{iff } \sim \mathbf{E}l(l, \neg l \in \Delta). \end{array}$$

Notice that incon(Γ) iff val(Γ) and con(Γ) iff inval(Γ).

Define

$$\begin{array}{l} v \models^{\mathbf{E}} \Gamma \text{ if } \mathbf{E}A \in \Gamma(v(A) = 0) \\ v \models^{\mathbf{E}} \Delta \text{ if } \mathbf{E}B \in \Delta(v(B) = 1) \\ v \models^{\mathbf{A}} \Gamma \text{ if } \mathbf{A}A \in \Gamma(v(A) = 0) \\ v \models^{\mathbf{A}} \Delta \text{ if } \mathbf{A}B \in \Delta(v(B) = 1). \end{array}$$

Proposition 4.1.2 *Let Γ, Δ be sets of literals. $\Gamma \Rightarrow \Delta$ is $\mathbf{G}^{\mathbf{EE}}$-valid if and only if* incon(Γ) *or* val(Δ) *or* $\Gamma \cap \Delta \neq \emptyset$.

Proof Assume that incon(Γ) or val(Δ) or $\Gamma \cap \Delta \neq \emptyset$. Then, (i) if there is a literal l such that $l, \neg l \in \Gamma$ then for any assignment v, either $v(l) = 0$, or $v(\neg l) = 0$, and $v \models^{\mathbf{E}} \Gamma$; (ii) if there is a literal m such that $m, \neg m \in \Delta$ then for any assignment v, either $v(m) = 1$, or $v(\neg m) = 1$, and $v \models^{\mathbf{E}} \Gamma$; and (iii) if there is a literal l such that $l \in \Gamma \cap \Delta$ then for any assignment v, either $v(l) = 0$, or $v(l) = 1$, and hence, either $v \models^{\mathbf{E}} \Gamma$, or $v \models^{\mathbf{E}} \Delta$.

Assume that con(Γ)&inval(Δ)&$\Gamma \cap \Delta = \emptyset$. We define an assignment v such that for any variable p,

$$v(p) = \begin{cases} 1 \text{ if } p \in \Gamma \text{ or } \neg p \in \Delta \\ 0 \text{ if } \neg p \in \Gamma \text{ or } p \in \Delta \\ 1 \text{ otherwise.} \end{cases}$$

Then, v is well-defined and $v \not\models^{\mathbf{EE}} \Gamma \Rightarrow \Delta$. □

Hence, we have the following axiom:

$$(\mathrm{A}^{\mathbf{EE}}) \ \frac{\mathrm{incon}(\Gamma) \text{ or } \mathrm{val}(\Delta) \text{ or } \Gamma \cap \Delta \neq \emptyset}{\Gamma \Rightarrow \Delta}$$

Proposition 4.1.3 *Let Γ, Δ be sets of literals. $\Gamma \Rightarrow \Delta$ is $\mathbf{G}^{\mathbf{EA}}$-valid if and only if* $\mathrm{incon}(\Gamma)$ *or* $\Delta \subseteq \Gamma$.

Proof Assume that $\mathrm{incon}(\Gamma)$ or $\Delta \subseteq \Gamma$. Then, for any assignment v, either $v(m) = 1$ for every $m \in \Delta$ (i.e., $v \models^{\mathbf{A}} \Delta$), or $v(m) = 0$ for some $m \in \Delta \subseteq \Gamma$, i.e., $v \models^{\mathbf{E}} \Gamma$.

Assume that $\mathrm{incon}(\Gamma) \& \Delta \not\subseteq \Gamma$. Let $l \in \Delta - \Gamma$. We define an assignment v such that for any variable p,

$$v(p) = \begin{cases} 1 \text{ if } p \in \Gamma \\ 0 \text{ if } \neg p \in \Gamma \\ 0 \text{ if } p = l \\ 1 \text{ if } p = \neg l \\ 1 \text{ otherwise.} \end{cases}$$

Then, v is well-defined and $v \not\models^{\mathbf{EA}} \Gamma \Rightarrow \Delta$. □

Hence, we have the following axiom:

$$(\mathrm{A}^{\mathbf{EA}}) \ \frac{\mathrm{incon}(\Gamma) \text{ or } \Delta \subseteq \Gamma}{\Gamma \Rightarrow \Delta}$$

Proposition 4.1.4 *Let Γ, Δ be sets of literals. $\Gamma \Rightarrow \Delta$ is $\mathbf{G}^{\mathbf{AE}}$-valid if and only if* $\mathrm{val}(\Delta)$ *or* $\Gamma \subseteq \Delta$.

Proof Assume that $\mathrm{val}(\Delta)$ or $\Gamma \subseteq \Delta$. Then, for any assignment v, either $v(l) = 0$ for every $l \in \Gamma$ (i.e., $v \models^{\mathbf{A}} \Gamma$), or $v(l) = 1$ for some $l \in \Gamma \subseteq \Delta$, i.e., $v \models^{\mathbf{E}} \Delta$.

Assume that $\mathrm{inval}(\Delta) \& \Gamma \not\subseteq \Delta$. Let $l \in \Gamma - \Delta$. We define an assignment v such that for any variable p,

$$v(p) = \begin{cases} 0 \text{ if } p \in \Delta \\ 1 \text{ if } \neg p \in \Delta \\ 1 \text{ if } p = l \\ 0 \text{ if } p = \neg l \\ 0 \text{ otherwise.} \end{cases}$$

Then, v is well-defined and $v \not\models^{\mathbf{AE}} \Gamma \Rightarrow \Delta$. □

Hence, we have the following axiom:

$$(\mathrm{A}^{\mathbf{AE}})\ \frac{\mathrm{val}(\Delta) \text{ or } \Gamma \subseteq \Delta}{\Gamma \Rightarrow \Delta}$$

Proposition 4.1.5 *Let* Γ, Δ *be sets of literals.* $\Gamma \Rightarrow \Delta$ *is* $\mathbf{G}^{\mathbf{AA}}$*-valid if and only if* $\Gamma = \emptyset$ *or* $\Delta = \emptyset$ *or* $\Gamma = \Delta = \{l\}$.

Proof Assume that $\Gamma = \emptyset$ or $\Delta = \emptyset$ or $\Gamma = \Delta = \{l\}$. Then, (i) if $\Gamma = \emptyset$ then for any assignment v, $v \models^{\mathbf{A}} \Gamma$; (ii) if $\Delta = \emptyset$ then for any assignment v, $v \models^{\mathbf{A}} \Delta$; and (iii) if $\Gamma = \Delta = \{l\}$ then for any assignment v, either $v(l) = 0$ for every $l \in \Gamma$ (i.e., $v \models^{\mathbf{A}} \Gamma$), or $v(m) = 1$ for every $m \in \Delta$, i.e., $v \models^{\mathbf{A}} \Delta$.

Assume that $\Gamma \neq \emptyset \neq \Delta \& (\Gamma \neq \{l\}$ or $\Delta \neq \{l\})$. Let $l \in \Gamma$ and $l \neq l' \in \Gamma - \Delta$. We define an assignment v such that for any variable p,

$$v(p) = \begin{cases} 1 \text{ if } p = l' \\ 0 \text{ if } p = \neg l' \\ 0 \text{ if } p = l \\ 1 \text{ if } p = \neg l \\ 0 \text{ otherwise.} \end{cases}$$

Then, v is well-defined and $v \not\models^{\mathbf{AA}} \Gamma \Rightarrow \Delta$. Similar for $l \neq m' \in \Delta - \Gamma$. □

Hence, we have the following axiom:

$$(\mathrm{A}^{\mathbf{AA}})\ \frac{\Gamma = \emptyset \text{ or } \Delta = \emptyset \text{ or } \Gamma = \Delta = \{l\}}{\Gamma \Rightarrow \Delta}$$

4.1.2 Deduction Rules

$\mathbf{G}^{L0}$ consists of the following deduction rules:

$$(\neg\neg^L)\ \frac{\Gamma, A \Rightarrow \Delta}{\Gamma, \neg\neg A \Rightarrow \Delta}$$

$$(\wedge^L)\ \frac{\begin{cases}\Gamma, A_1 \Rightarrow \Delta \\ \Gamma, A_2 \Rightarrow \Delta\end{cases}}{\Gamma, A_1 \wedge A_2 \Rightarrow \Delta} \qquad (\vee^L)\ \frac{\begin{bmatrix}\Gamma, A_1 \Rightarrow \Delta \\ \Gamma, A_2 \Rightarrow \Delta\end{bmatrix}}{\Gamma, A_1 \vee A_2 \Rightarrow \Delta}$$

$$(\neg\wedge^L)\ \frac{\begin{bmatrix}\Gamma, \neg A_1 \Rightarrow \Delta \\ \Gamma, \neg A_2 \Rightarrow \Delta\end{bmatrix}}{\Gamma, \neg(A_1 \vee A_2) \Rightarrow \Delta} \qquad (\neg\vee^L)\ \frac{\begin{cases}\Gamma, \neg A_1 \Rightarrow \Delta \\ \Gamma, \neg A_2 \Rightarrow \Delta\end{cases}}{\Gamma, \neg(A_1 \vee A_2) \Rightarrow \Delta}$$

$\mathbf{G}^{R1}$ consists of the following deduction rules:

$$(\neg\neg^R)\ \frac{\Gamma \Rightarrow B, \Delta}{\Gamma \Rightarrow \neg\neg B, \Delta}$$

$$(\wedge^R)\ \frac{\left[\begin{array}{l}\Gamma \Rightarrow B_1, \Delta \\ \Gamma \Rightarrow B_2, \Delta\end{array}\right.}{\Gamma \Rightarrow B_1 \wedge B_2, \Delta} \qquad (\vee^R)\ \frac{\left\{\begin{array}{l}\Gamma \Rightarrow B_1, \Delta \\ \Gamma \Rightarrow B_2, \Delta\end{array}\right.}{\Gamma \Rightarrow B_1 \vee B_2, \Delta}$$

$$(\neg\wedge^R)\ \frac{\left\{\begin{array}{l}\Gamma \Rightarrow \neg B_1, \Delta \\ \Gamma \Rightarrow \neg B_2, \Delta\end{array}\right.}{\Gamma \Rightarrow \neg(B_1 \wedge B_2), \Delta} \qquad (\neg\vee^R)\ \frac{\left[\begin{array}{l}\Gamma \Rightarrow \neg B_1, \Delta \\ \Gamma \Rightarrow \neg B_2, \Delta\end{array}\right.}{\Gamma \Rightarrow \neg(B_1 \vee B_2), \Delta}$$

4.1.3 Deduction Systems

Let $Q_1, Q_2 \in \{\mathbf{E}, \mathbf{A}\}$ and

$$\mathbf{G}^{Q_1Q_2} = \mathrm{A}^{Q_1Q_2} + \mathbf{G}^{L0} + \mathbf{G}^{R1}$$

Definition 4.1.6 A sequent $\Gamma \Rightarrow \Delta$ is provable in $\mathbf{G}^{Q_1Q_2}$, denoted by $\vdash^{Q_1Q_2} \Gamma \Rightarrow \Delta$, if there is a sequence $\{\Gamma_1 \Rightarrow \Delta_1, \ldots, \Gamma_n \Rightarrow \Delta_n\}$ of sequents such that $\Gamma_n \Rightarrow \Delta_n = \Gamma \Rightarrow \Delta$, and for each $1 \leq i \leq n$, $\Gamma_i \Rightarrow \Delta_i$ is an axiom or is deduced from the previous sequents by one of the deduction rules in $\mathbf{G}^{Q_1Q_2}$.

Theorem 4.1.7 (Soundness and completeness theorem) *Let* $Q_1, Q_2 \in \{\mathbf{E}, \mathbf{A}\}$. *For any sequent* $\Gamma \Rightarrow \Delta$,

$$\vdash^{Q_1Q_2} \Gamma \Rightarrow \Delta \textit{ iff } \models^{Q_1Q_2} \Gamma \Rightarrow \Delta.$$

□

4.2 R-Calculi $\mathbf{R}^{Q_1Q_2}$

Let $Q_1, Q_2 \in \{\mathbf{E}, \mathbf{A}\}$.

Given a sequent $\Gamma \Rightarrow \Delta$ and pair (A, B) of formulas such that $A \in \Gamma$ and $B \in \Delta$, the result of $\Gamma \Rightarrow \Delta$ $\mathbf{G}^{Q_1Q_2}$-revising (A, B) is sequent $\Gamma' \Rightarrow \Delta'$, denoted by $\models^{Q_1Q_2} \Gamma \Rightarrow \Delta|(A, B) \rightrightarrows \Gamma' \Rightarrow \Delta'$, if

$$\Gamma' = \begin{cases} \Gamma, A \text{ if } \models^{Q_1Q_2} \Gamma, A \Rightarrow \Delta \\ \Gamma \quad \text{otherwise;} \end{cases}$$
$$\Delta' = \begin{cases} \Delta, B \text{ if } \models^{Q_1Q_2} \Gamma' \Rightarrow \Delta, B \\ \Delta \quad \text{otherwise.} \end{cases}$$

4.2.1 Deduction Rules

Let $\mathbf{X} = \Gamma \Rightarrow \Delta$ and

$$\begin{array}{l}\mathbf{X}[A] = \Gamma \Rightarrow \Delta|(A,B) \rightrightarrows \Gamma - \{A\} \Rightarrow \Delta|B \\ \mathbf{X}(A) = \Gamma \Rightarrow \Delta|(A,B) \rightrightarrows \Gamma, A \Rightarrow \Delta|B \\ \mathbf{X}[B] = \Gamma' \Rightarrow \Delta|B \rightrightarrows \Gamma' \Rightarrow \Delta - \{B\} \\ \mathbf{X}(B) = \Gamma' \Rightarrow \Delta|B \rightrightarrows \Gamma' \Rightarrow \Delta, B.\end{array}$$

There are four basic sets of deduction rules:

- $\mathbf{S}^{LE}$: where $A \in \Gamma$.

$$(\neg\neg_0^L)\ \frac{\mathbf{X}|A \rightrightarrows \mathbf{X}}{\mathbf{X}|\neg\neg A \rightrightarrows \mathbf{X}} \qquad (\neg\neg_-^L)\ \frac{\mathbf{X}|A \rightrightarrows \mathbf{X}[A]}{\mathbf{X}|\neg\neg A \rightrightarrows \mathbf{X}[\neg\neg A]}$$

$$(\wedge_0^L)\ \frac{\left[\begin{array}{l}\mathbf{X}|A_1 \rightrightarrows \mathbf{X} \\ \mathbf{X}|A_2 \rightrightarrows \mathbf{X}\end{array}\right.}{\mathbf{X}|A_1 \wedge A_2 \rightrightarrows \mathbf{X}} \qquad (\wedge_-^L)\ \frac{\left\{\begin{array}{l}\mathbf{X}|A_1 \rightrightarrows \mathbf{X}[A_1] \\ \mathbf{X}|A_2 \rightrightarrows \mathbf{X}[A_2]\end{array}\right.}{\mathbf{X}|A_1 \wedge A_2 \rightrightarrows \mathbf{X}[A_1 \wedge A_2]}$$

$$(\vee_0^L)\ \frac{\left\{\begin{array}{l}\mathbf{X}|A_1 \rightrightarrows \mathbf{X} \\ \mathbf{X}[A_1]|A_2 \rightrightarrows \mathbf{X}[A_1]\end{array}\right.}{\mathbf{X}|A_1 \vee A_2 \rightrightarrows \mathbf{X}} \qquad (\vee_-^L)\ \frac{\left[\begin{array}{l}\mathbf{X}|A_1 \rightrightarrows \mathbf{X}[A_1] \\ \mathbf{X}[A_1]|A_2 \rightrightarrows \mathbf{X}[A_1, A_2]\end{array}\right.}{\mathbf{X}|A_1 \vee A_2 \rightrightarrows \mathbf{X}[A_1 \vee A_2]}$$

and

$$(\neg\wedge_0^L)\ \frac{\left\{\begin{array}{l}\mathbf{X}|\neg A_1 \rightrightarrows \mathbf{X} \\ \mathbf{X}[\neg A_1]|\neg A_2 \rightrightarrows \mathbf{X}[\neg A_1]\end{array}\right.}{\mathbf{X}|\neg(A_1 \wedge A_2) \rightrightarrows \mathbf{X}}$$

$$(\neg\wedge_-^L)\ \frac{\left[\begin{array}{l}\mathbf{X}|\neg A_1 \rightrightarrows \mathbf{X}[\neg A_1] \\ \mathbf{X}[\neg A_1]|\neg A_2 \rightrightarrows \mathbf{X}[\neg A_1, \neg A_2]\end{array}\right.}{\neg(A_1 \wedge A_2) \rightrightarrows \mathbf{X}[\neg(A_1 \wedge A_2)]}$$

$$(\neg\vee_0^L)\ \frac{\left[\begin{array}{l}\mathbf{X}|\neg A_1 \rightrightarrows \mathbf{X} \\ \mathbf{X}|\neg A_2 \rightrightarrows \mathbf{X}\end{array}\right.}{\neg(A_1 \vee A_2) \rightrightarrows \mathbf{X}}$$

$$(\neg\vee_-^L)\ \frac{\left\{\begin{array}{l}\mathbf{X}|\neg A_1 \rightrightarrows \mathbf{X}[\neg A_1] \\ \mathbf{X}|\neg A_2 \rightrightarrows \mathbf{X}[\neg A_2]\end{array}\right.}{\mathbf{X}|\neg(A_1 \vee A_2) \rightrightarrows \mathbf{X}[\neg(A_1 \vee A_2)}$$

A formula $A_1 \wedge A_2$ is extractable from Γ, if either A_1 or A_2 is extractable from Γ; and $A_1 \vee A_2$ is extractable from Γ if A_1 is extractable from Γ and A_2 is extractable from $\Gamma - \{A_1\}$.

A formula $A_1 \wedge A_2$ is not extractable from Γ if both A_1 and A_2 are not extractable from Γ; and $A_1 \vee A_2$ is not extractable from Γ if either A_1 is not extractable from Γ or A_2 is not extractable from $\Gamma - \{A_1\}$.

- $\mathbf{S}^{RE}$: where $B \in \Delta$

$$(\neg\neg_0^R)\ \frac{\mathbf{X}|B \rightrightarrows \mathbf{X}}{\mathbf{X}|\neg\neg B \rightrightarrows \mathbf{X}} \qquad (\neg\neg_-^R)\ \frac{\mathbf{X}|B \rightrightarrows \mathbf{X}[B]}{\mathbf{X}|\neg\neg B \rightrightarrows \mathbf{X}[\neg\neg B]}$$

$$(\wedge_0^R)\ \frac{\begin{cases}\mathbf{X}|B_1 \rightrightarrows \mathbf{X}\\ \mathbf{X}[B_1]|B_2 \rightrightarrows \mathbf{X}[B_1]\end{cases}}{\mathbf{X}|B_1 \wedge B_2 \rightrightarrows \mathbf{X}} \qquad (\wedge_-^R)\ \frac{\begin{bmatrix}\mathbf{X}|B_1 \rightrightarrows \mathbf{X}[B_1]\\ \mathbf{X}[B_1]|B_2 \rightrightarrows \mathbf{X}[B_1, B_2]\end{bmatrix}}{\mathbf{X}|B_1 \wedge B_2 \rightrightarrows \mathbf{X}[B_1 \wedge B_2]}$$

$$(\vee_0^R)\ \frac{\begin{bmatrix}\mathbf{X}|B_1 \rightrightarrows \mathbf{X}\\ \mathbf{X}|B_2 \rightrightarrows \mathbf{X}\end{bmatrix}}{\mathbf{X}|B_1 \vee B_2 \rightrightarrows \mathbf{X}} \qquad (\vee_-^R)\ \frac{\begin{cases}\mathbf{X}|B_1 \rightrightarrows \mathbf{X}[B_1]\\ \mathbf{X}|B_2 \rightrightarrows \mathbf{X}[B_2]\end{cases}}{\mathbf{X}|B_1 \vee B_2 \rightrightarrows \mathbf{X}[B_1 \vee B_2]}$$

and

$$(\neg\wedge_0^R)\ \frac{\begin{bmatrix}\mathbf{X}|\neg B_1 \rightrightarrows \mathbf{X}\\ \mathbf{X}|\neg B_2 \rightrightarrows \mathbf{X}\end{bmatrix}}{\mathbf{X}|\neg(B_1 \wedge B_2) \rightrightarrows \mathbf{X}}$$

$$(\neg\wedge_-^R)\ \frac{\begin{cases}\mathbf{X}|\neg B_1 \rightrightarrows \mathbf{X}[\neg B_1]\\ \mathbf{X}|\neg B_2 \rightrightarrows \mathbf{X}[\neg B_2]\end{cases}}{\mathbf{X}|\neg(B_1 \wedge B_2) \rightrightarrows \mathbf{X}[\neg(B_1 \wedge B_2)}$$

$$(\neg\vee_0^R)\ \frac{\begin{cases}\mathbf{X}|\neg B_1 \rightrightarrows \mathbf{X}\\ \mathbf{X}[\neg B_1]|\neg B_2 \rightrightarrows \mathbf{X}[\neg B_1]\end{cases}}{\mathbf{X}|\neg(B_1 \vee B_2) \rightrightarrows \mathbf{X}}$$

$$(\neg\vee_-^R)\ \frac{\begin{bmatrix}\mathbf{X}|\neg B_1 \rightrightarrows \mathbf{X}[\neg B_1]\\ \mathbf{X}[\neg B_1]|\neg B_2 \rightrightarrows \mathbf{X}[\neg B_1, \neg B_2]\end{bmatrix}}{\mathbf{X}|\neg(B_1 \vee B_2) \rightrightarrows \mathbf{X}[\neg(B_1 \vee B_2)]}$$

A formula $B_1 \wedge B_2$ is extractable from Δ, if B_1 is extractable from Δ and B_2 is extractable from $\Delta - \{B_1\}$; and $B_1 \vee B_2$ is extractable from Δ if either B_1 or B_2 is extractable from Δ.
A formula $B_1 \wedge B_2$ is not extractable from Δ if either B_1 is not extractable from Δ or B_2 are not extractable from $\Delta - \{B_1\}$; and $B_1 \vee B_2$ is not extractable from Δ if both B_1 and B_2 are not extractable from Δ.

- $\mathbf{S}^{LA}$:

$$(\neg\neg_0^L)\ \frac{\mathbf{X}|A \rightrightarrows \mathbf{X}}{\mathbf{X}|\neg\neg A \rightrightarrows \mathbf{X}} \qquad (\neg\neg_-^L)\ \frac{\mathbf{X}|A \rightrightarrows \mathbf{X}(A)}{\mathbf{X}|\neg\neg A \rightrightarrows \mathbf{X}(\neg\neg A)}$$

$$(\wedge_0^L)\ \frac{\begin{bmatrix}\mathbf{X}|A_1 \rightrightarrows \mathbf{X}\\ \mathbf{X}|A_2 \rightrightarrows \mathbf{X}\end{bmatrix}}{\mathbf{X}|A_1 \wedge A_2 \rightrightarrows \mathbf{X}} \qquad (\wedge_-^L)\ \frac{\begin{cases}\mathbf{X}|A_1 \rightrightarrows \mathbf{X}(A_1)\\ \mathbf{X}|A_2 \rightrightarrows \mathbf{X}(A_2)\end{cases}}{\mathbf{X}|A_1 \wedge A_2 \rightrightarrows \mathbf{X}(A_1 \wedge A_2)}$$

$$(\vee_0^L)\ \frac{\begin{cases}\mathbf{X}|A_1 \rightrightarrows \mathbf{X}\\ \mathbf{X}(A_1)|A_2 \rightrightarrows \mathbf{X}(A_1)\end{cases}}{\mathbf{X}|A_1 \vee A_2 \rightrightarrows \mathbf{X}} \qquad (\vee_-^L)\ \frac{\begin{bmatrix}\mathbf{X}|A_1 \rightrightarrows \mathbf{X}(A_1)\\ \mathbf{X}(A_1)|A_2 \rightrightarrows \mathbf{X}(A_1, A_2)\end{bmatrix}}{\mathbf{X}|A_1 \vee A_2 \rightrightarrows \mathbf{X}(A_1 \vee A_2)}$$

and

$$(\neg\wedge_0^L)\ \frac{\left\{\begin{array}{l}\mathbf{X}|\neg A_1 \rightrightarrows \mathbf{X}\\ \mathbf{X}(\neg A_1)|\neg A_2 \rightrightarrows \mathbf{X}(\neg A_1)\end{array}\right.}{\mathbf{X}|\neg(A_1 \wedge A_2) \rightrightarrows \mathbf{X}}$$

$$(\neg\wedge_-^L)\ \frac{\left[\begin{array}{l}\mathbf{X}|\neg A_1 \rightrightarrows \mathbf{X}(\neg A_1)\\ \mathbf{X}(\neg A_1)|\neg A_2 \rightrightarrows \mathbf{X}(\neg A_1, \neg A_2)\end{array}\right.}{\neg(A_1 \wedge A_2) \rightrightarrows \mathbf{X}(\neg(A_1 \wedge A_2))}$$

$$(\neg\vee_0^L)\ \frac{\left[\begin{array}{l}\mathbf{X}|\neg A_1 \rightrightarrows \mathbf{X}\\ \mathbf{X}|\neg A_2 \rightrightarrows \mathbf{X}\end{array}\right.}{\neg(A_1 \vee A_2) \rightrightarrows \mathbf{X}}$$

$$(\neg\vee_-^L)\ \frac{\left\{\begin{array}{l}\mathbf{X}|\neg A_1 \rightrightarrows \mathbf{X}(\neg A_1)\\ \mathbf{X}|\neg A_2 \rightrightarrows \mathbf{X}(\neg A_2)\end{array}\right.}{\mathbf{X}|\neg(A_1 \vee A_2) \rightrightarrows \mathbf{X}(\neg(A_1 \vee A_2)}$$

A formula $A_1 \wedge A_2$ is enumerable from Γ, if either A_1 or A_2 is enumerable from Γ; and $A_1 \vee A_2$ is enumerable from Γ if A_1 is enumerable from Γ and A_2 is enumerable from Γ, A_1.

A formula $A_1 \wedge A_2$ is not enumerable from Γ if both A_1 and A_2 are not enumerable from Γ; and $A_1 \vee A_2$ is not enumerable from Γ if either A_1 is not enumerable from Γ or A_2 is not enumerable from Γ, A_1.

- $\mathbf{S}^{R\mathbf{A}}$:

$$(\neg\neg_0^R)\ \frac{\mathbf{X}|B \rightrightarrows \mathbf{X}}{\mathbf{X}|\neg\neg B \rightrightarrows \mathbf{X}} \qquad (\neg\neg_-^R)\ \frac{\mathbf{X}|B \rightrightarrows \mathbf{X}(B)}{\mathbf{X}|\neg\neg B \rightrightarrows \mathbf{X}(\neg\neg B)}$$

$$(\wedge_0^R)\ \frac{\left\{\begin{array}{l}\mathbf{X}|B_1 \rightrightarrows \mathbf{X}\\ \mathbf{X}(B_1)|B_2 \rightrightarrows \mathbf{X}(B_1)\end{array}\right.}{\mathbf{X}|B_1 \wedge B_2 \rightrightarrows \mathbf{X}} \quad (\wedge_-^R)\ \frac{\left[\begin{array}{l}\mathbf{X}|B_1 \rightrightarrows \mathbf{X}(B_1)\\ \mathbf{X}(B_1)|B_2 \rightrightarrows \mathbf{X}(B_1, B_2)\end{array}\right.}{\mathbf{X}|B_1 \wedge B_2 \rightrightarrows \mathbf{X}(B_1 \wedge B_2)}$$

$$(\vee_0^R)\ \frac{\left[\begin{array}{l}\mathbf{X}|B_1 \rightrightarrows \mathbf{X}\\ \mathbf{X}|B_2 \rightrightarrows \mathbf{X}\end{array}\right.}{\mathbf{X}|B_1 \vee B_2 \rightrightarrows \mathbf{X}} \qquad (\vee_-^R)\ \frac{\left\{\begin{array}{l}\mathbf{X}|B_1 \rightrightarrows \mathbf{X}(B_1)\\ \mathbf{X}|B_2 \rightrightarrows \mathbf{X}(B_2)\end{array}\right.}{\mathbf{X}|B_1 \vee B_2 \rightrightarrows \mathbf{X}(B_1 \vee B_2)}$$

and

$$(\neg\wedge_0^R)\ \frac{\left[\begin{array}{l}\mathbf{X}|\neg B_1 \rightrightarrows \mathbf{X}\\ \mathbf{X}|\neg B_2 \rightrightarrows \mathbf{X}\end{array}\right.}{\mathbf{X}|\neg(B_1 \wedge B_2) \rightrightarrows \mathbf{X}}$$

$$(\neg\wedge_-^R)\ \frac{\left\{\begin{array}{l}\mathbf{X}|\neg B_1 \rightrightarrows \mathbf{X}(\neg B_1)\\ \mathbf{X}|\neg B_2 \rightrightarrows \mathbf{X}(\neg B_2)\end{array}\right.}{\mathbf{X}|\neg(B_1 \wedge B_2) \rightrightarrows \mathbf{X}(\neg(B_1 \wedge B_2)}$$

$$(\neg\vee_0^R)\ \frac{\left\{\begin{array}{l}\mathbf{X}|\neg B_1 \rightrightarrows \mathbf{X}\\ \mathbf{X}(\neg B_1)|\neg B_2 \rightrightarrows \mathbf{X}(\neg B_1)\end{array}\right.}{\mathbf{X}|\neg(B_1 \vee B_2) \rightrightarrows \mathbf{X}}$$

$$(\neg\vee_-^R)\ \frac{\left[\begin{array}{l}\mathbf{X}|\neg B_1 \rightrightarrows \mathbf{X}(\neg B_1)\\ \mathbf{X}(\neg B_1)|\neg B_2 \rightrightarrows \mathbf{X}(\neg B_1, \neg B_2)\end{array}\right.}{\mathbf{X}|\neg(B_1 \vee B_2) \rightrightarrows \mathbf{X}(\neg(B_1 \vee B_2))}$$

A formula $B_1 \wedge B_2$ is enumerable from Δ, if B_1 is enumerable from Δ and B_2 is enumerable from Δ, B_1; and $B_1 \vee B_2$ is enumerable from Δ if either B_1 or B_2 is enumerable from Δ.

A formula $B_1 \wedge B_2$ is not enumerable from Δ if either B_1 is not enumerable from Δ or B_2 are not enumerable from Δ, B_1; and $B_1 \vee B_2$ is not enumerable from Δ if both B_1 and B_2 are not enumerable from Δ.

4.2.2 Axioms

If a reduction $\Gamma \Rightarrow \Delta|(l, m) \Rightarrow \Gamma' \Rightarrow \Delta'$ is $\mathbf{G}^{Q_1Q_2}$-valid for literals l, m, then the reduction is called an axiom, where Γ', Δ' are sets of literals. We use deduction rules in $\mathbf{G}^{Q_1Q_2}$ to reduce $\Gamma \Rightarrow \Delta$ into literal sets.

Let Γ, Δ be sets of literals and l, m be literals such that $l \in \Gamma$ and $m \in \Delta$.

$$(\mathrm{A}^{Q_1Q_2})\ \Gamma \Rightarrow \Delta|(l, m) \rightrightarrows \Gamma' \Rightarrow \Delta'$$

if

$$\Gamma' = \begin{cases} \Gamma \pm l \text{ if} \vdash^{Q_1Q_2} \Gamma \pm l \Rightarrow \Delta \\ \Gamma \quad\quad \text{otherwise} \end{cases}$$
$$\Delta' = \begin{cases} \Delta \pm m \text{ if} \vdash^{Q_1Q_2} \Gamma' \Rightarrow \Delta \pm m \\ \Delta \quad\quad \text{otherwise} \end{cases}$$

where

$$\pm_1 l = \begin{cases} \cup\{l\} \text{ if } Q_1 = \mathbf{A} \\ -\{l\} \text{ if } Q_1 = \mathbf{E}, \end{cases} \quad \pm_2 m = \begin{cases} \cup\{m\} \text{ if } Q_2 = \mathbf{A} \\ -\{m\} \text{ if } Q_2 = \mathbf{E}. \end{cases}$$

In detail, we have the following axioms:

$$(\mathrm{A}^{\mathbf{EE}}_{-L})\ \frac{\begin{cases} \mathbf{E}l' \neq l(l', \neg l' \in \Gamma) \\ \mathbf{E}l' \neq l(l' \in \Gamma \& l' \in \Delta) \end{cases}}{\Gamma \Rightarrow \Delta|(l, m) \rightrightarrows \Gamma - \{l\} \Rightarrow \Delta|m}$$

$$(\mathrm{A}^{\mathbf{EE}}_{0L})\ \frac{\left[\begin{array}{l} \sim \mathbf{E}l' \neq l(l', \neg l' \in \Gamma) \\ \sim \mathbf{E}l' \neq l(l' \in \Gamma \& l' \in \Delta) \end{array}\right.}{\Gamma \Rightarrow \Delta|(l, m) \rightrightarrows \Gamma \Rightarrow \Delta|m}$$

$$(\mathrm{A}^{\mathbf{EE}}_{-R})\ \frac{\begin{cases} \mathbf{E}m' \neq m(m', \neg m' \in \Delta) \\ \mathbf{E}m' \neq m(m' \in \Gamma' \& m' \in \Delta) \end{cases}}{\Gamma' \Rightarrow \Delta|m \rightrightarrows \Gamma' \Rightarrow \Delta - \{m\}}$$

$$(\mathrm{A}^{\mathbf{EE}}_{0R})\ \frac{\left[\begin{array}{l} \sim \mathbf{E}m' \neq m(m', \neg m' \in \Delta) \\ \sim \mathbf{E}m' \neq m(m' \in \Gamma' \& m' \in \Delta) \end{array}\right.}{\Gamma' \Rightarrow \Delta|m \rightrightarrows \Gamma' \Rightarrow \Delta}$$

where $l \in \Gamma$ and $m \in \Delta$, and

$$(\mathtt{A}^{\mathbf{EA}}_{-L}) \frac{\begin{cases} \mathbf{E}l' \neq l(l', \neg l' \in \Gamma) \\ \Delta \subseteq \Gamma - \{l\} \end{cases}}{\Gamma \Rightarrow \Delta|(l, m) \rightrightarrows \Gamma - \{l\} \Rightarrow \Delta|m}$$

$$(\mathtt{A}^{\mathbf{EA}}_{0L}) \frac{\begin{bmatrix} \sim \mathbf{E}l' \neq l(l', \neg l' \in \Gamma) \\ \Delta \nsubseteq \Gamma - \{l\} \end{bmatrix}}{\Gamma \Rightarrow \Delta|(l, m) \rightrightarrows \Gamma \Rightarrow \Delta|m}$$

$$(\mathtt{A}^{\mathbf{EA}}_{+R}) \frac{\Delta \cup \{m\} \subseteq \Gamma'}{\Gamma' \Rightarrow \Delta|m \rightrightarrows \Gamma' \Rightarrow \Delta, m}$$

$$(\mathtt{A}^{\mathbf{EA}}_{0R}) \frac{\Delta \cup \{m\} \nsubseteq \Gamma'}{\Gamma' \Rightarrow \Delta|m \rightrightarrows \Gamma' \Rightarrow \Delta}$$

where $l \in \Gamma$, and

$$(\mathtt{A}^{\mathbf{AE}}_{+L}) \frac{\Gamma \cup \{l\} \subseteq \Delta}{\Gamma \Rightarrow \Delta|(l, m) \rightrightarrows \Gamma, l \Rightarrow \Delta|m}$$

$$(\mathtt{A}^{\mathbf{AE}}_{0L}) \frac{\Gamma \cup \{l\} \nsubseteq \Delta}{\Gamma \Rightarrow \Delta|(l, m) \rightrightarrows \Gamma \Rightarrow \Delta|m}$$

$$(\mathtt{A}^{\mathbf{AE}}_{-R}) \frac{\begin{cases} \mathbf{E}m' \neq m(m', \neg m' \in \Delta) \\ \Gamma' \subseteq \Delta - \{m\} \end{cases}}{\Gamma' \Rightarrow \Delta|m \rightrightarrows \Gamma' \Rightarrow \Delta - \{m\}}$$

$$(\mathtt{A}^{\mathbf{AE}}_{+R}) \frac{\begin{bmatrix} \sim \mathbf{E}m' \neq m(m', \neg m' \in \Delta) \\ \Gamma' \nsubseteq \Delta - \{m\} \end{bmatrix}}{\Gamma' \Rightarrow \Delta|m \rightrightarrows \Gamma' \Rightarrow \Delta}$$

where $m \in \Delta$, and

$$(\mathtt{A}^{\mathbf{AA}}_{+L}) \frac{l \in \Gamma}{\Gamma \Rightarrow \Delta|(l, m) \rightrightarrows \Gamma, l \Rightarrow \Delta|m}$$

$$(\mathtt{A}^{\mathbf{AA}}_{0L}) \frac{l \notin \Gamma}{\Gamma \Rightarrow \Delta|(l, m) \rightrightarrows \Gamma \Rightarrow \Delta|m}$$

$$(\mathtt{A}^{\mathbf{AA}}_{+R}) \frac{m \in \Delta}{\Gamma' \Rightarrow \Delta|m \rightrightarrows \Gamma' \Rightarrow \Delta, m}$$

$$(\mathtt{A}^{\mathbf{AA}}_{0R}) \frac{m \notin \Delta}{\Gamma' \Rightarrow \Delta|m \rightrightarrows \Gamma' \Rightarrow \Delta}$$

Proposition 4.2.1 *Let $\Gamma \Rightarrow \Delta$ be literal. $\Gamma \Rightarrow \Delta|(A, B) \Rightarrow \Gamma' \Rightarrow \Delta'$ is $\mathbf{R}^{Q_1 Q_2}$-valid if and only if $\Gamma \Rightarrow \Delta|(A, B) \rightrightarrows \Gamma' \Rightarrow \Delta'$ is the conclusion of an axiom.* □

4.2.3 Deduction Systems $\mathbf{R}^{Q_1Q_2}$

Define

$$\mathbf{R}^{Q_1Q_2} = \mathtt{A}^{Q_1Q_2} + \mathbf{S}^{LQ_1} + \mathbf{S}^{RQ_2}.$$

Definition 4.2.2 A reduction $\delta = \Gamma \Rightarrow \Delta|(A, B) \Rightarrow \Gamma' \Rightarrow \Delta'$ is provable in $\mathbf{R}^{Q_1Q_2}$, denoted by $\vdash_{Q_1Q_2} \delta$, if there is a sequence $\{\delta_1, \ldots, \delta_n\}$ of reductions such that $\delta_n = \delta$, and for each $1 \leq i \leq n$, δ_i is either an axiom or deduced from the previous reductions by one of the deduction rules in $\mathbf{R}^{Q_1Q_2}$.

Theorem 4.2.3 (Soundness and completeness theorem) *For any $Q_1, Q_2 \in \{\mathbf{A}, \mathbf{E}\}$ and reduction $\delta = \Gamma \Rightarrow \Delta|(A, B) \rightrightarrows \Gamma' \Rightarrow \Delta'$, δ is $\mathbf{R}^{Q_1Q_2}$-valid if and only if δ is provable in $\mathbf{R}^{Q_1Q_2}$. That is,*

$$\vdash^{Q_1Q_2} \delta \textit{ iff } \models^{Q_1Q_2} \delta.$$

Proof Here we give the proof of completeness part for $Q_1 = Q_2 = \mathbf{E}$. Given a reduction $\Gamma \Rightarrow \Delta|(A, B) \rightrightarrows \Gamma \pm \{A\} \Rightarrow \Delta|B$, we construct a trees T such that either

(i) T is a proof tree of $\Gamma \Rightarrow \Delta|(A, B) \rightrightarrows \Gamma \pm \{A\} \Rightarrow \Delta|B$, i.e., for each branch ξ of T, there is a reduction γ at the leaf of ξ which is an axiom of form ($\mathtt{A}^{\mathtt{t}}_{-}$), or
(ii) there is a branch $\xi \in T$ such that ξ is a proof of $\Gamma \Rightarrow \Delta|(A, B) \rightrightarrows \Gamma \Rightarrow \Delta|B$.

T is constructed as follows:

- the root of T is $\Gamma \Rightarrow \Delta|(A, B) \rightrightarrows \Gamma \pm \{A\} \Rightarrow \Delta|B$;
- for a node ξ, if each reduction at ξ is of form $\Gamma' \Rightarrow \Delta'|(l', B) \rightrightarrows \Gamma - \{l'\} \Rightarrow \Delta|B$ then the node is a leaf;
- otherwise, ξ has the direct child containing the following reductions:

$$\left\{\begin{array}{ll}
\Gamma_1 \Rightarrow \Delta_1|(A, B) \rightrightarrows \Gamma_1 - \{A\} \Rightarrow \Delta_1|B & \text{if } \Gamma_1(\neg\neg A, B) \in \xi \\
\left\{\begin{array}{l}\Gamma_1 \Rightarrow \Delta_1|(A_1, B) \rightrightarrows \Gamma_1 - \{A_1\} \Rightarrow \Delta_1|B \\ \Gamma_1 \Rightarrow \Delta_1|(A_2, B) \rightrightarrows \Gamma_1 - \{A_2\} \Rightarrow \Delta_1|B\end{array}\right. & \text{if } \Gamma_1(A_1 \wedge A_2, B) \in \xi \\
\left[\begin{array}{l}\Gamma_1 \Rightarrow \Delta_1|(\neg A_1, B) \rightrightarrows \Gamma_1 - \{\neg A_1\} \Rightarrow \Delta_1|B \\ \Gamma_1 - \{\neg A_1\} \Rightarrow \Delta_1|(\neg A_2, B) \rightrightarrows \Gamma_1 - \{\neg A_1, \neg A_2\} \Rightarrow \Delta_1|B\end{array}\right. & \text{if } \Gamma_1(\neg(A_1 \wedge A_2), B) \in \xi \\
\left[\begin{array}{l}\Gamma_1 \Rightarrow \Delta_1|(A_1, B) \rightrightarrows \Gamma_1 - \{A_1\} \Rightarrow \Delta_1|B \\ \Gamma_1 - \{A_1\} \Rightarrow \Delta_1|(A_2, B) \rightrightarrows \Gamma_1 - \{A_1, A_2\} \Rightarrow \Delta_1|B\end{array}\right. & \text{if } \Gamma_1(A_1 \vee A_2, B) \in \xi \\
\left\{\begin{array}{l}\Gamma_1 \Rightarrow \Delta_1|(\neg A_1, B) \rightrightarrows \Gamma_1 - \{\neg A_1\} \Rightarrow \Delta_1|B \\ \Gamma_1 \Rightarrow \Delta_1|(\neg A_2, B) \rightrightarrows \Gamma_1 - \{\neg A_2\} \Rightarrow \Delta_1|B\end{array}\right. & \text{if } \Gamma_1(\neg(A_1 \vee A_2), B) \in \xi
\end{array}\right.$$

where $\Gamma_1(A, B) = \Gamma_1 \Rightarrow \Delta_1|(A, B) \rightrightarrows \Gamma_1 - \{A\} \Rightarrow \Delta_1|B$.

Lemma 4.2.4 *If for each branch $\xi \subseteq T$, there is a reduction $\Gamma_1 \Rightarrow \Delta_1|(A, B) \rightrightarrows \Gamma_1 - \{A\} \Rightarrow \Delta|B \in \xi$ which is the conclusion of an axiom in $\mathbf{R}^{\mathbf{EE}}$ then T is a proof tree of $\Gamma \Rightarrow \Delta|(A, B) \rightrightarrows \Gamma - \{A\} \Rightarrow \Delta|B$ in $\mathbf{R}^{\mathbf{EE}}$.*

Proof By the definition of T, T is a proof tree of $\Gamma \Rightarrow \Delta|(A, B) \rightrightarrows \Gamma - \{A\} \Rightarrow \Delta|B$.
□

Lemma 4.2.5 *If there is a branch $\xi \subseteq T$ such that each reduction $\Gamma_1 \Rightarrow \Delta_1|(A', B) \rightrightarrows \Gamma_1 - \{A'\} \Rightarrow \Delta|B \in \xi$ is not the conclusion of an axioms in $\mathbf{R}^{\mathbf{EE}}$ then ξ is a proof of $\Gamma \Rightarrow \Delta|(A, B) \rightrightarrows \Gamma \Rightarrow \Delta|B$.*

Proof Let ξ be a branch of T such that each reduction $\Gamma_1 \Rightarrow \Delta_1|(A', B) \rightrightarrows \Gamma_1 - \{A'\} \Rightarrow \Delta|B \in \xi$ is not the conclusion of an axiom in $\mathbf{R}^{\mathbf{EE}}$.

We proved by induction on node $\eta \in \xi$ that for each reduction $\Gamma' \Rightarrow \Delta|(A', B) \rightrightarrows \Gamma' \Rightarrow \Delta|B \in \eta$, $\vdash^{\mathtt{t}} \Gamma' \Rightarrow \Delta|(A', B) \rightrightarrows \Gamma' \Rightarrow \Delta|B$.

Case $\Gamma' \Rightarrow \Delta'|(l, m) \rightrightarrows \Gamma' - \{l\} \Rightarrow \Delta'|m \in \eta$ is a leaf node of ξ. Then, there is no literal $l' \neq l$ such that

$$l', \neg l' \in \Gamma' \text{ or } l' \in \Gamma' \cap \Delta'.$$

$\vdash^{\mathtt{t}} \Gamma' \Rightarrow \Delta'|(l, m) \rightrightarrows \Gamma' \Rightarrow \Delta'|m$.

Case $\delta = \Gamma_2 \Rightarrow \Delta_2|(\neg\neg A, B) \rightrightarrows \Gamma_2 - \{\neg\neg A\} \Rightarrow \Delta_2|B \in \eta$. Then, η has a child node $\in \xi$ containing reduction $\Gamma_2 \Rightarrow \Delta_2|(A, B) \rightrightarrows \Gamma_2 - \{A\} \Rightarrow \Delta_2|B$. By induction assumption, $\vdash^{\mathtt{t}} \Gamma_2 \Rightarrow \Delta_2|(A, B) \rightrightarrows \Gamma_2 \Rightarrow \Delta_2|B$, and by $(\neg\neg^0)$, $\vdash^{\mathtt{t}} \Gamma_2 \Rightarrow \Delta_2|(\neg\neg A, B) \rightrightarrows \Gamma_2 \Rightarrow \Delta_2|B$.

Case $\delta = \Gamma_2 \Rightarrow \Delta_2|(A_1 \wedge A_2, B) \rightrightarrows \Gamma_2 - \{A_1 \wedge A_2\} \Rightarrow \Delta_2|B \in \eta$. Then, η has a child node $\in \xi$ containing reductions $\Gamma_2 \Rightarrow \Delta_2|(A_1, B) \rightrightarrows \Gamma_2 - \{A_1\} \Rightarrow \Delta_2|B$ and $\Gamma_2 \Rightarrow \Delta_2|(A_2, B) \rightrightarrows \Gamma_2 - \{A_2\} \Rightarrow \Delta_2|B$. By induction assumption, $\vdash^{\mathtt{t}} \Gamma_2 \Rightarrow \Delta_2|(A_1, B) \rightrightarrows \Gamma_2 \Rightarrow \Delta_2|B$, and $\vdash^{\mathtt{t}} \Gamma_2 \Rightarrow \Delta_2|(A_2, B) \rightrightarrows \Gamma_2 \Rightarrow \Delta_2|B$. By $(\wedge^0)$, we have $\vdash^{\mathtt{t}} \Gamma_2 \Rightarrow \Delta_2|(A_1 \wedge A_2, B) \rightrightarrows \Gamma_2 \Rightarrow \Delta_2|B$.

Case $\delta = \Gamma_2 \Rightarrow \Delta_2|(A_1 \vee A_2, B) \rightrightarrows \Gamma_2 \Rightarrow \Delta_2|B \in \eta$. Then, η has a child node $\in \xi$ containing reduction either $\Gamma_2 \Rightarrow \Delta_2|(A_1, B) \rightrightarrows \Gamma_2 - \{A_1\} \Rightarrow \Delta_2|B$ or $\Gamma_2 - \{A_1\} \Rightarrow \Delta_2|(A_2, B) \rightrightarrows \Gamma_2 - \{A_1, A_2\} \Rightarrow \Delta_2|B$. By induction assumption, either $\vdash^{\mathtt{t}} \Gamma_2 \Rightarrow \Delta_2|(A_1, B) \rightrightarrows \Gamma_2 \Rightarrow \Delta_2|B$ or $\vdash^{\mathtt{t}} \Gamma_2 - \{A_1\} \Rightarrow \Delta_2|(A_2, B) \rightrightarrows \Gamma_2 - \{A_1\} \Rightarrow \Delta_2|B$. By $(\vee^0)$, we have $\vdash^{\mathtt{t}} \Gamma_2 \Rightarrow \Delta_2|(A_1 \vee A_2, B) \rightrightarrows \Gamma_2 \Rightarrow \Delta_2|B$.

Similar for other cases. □

Given a reduction $\Gamma \Rightarrow \Delta|B \rightrightarrows \Gamma \Rightarrow \Delta'$, we construct a trees T' such that either

(i) T' is a proof tree of $\Gamma \Rightarrow \Delta|B \rightrightarrows \Gamma \Rightarrow \Delta - \{B\}$, i.e., for each branch ξ of T', there is a reduction γ at the leaf of ξ which is an axiom of form $(\mathtt{A}^{\mathtt{t}}_{-})$.

(ii) There is a branch $\xi \in T'$ such that ξ is a proof of $\Gamma \Rightarrow \Delta|B \rightrightarrows \Gamma \Rightarrow \Delta$.

T' is constructed as follows:

- The root of T' is $\Gamma \Rightarrow \Delta|B \rightrightarrows \Gamma \Rightarrow \Delta - \{B\}$.
- For a node ξ, if each reduction at ξ is of form $\Gamma \Rightarrow \Delta|m \rightrightarrows \Gamma \Rightarrow \Delta - \{m\}$ then the node is a leaf.
- Otherwise, ξ has the direct child containing the following reductions:

$$\begin{cases} \Gamma_1 \Rightarrow \Delta_1|B \rightrightarrows \Gamma_1 \Rightarrow \Delta_1 - \{B\} & \text{if } \Gamma_1(\neg\neg B) \in \xi \\ \begin{cases} \Gamma_1 \Rightarrow \Delta_1|B_1 \rightrightarrows \Gamma_1 \Rightarrow \Delta_1 - \{B_1\} \\ \Gamma_1 \Rightarrow \Delta_1 - \{B_1\}|B_2 \rightrightarrows \Gamma_1 \Rightarrow \Delta_1 - \{B_1, B_2\} \end{cases} & \text{if } \Gamma_1(B_1 \wedge B_2) \in \xi \\ \left[\begin{array}{l} \Gamma_1 \Rightarrow \Delta_1|\neg B_1 \rightrightarrows \Gamma_1 \Rightarrow \Delta_1 - \{\neg B_1\} \\ \Gamma_1 \Rightarrow \Delta_1|\neg B_2 \rightrightarrows \Gamma_1 \Rightarrow \Delta_1 - \{\neg B_2\} \end{array}\right. & \text{if } \Gamma_1(\neg(B_1 \wedge B_2)) \in \xi \\ \left[\begin{array}{l} \Gamma_1 \Rightarrow \Delta_1|B_1 \rightrightarrows \Gamma_1 \Rightarrow \Delta_1 - \{B_1\} \\ \Gamma_1 \Rightarrow \Delta_1|B_2 \rightrightarrows \Gamma_1 \Rightarrow \Delta_1 - \{B_2\} \end{array}\right. & \text{if } \Gamma_1(B_1 \vee B_2) \in \xi \\ \begin{cases} \Gamma_1 \Rightarrow \Delta_1|\neg B_1 \rightrightarrows \Gamma_1 \Rightarrow \Delta_1 - \{\neg B_1\} \\ \Gamma_1 \Rightarrow \Delta_1 - \{\neg B_1\}|\neg B_2 \rightrightarrows \Gamma_1 \Rightarrow \Delta_1 - \{\neg B_1, \neg B_2\} \end{cases} & \text{if } \Gamma_1(\neg(B_1 \vee B_2)) \in \xi \end{cases}$$

where $\Gamma_1(B) = \Gamma_1 \Rightarrow \Delta_1|B \rightrightarrows \Gamma_1 \Rightarrow \Delta_1 - \{B\}$.

Lemma 4.2.6 *If for each branch $\xi \subseteq T'$, there is a reduction $\Gamma_1 \Rightarrow \Delta_1|B' \rightrightarrows \Gamma_1 \Rightarrow \Delta - \{B'\} \in \xi$ which is the conclusion of an axiom in $\mathbf{R}^{\mathbf{EE}}$ then T' is a proof tree of $\Gamma \Rightarrow \Delta|B \rightrightarrows \Gamma \Rightarrow \Delta - \{B\}$ in $\mathbf{R}^{\mathbf{EE}}$.*

Proof By the definition of T', T' is a proof tree of $\Gamma \Rightarrow \Delta|B \rightrightarrows \Gamma \Rightarrow \Delta - \{B\}$ in $\mathbf{R}^{\mathbf{EE}}$. □

Lemma 4.2.7 *If there is a branch $\xi \subseteq T'$ such that each reduction $\Gamma_1 \Rightarrow \Delta_1|B' \rightrightarrows \Gamma_1 \Rightarrow \Delta - \{B'\} \in \xi$ is not the conclusion of an axiom in $\mathbf{R}^{\mathbf{EE}}$ then ξ is a proof of $\Gamma \Rightarrow \Delta|B \rightrightarrows \Gamma \Rightarrow \Delta$.*

Proof Let ξ be a branch of T' such that any reduction $\Gamma_1 \Rightarrow \Delta_1|B' \rightrightarrows \Gamma_1 \Rightarrow \Delta_1 - \{B'\} \in \xi$ is not an axiom in $\mathbf{R}^{\mathbf{EE}}$.

We proved by induction on node $\eta \in \xi$ that for each reduction $\Gamma' \Rightarrow \Delta|B' \rightrightarrows \Gamma' \Rightarrow \Delta - \{B'\} \in \eta$, $\vdash^{\mathrm{t}} \Gamma' \Rightarrow \Delta|B' \rightrightarrows \Gamma' \Rightarrow \Delta$.

Case $\Gamma' \Rightarrow \Delta|m \rightrightarrows \Gamma' \Rightarrow \Delta - \{m\} \in \eta$ is a leaf node of ξ. Then, there is no literal $m' \neq m$ such that

$$m', \neg m' \in \Delta \text{ or } m' \in \Gamma' \cap \Delta.$$

$\vdash^{\mathrm{t}} \Gamma' \Rightarrow \Delta|m \rightrightarrows \Gamma' \Rightarrow \Delta$.

Case $\Gamma_2 \Rightarrow \Delta_2|\neg\neg B \rightrightarrows \Gamma_2 \Rightarrow \Delta_2 - \{\neg\neg B\} \in \eta$. Then, η has a child node $\in \xi$ containing reduction $\Gamma_2 \Rightarrow \Delta_2|B \rightrightarrows \Gamma_2 \Rightarrow \Delta_2 - \{B\}$. By induction assumption, $\vdash^{\mathrm{t}} \Gamma_2 \Rightarrow \Delta_2|B \rightrightarrows \Gamma_2 \Rightarrow \Delta_2$, and by $(\neg\neg^0)$, $\vdash^{\mathrm{t}} \Gamma_2 \Rightarrow \Delta_2|\neg\neg B \rightrightarrows \Gamma_2 \Rightarrow \Delta_2$.

Case $\delta = \Gamma_2 \Rightarrow \Delta_2|B_1 \wedge B_2 \rightrightarrows \Gamma_2 \Rightarrow \Delta_2 - \{B_1 \wedge B_2\} \in \eta$. Then, η has a child node $\in \xi$ containing reduction either $\Gamma_2 \Rightarrow \Delta_2|B_1 \rightrightarrows \Gamma_2 \Rightarrow \Delta_2 - \{B_1\}$ or $\Gamma_2 \Rightarrow \Delta_2 - \{B_1\}|B_2 \rightrightarrows \Gamma_2 \Rightarrow \Delta_2 - \{B_1, B_2\}$. By induction assumption, either $\vdash^{\mathrm{t}} \Gamma_2 \Rightarrow \Delta_2|B_1 \rightrightarrows \Gamma_2 \Rightarrow \Delta_2$, or $\vdash^{\mathrm{t}} \Gamma_2 \Rightarrow \Delta_2 - \{B_1\}|B_2 \rightrightarrows \Gamma_2 \Rightarrow \Delta_2 - \{B_1\}$. By $(\wedge^0)$, we have $\vdash^{\mathrm{t}} \Gamma_2 \Rightarrow \Delta_2|B_1 \wedge B_2 \rightrightarrows \Gamma_2 \Rightarrow \Delta_2$.

Case $\delta = \Gamma_2 \Rightarrow \Delta_2|B_1 \vee B_2 \rightrightarrows \Gamma_2 \Rightarrow \Delta_2 - \{B_1 \vee B_2\} \in \eta$. Then, η has a child node $\in \xi$ containing reductions $\Gamma_2 \Rightarrow \Delta_2|B_1 \rightrightarrows \Gamma_2 \Rightarrow \Delta_2 - \{B_1\}$ and $\Gamma_2 \Rightarrow \Delta_2|B_2 \rightrightarrows \Gamma_2 \Rightarrow \Delta_2 - \{B_2\}$. By induction assumption, $\vdash^{\mathrm{t}} \Gamma_2 \Rightarrow \Delta_2|B_1 \rightrightarrows \Gamma_2 \Rightarrow \Delta_2$ and $\vdash^{\mathrm{t}} \Gamma_2 \Rightarrow \Delta_2|B_2 \rightrightarrows \Gamma_2 \Rightarrow \Delta_2$. By $(\vee^0)$, we have $\vdash^{\mathrm{t}} \Gamma_2 \Rightarrow \Delta_2|B_1 \vee B_2 \rightrightarrows \Gamma_2 \Rightarrow \Delta_2$.

Similar for other cases. □

4.3 Gentzen Deduction Systems $\mathbf{G}_{Q_1Q_2}$

Definition 4.3.1 A co-sequent $\Gamma \mapsto \Delta$ is $\mathbf{G}_{Q_1Q_2}$-valid, denoted by $\models_{Q_1Q_2} \Gamma \mapsto \Delta$, if there is an assignment v such that $Q_1A \in \Gamma(v(A) = 1)$ and $Q_2B \in \Delta(v(B) = 0)$.

4.3.1 Axioms

Lemma 4.3.2 *Given two sets* Γ, Δ *of literals,* $\models_{\mathbf{AA}} \Gamma \mapsto \Delta$ *if and only if* $\mathrm{con}(\Gamma)\&\mathrm{inval}(\Delta)\&\Gamma \cap \Delta = \emptyset$. □

Hence, we have the following axiom:

$$(\mathbb{A}_{\mathbf{AA}})\ \frac{\mathrm{con}(\Gamma)\&\mathrm{inval}(\Delta)\&\Gamma \cap \Delta = \emptyset}{\Gamma \mapsto \Delta}$$

Proposition 4.3.3 *Let* Γ, Δ *be sets of literals.* $\models_{\mathbf{EA}} \Gamma \mapsto \Delta$ *if and only if* $\mathrm{inval}(\Delta)\&\Gamma \not\subseteq \Delta$. □

Hence, we have the following axiom:

$$(\mathbb{A}_{\mathbf{EA}})\ \frac{\mathrm{inval}(\Delta)\&\Gamma \not\subseteq \Delta}{\Gamma \mapsto \Delta}$$

Proposition 4.3.4 *Let* Γ, Δ *be sets of literals.* $\models_{\mathbf{AE}} \Gamma \mapsto \Delta$ *if and only if* $\mathrm{con}(\Gamma)\&\Delta \not\subseteq \Gamma$. □

Hence, we have the following axiom:

$$(\mathbb{A}_{\mathbf{AE}})\ \frac{\mathrm{con}(\Gamma)\&\Delta \not\subseteq \Gamma}{\Gamma \mapsto \Delta}$$

Proposition 4.3.5 *Let* Γ, Δ *be sets of literals.* $\models_{\mathbf{EE}} \Gamma \mapsto \Delta$ *if and only if* $\Gamma \neq \emptyset \neq \Delta\&\Delta \neq \Gamma \neq \{l\}$ *for any literal* l. □

Hence, we have the following axiom:

$$(\mathbb{A}_{\mathbf{EE}})\ \frac{\Gamma \neq \emptyset \neq \Delta\&\Delta \neq \Gamma \neq \{l\}}{\Gamma \mapsto \Delta}$$

4.3.2 Deduction Rules

$\mathbf{G}_{L1}$ consists of the following deduction rules:

$$
\begin{array}{ll}
(\neg\neg^L)\ \dfrac{\Gamma, A \mapsto \Delta}{\Gamma, \neg\neg A \mapsto \Delta} & \\
(\wedge^L)\ \dfrac{\left[\begin{array}{l}\Gamma, A_1 \mapsto \Delta \\ \Gamma, A_2 \mapsto \Delta\end{array}\right.}{\Gamma, A_1 \wedge A_2 \mapsto \Delta} & (\vee^L)\ \dfrac{\left\{\begin{array}{l}\Gamma, A_1 \mapsto \Delta \\ \Gamma, A_2 \mapsto \Delta\end{array}\right.}{\Gamma, A_1 \vee A_2 \mapsto \Delta} \\
(\neg\wedge^L)\ \dfrac{\left\{\begin{array}{l}\Gamma, \neg A_1 \mapsto \Delta \\ \Gamma, \neg A_2 \mapsto \Delta\end{array}\right.}{\Gamma, \neg(A_1 \vee A_2) \mapsto \Delta} & (\neg\vee^L)\ \dfrac{\left[\begin{array}{l}\Gamma, \neg A_1 \mapsto \Delta \\ \Gamma, \neg A_2 \mapsto \Delta\end{array}\right.}{\Gamma, \neg(A_1 \vee A_2) \mapsto \Delta}
\end{array}
$$

$\mathbf{G}_{R0}$ consists of the following deduction rules:

$$
\begin{array}{ll}
(\neg\neg^R)\ \dfrac{\Gamma \mapsto B, \Delta}{\Gamma \mapsto \neg\neg B, \Delta} & \\
(\wedge^R)\ \dfrac{\left\{\begin{array}{l}\Gamma \mapsto B_1, \Delta \\ \Gamma \mapsto B_2, \Delta\end{array}\right.}{\Gamma \mapsto B_1 \wedge B_2, \Delta} & (\vee^R)\ \dfrac{\left[\begin{array}{l}\Gamma \mapsto B_1, \Delta \\ \Gamma \mapsto B_2, \Delta\end{array}\right.}{\Gamma \mapsto B_1 \vee B_2, \Delta} \\
(\neg\wedge^R)\ \dfrac{\left[\begin{array}{l}\Gamma \mapsto \neg B_1, \Delta \\ \Gamma \mapsto \neg B_2, \Delta\end{array}\right.}{\Gamma \mapsto \neg(B_1 \wedge B_2), \Delta} & (\neg\vee^R)\ \dfrac{\left\{\begin{array}{l}\Gamma \mapsto \neg B_1, \Delta \\ \Gamma \mapsto \neg B_2, \Delta\end{array}\right.}{\Gamma \mapsto \neg(B_1 \vee B_2), \Delta}
\end{array}
$$

4.3.3 Deduction Systems

For $Q_1, Q_2 \in \{\mathbf{A}, \mathbf{E}\}$, let

$$\mathbf{G}_{Q_1Q_2} = \mathbb{A}_{Q_1Q_2} + \mathbf{G}_{L1} + \mathbf{G}_{R0}.$$

Definition 4.3.6 A co-sequent $\Gamma \mapsto \Delta$ is provable in $\mathbf{G}_{Q_1Q_2}$, denoted by $\vdash_{Q_1Q_2} \Gamma \mapsto \Delta$, if there is a sequence $\{\Gamma_1 \mapsto \Delta_1, \ldots, \Gamma_n \mapsto \Delta_n\}$ of co-sequents such that $\Gamma_n \mapsto \Delta_n = \Gamma \mapsto \Delta$, and for each $1 \leq i \leq n$, $\Gamma_i \mapsto \Delta_i$ is an axiom or is deduced from the previous co-sequents by one of the deduction rules in $\mathbf{G}_{Q_1Q_2}$.

Theorem 4.3.7 (Soundness and completeness theorem) *Let $Q_1, Q_2 \in \{\mathbf{E}, \mathbf{A}\}$. For any co-sequent $\Gamma \mapsto \Delta$,*

$$\vdash_{Q_1Q_2} \Gamma \mapsto \Delta \textit{ if and only if } \models_{Q_1Q_2} \Gamma \mapsto \Delta.$$

□

4.4 R-Calculi $\mathbf{R}_{Q_1Q_2}$

Let $Q_1, Q_2 \in \{\mathbf{E}, \mathbf{A}\}$.

Given a co-sequent $\Gamma \mapsto \Delta$ and pair (A, B) of formulas, the result of $\Gamma \mapsto \Delta$ $\mathbf{G}_{Q_1Q_2}$-revising (A, B) is denoted by

$$\models_{Q_1Q_2} \Gamma \mapsto \Delta|(A, B) \rightrightarrows \Gamma' \mapsto \Delta',$$

if

$$\Gamma' = \begin{cases} \Gamma \pm \{A\} \text{ if } \vdash_{Q_1Q_2} \Gamma \pm \{A\} \mapsto \Delta \\ \Gamma \qquad \text{otherwise;} \end{cases}$$
$$\Delta' = \begin{cases} \Delta \pm \{B\} \text{ if } \vdash_{Q_1Q_2} \Gamma' \mapsto \Delta \pm \{B\} \\ \Delta \qquad \text{otherwise,} \end{cases}$$

where

$$\pm_1 l = \begin{cases} \cup\{l\} \text{ if } Q_1 = \mathbf{A} \\ -\{l\} \text{ if } Q_1 = \mathbf{E} \text{ and } l \in \Gamma, \end{cases}$$
$$\pm_2 m = \begin{cases} \cup\{m\} \text{ if } Q_2 = \mathbf{A} \\ -\{m\} \text{ if } Q_2 = \mathbf{E} \text{ and } m \in \Delta. \end{cases}$$

For example, let $Q_1 = \mathbf{A}$, $Q_2 = \mathbf{E}$ and $B \in \Delta$. Then,

$$\Gamma' = \begin{cases} \Gamma, A \text{ if } \vdash_{\mathbf{AE}} \Gamma, A \mapsto \Delta \\ \Gamma \quad \text{otherwise;} \end{cases}$$
$$\Delta' = \begin{cases} \Delta - \{B\} \text{ if } \vdash_{\mathbf{AE}} \Gamma' \mapsto \Delta - \{B\} \\ \Delta \qquad \text{otherwise.} \end{cases}$$

4.4.1 Axioms

A reduction $\Gamma \mapsto \Delta|(A, B) \rightrightarrows \Gamma' \mapsto \Delta'$ is reduced to literals for A and B, that is, Γ', Δ' are sets of literals. We use deduction rules in $\mathbf{G}_{Q_1Q_2}$ to reduce $\Gamma \mapsto \Delta$ into literal ones, say $\Gamma' \mapsto \Delta'$.

Let Γ, Δ be sets of literals and l, m be literals.

$$(\mathtt{A}_{Q_1Q_2})\ \Gamma \Rightarrow \Delta|(l, m) \rightrightarrows \Gamma' \Rightarrow \Delta'$$

if

$$\Gamma' = \begin{cases} \Gamma \pm l \text{ if } \vdash^{Q_1Q_2} \Gamma \pm l \Rightarrow \Delta \\ \Gamma \qquad \text{otherwise} \end{cases}$$
$$\Delta' = \begin{cases} \Delta \pm m \text{ if } \vdash^{Q_1Q_2} \Gamma' \Rightarrow \Delta \pm m \\ \Delta \qquad \text{otherwise.} \end{cases}$$

In detail, we have the following axioms: Let Γ, Δ be sets of literals.

$$(\mathtt{A}^{+L}_{\mathbf{AA}})\ \frac{\left[\begin{array}{l}\neg l \notin \Gamma \\ l \notin \Delta\end{array}\right.}{\Gamma \mapsto \Delta|(l,m) \rightrightarrows \Gamma, l \mapsto \Delta|m}$$

$$(\mathtt{A}^{0L}_{\mathbf{AA}})\ \frac{\left\{\begin{array}{l}\neg l \in \Gamma \\ l \in \Delta\end{array}\right.}{\Gamma \mapsto \Delta|(l,m) \rightrightarrows \Gamma \mapsto \Delta|m}$$

$$(\mathtt{A}^{+R}_{\mathbf{AA}})\ \frac{\left[\begin{array}{l}m \notin \Gamma' \\ \neg m \notin \Delta\end{array}\right.}{\Gamma' \mapsto \Delta|m \rightrightarrows \Gamma' \mapsto \Delta, m}$$

$$(\mathtt{A}^{0R}_{\mathbf{AA}})\ \frac{\left\{\begin{array}{l}m \in \Gamma' \\ \neg m \in \Delta\end{array}\right.}{\Gamma' \mapsto \Delta|m \rightrightarrows \Gamma' \mapsto \Delta}$$

and

$$(\mathtt{A}^{-L}_{\mathbf{EA}})\ \frac{\Gamma, l \nsubseteq \Delta}{\Gamma \mapsto \Delta|(l,m) \rightrightarrows \Gamma - \{l\} \mapsto \Delta|m}$$

$$(\mathtt{A}^{0L}_{\mathbf{EA}})\ \frac{\Gamma, l \subseteq \Delta}{\Gamma \mapsto \Delta|(l,m) \rightrightarrows \Gamma \mapsto \Delta|m}$$

$$(\mathtt{A}^{+R}_{\mathbf{EA}})\ \frac{\neg m \notin \Delta}{\Gamma' \mapsto \Delta|m \rightrightarrows \Gamma' \mapsto \Delta - \{m\}}$$

$$(\mathtt{A}^{0R}_{\mathbf{EA}})\ \frac{\neg m \in \Delta}{\Gamma' \mapsto \Delta|m \rightrightarrows \Gamma' \mapsto \Delta}$$

where $l \in \Gamma$, and

$$(\mathtt{A}^{+L}_{\mathbf{AE}})\ \frac{\neg l \notin \Gamma}{\Gamma \mapsto \Delta|(l,m) \rightrightarrows \Gamma, l \mapsto \Delta|m}$$

$$(\mathtt{A}^{0L}_{\mathbf{AA}})\ \frac{\neg l \in \Gamma}{\Gamma \mapsto \Delta|(l,m) \rightrightarrows \Gamma \mapsto \Delta|m}$$

$$(\mathtt{A}^{-R}_{\mathbf{AA}})\ \frac{\Delta - \{m\} \nsubseteq \Gamma}{\Gamma' \mapsto \Delta|m \rightrightarrows \Gamma' \mapsto \Delta - \{m\}}$$

$$(\mathtt{A}^{0R}_{\mathbf{AA}})\ \frac{\Delta - \{m\} \subseteq \Gamma}{\Gamma' \mapsto \Delta|m \rightrightarrows \Gamma' \mapsto \Delta}$$

where $m \in \Delta$, and

$$(\mathtt{A}^{-L}_{\mathbf{EE}})\ \frac{\Gamma - \{l\} \neq \{l'\} \neq \Delta}{\Gamma \mapsto \Delta|(l,m) \rightrightarrows \Gamma - \{l\} \mapsto \Delta|m}$$

$$(\mathtt{A}^{0L}_{\mathbf{EE}})\ \frac{\Gamma - \{l\} = \{l'\} = \Delta}{\Gamma \mapsto \Delta|(l,m) \rightrightarrows \Gamma \mapsto \Delta|m}$$

$$(\mathtt{A}^{-R}_{\mathbf{EE}})\ \frac{\Gamma' \neq \{l'\} \neq \Delta - \{m\}}{\Gamma' \mapsto \Delta|m \rightrightarrows \Gamma' \mapsto \Delta - \{m\}}$$

$$(\mathtt{A}^{0R}_{\mathbf{EE}})\ \frac{\Gamma' = \{l'\} = \Delta - \{m\}}{\Gamma' \mapsto \Delta|m \rightrightarrows \Gamma' \mapsto \Delta}$$

where $l \in \Gamma$ and $m \in \Delta$.

Proposition 4.4.1 *Let $\Gamma' \mapsto \Delta'$ be literal. $\Gamma \mapsto \Delta|(A,B) \rightrightarrows \Gamma' \mapsto \Delta'$ is $\mathbf{R}_{Q_1Q_2}$-valid if and only if $\Gamma \mapsto \Delta|(A,B) \rightrightarrows \Gamma' \mapsto \Delta'$ is the conclusion of an axiom.* □

4.4.2 Deduction Rules

Let $\mathbf{X} = \Gamma \mapsto \Delta$ and

$$\begin{aligned}
\mathbf{X}[A] &= \Gamma \mapsto \Delta|(A,B) \rightrightarrows \Gamma - \{A\} \mapsto \Delta|B\\
\mathbf{X}(A) &= \Gamma \mapsto \Delta|(A,B) \rightrightarrows \Gamma, A \mapsto \Delta|B\\
\mathbf{X}[B] &= \Gamma' \mapsto \Delta|B \rightrightarrows \Gamma' \mapsto \Delta - \{B\}\\
\mathbf{X}(B) &= \Gamma' \mapsto \Delta|B \rightrightarrows \Gamma' \mapsto \Delta, B.
\end{aligned}$$

- $\mathbf{S}_{L\mathbf{A}}$:

$$(\neg\neg_+^L)\ \frac{\mathbf{X}|A \rightrightarrows \mathbf{X}(A)}{\mathbf{X}|\neg\neg A \rightrightarrows \mathbf{X}(\neg\neg A_1)} \qquad (\neg\neg_0^L)\ \frac{\mathbf{X}|A \rightrightarrows \mathbf{X}}{\mathbf{X}|\neg\neg A \rightrightarrows \mathbf{X}}$$

$$(\wedge_+^L)\ \frac{\left[\begin{array}{l}\mathbf{X}|A_1 \rightrightarrows \mathbf{X}(A_1)\\ \mathbf{X}(A_1)|A_2 \rightrightarrows \mathbf{X}(A_1,A_2)\end{array}\right.}{\mathbf{X}|A_1 \wedge A_2 \rightrightarrows \mathbf{X}(A_1 \wedge A_2)} \qquad (\wedge_0^L)\ \frac{\left\{\begin{array}{l}\mathbf{X}|A_1 \rightrightarrows \mathbf{X}\\ \mathbf{X}(A_1)|A_2 \rightrightarrows \mathbf{X}(A_1)\end{array}\right.}{\mathbf{X}|A_1 \wedge A_2 \rightrightarrows \mathbf{X}}$$

$$(\vee_+^L)\ \frac{\left\{\begin{array}{l}\mathbf{X}|A_1 \rightrightarrows \mathbf{X}(A_1)\\ \mathbf{X}|A_2 \rightrightarrows \mathbf{X}(A_2)\end{array}\right.}{\mathbf{X}|A_1 \vee A_2 \rightrightarrows \mathbf{X}(A_1 \vee A_2)} \qquad (\vee_0^L)\ \frac{\left[\begin{array}{l}\mathbf{X}|A_1 \rightrightarrows \mathbf{X}\\ \mathbf{X}|A_2 \rightrightarrows \mathbf{X}\end{array}\right.}{\mathbf{X}|A_1 \vee A_2 \rightrightarrows \mathbf{X}}$$

and

$$(\neg\wedge_+^L)\ \frac{\left\{\begin{array}{l}\mathbf{X}|\neg A_1 \rightrightarrows \mathbf{X}(\neg A_1)\\ \mathbf{X}|\neg A_2 \rightrightarrows \mathbf{X}(\neg A_2)\end{array}\right.}{\mathbf{X}|\neg(A_1 \wedge A_2) \rightrightarrows \mathbf{X}(\neg(A_1 \wedge A_2))}$$

$$(\neg\wedge_0^L)\ \frac{\left[\begin{array}{l}\mathbf{X}|\neg A_1 \rightrightarrows \mathbf{X}\\ \mathbf{X}|\neg A_2 \rightrightarrows \mathbf{X}\end{array}\right.}{\mathbf{X}|\neg(A_1 \wedge A_2) \rightrightarrows \mathbf{X}}$$

$$(\neg\vee_+^L)\ \frac{\left[\begin{array}{l}\mathbf{X}|\neg A_1 \rightrightarrows \mathbf{X}(\neg A_1)\\ \mathbf{X}(\neg A_1)|\neg A_2 \rightrightarrows \mathbf{X}(\neg A_1, \neg A_2)\end{array}\right.}{\mathbf{X}|\neg(A_1 \vee A_2) \rightrightarrows \mathbf{X}(\neg(A_1 \vee A_2))}$$

$$(\neg\vee_0^L)\ \frac{\left\{\begin{array}{l}\mathbf{X}|\neg A_1 \rightrightarrows \mathbf{X}\\ \mathbf{X}(\neg A_1)|\neg A_2 \rightrightarrows \mathbf{X}(\neg A_1)\end{array}\right.}{\mathbf{X}|\neg(A_1 \vee A_2) \rightrightarrows \mathbf{X}}$$

A formula $A_1 \wedge A_2$ is enumerable into Γ, if either A_1 or A_2 is enumerable into Γ; and $A_1 \vee A_2$ is enumerable into Γ if A_1 is enumerable into Γ and A_2 is enumerable into $\Gamma \cup \{A_1\}$.

$A_1 \wedge A_2$ is not enumerable into Γ, if both A_1 and A_2 are not enumerable into Γ; and $A_1 \vee A_2$ is not enumerable into Γ if either A_1 is not enumerable into Γ or A_2 is not enumerable into $\Gamma \cup \{A_1\}$.

- $\mathbf{S}_{R\mathbf{A}}$:

$$(\neg\neg_+^R)\ \frac{\mathbf{X}|B \rightrightarrows \mathbf{X}(B)}{\mathbf{X}|\neg\neg B \rightrightarrows \mathbf{X}(\neg\neg B)} \qquad (\neg\neg_0^R)\ \frac{\mathbf{X}|B \rightrightarrows \mathbf{X}}{\mathbf{X}|\neg\neg B \rightrightarrows \mathbf{X}}$$

$$(\wedge_+^R)\ \frac{\left\{\begin{array}{l}\mathbf{X}|B_1 \rightrightarrows \mathbf{X}(B_1)\\ \mathbf{X}|B_2 \rightrightarrows \mathbf{X}(B_2)\end{array}\right.}{\mathbf{X}|B_1 \wedge B_2 \rightrightarrows \mathbf{X}(B_1 \wedge B_2)} \qquad (\wedge_0^R)\ \frac{\left[\begin{array}{l}\mathbf{X}|B_1 \rightrightarrows \mathbf{X}\\ \mathbf{X}|B_2 \rightrightarrows \mathbf{X}\end{array}\right.}{\mathbf{X}|B_1 \wedge B_2 \rightrightarrows \mathbf{X}}$$

$$(\vee_+^R)\ \frac{\left[\begin{array}{l}\mathbf{X}|B_1 \rightrightarrows \mathbf{X}(B_1)\\ \mathbf{X}(B_1)|B_2 \rightrightarrows \mathbf{X}(B_1, B_2)\end{array}\right.}{\mathbf{X}|B_1 \vee B_2 \rightrightarrows \mathbf{X}(B_1 \vee B_2)} \qquad (\vee_0^R)\ \frac{\left\{\begin{array}{l}\mathbf{X}|B_1 \rightrightarrows \mathbf{X}\\ \mathbf{X}(B_1)|B_2 \rightrightarrows \mathbf{X}(B_1)\end{array}\right.}{\mathbf{X}|B_1 \vee B_2 \rightrightarrows \mathbf{X}}$$

and

$$(\neg\wedge_+^R)\ \frac{\left[\begin{array}{l}\mathbf{X}|\neg B_1 \rightrightarrows \mathbf{X}(\neg B_1)\\ \mathbf{X}(\neg B_1)|\neg B_2 \rightrightarrows \mathbf{X}(\neg B_1, \neg B_2)\end{array}\right.}{\mathbf{X}|\neg(B_1 \wedge B_2) \rightrightarrows \mathbf{X}(\neg(B_1 \wedge B_2))}$$

$$(\neg\wedge_0^R)\ \frac{\left\{\begin{array}{l}\mathbf{X}|\neg B_1 \rightrightarrows \mathbf{X}\\ \mathbf{X}(\neg B_1)|\neg B_2 \rightrightarrows \mathbf{X}(\neg B_1)\end{array}\right.}{\mathbf{X}|\neg(B_1 \wedge B_2) \rightrightarrows \mathbf{X}}$$

$$(\neg\vee_+^R)\ \frac{\left\{\begin{array}{l}\mathbf{X}|\neg B_1 \rightrightarrows \mathbf{X}(\neg B_1)\\ \mathbf{X}|\neg B_2 \rightrightarrows \mathbf{X}(\neg B_2)\end{array}\right.}{\mathbf{X}|\neg(B_1 \vee B_2) \rightrightarrows \mathbf{X}\neg(B_1 \vee B_2))}$$

$$(\neg\vee_0^R)\ \frac{\left[\begin{array}{l}\mathbf{X}|\neg B_1 \rightrightarrows \mathbf{X}\\ \mathbf{X}|\neg B_2 \rightrightarrows \mathbf{X}\end{array}\right.}{\mathbf{X}|\neg(B_1 \vee B_2) \rightrightarrows \mathbf{X}}$$

A formula $B_1 \wedge B_2$ is enumerable into Δ, if B_1 or B_2 is enumerable into Δ; and $B_1 \vee B_2$ is enumerable into Δ if B_1 is enumerable into Δ and B_2 is enumerable into $\Delta \cup \{B_1\}$.

Formula $B_1 \wedge B_2$ is not enumerable into Δ, if both B_1 and B_2 are not enumerable into Δ; and $B_1 \vee B_2$ is not enumerable into Δ if either B_1 is not enumerable into Δ, or B_2 is not enumerable into $\Delta \cup \{B_1\}$.

- $\mathbf{S}_{LE}$:

$$(\neg\neg_+^L)\ \frac{\mathbf{X}|A \rightrightarrows \mathbf{X}[A]}{\mathbf{X}|\neg\neg A \rightrightarrows \mathbf{X}[\neg\neg A_1]}$$

$$(\neg\neg_0^L)\ \frac{\mathbf{X}|A \rightrightarrows \mathbf{X}}{\mathbf{X}|\neg\neg A \rightrightarrows \mathbf{X}}$$

$$(\wedge_+^L)\ \frac{\left[\begin{array}{l}\mathbf{X}|A_1 \rightrightarrows \mathbf{X}[A_1]\\ \mathbf{X}[A_1]|A_2 \rightrightarrows \mathbf{X}[A_1, A_2]\end{array}\right.}{\mathbf{X}|A_1 \wedge A_2 \rightrightarrows \mathbf{X}[A_1 \wedge A_2]}$$

$$(\wedge_0^L)\ \frac{\left\{\begin{array}{l}\mathbf{X}|A_1 \rightrightarrows \mathbf{X}\\ \mathbf{X}[A_1]|A_2 \rightrightarrows \mathbf{X}[A_1]\end{array}\right.}{\mathbf{X}|A_1 \wedge A_2 \rightrightarrows \mathbf{X}}$$

$$(\vee_+^L)\ \frac{\left\{\begin{array}{l}\mathbf{X}|A_1 \rightrightarrows \mathbf{X}[A_1]\\ \mathbf{X}|A_2 \rightrightarrows \mathbf{X}[A_2]\end{array}\right.}{\mathbf{X}|A_1 \vee A_2 \rightrightarrows \mathbf{X}[A_1 \vee A_2]}$$

$$(\vee_0^L)\ \frac{\left[\begin{array}{l}\mathbf{X}|A_1 \rightrightarrows \mathbf{X}\\ \mathbf{X}|A_2 \rightrightarrows \mathbf{X}\end{array}\right.}{\mathbf{X}|A_1 \vee A_2 \rightrightarrows \mathbf{X}}$$

and

$$(\neg\wedge_+^L)\ \frac{\left\{\begin{array}{l}\mathbf{X}|\neg A_1 \rightrightarrows \mathbf{X}[\neg A_1]\\ \mathbf{X}|\neg A_2 \rightrightarrows \mathbf{X}[\neg A_2]\end{array}\right.}{\mathbf{X}|\neg(A_1 \wedge A_2) \rightrightarrows \mathbf{X}[\neg(A_1 \wedge A_2)]}$$

$$(\neg\wedge_0^L)\ \frac{\left[\begin{array}{l}\mathbf{X}|\neg A_1 \rightrightarrows \mathbf{X}\\ \mathbf{X}|\neg A_2 \rightrightarrows \mathbf{X}\end{array}\right.}{\mathbf{X}|\neg(A_1 \wedge A_2) \rightrightarrows \mathbf{X}}$$

$$(\neg\vee_+^L)\ \frac{\left[\begin{array}{l}\mathbf{X}|\neg A_1 \rightrightarrows \mathbf{X}[\neg A_1]\\ \mathbf{X}[\neg A_1]|\neg A_2 \rightrightarrows \mathbf{X}[\neg A_1, \neg A_2]\end{array}\right.}{\mathbf{X}|\neg(A_1 \vee A_2) \rightrightarrows \mathbf{X}[\neg(A_1 \vee A_2)]}$$

$$(\neg\vee_0^L)\ \frac{\left\{\begin{array}{l}\mathbf{X}|\neg A_1 \rightrightarrows \mathbf{X}\\ \mathbf{X}[\neg A_1]|\neg A_2 \rightrightarrows \mathbf{X}[\neg A_1]\end{array}\right.}{\mathbf{X}|\neg(A_1 \vee A_2) \rightrightarrows \mathbf{X}}$$

A formula $A_1 \wedge A_2$ is extractable into Γ, if either A_1 or A_2 is extractable into Γ; and $A_1 \vee A_2$ is extractable into Γ if A_1 is extractable into Γ and A_2 is extractable into $\Gamma \cup \{A_1\}$.

$A_1 \wedge A_2$ is not extractable into Γ, if both A_1 and A_2 are not extractable into Γ; and $A_1 \vee A_2$ is not extractable into Γ if either A_1 is not extractable into Γ or A_2 is not extractable into $\Gamma \cup \{A_1\}$.

- $\mathbf{S}_{R\mathbf{E}}$:

$$(\neg\neg_{+}^{R})\ \frac{\mathbf{X}|B \rightrightarrows \mathbf{X}[B]}{\mathbf{X}|\neg\neg B \rightrightarrows \mathbf{X}[\neg\neg B]}$$

$$(\neg\neg_{0}^{R})\ \frac{\mathbf{X}|B \rightrightarrows \mathbf{X}}{\mathbf{X}|\neg\neg B \rightrightarrows \mathbf{X}}$$

$$(\wedge_{+}^{R})\ \frac{\left\{\begin{array}{l}\mathbf{X}|B_1 \rightrightarrows \mathbf{X}[B_1]\\ \mathbf{X}|B_2 \rightrightarrows \mathbf{X}[B_2]\end{array}\right.}{\mathbf{X}|B_1 \wedge B_2 \rightrightarrows \mathbf{X}[B_1 \wedge B_2]}$$

$$(\wedge_{0}^{R})\ \frac{\left[\begin{array}{l}\mathbf{X}|B_1 \rightrightarrows \mathbf{X}\\ \mathbf{X}|B_2 \rightrightarrows \mathbf{X}\end{array}\right.}{\mathbf{X}|B_1 \wedge B_2 \rightrightarrows \mathbf{X}}$$

$$(\vee_{+}^{R})\ \frac{\left[\begin{array}{l}\mathbf{X}|B_1 \rightrightarrows \mathbf{X}[B_1]\\ \mathbf{X}[B_1]|B_2 \rightrightarrows \mathbf{X}[B_1, B_2]\end{array}\right.}{\mathbf{X}|B_1 \vee B_2 \rightrightarrows \mathbf{X}[B_1 \vee B_2]}$$

$$(\vee_{0}^{R})\ \frac{\left\{\begin{array}{l}\mathbf{X}|B_1 \rightrightarrows \mathbf{X}\\ \mathbf{X}[B_1]|B_2 \rightrightarrows \mathbf{X}[B_1]\end{array}\right.}{\mathbf{X}|B_1 \vee B_2 \rightrightarrows \mathbf{X}}$$

and

$$(\neg\wedge_{+}^{R})\ \frac{\left[\begin{array}{l}\mathbf{X}|\neg B_1 \rightrightarrows \mathbf{X}[\neg B_1]\\ \mathbf{X}[\neg B_1]|\neg B_2 \rightrightarrows \mathbf{X}[\neg B_1, \neg B_2]\end{array}\right.}{\mathbf{X}|\neg(B_1 \wedge B_2) \rightrightarrows \mathbf{X}[\neg(B_1 \wedge B_2)]}$$

$$(\neg\wedge_{0}^{R})\ \frac{\left\{\begin{array}{l}\mathbf{X}|\neg B_1 \rightrightarrows \mathbf{X}\\ \mathbf{X}[\neg B_1]|\neg B_2 \rightrightarrows \mathbf{X}[\neg B_1]\end{array}\right.}{\mathbf{X}|\neg(B_1 \wedge B_2) \rightrightarrows \mathbf{X}}$$

$$(\neg\vee_{+}^{R})\ \frac{\left\{\begin{array}{l}\mathbf{X}|\neg B_1 \rightrightarrows \mathbf{X}[\neg B_1]\\ \mathbf{X}|\neg B_2 \rightrightarrows \mathbf{X}[\neg B_2]\end{array}\right.}{\mathbf{X}|\neg(B_1 \vee B_2) \rightrightarrows \mathbf{X}\neg(B_1 \vee B_2)]}$$

$$(\neg\vee_{0}^{R})\ \frac{\left[\begin{array}{l}\mathbf{X}|\neg B_1 \rightrightarrows \mathbf{X}\\ \mathbf{X}|\neg B_2 \rightrightarrows \mathbf{X}\end{array}\right.}{\mathbf{X}|\neg(B_1 \vee B_2) \rightrightarrows \mathbf{X}}$$

A formula $B_1 \wedge B_2$ is extractable into Δ, if B_1 or B_2 is extractable into Δ; and $B_1 \vee B_2$ is extractable into Δ if B_1 is extractable into Δ and B_2 is extractable into $\Delta \cup \{B_1\}$.

Formula $B_1 \wedge B_2$ is not extractable into Δ, if both B_1 and B_2 are not extractable into Δ; and $B_1 \vee B_2$ is not extractable into Δ if either B_1 is not extractable into Δ, or B_2 is not extractable into $\Delta \cup \{B_1\}$.

4.4.3 Deduction Systems $\mathbf{R}_{Q_1Q_2}$

For $Q_1, Q_2 \in \{\mathbf{A}, \mathbf{E}\}$, define

$$\mathbf{R}_{Q_1Q_2} = \mathbb{A}_{Q_1Q_2} + \mathbf{S}_{LQ_1} + \mathbf{S}_{RQ_2}.$$

Theorem 4.4.2 (Soundness and completeness theorem) *For any $Q_1, Q_2 \in \{\mathbf{A}, \mathbf{E}\}$ and reduction $\delta = \Gamma \mapsto \Delta|(A, B) \rightrightarrows \Gamma' \mapsto \Delta'$, δ is $\mathbf{R}_{Q_1Q_2}$-valid if and only if δ is provable in $\mathbf{R}_{Q_1Q_2}$. That is,*

$$\vdash_{Q_1Q_2} \delta \text{ iff } \models_{Q_1Q_2} \delta. \qquad \square$$

4.5 Conclusions

It is true that

$$\begin{aligned}\not\models^{Q_1Q_2} \Gamma \Rightarrow \Delta \text{ iff } \models_{\overline{Q_1Q_2}} \Gamma \mapsto \Delta \\ \nvdash^{Q_1Q_2} \Gamma \Rightarrow \Delta \text{ iff } \vdash_{\overline{Q_1Q_2}} \Gamma \mapsto \Delta\end{aligned}$$

and

$$\begin{aligned}\not\models^{Q_1Q_2} \Gamma \Rightarrow \Delta|(A, B) \rightrightarrows \Gamma' \Rightarrow \Delta' \text{ iff } \models_{\overline{Q_1Q_2}} \Gamma \mapsto \Delta|(A, B) \rightrightarrows \Gamma' \mapsto \Delta' \\ \nvdash^{Q_1Q_2} \Gamma \Rightarrow \Delta|(A, B) \rightrightarrows \Gamma' \Rightarrow \Delta' \text{ iff } \vdash_{\overline{Q_1Q_2}} \Gamma \mapsto \Delta|(A, B) \rightrightarrows \Gamma' \mapsto \Delta',\end{aligned}$$

where $\overline{\mathbf{E}} = \mathbf{A}$ and $\overline{\mathbf{A}} = \mathbf{E}$.

There are eight Gentzen deduction systems:

$\mathbf{A}v$	$\mathbf{G}^{\mathbf{EE}}$	$\mathbf{G}^{\mathbf{EA}}$	$\mathbf{G}^{\mathbf{AE}}$	$\mathbf{G}^{\mathbf{AA}}$
$\mathbf{Q}v$	$\mathbf{G}_{\mathbf{AA}}$	$\mathbf{G}_{\mathbf{AE}}$	$\mathbf{G}_{\mathbf{EA}}$	$\mathbf{G}_{\mathbf{EE}}$

where

$$\begin{aligned}\mathbf{G}^{Q_1Q_2} &= \mathbb{A}^{Q_1Q_2} + \mathbf{G}^{L0} + \mathbf{G}^{R1} \\ \mathbf{G}_{Q_1Q_2} &= \mathbb{A}_{Q_1Q_2} + \mathbf{G}_{L1} + \mathbf{G}_{R0} \\ \mathbf{G}^{L0} &\equiv \mathbf{G}_{L0}, \mathbf{G}^{L1} \equiv \mathbf{G}_{L1}, \\ \mathbf{G}^{R0} &\equiv \mathbf{G}_{R0}, \mathbf{G}^{R1} \equiv \mathbf{G}_{R1}.\end{aligned}$$

Moreover, $\mathbf{G}^{Q_1Q_2}/\mathbf{G}_{Q_1Q_2}$ is monotonic in Γ if and only if $Q_1 = \mathbf{E}$; and nonmonotonic in Δ if and only if $Q_2 = \mathbf{A}$.

Correspondingly there are eight R-calculi:

$\mathbf{A}v$	$\mathbf{R}^{\mathbf{EE}}$	$\mathbf{R}^{\mathbf{EA}}$	$\mathbf{R}^{\mathbf{AE}}$	$\mathbf{R}^{\mathbf{AA}}$
$\mathbf{E}v$	$\mathbf{R}_{\mathbf{AA}}$	$\mathbf{R}_{\mathbf{AE}}$	$\mathbf{R}_{\mathbf{EA}}$	$\mathbf{R}_{\mathbf{EE}}$

where

$$\begin{aligned}
&\mathbf{R}^{Q_1Q_2} = \mathbb{A}^{Q_1Q_2} + \mathbf{R}^{LQ_1} + \mathbf{R}^{RQ_2}\\
&\mathbf{R}_{Q_1Q_2} = \mathbb{A}_{Q_1Q_2} + \mathbf{R}_{LQ_1} + \mathbf{R}_{RQ_2};\\
&\mathbf{R}^{LQ_1} \equiv \mathbf{R}_{RQ_1},\\
&\mathbf{R}^{RQ_2} \equiv \mathbf{R}_{LQ_2}.
\end{aligned}$$

References

Alchourrón, C.E., Gärdenfors, P., Makinson, D.: On the logic of theory change: partial meet contraction and revision functions. J. Symbolic Logic **50**, 510–530 (1985)

Cao, C., Sui, Y., Wang, Y.: The nonmonotonic propositional logics. Artif. Intell. Res. **5**, 111–120 (2016)

Darwiche, A., Pearl, J.: On the logic of iterated belief revision. Artif. Intell. **89**, 1–29 (1997)

Fermé, E., Hansson, S.O.: AGM 25 years, twenty-five years of research in belief change. J. Philos. Logic **40**, 295–331 (2011)

Gärdenfors, P., Rott, H.: Belief revision. In: Handbook of Logic in Artificial Intelligence and Logic Programming, vol. 4, pp. 35–132. Oxford University Press, Oxford (1995)

Ginsberg, M.L. (ed.): Readings in Nonmonotonic Reasoning. Morgan Kaufmann, San Francisco (1987)

Li, W.: R-calculus: an inference system for belief revision. Comput. J. **50**, 378–390 (2007)

Li, W.: Mathematical logic, foundations for information science. In: Progress in Computer Science and Applied Logic, vol. 25. Birkhäuser (2010)

Reiter, R.: A logic for default reasoning. Artif. Intell. **13**, 81–132 (1980)

Takeuti, G.: Proof theory. In: Barwise, J. (ed.), Handbook of Mathematical Logic. Studies in Logic and the Foundations of Mathematics. North-Holland, Amsterdam, NL (1987)

Chapter 5
R-Calculi $\mathbf{R}^{Q_1 i Q_2 j}/\mathbf{R}_{Q_1 i Q_2 j}$

We consider sequents of form $\mathbf{G}^{Q_1 i Q_2 j}$ and co-sequents of form $\mathbf{G}_{Q_1 i Q_2 j}$ (Li 2010; Takeuti and Barwise 1987), and corresponding R-calculi $\mathbf{R}^{Q_1 i Q_2 j}$ and $\mathbf{R}_{Q_1 i Q_2 j}$ (Li 2007), where $Q_1, Q_2 \in \mathbf{A}, \mathbf{E}$ and $i, j \in 0, 1$.

Let $Q_1, Q_2 \in \mathbf{A}, \mathbf{E}$ and $i, j \in 0, 1$.

- $\mathbf{G}^{Q_1 i Q_2 j}$: A sequent $\Gamma \Rightarrow \Delta$ is $\mathbf{G}^{Q_1 i Q_2 j}$-valid, denoted by $\models^{Q_1 i Q_2 j} \Gamma \Rightarrow \Delta$, if for any assignment v, either $Q_1 A \in \Gamma(v(A) = i)$ or $Q_2 B \in \Delta(v(B) = j)$.
- $\mathbf{G}_{Q_1 i Q_2 j}$: A co-sequent $\Gamma \mapsto \Delta$ is $\mathbf{G}_{Q_1 i Q_2 j}$-valid, denoted by $\models_{Q_1 i Q_2 j} \Gamma \mapsto \Delta$, if there is an assignment v such that $Q_1 A \in \Gamma(v(A) = i)$ and $Q_2 B \in \Delta(v(B) = j)$. There are 32 kinds of sequents and co-sequents:

sequents	$\mathbf{G}^{\mathrm{E0E0}}$	$\mathbf{G}^{\mathrm{E0A0}}$	$\mathbf{G}^{\mathrm{A0E0}}$	$\mathbf{G}^{\mathrm{A0A0}}$
	$\mathbf{G}^{\mathrm{E0E1}}$	$\mathbf{G}^{\mathrm{E0A1}}$	$\mathbf{G}^{\mathrm{A0E1}}$	$\mathbf{G}^{\mathrm{A0A1}}$
	$\mathbf{G}^{\mathrm{E1E0}}$	$\mathbf{G}^{\mathrm{E1A0}}$	$\mathbf{G}^{\mathrm{A1E0}}$	$\mathbf{G}^{\mathrm{A1A0}}$
	$\mathbf{G}^{\mathrm{E1E1}}$	$\mathbf{G}^{\mathrm{E1A1}}$	$\mathbf{G}^{\mathrm{A1E1}}$	$\mathbf{G}^{\mathrm{A1A1}}$
co-sequents	$\mathbf{G}_{\mathrm{A1A1}}$	$\mathbf{G}_{\mathrm{A1E1}}$	$\mathbf{G}_{\mathrm{E1A1}}$	$\mathbf{G}_{\mathrm{E1E1}}$
	$\mathbf{G}_{\mathrm{A1A0}}$	$\mathbf{G}_{\mathrm{A1E0}}$	$\mathbf{G}_{\mathrm{E1A0}}$	$\mathbf{G}_{\mathrm{E1E0}}$
	$\mathbf{G}_{\mathrm{A0A1}}$	$\mathbf{G}_{\mathrm{A0E1}}$	$\mathbf{G}_{\mathrm{E0A1}}$	$\mathbf{G}_{\mathrm{E0E1}}$
	$\mathbf{G}_{\mathrm{A0A0}}$	$\mathbf{G}_{\mathrm{A0E0}}$	$\mathbf{G}_{\mathrm{E0A0}}$	$\mathbf{G}_{\mathrm{E0E0}}$

where

$$\mathbf{G}^{\mathrm{E0E1}} = \mathbf{G}^{\mathrm{EE}},\ \mathbf{G}^{\mathrm{E0A1}} = \mathbf{G}^{\mathrm{EA}},$$
$$\mathbf{G}^{\mathrm{A0E1}} = \mathbf{G}^{\mathrm{AE}},\ \mathbf{G}^{\mathrm{A0A1}} = \mathbf{G}^{\mathrm{AA}}.$$

- $\mathbf{R}^{Q_1 i Q_2 j}$: Given a sequent $\Gamma \Rightarrow \Delta$ and pair (A, B) of formulas, the result of $\Gamma \Rightarrow \Delta$ $\mathbf{G}^{Q_1 i Q_2 j}$-revising (A, B) is $\Gamma' \Rightarrow \Delta'$, denoted by

$$\models^{Q_1 i Q_2 j} \Gamma \Rightarrow \Delta | (A, B) \rightrightarrows \Gamma' \Rightarrow \Delta',$$

W. Li and Y. Sui, *R-Calculus, IV: Propositional Logic*,
Perspectives in Formal Induction, Revision and Evolution,
https://doi.org/10.1007/978-981-19-8633-8_5

where

$$\Gamma' \Rightarrow \Delta' = \begin{cases} \Gamma \pm_1 A \Rightarrow \Delta \pm_2 B & \text{if } \models^{Q_1 i Q_2 j} \Gamma \pm_1 A \Rightarrow \Delta \pm_2 B \\ \Gamma \pm_1 A \Rightarrow \Delta & \text{if } \models^{Q_1 i Q_2 j} \Gamma \pm_1 A \Rightarrow \Delta \\ \Gamma \Rightarrow \Delta \pm_2 B & \text{if } \models^{Q_1 i Q_2 j} \Gamma \Rightarrow \Delta \pm_2 B \\ \Gamma \Rightarrow \Delta & \text{otherwise,} \end{cases}$$

where

$$\pm_1 A = \begin{cases} \cup\{A\} \text{ if } Q_1 = \mathbf{A} \\ -\{A\} \text{ if } Q_1 = \mathbf{E}, \end{cases} \qquad \pm_2 B = \begin{cases} \cup\{B\} \text{ if } Q_2 = \mathbf{A} \\ -\{B\} \text{ if } Q_2 = \mathbf{E}. \end{cases}$$

- $\mathbf{R}_{Q_1 i Q_2 j}$: Given a co-sequent $\Gamma \mapsto \Delta$ and pair (A, B), the result of $\Gamma \mapsto \Delta$ $\mathbf{G}_{Q_1 i Q_2 j}$-revising (A, B) is $\Gamma' \mapsto \Delta'$, denoted by $\models_{Q_1 i Q_2 j} \Gamma \mapsto \Delta|(A, B) \rightrightarrows \Gamma' \mapsto \Delta'$, where

$$\Gamma' \mapsto \Delta' = \begin{cases} \Gamma \pm_1 A \mapsto \Delta \pm_2 B & \text{if } \Gamma \pm_1 A \mapsto \Delta \pm_2 B \text{is} \mathbf{G}_{Q_1 i Q_2 j}-\text{valid} \\ \Gamma \pm_1 A \mapsto \Delta & \text{otherwise, if } \Gamma \pm_1 A \mapsto \Delta \text{is} \mathbf{G}_{Q_1 i Q_2 j}-\text{valid} \\ \Gamma \mapsto \Delta \pm_2 B & \text{otherwise, if } \Gamma \mapsto \Delta \pm_2 B \text{is} \mathbf{G}_{Q_1 i Q_2 j}-\text{valid} \\ \Gamma \mapsto \Delta & \text{otherwise;} \end{cases}$$

Correspondingly, there are 32 kinds of R-calculi:

sequents	$\mathbf{R}^{\mathbf{E0E0}}$	$\mathbf{R}^{\mathbf{E0A0}}$	$\mathbf{R}^{\mathbf{A0E0}}$	$\mathbf{R}^{\mathbf{A0A0}}$
	$\mathbf{R}^{\mathbf{E0E1}}$	$\mathbf{R}^{\mathbf{E0A1}}$	$\mathbf{R}^{\mathbf{A0E1}}$	$\mathbf{R}^{\mathbf{A0A1}}$
	$\mathbf{R}^{\mathbf{E1E0}}$	$\mathbf{R}^{\mathbf{E1A0}}$	$\mathbf{R}^{\mathbf{A1E0}}$	$\mathbf{R}^{\mathbf{A1A0}}$
	$\mathbf{R}^{\mathbf{E1E1}}$	$\mathbf{R}^{\mathbf{E1A1}}$	$\mathbf{R}^{\mathbf{A1E1}}$	$\mathbf{R}^{\mathbf{A1A1}}$
co-sequents	$\mathbf{R}_{\mathbf{A1A1}}$	$\mathbf{R}_{\mathbf{A1E1}}$	$\mathbf{R}_{\mathbf{E1A1}}$	$\mathbf{R}_{\mathbf{E1E1}}$
	$\mathbf{R}_{\mathbf{A1A0}}$	$\mathbf{R}_{\mathbf{A1E0}}$	$\mathbf{R}_{\mathbf{E1A0}}$	$\mathbf{R}_{\mathbf{E1E0}}$
	$\mathbf{R}_{\mathbf{A0A1}}$	$\mathbf{R}_{\mathbf{A0E1}}$	$\mathbf{R}_{\mathbf{E0A1}}$	$\mathbf{R}_{\mathbf{E0E1}}$
	$\mathbf{R}_{\mathbf{A0A0}}$	$\mathbf{R}_{\mathbf{A0E0}}$	$\mathbf{R}_{\mathbf{E0A0}}$	$\mathbf{R}_{\mathbf{E0E0}}$

where

$$\mathbf{R}^{\mathbf{E0E1}} = \mathbf{R}^{\mathbf{EE}}, \mathbf{R}^{\mathbf{E0A1}} = \mathbf{R}^{\mathbf{EA}}, \\ \mathbf{R}^{\mathbf{A0E1}} = \mathbf{R}^{\mathbf{AE}}, \mathbf{R}^{\mathbf{A0A1}} = \mathbf{R}^{\mathbf{AA}}.$$

We have the following equivalences:

$$\mathbf{G}^{\mathbf{EE}} = \mathbf{G}^{\mathbf{E0E1}}, \mathbf{G}^{\mathbf{AA}} = \mathbf{G}^{\mathbf{A0A1}}; \\ \mathbf{G}_{\mathbf{AA}} = \mathbf{G}_{\mathbf{A1A0}}, \mathbf{G}_{\mathbf{EE}} = \mathbf{G}_{\mathbf{E1E0}}.$$

About the monotonicity (Cao et al. 2016; Ginsberg 1987; Reiter 1980), we have the following conclusions:

- $\mathbf{G}^{Q_1 i Q_2 j}$ is monotonic in Γ iff $Q_1 = \mathbf{E}$, and monotonic in Δ iff $Q_2 = \mathbf{E}$;
- $\mathbf{G}_{Q_1 i Q_2 j}$ is nonmonotonic in Γ iff $Q_1 = \mathbf{A}$, and nonmonotonic in Δ iff $Q_2 = \mathbf{A}$;

- $\mathbf{R}^{Q_1 i Q_2 j}(+/-)$ is monotonic in Γ iff $Q_1 = \mathbf{E}$, and monotonic in Δ iff $Q_2 = \mathbf{E}$; and
- $\mathbf{R}_{Q_1 i Q_2 j}(+/-)$ is nonmonotonic in Γ iff $Q_1 = \mathbf{E}$, and nonmonotonic in Δ iff $Q_2 = \mathbf{E}$.

5.1 Gentzen Deduction Systems $\mathbf{G}^{\mathbf{E0}*}/\mathbf{G}_{\mathbf{E1}*}$

Let $* \in \{\mathbf{A}1, \mathbf{A}0, \mathbf{E}1, \mathbf{E}0\}$.

Definition 5.1.1 A sequent $\Gamma \Rightarrow \Delta$ is $\mathbf{G}^{\mathbf{E0}*}$-valid, denoted by $\models^{\mathbf{E0}*} \Gamma \Rightarrow \Delta$, if for any assignment v, either $v(A) = 0$ for some $A \in \Gamma$ or

$$\begin{cases} v(B) = 0 \text{ for some} B \in \Delta \text{ if } * = \mathbf{E}0 \\ v(B) = 0 \text{ for every} B \in \Delta \text{ if } * = \mathbf{A}0 \\ v(B) = 1 \text{ for some} B \in \Delta \text{ if } * = \mathbf{E}1 \\ v(B) = 1 \text{ for every} B \in \Delta \text{ if } * = \mathbf{A}1 \end{cases}$$

Definition 5.1.2 A co-sequent $\Gamma \mapsto \Delta$ is $\mathbf{G}_{\mathbf{A}1*}$-valid, denoted by $\models_{\mathbf{A}1*} \Gamma \mapsto \Delta$, if there is an assignment v such that $v(A) = 1$ for every $A \in \Gamma$, and

$$\begin{cases} v(B) = 1 \text{ for every } B \in \Delta \text{ if } * = \mathbf{A}1 \\ v(B) = 1 \text{ for some } B \in \Delta \text{ if } * = \mathbf{E}1 \\ v(B) = 0 \text{ for every } B \in \Delta \text{ if } * = \mathbf{A}0 \\ v(B) = 0 \text{ for some } B \in \Delta \text{ if } * = \mathbf{E}0 \end{cases}$$

5.1.1 Axioms

Proposition 5.1.3 *Let* Γ, Δ *be sets of literals.* $\Gamma \Rightarrow \Delta$ *is* $\mathbf{G}^{\mathbf{E0E1}}$*-valid if and only if* $\text{incon}(\Gamma)$ *or* $\text{val}(\Delta)$ *or* $\Gamma \cap \Delta \neq \emptyset$.

Proof Because $\Gamma \Rightarrow \Delta$ is $\mathbf{G}^{\mathbf{E0E1}}$-valid iff $\Gamma \Rightarrow \Delta$ is $\mathbf{G}^{\mathbf{EE}}$-valid. □

Hence, we have the following axioms:

$$(\mathbb{A}^{\mathbf{E0E1}}) \frac{\text{incon}(\Gamma) \text{ or val}(\Delta) \text{ or d}\Gamma \cap \Delta \neq \emptyset}{\Gamma \Rightarrow \Delta}$$

$$(\mathbb{A}_{\mathbf{A1A0}}) \frac{\text{con}(\Gamma)\&\text{inval}(\Delta)\&\Gamma \cap \Delta = \emptyset}{\Gamma \mapsto \Delta}$$

Proposition 5.1.4 *Let* Γ, Δ *be sets of literals.* $\Gamma \Rightarrow \Delta$ *is* $\mathbf{G}_{\mathbf{E0E0}}$*-valid if and only if* $\text{incon}(\Gamma)$ *or* $\text{val}(\Delta)$ *or* $\Gamma \cap \neg\Delta \neq \emptyset$.

Proof Assume that incon(Γ) or val(Δ) or $\Gamma \cap \neg\Delta \neq \emptyset$, and $\Gamma \cap \neg\Delta \neq \emptyset$. Then, if incon$(\Gamma)$ or val(Δ) then for any assignment v, $v \models^{\mathbf{E}} \Gamma$ or $v \models^{\mathbf{E}} \Delta$; and if $\Gamma \cap \neg\Delta \neq \emptyset$ then there is a literal $l \in \Gamma \cap \neg\Delta$, and for any assignment v, either $v(l) = 0$ or $v(l) = 1$. If $v(l) = 0$ then $v \models^{\mathbf{E}} \Gamma$; and if $v(l) = 1$ then $v(\neg l) = 0, \neg l \in \Delta$, $v \models^{\mathbf{E}} \Delta$.

Assume that con(Γ)&inval(Δ)&$\Gamma \cap \neg\Delta = \emptyset$. We define an assignment v such that for any variable p,

$$v(p) = \begin{cases} 1 \text{ if } p \in \Gamma \text{ or } \neg p \in \Delta \\ 0 \text{ if } \neg p \in \Gamma \text{ or } p \in \Delta \\ 1 \text{ otherwise.} \end{cases}$$

Then, v is well-defined and $v \not\models^{\mathbf{E0E0}} \Gamma \Rightarrow \Delta$. □

Hence, we have the following axioms:

$$(\mathbb{A}^{\mathbf{E0E0}}) \frac{\text{incon}(\Gamma) \text{ or val}(\Delta) \text{ or } \Gamma \cap \neg\Delta \neq \emptyset}{\Gamma \Rightarrow \Delta}$$
$$(\mathbb{A}_{\mathbf{A1A1}}) \frac{\text{con}(\Gamma)\&\text{inval}(\Delta)\&\Gamma \cap \neg\Delta = \emptyset}{\Gamma \mapsto \Delta}$$

Proposition 5.1.5 *Let Γ, Δ be sets of literals.* $\models^{\mathbf{E0A1}} \Gamma \Rightarrow \Delta$ *if and only if*

$$\Delta \subseteq \Gamma \textit{ or } \text{incon}(\Gamma).$$

Proof Assume that $\Delta \subseteq \Gamma$. Then, for any assignment v, either $v \models^{\mathbf{A}} \Delta$, or there is a formula $B \in \Delta$ such that $v(B) = 0$, and by assumption, $B \in \Gamma$, i.e., $v \models^{\mathbf{E}} \Gamma$. Hence, $v \models^{\mathbf{E0A1}} \Gamma \Rightarrow \Delta$.

Conversely, assume that $\Delta \not\subseteq \Gamma$ and con(Γ). There is a literal $l \in \Delta - \Gamma$. Define an assignment v such that for any variable p,

$$v(p) = \begin{cases} 1 \text{ if } p \in \Gamma \\ 0 \text{ if } \neg p \in \Gamma \\ 0 \text{ if } p = l \\ 1 \text{ if } p = \neg l \\ 0 \text{ otherwise.} \end{cases}$$

Then, $v \models_{\mathbf{A}} \Gamma, v \models_{\mathbf{E}} \Delta$, and $v \models_{\mathbf{A1E0}} \Gamma \mapsto \Delta$. □

Hence, we have the following axioms:

$$(\mathbb{A}^{\mathbf{E0A1}}) \frac{\Delta \subseteq \Gamma \text{ or incon}(\Gamma)}{\Gamma \Rightarrow \Delta}$$
$$(\mathbb{A}_{\mathbf{A1E0}}) \frac{\Delta \not\subseteq \Gamma\&\text{con}(\Gamma)}{\Gamma \mapsto \Delta}$$

Proposition 5.1.6 *Let* Γ, Δ *be sets of literals.* $\models^{\mathbf{E0A0}} \Gamma \Rightarrow \Delta$ *if and only if*

$$\neg\Delta \subseteq \Gamma \text{ or } \mathrm{incon}(\Gamma).$$

Proof Assume that $\neg\Delta \subseteq \Gamma$. Then, for any assignment v, either $v \models^{\mathbf{E}} \Delta$, or there is a formula $B \in \Delta$ such that $v(B) = 1$, and by assumption, $\neg B \in \Gamma$ and $v(\neg B) = 0$, i.e., $v \models^{\mathbf{A}} \Gamma$. Hence, $v \models^{\mathbf{E0A0}} \Gamma \Rightarrow \Delta$.

Conversely, assume that $\neg\Delta \not\subseteq \Gamma$ and $\mathrm{con}(\Gamma)$. There is a literal $l \in \neg\Delta - \Gamma$. Define an assignment v such that for any variable p,

$$v(p) = \begin{cases} 1 \text{ if } p \in \Gamma \\ 0 \text{ if } \neg p \in \Gamma \\ 0 \text{ if } p = l \\ 1 \text{ if } p = \neg l \\ 0 \text{ otherwise.} \end{cases}$$

Then, $v \models_{\mathbf{A}} \Gamma$, $v \models_{\mathbf{E}} \Delta$, and $v \models_{\mathbf{A1E0}} \Gamma \mapsto \Delta$. □

Hence, we have the following axioms:

$$(\mathbb{A}^{\mathbf{E0A1}})\ \frac{\neg\Delta \subseteq \Gamma \text{ or } \mathrm{incon}(\Gamma)}{\Gamma \Rightarrow \Delta}$$

$$(\mathbb{A}_{\mathbf{A1E0}})\ \frac{\neg\Delta \not\subseteq \Gamma \& \mathrm{con}(\Gamma)}{\Gamma \mapsto \Delta}$$

5.1.2 Deduction Rules

There are four kinds of deduction rules.

$\mathbf{G}^{L0}$ consists of the following deduction rules:

$$(\neg\neg^L)\ \frac{\Gamma, A \Rightarrow \Delta}{\Gamma, \neg\neg A \Rightarrow \Delta}$$

$$(\wedge^L)\ \frac{\begin{cases}\Gamma, A_1 \Rightarrow \Delta \\ \Gamma, A_2 \Rightarrow \Delta\end{cases}}{\Gamma, A_1 \wedge A_2 \Rightarrow \Delta} \qquad (\vee^L)\ \frac{\left[\begin{array}{l}\Gamma, A_1 \Rightarrow \Delta \\ \Gamma, A_2 \Rightarrow \Delta\end{array}\right.}{\Gamma, A_1 \vee A_2 \Rightarrow \Delta}$$

$$(\neg\wedge^L)\ \frac{\left[\begin{array}{l}\Gamma, \neg A_1 \Rightarrow \Delta \\ \Gamma, \neg A_2 \Rightarrow \Delta\end{array}\right.}{\Gamma, \neg(A_1 \vee A_2) \Rightarrow \Delta} \qquad (\neg\vee^L)\ \frac{\begin{cases}\Gamma, \neg A_1 \Rightarrow \Delta \\ \Gamma, \neg A_2 \Rightarrow \Delta\end{cases}}{\Gamma, \neg(A_1 \vee A_2) \Rightarrow \Delta}$$

$\mathbf{G}^{R0}$ consists of the following deduction rules:

$$(\neg\neg^R)\ \frac{\Gamma \Rightarrow B, \Delta}{\Gamma \Rightarrow \neg\neg B, \Delta}$$

$$(\wedge^R)\ \frac{\left\{\begin{array}{l}\Gamma \Rightarrow B_1, \Delta\\ \Gamma \Rightarrow B_2, \Delta\end{array}\right.}{\Gamma \Rightarrow B_1 \wedge B_2, \Delta} \qquad (\vee^R)\ \frac{\left[\begin{array}{l}\Gamma \Rightarrow B_1, \Delta\\ \Gamma \Rightarrow B_2, \Delta\end{array}\right.}{\Gamma \Rightarrow B_1 \vee B_2, \Delta}$$

$$(\neg\wedge^R)\ \frac{\left[\begin{array}{l}\Gamma \Rightarrow \neg B_1, \Delta\\ \Gamma \Rightarrow \neg B_2, \Delta\end{array}\right.}{\Gamma \Rightarrow \neg(B_1 \wedge B_2), \Delta} \qquad (\neg\vee^R)\ \frac{\left\{\begin{array}{l}\Gamma \Rightarrow \neg B_1, \Delta\\ \Gamma \Rightarrow \neg B_2, \Delta\end{array}\right.}{\Gamma \Rightarrow \neg(B_1 \vee B_2), \Delta}$$

$\mathbf{G}^{R1}$ consists of the following deduction rules:

$$(\neg\neg^R)\ \frac{\Gamma \Rightarrow B, \Delta}{\Gamma \Rightarrow \neg\neg B, \Delta}$$

$$(\wedge^R)\ \frac{\left[\begin{array}{l}\Gamma \Rightarrow B_1, \Delta\\ \Gamma \Rightarrow B_2, \Delta\end{array}\right.}{\Gamma \Rightarrow B_1 \wedge B_2, \Delta} \qquad (\vee^R)\ \frac{\left\{\begin{array}{l}\Gamma \Rightarrow B_1, \Delta\\ \Gamma \Rightarrow B_2, \Delta\end{array}\right.}{\Gamma \Rightarrow B_1 \vee B_2, \Delta}$$

$$(\neg\wedge^R)\ \frac{\left\{\begin{array}{l}\Gamma \Rightarrow \neg B_1, \Delta\\ \Gamma \Rightarrow \neg B_2, \Delta\end{array}\right.}{\Gamma \Rightarrow \neg(B_1 \wedge B_2), \Delta} \qquad (\neg\vee^R)\ \frac{\left[\begin{array}{l}\Gamma \Rightarrow \neg B_1, \Delta\\ \Gamma \Rightarrow \neg B_2, \Delta\end{array}\right.}{\Gamma \Rightarrow \neg(B_1 \vee B_2), \Delta}$$

$\mathbf{G}_{L1}$ consists of the following deduction rules:

$$(\neg\neg^L)\ \frac{\Gamma, A \mapsto \Delta}{\Gamma, \neg\neg A \mapsto \Delta}$$

$$(\wedge^L)\ \frac{\left[\begin{array}{l}\Gamma, A_1 \mapsto \Delta\\ \Gamma, A_2 \mapsto \Delta\end{array}\right.}{\Gamma, A_1 \wedge A_2 \mapsto \Delta} \qquad (\vee^L)\ \frac{\left\{\begin{array}{l}\Gamma, A_1 \mapsto \Delta\\ \Gamma, A_2 \mapsto \Delta\end{array}\right.}{\Gamma, A_1 \vee A_2 \mapsto \Delta}$$

$$(\neg\wedge^L)\ \frac{\left\{\begin{array}{l}\Gamma, \neg A_1 \mapsto \Delta\\ \Gamma, \neg A_2 \mapsto \Delta\end{array}\right.}{\Gamma, \neg(A_1 \vee A_2) \mapsto \Delta} \qquad (\neg\vee^L)\ \frac{\left[\begin{array}{l}\Gamma, \neg A_1 \mapsto \Delta\\ \Gamma, \neg A_2 \mapsto \Delta\end{array}\right.}{\Gamma, \neg(A_1 \vee A_2) \mapsto \Delta}$$

$\mathbf{G}_{R0}$ consists of the following deduction rules:

$$(\neg\neg^R)\ \frac{\Gamma \mapsto B, \Delta}{\Gamma \Rightarrow \neg\neg B, \Delta}$$

$$(\wedge^R)\ \frac{\left\{\begin{array}{l}\Gamma \mapsto B_1, \Delta\\ \Gamma \mapsto B_2, \Delta\end{array}\right.}{\Gamma \mapsto B_1 \wedge B_2, \Delta} \qquad (\vee^R)\ \frac{\left[\begin{array}{l}\Gamma \mapsto B_1, \Delta\\ \Gamma \mapsto B_2, \Delta\end{array}\right.}{\Gamma \mapsto B_1 \vee B_2, \Delta}$$

$$(\neg\wedge^R)\ \frac{\left[\begin{array}{l}\Gamma \mapsto \neg B_1, \Delta\\ \Gamma \mapsto \neg B_2, \Delta\end{array}\right.}{\Gamma \mapsto \neg(B_1 \wedge B_2), \Delta} \qquad (\neg\vee^R)\ \frac{\left\{\begin{array}{l}\Gamma \mapsto \neg B_1, \Delta\\ \Gamma \mapsto \neg B_2, \Delta\end{array}\right.}{\Gamma \mapsto \neg(B_1 \vee B_2), \Delta}$$

$\mathbf{G}_{R1}$ consists of the following deduction rules:

$$(\neg\neg^R)\ \frac{\Gamma \mapsto B, \Delta}{\Gamma \Rightarrow \neg\neg B, \Delta}$$

$$(\wedge^R)\ \frac{\left[\begin{array}{l}\Gamma \mapsto B_1, \Delta \\ \Gamma \mapsto B_2, \Delta\end{array}\right.}{\Gamma \mapsto B_1 \wedge B_2, \Delta} \qquad (\vee^R)\ \frac{\left\{\begin{array}{l}\Gamma \mapsto B_1, \Delta \\ \Gamma \mapsto B_2, \Delta\end{array}\right.}{\Gamma \mapsto B_1 \vee B_2, \Delta}$$

$$(\neg\wedge^R)\ \frac{\left\{\begin{array}{l}\Gamma \mapsto \neg B_1, \Delta \\ \Gamma \mapsto \neg B_2, \Delta\end{array}\right.}{\Gamma \mapsto \neg(B_1 \wedge B_2), \Delta} \qquad (\neg\vee^R)\ \frac{\left[\begin{array}{l}\Gamma \mapsto \neg B_1, \Delta \\ \Gamma \mapsto \neg B_2, \Delta\end{array}\right.}{\Gamma \mapsto \neg(B_1 \vee B_2), \Delta}$$

It is easy to see that

$$\begin{array}{l}\mathbf{G}^{L1} = \mathbf{G}_{L1}\ \mathbf{G}^{L0} = \mathbf{G}_{L0} \\ \mathbf{G}^{R1} = \mathbf{G}_{R1}\ \mathbf{G}^{R0} = \mathbf{G}_{R0}.\end{array}$$

5.1.3 Deduction Systems

Let

$$\begin{array}{l}\mathbf{G}^{\mathbf{E0E0}} = \mathbb{A}^{\mathbf{E0E0}} + \mathbf{G}^{L0} + \mathbf{G}^{R0} \\ \mathbf{G}^{\mathbf{E0E1}} = \mathbb{A}^{\mathbf{E0E1}} + \mathbf{G}^{L0} + \mathbf{G}^{R1} \\ \mathbf{G}^{\mathbf{E0A0}} = \mathbb{A}^{\mathbf{E0A0}} + \mathbf{G}^{L0} + \mathbf{G}^{R0} \\ \mathbf{G}^{\mathbf{E0A1}} = \mathbb{A}^{\mathbf{E0A1}} + \mathbf{G}^{L0} + \mathbf{G}^{R1}\end{array}$$

and

$$\begin{array}{l}\mathbf{G}_{\mathbf{A1E0}} = \mathbb{A}_{\mathbf{A1E0}} + \mathbf{G}_{L1} + \mathbf{G}_{R0} \\ \mathbf{G}_{\mathbf{A1E1}} = \mathbb{A}_{\mathbf{A1E1}} + \mathbf{G}_{L1} + \mathbf{G}_{R1} \\ \mathbf{G}_{\mathbf{A1A0}} = \mathbb{A}_{\mathbf{A1A0}} + \mathbf{G}_{L1} + \mathbf{G}_{R0} \\ \mathbf{G}_{\mathbf{A1A1}} = \mathbb{A}_{\mathbf{A1A1}} + \mathbf{G}_{L1} + \mathbf{G}_{R1}\end{array}$$

Let $* \in \{\mathbf{E}0, \mathbf{E}1, \mathbf{A}0, \mathbf{A}1\}$.

Definition 5.1.7 A sequent $\Gamma \Rightarrow \Delta$ is provable in $\mathbf{G}^{\mathbf{E0}*}$, denoted by $\vdash^{\mathbf{E0}*} \Gamma \Rightarrow \Delta$, if there is a sequence $\{\Gamma_1 \Rightarrow \Delta_1, ..., \Gamma_n \Rightarrow \Delta_n\}$ of sequents such that $\Gamma_n \Rightarrow \Delta_n = \Gamma \Rightarrow \Delta$, and for each $1 \le i \le n$, $\Gamma_i \Rightarrow \Delta_i$ is an axiom or is deduced from the previous sequents by one of the deduction rules in $\mathbf{G}^{\mathbf{E0}*}$.

Definition 5.1.8 A co-sequent $\Gamma \mapsto \Delta$ is provable in $\mathbf{G}_{\mathbf{E0}*}$, denoted by $\vdash_{\mathbf{E0}*} \Gamma \mapsto \Delta$ if there is a sequence $\{\Gamma_1 \mapsto \Delta_1, ..., \Gamma_n \mapsto \Delta_n\}$ of co-sequents such that $\Gamma_n \mapsto \Delta_n = \Gamma \mapsto \Delta$, and for each $1 \le i \le n$, $\Gamma_i \mapsto \Delta_i$ is an axiom or is deduced from the previous co-sequents by one of the deduction rules in $\mathbf{G}_{\mathbf{E0}*}$.

Theorem 5.1.9 (*Soundness and completeness theorem) For any sequent* $\Gamma \Rightarrow \Delta$ *and co-sequent* $\Gamma \mapsto \Delta$,

$$\vdash^{\mathbf{E0}*} \Gamma \Rightarrow \Delta \textit{iff} \models^{\mathbf{A1}*} \Gamma \Rightarrow \Delta$$
$$\vdash_{\mathbf{A1}*} \Gamma \mapsto \Delta \textit{iff} \models_{\mathbf{A1}*} \Gamma \mapsto \Delta.$$

□

Notice that

	negation
$\mathbf{G}^{\mathbf{E0}*}$	$\mathbf{G}_{\mathbf{A1}\overline{*}}$,
$\mathbf{G}_{\mathbf{A1}*}$	$\mathbf{G}^{\mathbf{E0}\overline{*}}$,

where $\overline{*} = \begin{cases} \mathbf{A1} \text{ if } * = \mathbf{E0} \\ \mathbf{A0} \text{ if } * = \mathbf{E1} \\ \mathbf{E0} \text{ if } * = \mathbf{A1} \\ \mathbf{E1} \text{ if } * = \mathbf{A0} \end{cases}$

5.1.4 Monotonicity of $\mathbf{G}^{\mathbf{E0}*}$ and $\mathbf{G}_{\mathbf{A1}*}$

Definition 5.1.10 A deduction system $\overline{\mathbf{X}}$ is monotonic in Γ if for any formula sets Γ, Γ' and Δ,

$$\vdash^{\mathbf{X}} \Gamma \Rightarrow \Delta \& \Gamma' \supseteq \Gamma \text{imply} \vdash^{\mathbf{X}} \Gamma' \Rightarrow \Delta.$$

$\mathbf{X}$ is monotonic in Δ if for any formula sets Γ, Δ and Δ',

$$\vdash^{\mathbf{X}} \Gamma \Rightarrow \Delta \& \Delta' \supseteq \Delta \text{imply} \vdash^{\mathbf{X}} \Gamma \Rightarrow \Delta'.$$

Theorem 5.1.11 (*Monotonicity theorem*) $\mathbf{G}^{\mathbf{E0E}i}$ *is monotonic in both* Γ *and* Δ, *that is, for any formula sets* Γ, Γ', Δ *and* Δ',

$$\Gamma \subseteq \Gamma' \& \vdash^{\mathbf{E0E}i} \Gamma \Rightarrow \Delta \textit{imply} \vdash^{\mathbf{E0E}i} \Gamma' \Rightarrow \Delta;$$
$$\Delta \subseteq \Delta' \& \vdash^{\mathbf{E0E}i} \Gamma \Rightarrow \Delta \textit{imply} \vdash^{\mathbf{E0E}i} \Gamma \Rightarrow \Delta';$$

and $\mathbf{G}^{\mathbf{E0A}i}$ *is monotonic in* Γ *and nonmonotonic in* Δ, *that is, for any formula sets* Γ, Γ', Δ *and* Δ',

$$\Gamma \subseteq \Gamma' \& \vdash^{\mathbf{E0A}i} \Gamma \Rightarrow \Delta \textit{imply} \vdash^{\mathbf{E0A}i} \Gamma' \Rightarrow \Delta;$$
$$\Delta \subseteq \Delta' \& \vdash^{\mathbf{E0A}i} \Gamma \Rightarrow \Delta \textit{may not imply} \vdash^{\mathbf{E0A}i} \Gamma \Rightarrow \Delta'.$$

Definition 5.1.12 A deduction system $\underline{\mathbf{X}}$ is nonmonotonic in Γ if for any formula sets Γ, Γ' and Δ,

$$\vdash_{\mathbf{X}} \Gamma \mapsto \Delta \& \Gamma' \supseteq \Gamma \text{may not imply} \vdash_{\mathbf{X}} \Gamma' \mapsto \Delta.$$

$\mathbf{X}$ is nonmonotonic in Δ if for any formula sets Γ, Δ and Δ',

$$\vdash_{\mathbf{X}} \Gamma \mapsto \Delta \& \Delta' \supseteq \Delta \text{may not imply} \vdash_{\mathbf{X}} \Gamma \mapsto \Delta'.$$

Theorem 5.1.13 (*Nonmonotonicity theorem*) *(i)* $\mathbf{G}_{\mathbf{A1E}i}$ *is nonmonotonic in* Γ *and and monotonic in* Δ, *that is, for any formula sets* Γ, Γ', Δ *and* Δ',

$$\Gamma \subseteq \Gamma' \& \vdash_{\mathbf{G}_{\mathbf{A1E}i}} \Gamma \mapsto \Delta \textit{ may not imply} \vdash_{\mathbf{G}_{\mathbf{A1E}i}} \Gamma' \mapsto \Delta;$$
$$\Delta \subseteq \Delta' \& \vdash_{\mathbf{G}_{\mathbf{A1E}i}} \Gamma \mapsto \Delta \textit{ imply} \vdash_{\mathbf{G}_{\mathbf{A1E}i}} \Gamma \mapsto \Delta'.$$

(ii) $\mathbf{G}_{\mathbf{A1A}i}$ *is nonmonotonic in both* Γ *and* Δ, *that is, for any formula sets* Γ, Γ', Δ *and* Δ',

$$\Gamma \subseteq \Gamma' \& \vdash_{\mathbf{G}_{\mathbf{A1A}i}} \Gamma \mapsto \Delta \textit{ may not imply} \vdash_{\mathbf{G}_{\mathbf{A1A}i}} \Gamma' \mapsto \Delta;$$
$$\Delta \subseteq \Delta' \& \vdash_{\mathbf{G}_{\mathbf{A1A}i}} \Gamma \mapsto \Delta \textit{ may not imply} \vdash_{\mathbf{G}_{\mathbf{A1A}i}} \Gamma \mapsto \Delta'.$$

Proof We prove that the axiom is nonmonotonic and each deduction rule preserves the monotonicity.

Assume that $\mathrm{con}(\Gamma)$, $\mathrm{inval}(\Delta)$ and $\Gamma \cap \Delta = \emptyset$. There is a superset $\Gamma' \supseteq \Gamma$ such that Γ' is inconsistent; and there is a superset $\Delta' \supseteq \Delta$ such that Δ' is inconsistent; and there are supersets $\Gamma' \supseteq \Gamma$ and $\Delta' \supseteq \Delta$ such that $\Gamma' \cap \Delta' \neq \emptyset$. Hence, $\mathbf{G}_{\mathbf{A1A}i}$ is nonmonotonic in both Γ and Δ.

To show that $(\wedge^L)$ preserves the monotonicity of Γ, assume that $\Gamma, A_1 \mapsto \Delta$ and $\Gamma, A_1 \mapsto \Delta$ are monotonic with respect to Γ. By $(\wedge^L)$, from $\Gamma, A_1 \mapsto \Delta$ and $\Gamma, A_2 \mapsto \Delta$, we infer $\Gamma, A_1 \wedge A_2 \mapsto \Delta$. Then, for any $\Gamma' \supseteq \Gamma, \Gamma', A_1 \mapsto \Delta$ and $\Gamma, A_2 \mapsto \Delta$ follows; and by $(\wedge^L)$, from $\Gamma', A_1 \mapsto \Delta$ and $\Gamma', A_2 \mapsto \Delta$, we infer $\Gamma', A_1 \wedge A_2 \mapsto \Delta$. Hence, $\Gamma, A_1 \wedge A_2 \mapsto \Delta$ implies $\Gamma', A_1 \wedge A_2 \mapsto \Delta$, that is, $\Gamma, A_1 \wedge A_2 \mapsto \Delta$ is monotonic with respect to Γ.

To show that $(\wedge^L)$ preserves the nonmonotonicity of Γ, assume that $\Gamma, A_1 \mapsto \Delta$ and $\Gamma, A_2 \mapsto \Delta$ are nonmonotonic with respect to Γ. By $(\wedge^L)$, from $\Gamma, A_1 \mapsto \Delta$ and $\Gamma, A_2 \mapsto \Delta$, we infer $\Gamma, A_1 \wedge A_2 \mapsto \Delta$. Then, for some $\Gamma' \supseteq \Gamma$,

$$\Gamma, A_1 \mapsto \Delta \text{may not imply} \Gamma', A_1 \mapsto \Delta;$$
$$\Gamma, A_2 \mapsto \Delta \text{may not imply} \Gamma', A_2 \mapsto \Delta;$$

and by $(\wedge^L)$, $\Gamma, A_1 \wedge A_2 \mapsto \Delta$ may not imply $\Gamma', A_1 \wedge A_2 \mapsto \Delta$, that is, $\Gamma, A_1 \wedge A_2 \mapsto \Delta$ is nonmonotonic with respect to Γ.

To show that $(\wedge^L)$ preserves the monotonicity of Δ, assume that $\Gamma, A_1 \mapsto \Delta$ and $\Gamma, A_2 \mapsto \Delta$ are monotonic with respect to Δ. By $(\wedge^L)$, from $\Gamma, A_1, \mapsto \Delta$ and $\Gamma, A_2 \mapsto \Delta$, we infer $\Gamma, A_1 \wedge A_2 \mapsto \Delta$. Then, for any $\Delta' \supseteq \Delta, \Gamma, A_1 \mapsto \Delta'$ and $\Gamma, A_2 \mapsto \Delta'$ follow; and by $(\wedge^L)$, from $\Gamma, A_1 \mapsto \Delta'$ and $\Gamma, A_2 \mapsto \Delta'$, we infer $\Gamma, A_1 \wedge A_2 \mapsto \Delta'$. Hence, $\Gamma, A_1 \wedge A_2 \mapsto \Delta$ implies $\Gamma, A_1 \wedge A_2 \mapsto \Delta'$, that is, $\Gamma, A_1 \wedge A_2 \mapsto \Delta$ is monotonic with respect to Δ.

To show that $(\wedge^L)$ preserves the nonmonotonicity of Δ, assume that $\Gamma, A_1 \mapsto \Delta$ and $\Gamma, A_2 \mapsto \Delta$ are nonmonotonic with respect to Δ'. By $(\wedge^L)$, from $\Gamma, A_1 \mapsto \Delta$ and $\Gamma, A_2 \mapsto \Delta$, we infer $\Gamma, A_1 \wedge A_2 \mapsto \Delta$. Then, for some $\Delta' \supseteq \Delta$,

$$\Gamma, A_1 \mapsto \Delta \text{may not imply} \Gamma, A_1 \mapsto \Delta';$$
$$\Gamma, A_2 \mapsto \Delta \text{may not imply} \Gamma, A_2 \mapsto \Delta';$$

and by $(\wedge^L)$, $\Gamma, A_1 \wedge A_2 \mapsto \Delta$ may not imply $\Gamma, A_1 \wedge A_2 \mapsto \Delta'$, that is, $\Gamma, A_1 \wedge A_2 \mapsto \Delta$ is nonmonotonic with respect to Δ.

Similar to show that other deduction rules preserves the monotonicity and non-monotonicity with respect to Γ and Δ. □

Generally we have the following table for the monotonicity of Getzen deduction systems $\mathbf{G}^{\mathbf{E}0*}/\mathbf{G}_{\mathbf{A}1*}$:

System	Axiom	mono Γ	mono Δ
$\mathbf{G}^{\mathbf{E0E1}}$	$\text{incon}(\Gamma)$ or $\text{val}(\Delta)$ or $\Gamma \cap \Delta \neq \emptyset$	Y	Y
$\mathbf{G}_{\mathbf{A1A0}}$	$\text{con}(\Gamma)\&\text{inval}(\Delta)\&\Gamma \cap \Delta = \emptyset$	N	N
$\mathbf{G}^{\mathbf{E0E0}}$	$\text{incon}(\Gamma)$ or $\text{val}(\Delta)$ or $\Gamma \cap \neg\Delta \neq \emptyset$	Y	Y
$\mathbf{G}_{\mathbf{A1A1}}$	$\text{con}(\Gamma)\&\text{inval}(\Delta)\&\Gamma \cap \neg\Delta = \emptyset$	N	N
$\mathbf{G}^{\mathbf{E0A1}}$	$\Delta \subseteq \Gamma$ or $\text{incon}(\Gamma)$	Y	N
$\mathbf{G}_{\mathbf{A1E0}}$	$\Delta \nsubseteq \Gamma\&\text{con}(\Gamma)$	N	Y
$\mathbf{G}^{\mathbf{E0A0}}$	$\neg\Delta \subseteq \Gamma$ or $\text{incon}(\Gamma)$	Y	N
$\mathbf{G}_{\mathbf{A1E1}}$	$\neg\Delta \nsubseteq \Gamma\&\text{con}(\Gamma)$	N	Y.

5.2 Gentzen Deduction Systems $\mathbf{G}^{Q_1 i\, Q_2 j}$

Definition 5.2.1 A sequent $\Gamma \Rightarrow \Delta$ is $\mathbf{G}^{Q_1 i Q_2 j}$-valid, denoted by $\models^{Q_1 i Q_2 j} \Gamma \Rightarrow \Delta$, if for any assignment v, either

$$\begin{cases} v(A) = 1 \text{ for every} A \in \Gamma \text{ if } Q_1 i = \mathbf{A}1 \\ v(A) = 0 \text{ for every} A \in \Gamma \text{ if } Q_1 i = \mathbf{A}0 \\ v(A) = 1 \text{ for some} A \in \Gamma \text{ if } Q_1 i = \mathbf{E}1 \\ v(A) = 0 \text{ for some} A \in \Gamma \text{ if } Q_1 i = \mathbf{E}0 \end{cases}$$

or

$$\begin{cases} v(B) = 1 \text{ for every} B \in \Delta \text{ if } Q_2 j = \mathbf{A}1 \\ v(B) = 0 \text{ for every} B \in \Delta \text{ if } Q_2 j = \mathbf{A}0 \\ v(B) = 1 \text{ for some} B \in \Delta \text{ if } Q_2 j = \mathbf{E}1 \\ v(B) = 0 \text{ for some} B \in \Delta \text{ if } Q_2 j = \mathbf{E}0 \end{cases}$$

5.2.1 Axioms

The axioms are classified into four classes: $\mathbf{E}0Q_2 j, \mathbf{E}1Q_2 j, \mathbf{A}0Q_2 j, \mathbf{A}1Q_2 j$.

- $\mathbf{E}0Q_2 j$:

Proposition 5.2.2 *Let* Γ, Δ *be sets of literals.* $\Gamma \Rightarrow \Delta$ *is* $\mathbf{G}^{\mathbf{E0E0}}$*-valid if and only if* $\text{incon}(\Gamma)$ *or* $\text{val}(\Delta)$ *or* $\Gamma \cap \neg\Delta \neq \emptyset$.

Proof Same as Proposition 5.1.5. □

Hence, we have the following axioms:

$$(\mathrm{A}^{\mathbf{E0E0}})\ \frac{\text{incon}(\Gamma) \text{ or val}(\Delta) \text{ or } \Gamma \cap \neg\Delta \neq \emptyset}{\Gamma \Rightarrow \Delta}$$

Proposition 5.2.3 *Let* Γ, Δ *be sets of literals.* $\Gamma \Rightarrow \Delta$ *is* $\mathbf{G}^{\mathbf{E0E1}}$*-valid if and only if* incon(Γ) *or* val(Δ) *or* $\Gamma \cap \Delta \neq \emptyset$.

Proof $\Gamma \Rightarrow \Delta$ is $\mathbf{G}^{\mathbf{E0E1}}$-valid iff $\Gamma \Rightarrow \Delta$ is $\mathbf{G}^{\mathbf{EE}}$-valid. □

Hence, we have the following axiom:

$$(\mathrm{A}^{\mathbf{E0E1}})\ \frac{\text{incon}(\Gamma)\text{orval}(\Delta) \text{ or } \Gamma \cap \Delta \neq \emptyset}{\Gamma \Rightarrow \Delta}$$

Proposition 5.2.4 *Let* Γ, Δ *be sets of literals.* $\Gamma \Rightarrow \Delta$ *is* $\mathbf{G}^{\mathbf{E0A1}}$*-valid if and only if* $\Delta \subseteq \Gamma$ *or* incon(Γ).

Proof $\Gamma \Rightarrow \Delta$ is $\mathbf{G}^{\mathbf{E0A1}}$-valid iff $\Gamma \Rightarrow \Delta$ is $\mathbf{G}^{\mathbf{EA}}$-valid. □

Hence, we have the following axiom:

$$(\mathrm{A}^{\mathbf{E0A1}})\ \frac{\Delta \subseteq \Gamma \text{ or incon}(\Gamma)}{\Gamma \Rightarrow \Delta}$$

Proposition 5.2.5 *Let* Γ, Δ *be sets of literals.* $\Gamma \Rightarrow \Delta$ *is* $\mathbf{G}^{\mathbf{E0A0}}$*-valid if and only if* $\neg\Delta \subseteq \Gamma$ *or* incon(Γ).

Proof Same as Proposition 5.1.8. □

Hence, we have the following axiom:

$$(\mathrm{A}^{\mathbf{E0A0}})\ \frac{\neg\Delta \subseteq \Gamma \text{ or incon}(\Gamma)}{\Gamma \Rightarrow \Delta}$$

- $\mathbf{E}1Q_2 j$:

Proposition 5.2.6 *Let* Γ, Δ *be sets of literals.* $\Gamma \Rightarrow \Delta$ *is* $\mathbf{G}^{\mathbf{E1E1}}$*-valid if and only if* incon(Γ) *or* val(Δ) *or* $\Gamma \cap \neg\Delta \neq \emptyset$.

Proof $\Gamma \Rightarrow \Delta$ is $\mathbf{G}^{\mathbf{E1E1}}$-valid iff $\neg\Delta \Rightarrow \neg\Gamma$ is $\mathbf{G}^{\mathbf{E0E0}}$-valid. □

Hence, we have the following axiom:

$$(\mathrm{A}^{\mathbf{E1E1}})\ \frac{\text{incon}(\Gamma) \text{ or val}(\Delta) \text{ or } \Gamma \cap \neg\Delta \neq \emptyset}{\Gamma \Rightarrow \Delta}$$

Proposition 5.2.7 *Let* Γ, Δ *be sets of literals.* $\Gamma \Rightarrow \Delta$ *is* $\mathbf{G}^{\mathbf{E1E0}}$*-valid if and only if* incon(Γ) *or* val(Δ) *or* $\Gamma \cap \Delta \neq \emptyset$.

Proof $\Gamma \Rightarrow \Delta$ is $\mathbf{G}^{\mathbf{E1E0}}$-valid iff $\neg\Delta \Rightarrow \neg\Gamma$ is $\mathbf{G}^{\mathbf{E0E1}}$-valid. □

Hence, we have the following axiom:

$$(\mathrm{A}^{\mathbf{E1E0}}) \frac{\mathrm{incon}(\Gamma) \text{ or } \mathrm{val}(\Delta) \text{ or } \Gamma \cap \Delta \neq \emptyset}{\Gamma \Rightarrow \Delta}$$

Proposition 5.2.8 *Let* Γ, Δ *be sets of literals.* $\models^{\mathbf{E1A1}} \Gamma \Rightarrow \Delta$ *if and only if*

$$\neg\Delta \subseteq \Gamma \text{ or } \mathrm{incon}(\Gamma).$$

Proof Assume that $\neg\Delta \subseteq \Gamma$. Then, for any assignment v, either $v \models^{\mathbf{A}} \Delta$, or there is a formula $B \in \Delta$ such that $v(B) = 0$, and by assumption, $\neg B \in \Gamma$ and $v(\neg B) = 1$, i.e., $v \models^{\mathbf{E}} \Gamma$. Hence, $v \models^{\mathbf{E1A1}} \Gamma \Rightarrow \Delta$.

Conversely, assume that $\neg\Delta \nsubseteq \Gamma$ and $\mathrm{con}(\Gamma)$. There is a literal $l \in \neg\Delta - \Gamma$. Define an assignment v such that for any variable p,

$$v(p) = \begin{cases} 1 \text{ if } p \in \Gamma \\ 0 \text{ if } \neg p \in \Gamma \\ 1 \text{ if } p = l \\ 0 \text{ if } p = \neg l \\ 0 \text{ otherwise.} \end{cases}$$

Then, $v \models_{\mathbf{A}} \Gamma$, $v \models_{\mathbf{E}} \Delta$, and $v \models_{\mathbf{A1E0}} \Gamma \mapsto \Delta$. □

Hence, we have the following axioms:

$$(\mathrm{A}^{\mathbf{E1A1}}) \frac{\neg\Delta \subseteq \Gamma \text{ or } \mathrm{incon}(\Gamma)}{\Gamma \Rightarrow \Delta}$$

Proposition 5.2.9 *Let* Γ, Δ *be sets of literals.* $\models^{\mathbf{E1A0}} \Gamma \Rightarrow \Delta$ *if and only if*

$$\Delta \subseteq \Gamma \text{ or } \mathrm{incon}(\Gamma).$$

Proof Assume that $\Delta \subseteq \Gamma$. Then, for any assignment v, either $v \models \Delta$, or there is a formula $B \in \Delta$ such that $v(B) = 1$, and by assumption, $B \in \Gamma$, i.e., $v \models \Gamma$. Hence, $v \models^{\mathbf{E1A0}} \Gamma \Rightarrow \Delta$.

Conversely, assume that $\Delta \nsubseteq \Gamma$ and $\mathrm{con}(\Gamma)$. There is a literal $l \in \Delta - \Gamma$. Define an assignment v such that for any variable p,

$$v(p) = \begin{cases} 0 \text{ if } p \in \Gamma \\ 1 \text{ if } \neg p \in \Gamma \\ 1 \text{ if } p = l \\ 0 \text{ if } p = \neg l \\ 0 \text{ otherwise.} \end{cases}$$

Then, $v \models \Gamma$ and $v \models_{\mathbf{A0E1}} \Gamma \mapsto \Delta$. □

Hence, we have the following axioms:

$$(\mathbb{A}^{\mathbf{E1A0}})\ \frac{\Delta \subseteq \Gamma \text{ or incon}(\Gamma)}{\Gamma \Rightarrow \Delta}$$

- $\mathbf{A}0Q_2 j$:

Proposition 5.2.10 *Let* Γ, Δ *be sets of literals.* $\Gamma \Rightarrow \Delta$ *is* $\mathbf{G}^{\mathbf{A0E0}}$*-valid if and only if* $\text{val}(\Delta)$ *or* $\neg\Gamma \subseteq \Delta$.

Proof Assume that $\neg\Gamma \subseteq \Delta$ or $\text{val}(\Delta)$. Then, if $\text{val}(\Delta)$ then for any assignment v, $v \models^{\mathbf{E}} \Delta$; and if $\neg\Gamma \subseteq \Delta$ then for any assignment v, either $v(l) = 0$ for every literal $l \in \Gamma$ (hence, $v \models^{\mathbf{A}} \Gamma$) or $v(l) = 1$ for some $l \in \Gamma$, and $\neg l \in \Delta$, i.e., $v(\neg l) = 0$ and hence, $v \models^{\mathbf{E}} \Delta$, that is, $\models^{\mathbf{A0E0}} \Gamma \Rightarrow \Delta$.

Conversely, assume that $\neg\Gamma \nsubseteq \Delta$ and $\text{inval}(\Delta)$. There is a literal $l \in \neg\Gamma - \Delta$. Define an assignment v such that for any variable p,

$$v(p) = \begin{cases} 1 \text{ if } p \in \Delta \\ 0 \text{ if } \neg p \in \Delta \\ 0 \text{ if } p = l \\ 1 \text{ if } p = \neg l \\ 0 \text{ otherwise.} \end{cases}$$

Then, $v \models_{\mathbf{E}} \Gamma$, $v \models_{\mathbf{A}} \Delta$, and $v \models_{\mathbf{E1A1}} \Gamma \mapsto \Delta$. □

Hence, we have the following axioms:

$$(\mathbb{A}^{\mathbf{A0E0}})\ \frac{\text{val}(\Delta) \text{ or } \neg\Gamma \subseteq \Delta}{\Gamma \Rightarrow \Delta}$$

Proposition 5.2.11 *Let* Γ, Δ *be sets of literals.* $\Gamma \Rightarrow \Delta$ *is* $\mathbf{G}^{\mathbf{A0E1}}$*-valid if and only if* $\text{val}(\Delta)$ *or* $\Gamma \subseteq \Delta$.

Proof Assume that $\Gamma \subseteq \Delta$ or $\text{val}(\Delta)$. Then, if $\text{val}(\Delta)$ then for any assignment v, $v \models^{\mathbf{E}} \Delta$; and if $\Gamma \subseteq \Delta$ then for any assignment v, either $v(l) = 0$ for every literal $l \in \Gamma$ (hence, $v \models^{\mathbf{A}} \Gamma$) or $v(l) = 1$ for some $l \in \Gamma \subseteq \Delta$, $v \models^{\mathbf{E}} \Delta$, that is, $\models^{\mathbf{A0E1}} \Gamma \Rightarrow \Delta$.

Conversely, assume that $\Gamma \nsubseteq \Delta$ and $\text{inval}(\Delta)$. There is a literal $l \in \Gamma - \Delta$. Define an assignment v such that for any variable p,

$$v(p) = \begin{cases} 0 \text{ if } p \in \Delta \\ 1 \text{ if } \neg p \in \Delta \\ 1 \text{ if } p = l \\ 0 \text{ if } \neg p = l \\ 0 \text{ otherwise.} \end{cases}$$

Then, $v \models_{\mathbf{E}} \Gamma$, $v \models_{\mathbf{A}} \Delta$, and $v \models_{\mathbf{E1A0}} \Gamma \mapsto \Delta$. □

Hence, we have the following axioms:

$$(\mathbb{A}^{\mathbf{A0E1}})\ \frac{\Gamma \subseteq \Delta \text{ or val}(\Delta)}{\Gamma \Rightarrow \Delta}$$

Proposition 5.2.12 *Given two sets* Γ, Δ *of literals,* $\models^{\mathbf{A0A0}} \Gamma \Rightarrow \Delta$ *if and only if* $\Gamma = \emptyset$ *or* $\Delta = \emptyset$ *or* $\Gamma = \{l\} = \neg\Delta$.

Proof Assume that $\Gamma = \emptyset$ or $\Delta = \emptyset$ or $\Gamma = \{l\} = \neg\Delta$. Then, for any assignment v, either $v(l) = 0$ for every $l \in \Gamma$, or $v(m) = 0$ for every $m \in \Delta$.

Assume that $\Gamma \neq \emptyset \neq \Delta \& (\Gamma \neq \{l\}$ or $\{l\} \neq \neg\Delta)$. Let $l \neq l' \in \Gamma - \Delta$ and $l \in \Delta$. We define an assignment v such that for any variable p,

$$v(p) = \begin{cases} 1 \text{ if } p = l' \\ 0 \text{ if } p = \neg l' \\ 1 \text{ if } p = l \\ 0 \text{ if } p = \neg l \\ 1 \text{ otherwise.} \end{cases}$$

Then, v is well-defined and $v \models_{\mathbf{E1E1}} \Gamma \mapsto \Delta$. Similar for $l \in \Delta - \Gamma$.

□

Hence, we have the following axioms:

$$(\mathbb{A}^{\mathbf{A0A0}})\ \frac{\Gamma = \emptyset \text{ or } \Delta = \emptyset \text{ or } \Gamma = \{l\} = \neg\Delta}{\Gamma \Rightarrow \Delta}$$

Proposition 5.2.13 *Given two sets* Γ, Δ *of literals,* $\models^{\mathbf{A0A1}} \Gamma \Rightarrow \Delta$ *if and only if* $\Gamma = \emptyset$ *or* $\Delta = \emptyset$ *or* $\Gamma = \{l\} = \Delta$.

Proof $\models^{\mathbf{A0A1}} \Gamma \Rightarrow \Delta$ iff $\models^{\mathbf{AA}} \Gamma \Rightarrow \Delta$. □

Hence, we have the following axioms:

$$(\mathbb{A}^{\mathbf{A0A1}})\ \frac{\Gamma = \emptyset \text{ or } \Delta = \emptyset \text{ or } \Gamma = \{l\} = \Delta}{\Gamma \Rightarrow \Delta}$$

- $\mathbf{A1}Q_2 j$:

Proposition 5.2.14 *Let* Γ, Δ *be sets of literals.* $\Gamma \Rightarrow \Delta$ *is* $\mathbf{G}^{\mathbf{A}1\mathbf{E}0}$*-valid if and only if* $\text{val}(\Delta)$ *or* $\Gamma \subseteq \Delta$.

Proof Assume that $\Gamma \subseteq \Delta$ or $\text{val}(\Delta)$. Then, if $\text{val}(\Delta)$ then for any assignment v, $v \models^{\mathbf{E}} \Delta$; and if $\Gamma \subseteq \Delta$ then for any assignment v, either $v(l) = 1$ for every literal $l \in \Gamma$ or $v(l) = 0$ for some $l \in \Gamma \subseteq \Delta$, that is, $\models^{\mathbf{A}1\mathbf{E}0} \Gamma \Rightarrow \Delta$.

Conversely, assume that $\Gamma \not\subseteq \Delta$ and $\text{inval}(\Delta)$. There is a literal $l \in \Gamma - \Delta$. Define an assignment v such that for any variable p,

$$v(p) = \begin{cases} 0 \text{ if } p = l \\ 1 \text{ if } p = \neg l \\ 1 \text{ if } p \in \Delta \\ 0 \text{ if } \neg p \in \Delta \\ 1 \text{ otherwise.} \end{cases}$$

Then, $v \models_{\mathbf{E}} \Gamma$, $v \models_{\mathbf{A}} \Delta$, and $v \models_{\mathbf{E}0\mathbf{A}1} \Gamma \mapsto \Delta$. □

Hence, we have the following axioms:

$$(\mathbb{A}^{\mathbf{A}1\mathbf{E}0}) \ \frac{\text{val}(\Delta) \text{ or } \Gamma \subseteq \Delta}{\Gamma \Rightarrow \Delta}$$

Proposition 5.2.15 *Let* Γ, Δ *be sets of literals.* $\Gamma \Rightarrow \Delta$ *is* $\mathbf{G}^{\mathbf{A}1\mathbf{E}1}$*-valid if and only if* $\text{val}(\Delta)$ *or* $\neg\Gamma \subseteq \Delta$.

Proof Assume that $\neg\Gamma \subseteq \Delta$ or $\text{val}(\Delta)$. Then, if $\text{val}(\Delta)$ then for any assignment v, $v \models^{\mathbf{E}} \Delta$; and if $\neg\Gamma \subseteq \Delta$ then for any assignment v, either $v(l) = 1$ for every literal $l \in \Gamma$ (hence, $v \models^{\mathbf{A}} \Gamma$), or $v(l) = 0$ for some $l \in \Gamma$, i.e., $v(\neg l) = 1$. Then, $\neg l \in \Delta$ and $v \models^{\mathbf{E}} \Delta$. Hence, $\models^{\mathbf{A}1\mathbf{E}1} \Gamma \Rightarrow \Delta$.

Conversely, assume that $\neg\Gamma \not\subseteq \Delta$ and $\text{inval}(\Delta)$. There is a literal $l \in \neg\Gamma - \Delta$. Define an assignment v such that for any variable p,

$$v(p) = \begin{cases} 1 \text{ if } p = l \\ 0 \text{ if } p = \neg l \\ 0 \text{ if } p \in \Delta \\ 1 \text{ if } \neg p \in \Delta \\ 1 \text{ otherwise.} \end{cases}$$

Then, $v \models_{\mathbf{E}} \Gamma$, $v \models_{\mathbf{A}} \Delta$, and $v \models_{\mathbf{E}0\mathbf{A}0} \Gamma \mapsto \Delta$. □

Hence, we have the following axioms:

$$(\mathbb{A}^{\mathbf{A}1\mathbf{E}1}) \ \frac{\neg\Gamma \subseteq \Delta \text{ or } \text{val}(\Delta)}{\Gamma \Rightarrow \Delta}$$

Proposition 5.2.16 *Given two sets* Γ, Δ *of literals,* $\models^{\mathbf{A1A0}} \Gamma \Rightarrow \Delta$ *if and only if* $\Gamma = \emptyset$ *or* $\Delta = \emptyset$ *or* $\Gamma = \{l\} = \Delta$.

Proof Assume that $\Gamma = \emptyset$ or $\Delta = \emptyset$ or $\Gamma = \{l\} = \Delta$. Then, for any assignment v, either $v(l) = 1$ for every $l \in \Gamma$, or $v(m) = 0$ for every $m \in \Delta$.

Assume that $\Gamma \neq \emptyset \neq \Delta \& (\Gamma \neq \{l\} \neq \Delta)$. Let $l' \in \Gamma - \Delta$ and $l' \neq l \in \Delta$. We define an assignment v such that for any variable p,

$$v(p) = \begin{cases} 1 \text{ if } p = l \\ 0 \text{ if } p = \neg l \\ 0 \text{ if } p = l' \\ 1 \text{ if } p = \neg l' \\ 1 \text{ otherwise.} \end{cases}$$

Then, v is well-defined and $v \models_{\mathbf{E0E1}} \Gamma \mapsto \Delta$. Similar for $l \in \Delta - \Gamma$.

□

Hence, we have the following axioms:

$$(\mathrm{A}^{\mathbf{A1A0}}) \frac{\Gamma = \emptyset \text{ or } \Delta = \emptyset \text{ or } \Gamma = \{l\} = \Delta}{\Gamma \Rightarrow \Delta}$$

Proposition 5.2.17 *Given two sets* Γ, Δ *of literals,* $\models^{\mathbf{A1A1}} \Gamma \Rightarrow \Delta$ *if and only if* $\Gamma = \emptyset$ *or* $\Delta = \emptyset$ *or* $\Gamma = \{l\} = \neg\Delta$.

Proof Assume that $\Gamma = \emptyset$ or $\Delta = \emptyset$ or $\Gamma = \{l\} = \neg\Delta$. Then, for any assignment v, either $v(l) = 1$ for every $l \in \Gamma$, or $v(m) = 1$ for every $m \in \Delta$.

Assume that $\Gamma \neq \emptyset \neq \Delta \& (\Gamma \neq \{l\} \neq \neg\Delta)$. Let $l \neq l' \in \Gamma - \neg\Delta$ and $l \in \Delta$. We define an assignment v such that for any variable p,

$$v(p) = \begin{cases} 0 \text{ if } p = l \\ 1 \text{ if } p = \neg l \\ 0 \text{ if } p = l' \\ 1 \text{ if } p = \neg l' \\ 1 \text{ otherwise.} \end{cases}$$

Then, v is well-defined and $v \not\models^{\mathbf{A1A1}} \Gamma \Rightarrow \Delta$. Similar for $\neg l \neq \neg l' \in \Delta - \Gamma$.

□

Hence, we have the following axioms:

$$(\mathrm{A}^{\mathbf{A1A1}}) \frac{\Gamma = \emptyset \text{ or } \Delta = \emptyset \text{ or } \Gamma = \{l\} = \neg\Delta}{\Gamma \Rightarrow \Delta}$$

5.2.2 Deduction Rules

$\mathbf{G}^{LQ_1 0}$ consists of the following deduction rules:

$$(\neg\neg^L)\ \frac{\Gamma, A \Rightarrow \Delta}{\Gamma, \neg\neg A \Rightarrow \Delta}$$

$$(\wedge^L)\ \frac{\left\{\begin{array}{l}\Gamma, A_1 \Rightarrow \Delta\\ \Gamma, A_2 \Rightarrow \Delta\end{array}\right.}{\Gamma, A_1 \wedge A_2 \Rightarrow \Delta} \qquad (\vee^L)\ \frac{\left[\begin{array}{l}\Gamma, A_1 \Rightarrow \Delta\\ \Gamma, A_2 \Rightarrow \Delta\end{array}\right.}{\Gamma, A_1 \vee A_2 \Rightarrow \Delta}$$

$$(\neg\wedge^L)\ \frac{\left[\begin{array}{l}\Gamma, \neg A_1 \Rightarrow \Delta\\ \Gamma, \neg A_2 \Rightarrow \Delta\end{array}\right.}{\Gamma, \neg(A_1 \vee A_2) \Rightarrow \Delta} \qquad (\neg\vee^L)\ \frac{\left\{\begin{array}{l}\Gamma, \neg A_1 \Rightarrow \Delta\\ \Gamma, \neg A_2 \Rightarrow \Delta\end{array}\right.}{\Gamma, \neg(A_1 \vee A_2) \Rightarrow \Delta}$$

$\mathbf{G}^{LQ_1 1}$ consists of the following deduction rules:

$$(\neg\neg^L)\ \frac{\Gamma, A \Rightarrow \Delta}{\Gamma, \neg\neg A \Rightarrow \Delta}$$

$$(\wedge^L)\ \frac{\left[\begin{array}{l}\Gamma, A_1 \Rightarrow \Delta\\ \Gamma, A_2 \Rightarrow \Delta\end{array}\right.}{\Gamma, A_1 \wedge A_2 \Rightarrow \Delta} \qquad (\vee^L)\ \frac{\left\{\begin{array}{l}\Gamma, A_1 \Rightarrow \Delta\\ \Gamma, A_2 \Rightarrow \Delta\end{array}\right.}{\Gamma, A_1 \vee A_2 \Rightarrow \Delta}$$

$$(\neg\wedge^L)\ \frac{\left\{\begin{array}{l}\Gamma, \neg A_1 \Rightarrow \Delta\\ \Gamma, \neg A_2 \Rightarrow \Delta\end{array}\right.}{\Gamma, \neg(A_1 \vee A_2) \Rightarrow \Delta} \qquad (\neg\vee^L)\ \frac{\left[\begin{array}{l}\Gamma, \neg A_1 \Rightarrow \Delta\\ \Gamma, \neg A_2 \Rightarrow \Delta\end{array}\right.}{\Gamma, \neg(A_1 \vee A_2) \Rightarrow \Delta}$$

$\mathbf{G}^{RQ_2 1}$ consists of the following deduction rules:

$$(\neg\neg^R)\ \frac{\Gamma \Rightarrow B, \Delta}{\Gamma \Rightarrow \neg\neg B, \Delta}$$

$$(\wedge^R)\ \frac{\left[\begin{array}{l}\Gamma \Rightarrow B_1, \Delta\\ \Gamma \Rightarrow B_2, \Delta\end{array}\right.}{\Gamma \Rightarrow B_1 \wedge B_2, \Delta} \qquad (\vee^R)\ \frac{\left\{\begin{array}{l}\Gamma \Rightarrow B_1, \Delta\\ \Gamma \Rightarrow B_2, \Delta\end{array}\right.}{\Gamma \Rightarrow B_1 \vee B_2, \Delta}$$

$$(\neg\wedge^R)\ \frac{\left\{\begin{array}{l}\Gamma \Rightarrow \neg B_1, \Delta\\ \Gamma \Rightarrow \neg B_2, \Delta\end{array}\right.}{\Gamma \Rightarrow \neg(B_1 \wedge B_2), \Delta} \qquad (\neg\vee^R)\ \frac{\left[\begin{array}{l}\Gamma \Rightarrow \neg B_1, \Delta\\ \Gamma \Rightarrow \neg B_2, \Delta\end{array}\right.}{\Gamma \Rightarrow \neg(B_1 \vee B_2), \Delta}$$

$\mathbf{G}^{RQ_2 0}$ consists of the following deduction rules:

$$(\neg\neg^R)\ \frac{\Gamma \Rightarrow B, \Delta}{\Gamma \Rightarrow \neg\neg B, \Delta}$$

$$(\wedge^R)\ \frac{\left\{\begin{array}{l}\Gamma \Rightarrow B_1, \Delta\\ \Gamma \Rightarrow B_2, \Delta\end{array}\right.}{\Gamma \Rightarrow B_1 \wedge B_2, \Delta} \qquad (\vee^R)\ \frac{\left[\begin{array}{l}\Gamma \Rightarrow B_1, \Delta\\ \Gamma \Rightarrow B_2, \Delta\end{array}\right.}{\Gamma \Rightarrow B_1 \vee B_2, \Delta}$$

$$(\neg\wedge^R)\ \frac{\left[\begin{array}{l}\Gamma \Rightarrow \neg B_1, \Delta\\ \Gamma \Rightarrow \neg B_2, \Delta\end{array}\right.}{\Gamma \Rightarrow \neg(B_1 \wedge B_2), \Delta} \qquad (\neg\vee^R)\ \frac{\left\{\begin{array}{l}\Gamma \Rightarrow \neg B_1, \Delta\\ \Gamma \Rightarrow \neg B_2, \Delta\end{array}\right.}{\Gamma \Rightarrow \neg(B_1 \vee B_2), \Delta}$$

5.2.3 Deduction Systems

Let $Q_1, Q_2 \in \{\mathbf{E}, \mathbf{A}\}$ and $i, j \in \{0, 1\}$. Define

$$\mathbf{G}^{Q_1 i Q_2 j} = \mathrm{A}^{Q_1 i Q_2 j} + \mathbf{G}^{L Q_1 i} + \mathbf{G}^{R Q_2 j}.$$

Definition 5.2.18 A sequent $\Gamma \Rightarrow \Delta$ is provable in $\mathbf{G}^{Q_1 i Q_2 j}$, denoted by $\vdash^{Q_1 i Q_2 j} \Gamma \Rightarrow \Delta$, if there is a sequence $\{\Gamma_1 \Rightarrow \Delta_1, ..., \Gamma_n \Rightarrow \Delta_n\}$ of sequents such that $\Gamma_n \Rightarrow \Delta_n = \Gamma \Rightarrow \Delta$, and for each $1 \leq i \leq n$, $\Gamma_i \Rightarrow \Delta_i$ is an axiom or is deduced from the previous sequents by one of the deduction rules in $\mathbf{G}^{Q_1 i Q_2 j}$.

Theorem 5.2.19 (*Soundness and completeness theorem*) *For any sequent* $\Gamma \Rightarrow \Delta$,

$$\vdash^{Q_1 i Q_2 j} \Gamma \Rightarrow \Delta \textit{ iff } \models^{Q_1 i Q_2 j} \Gamma \Rightarrow \Delta.$$

□

Theorem 5.2.20 (*Monotonicity theorem*) $\mathbf{G}^{Q_1 i Q_2 j}$ *is monotonic in* Γ *if and only if* $Q_1 = \mathbf{E}$; *and nonmonotonic in* Δ *if and only if* $Q_2 = \mathbf{A}$.

5.3 R-Calculi $\mathbf{R}^{Q_1 i Q_2 j}$

Let $Q_1, Q_2 \in \mathbf{E}, \mathbf{A}$ and $i, j \in 0, 1$.

Given a sequent $\Gamma \Rightarrow^{Q_1 i Q_2 j} \Delta$ and pair (A, B) of formulas, the result of $\Gamma \Rightarrow \Delta$ $\mathbf{G}^{Q_1 i Q_2 j}$-revising (A, B) is denoted by

$$\models^{Q_1 i Q_2 j} \Gamma \Rightarrow \Delta | (A, B) \rightrightarrows \Gamma' \Rightarrow \Delta',$$

where

$$\Gamma' \Rightarrow \Delta' = \begin{cases} \Gamma \pm_1 A \Rightarrow \Delta \pm_2 B \text{ if } \models^{Q_1 i Q_2 j} \Gamma \pm_1 A \Rightarrow \Delta \pm_2 B \\ \Gamma \pm_1 A \Rightarrow \Delta \qquad \text{if } \models^{Q_1 i Q_2 j} \Gamma \pm_1 A \Rightarrow \Delta \\ \Gamma \Rightarrow \Delta \pm_2 B \qquad \text{if } \models^{Q_1 i Q_2 j} \Gamma \Rightarrow \Delta \pm_2 B \\ \Gamma \Rightarrow \Delta \qquad \text{otherwise,} \end{cases}$$

where

$$\pm_1 A = \begin{cases} \cup\{A\} \text{ if } Q_1 = \mathbf{A} \\ -\{A\} \text{ if } Q_1 = \mathbf{E}, \end{cases} \qquad \pm_2 B = \begin{cases} \cup\{B\} \text{ if } Q_2 = \mathbf{A} \\ -\{B\} \text{ if } Q_2 = \mathbf{E}. \end{cases}$$

5.3.1 Axioms

We reorder the axioms in Gentzen deduction systems as follows:

$$(\mathbb{A}^{\mathbf{E0E1}})\ \frac{\text{incon}(\Gamma) \text{ or val}(\Delta) \text{ or } \Gamma \cap \Delta \neq \emptyset}{\Gamma \Rightarrow \Delta}$$
$$(\mathbb{A}^{\mathbf{E1E0}})\ \frac{\text{incon}(\Gamma) \text{ or val}(\Delta) \text{ or } \Gamma \cap \Delta \neq \emptyset}{\Gamma \Rightarrow \Delta}$$
$$(\mathbb{A}^{\mathbf{E0E0}})\ \frac{\text{incon}(\Gamma) \text{ or val}(\Delta) \text{ or } \Gamma \cap \neg\Delta \neq \emptyset}{\Gamma \Rightarrow \Delta}$$
$$(\mathbb{A}^{\mathbf{E1E1}})\ \frac{\text{incon}(\Gamma) \text{ or val}(\Delta) \text{ or } \Gamma \cap \neg\Delta \neq \emptyset}{\Gamma \Rightarrow \Delta}$$

and

$$(\mathbb{A}^{\mathbf{E0A1}})\ \frac{\Delta \subseteq \Gamma \text{ or incon}(\Gamma)}{\Gamma \Rightarrow \Delta}$$
$$(\mathbb{A}^{\mathbf{E1A0}})\ \frac{\Delta \subseteq \Gamma \text{ or incon}(\Gamma)}{\Gamma \Rightarrow \Delta}$$
$$(\mathbb{A}^{\mathbf{E0A0}})\ \frac{\neg\Delta \subseteq \Gamma \text{ or incon}(\Gamma)}{\Gamma \Rightarrow \Delta}$$
$$(\mathbb{A}^{\mathbf{E1A1}})\ \frac{\neg\Delta \subseteq \Gamma \text{ or incon}(\Gamma)}{\Gamma \Rightarrow \Delta}$$

and

$$(\mathbb{A}^{\mathbf{A0E1}})\ \frac{\Gamma \subseteq \Delta \text{ or val}(\Delta)}{\Gamma \Rightarrow \Delta}$$
$$(\mathbb{A}^{\mathbf{A1E0}})\ \frac{\text{val}(\Delta) \text{ or } \Gamma \subseteq \Delta}{\Gamma \Rightarrow \Delta}$$
$$(\mathbb{A}^{\mathbf{A0E0}})\ \frac{\text{val}(\Delta) \text{ or } \neg\Gamma \subseteq \Delta}{\Gamma \Rightarrow \Delta}$$
$$(\mathbb{A}^{\mathbf{A1E1}})\ \frac{\neg\Gamma \subseteq \Delta \text{ or val}(\Delta)}{\Gamma \Rightarrow \Delta}$$

and

$$(\mathbb{A}^{\mathbf{A0A1}})\ \frac{\Gamma = \emptyset \text{ or } \Delta = \emptyset \text{ or } \Gamma = \{l\} = \Delta}{\Gamma \Rightarrow \Delta}$$
$$(\mathbb{A}^{\mathbf{A1A0}})\ \frac{\Gamma = \emptyset \text{ or } \Delta = \emptyset \text{ or } \Gamma = \{l\} = \Delta}{\Gamma \Rightarrow \Delta}$$
$$(\mathbb{A}^{\mathbf{A0A0}})\ \frac{\Gamma = \emptyset \text{ or } \Delta = \emptyset \text{ or } \Gamma = \{l\} = \neg\Delta}{\Gamma \Rightarrow \Delta}$$
$$(\mathbb{A}^{\mathbf{A1A1}})\ \frac{\Gamma = \emptyset \text{ or } \Delta = \emptyset \text{ or } \Gamma = \{l\} = \neg\Delta}{\Gamma \Rightarrow \Delta}$$

Correspondingly we have the following axioms for R-calculi:

- $Q_1 i Q_2 j = \mathbf{E}0\mathbf{E}1/\mathbf{E}1\mathbf{E}0$:

$$(\mathtt{A}_{-L}^{Q_1 i Q_2 j})\ \frac{\begin{cases}\mathbf{E}l' \neq l(l', \neg l' \in \Gamma) \\ \mathbf{E}l' \neq l(l' \in \Gamma \& l' \in \Delta)\end{cases}}{\Gamma \Rightarrow \Delta|(l, m) \rightrightarrows \Gamma - \{l\} \Rightarrow \Delta|m}$$

$$(\mathtt{A}_{0L}^{Q_1 i Q_2 j})\ \frac{\begin{bmatrix}\sim \mathbf{E}l' \neq l(l', \neg l' \in \Gamma) \\ \sim \mathbf{E}l' \neq l(l' \in \Gamma \& l' \in \Delta)\end{bmatrix}}{\Gamma \Rightarrow \Delta|(l, m) \rightrightarrows \Gamma \Rightarrow \Delta|m}$$

$$(\mathtt{A}_{-R}^{Q_1 i Q_2 j})\ \frac{\begin{cases}\mathbf{E}m' \neq m(m', \neg m' \in \Delta) \\ \mathbf{E}m' \neq m(m' \in \Gamma' \& m' \in \Delta)\end{cases}}{\Gamma' \Rightarrow \Delta|m \rightrightarrows \Gamma' \Rightarrow \Delta - \{m\}}$$

$$(\mathtt{A}_{0R}^{Q_1 i Q_2 j})\ \frac{\begin{bmatrix}\sim \mathbf{E}m' \neq m(m', \neg m' \in \Delta) \\ \sim \mathbf{E}m' \neq m(m' \in \Gamma' \& m' \in \Delta)\end{bmatrix}}{\Gamma' \Rightarrow \Delta|m \rightrightarrows \Gamma' \Rightarrow \Delta}$$

where $l \in \Gamma$ and $m \in \Delta$.

- $Q_1 i Q_2 j = \mathbf{E}0\mathbf{E}0/\mathbf{E}1\mathbf{E}1$:

$$(\mathtt{A}_{-L}^{Q_1 i Q_2 j})\ \frac{\begin{cases}\mathbf{E}l' \neq l(l', \neg l' \in \Gamma) \\ \mathbf{E}l' \neq l(l' \in \Gamma \& \neg l' \in \Delta)\end{cases}}{\Gamma \Rightarrow \Delta|(l, m) \rightrightarrows \Gamma - \{l\} \Rightarrow \Delta|m}$$

$$(\mathtt{A}_{0L}^{Q_1 i Q_2 j})\ \frac{\begin{bmatrix}\sim \mathbf{E}l' \neq l(l', \neg l' \in \Gamma) \\ \sim \mathbf{E}l' \neq l(l' \in \Gamma \& \neg l' \in \Delta)\end{bmatrix}}{\Gamma \Rightarrow \Delta|(l, m) \rightrightarrows \Gamma \Rightarrow \Delta|m}$$

$$(\mathtt{A}_{-R}^{Q_1 i Q_2 j})\ \frac{\begin{cases}\mathbf{E}m' \neq m(m', \neg m' \in \Delta) \\ \mathbf{E}m' \neq m(m' \in \Gamma' \& \neg m' \in \Delta)\end{cases}}{\Gamma' \Rightarrow \Delta|m \rightrightarrows \Gamma' \Rightarrow \Delta - \{m\}}$$

$$(\mathtt{A}_{0R}^{Q_1 i Q_2 j})\ \frac{\begin{bmatrix}\sim \mathbf{E}m' \neq m(m', \neg m' \in \Delta) \\ \sim \mathbf{E}m' \neq m(m' \in \Gamma' \& \neg m' \in \Delta)\end{bmatrix}}{\Gamma' \Rightarrow \Delta|m \rightrightarrows \Gamma' \Rightarrow \Delta}$$

where $l \in \Gamma$ and $m \in \Delta$.

- $Q_1 i Q_2 j = \mathbf{E}0\mathbf{A}1/\mathbf{E}1\mathbf{A}0$:

$$(\mathtt{A}_{-L}^{Q_1 i Q_2 j})\ \frac{\begin{cases}\mathbf{E}l' \neq l(l', \neg l' \in \Gamma) \\ \Delta \subseteq \Gamma - \{l\}\end{cases}}{\Gamma \Rightarrow \Delta|(l, m) \rightrightarrows \Gamma - \{l\} \Rightarrow \Delta|m}$$

$$(\mathtt{A}_{0L}^{Q_1 i Q_2 j})\ \frac{\begin{bmatrix}\sim \mathbf{E}l' \neq l(l', \neg l' \in \Gamma) \\ \Delta \nsubseteq \Gamma - \{l\}\end{bmatrix}}{\Gamma \Rightarrow \Delta|(l, m) \rightrightarrows \Gamma \Rightarrow \Delta|m}$$

$$(\mathtt{A}_{-R}^{Q_1 i Q_2 j})\ \frac{m \in \Gamma'}{\Gamma' \Rightarrow \Delta|m \rightrightarrows \Gamma' \Rightarrow \Delta, m}$$

$$(\mathtt{A}_{0R}^{Q_1 i Q_2 j})\ \frac{m \notin \Gamma'}{\Gamma' \Rightarrow \Delta|m \rightrightarrows \Gamma' \Rightarrow \Delta}$$

where $l \in \Gamma$.

- $Q_1 i Q_2 j = \mathbf{E}0\mathbf{A}0/\mathbf{E}1\mathbf{A}1$:

$$
\begin{array}{ll}
(\mathtt{A}_{-L}^{Q_1 i Q_2 j}) & \dfrac{\begin{cases}\mathbf{E}l' \neq l(l', \neg l' \in \Gamma) \\ \neg\Delta \subseteq \Gamma - \{l\}\end{cases}}{\Gamma \Rightarrow \Delta|(l, m) \rightrightarrows \Gamma - \{l\} \Rightarrow \Delta|m} \\
(\mathtt{A}_{0L}^{Q_1 i Q_2 j}) & \dfrac{\left[\begin{array}{l}\sim \mathbf{E}l' \neq l(l', \neg l' \in \Gamma) \\ \neg\Delta \not\subseteq \Gamma - \{l\}\end{array}\right.}{\Gamma \Rightarrow \Delta|(l, m) \rightrightarrows \Gamma \Rightarrow \Delta|m} \\
(\mathtt{A}_{-R}^{Q_1 i Q_2 j}) & \dfrac{\neg m \in \Gamma'}{\Gamma' \Rightarrow \Delta|m \rightrightarrows \Gamma' \Rightarrow \Delta, m} \\
(\mathtt{A}_{0R}^{Q_1 i Q_2 j}) & \dfrac{\neg m \notin \Gamma'}{\Gamma' \Rightarrow \Delta|m \rightrightarrows \Gamma' \Rightarrow \Delta}
\end{array}
$$

where $l \in \Gamma$.

- $Q_1 i Q_2 j = \mathbf{A}0\mathbf{E}1/\mathbf{A}1\mathbf{E}0$:

$$
\begin{array}{ll}
(\mathtt{A}_{-L}^{Q_1 i Q_2 j}) & \dfrac{l \in \Delta}{\Gamma \Rightarrow \Delta|(l, m) \rightrightarrows \Gamma, l \Rightarrow \Delta|m} \\
(\mathtt{A}_{0L}^{Q_1 i Q_2 j}) & \dfrac{l \notin \Delta}{\Gamma \Rightarrow \Delta|(l, m) \rightrightarrows \Gamma \Rightarrow \Delta|m} \\
(\mathtt{A}_{-R}^{Q_1 i Q_2 j}) & \dfrac{\begin{cases}\mathbf{E}m' \neq m(m', \neg m' \in \Delta) \\ \Gamma' \subseteq \Delta - \{m\}\end{cases}}{\Gamma' \Rightarrow \Delta|m \rightrightarrows \Gamma' \Rightarrow \Delta - \{m\}} \\
(\mathtt{A}_{0R}^{Q_1 i Q_2 j}) & \dfrac{\left[\begin{array}{l}\sim \mathbf{E}m' \neq m(m', \neg m' \in \Delta) \\ \Gamma' \not\subseteq \Delta - \{m\}\end{array}\right.}{\Gamma' \Rightarrow \Delta|m \rightrightarrows \Gamma' \Rightarrow \Delta}
\end{array}
$$

where $m \in \Delta$.

- $Q_1 i Q_2 j = \mathbf{A}0\mathbf{E}0/\mathbf{A}1\mathbf{E}1$:

$$
\begin{array}{ll}
(\mathtt{A}_{-L}^{Q_1 i Q_2 j}) & \dfrac{\neg l \in \Delta}{\Gamma \Rightarrow \Delta|(l, m) \rightrightarrows \Gamma, l \Rightarrow \Delta|m} \\
(\mathtt{A}_{0L}^{Q_1 i Q_2 j}) & \dfrac{\neg l \notin \Delta}{\Gamma \Rightarrow \Delta|(l, m) \rightrightarrows \Gamma \Rightarrow \Delta|m} \\
(\mathtt{A}_{-R}^{Q_1 i Q_2 j}) & \dfrac{\begin{cases}\mathbf{E}m' \neq m(m', \neg m' \in \Delta) \\ \neg\Gamma' \subseteq \Delta - \{m\}\end{cases}}{\Gamma' \Rightarrow \Delta|m \rightrightarrows \Gamma' \Rightarrow \Delta - \{m\}} \\
(\mathtt{A}_{0R}^{Q_1 i Q_2 j}) & \dfrac{\left[\begin{array}{l}\sim \mathbf{E}m' \neq m(m', \neg m' \in \Delta) \\ \neg\Gamma' \not\subseteq \Delta - \{m\}\end{array}\right.}{\Gamma' \Rightarrow \Delta|m \rightrightarrows \Gamma' \Rightarrow \Delta}
\end{array}
$$

where $m \in \Delta$.

- $Q_1 i Q_2 j = \mathbf{A}0\mathbf{A}1/\mathbf{A}1\mathbf{A}0$:

$$(\mathtt{A}_{+L}^{Q_1 i Q_2 j})\ \frac{l \in \Gamma}{\Gamma \Rightarrow \Delta|(l,m) \rightrightarrows \Gamma, l \Rightarrow \Delta|m}$$
$$(\mathtt{A}_{0L}^{Q_1 i Q_2 j})\ \frac{l \notin \Gamma}{\Gamma \Rightarrow \Delta|(l,m) \rightrightarrows \Gamma \Rightarrow \Delta|m}$$
$$(\mathtt{A}_{+R}^{Q_1 i Q_2 j})\ \frac{m \in \Delta}{\Gamma' \Rightarrow \Delta|m \rightrightarrows \Gamma' \Rightarrow \Delta, m}$$
$$(\mathtt{A}_{0R}^{Q_1 i Q_2 j})\ \frac{m \notin \Delta}{\Gamma' \Rightarrow \Delta|m \rightrightarrows \Gamma' \Rightarrow \Delta}$$

- $Q_1 i Q_2 j = \mathbf{A}0\mathbf{A}0/\mathbf{A}1\mathbf{A}1$:

$$(\mathtt{A}_{+L}^{Q_1 i Q_2 j})\ \frac{\neg l \in \Delta}{\Gamma \Rightarrow \Delta|(l,m) \rightrightarrows \Gamma, l \Rightarrow \Delta|m}$$
$$(\mathtt{A}_{0L}^{Q_1 i Q_2 j})\ \frac{\neg l \notin \Delta}{\Gamma \Rightarrow \Delta|(l,m) \rightrightarrows \Gamma \Rightarrow \Delta|m}$$
$$(\mathtt{A}_{+R}^{Q_1 i Q_2 j})\ \frac{\neg m \in \Gamma'}{\Gamma' \Rightarrow \Delta|m \rightrightarrows \Gamma' \Rightarrow \Delta, m}$$
$$(\mathtt{A}_{0R}^{Q_1 i Q_2 j})\ \frac{\neg m \notin \Gamma'}{\Gamma' \Rightarrow \Delta|m \rightrightarrows \Gamma' \Rightarrow \Delta}$$

Theorem 5.3.1 *Let $\Gamma \Rightarrow \Delta$ be literal and l, m be literals. $\Gamma \Rightarrow \Delta|(l,m) \Rightarrow \Gamma' \Rightarrow \Delta'$ is $\mathbf{R}^{Q_1 i Q_2 j}$-valid iff $\Gamma \Rightarrow \Delta|(l,m) \Rightarrow \Gamma' \Rightarrow \Delta'$ is the conclusion of an axiom.* □

5.3.2 Deduction Rules

Let $\mathbf{X} = \Gamma \Rightarrow \Delta$ and

$$\begin{aligned}\mathbf{X(A)} &= \Gamma, \mathbf{A} \Rightarrow \Delta \\ \mathbf{X(B)} &= \Gamma \Rightarrow \mathbf{B}, \Delta \\ \mathbf{X[A]} &= \Gamma - \{\mathbf{A}\} \Rightarrow \Delta \\ \mathbf{X[B]} &= \Gamma \Rightarrow \Delta - \{\mathbf{B}\}.\end{aligned}$$

There are four basic sets of deduction rules:

$$\mathbf{S}^{\mathbf{LQ_10}}, \mathbf{S}^{\mathbf{LQ_11}}, \mathbf{S}^{\mathbf{RQ_20}}, \mathbf{S}^{\mathbf{RQ_21}},$$

which are same as in the last chapter.

5.3.3 R-Calculi

Define

$$\mathbf{R}^{Q_1 i Q_2 j} = \mathrm{A}^{Q_1 i Q_2 j} + \mathbf{S}^{L Q_1 i} + \mathbf{S}^{R Q_2 j}.$$

Definition 5.3.2 A reduction $\delta = \Gamma \Rightarrow \Delta|(A, B) \rightrightarrows \Gamma' \Rightarrow \Delta'$ is provable in $\mathbf{R}^{Q_1 i Q_2 j}$, denoted by $\vdash^{Q_1 i Q_2 j} \delta$, if there is a sequence $\{\delta_1, ..., \delta_n\}$ of reductions such that $\delta_n = \delta$, and for each $1 \leq i \leq n$, δ_i is either an axiom or deduced from the previous reductions by one of the deduction rules in $\mathbf{R}^{Q_1 i Q_2 j}$.

Theorem 5.3.3 (*Soundness and completeness theorem*) *For any reduction* $\delta = \Gamma \mapsto \Delta|(A, B) \rightrightarrows \Gamma' \mapsto \Delta'$, δ *is* $\mathbf{R}^{Q_1 i Q_2 j}$*-valid if and only if* δ *is provable in* $\mathbf{R}^{Q_1 i Q_2 j}$. *That is,*

$$\vdash^{Q_1 i Q_2 j} \delta \mathit{iff} \models^{Q_1 i Q_2 j} \delta.$$

□

Theorem 5.3.4 (*Nonmonotonicity theorem*) $\mathbf{R}^{Q_1 i Q_2 j}$ *is nonmonotonic in* Γ *and in* Δ.

Proof $\mathbf{R}^{Q_1 i Q_2 j}$ is composed of two parts: the enumeration/elimination part (denoted by $\mathbf{R}_{\pm}^{Q_1 i Q_2 j}$) and the zero (doing nothing) part (denoted by $\mathbf{R}_0^{Q_1 i Q_2 j}$). Then, $\mathbf{R}_{\pm}^{Q_1 i Q_2 j}$ is nonmonotonic in Γ and in Δ; and $\mathbf{R}_0^{Q_1 i Q_2 j}$ is monotonic in Γ and in Δ. □

5.4 Gentzen Deduction Systems $\mathbf{G}_{Q_1 i Q_2 j}$

Definition 5.4.1 A co-sequent $\Gamma \mapsto \Delta$ is $\mathbf{G}_{Q_1 i Q_2 j}$-valid, denoted by $\models_{Q_1 i Q_2 j} \Gamma \mapsto \Delta$, if there is an assignment v such that

$$\begin{cases} v(A) = 1 \text{ for every } A \in \Gamma \text{ if } Q_1 i = \mathbf{A}1 \\ v(A) = 0 \text{ for every } A \in \Gamma \text{ if } Q_1 i = \mathbf{A}0 \\ v(A) = 1 \text{ for some } A \in \Gamma \text{ if } Q_1 i = \mathbf{E}1 \\ v(A) = 0 \text{ for some } A \in \Gamma \text{ if } Q_1 i = \mathbf{E}0 \end{cases}$$

and

$$\begin{cases} v(B) = 1 \text{ for some } B \in \Delta \text{ if } Q_2 j = \mathbf{A}1 \\ v(B) = 0 \text{ for some } B \in \Delta \text{ if } Q_2 j = \mathbf{A}0 \\ v(B) = 1 \text{ for every } B \in \Delta \text{ if } Q_2 j = \mathbf{E}1 \\ v(B) = 0 \text{ for every } B \in \Delta \text{ if } Q_2 j = \mathbf{E}0 \end{cases}$$

5.4.1 Axioms

The axioms are classified into the following four classes:

- $\mathbf{A}1Q_2j$:

Lemma 5.4.2 *Given two sets* Γ, Δ *of literals,* $\models_{\mathbf{A1A0}} \Gamma \mapsto \Delta$ *if and only if* $\mathrm{con}(\Gamma)\&\mathrm{inval}(\Delta)\&\Gamma \cap \Delta = \emptyset$. □

Hence, we have the following axiom:

$$(\mathbb{A}_{\mathbf{A1A0}})\ \frac{\mathrm{con}(\Gamma)\&\mathrm{inval}(\Delta)\&\Gamma \cap \Delta = \emptyset}{\Gamma \mapsto \Delta}$$

Lemma 5.4.3 *Given two sets* Γ, Δ *of literals,* $\models_{\mathbf{A1E1}} \Gamma \mapsto \Delta$ *if and only if* $\mathrm{con}(\Gamma)\&\mathrm{inval}(\Delta)\&\Gamma \cap \neg\Delta = \emptyset$. □

Hence, we have the following axiom:

$$(\mathbb{A}_{\mathbf{A1A1}})\ \frac{\mathrm{con}(\Gamma)\&\mathrm{inval}(\Delta)\&\Gamma \cap \neg\Delta = \emptyset}{\Gamma \mapsto \Delta}$$

Lemma 5.4.4 *Let* Γ, Δ *be sets of literals.* $\models_{\mathbf{A1E0}} \Gamma \Rightarrow \Delta$ *if and only if*

$$\Delta \not\subseteq \Gamma\&\mathrm{con}(\Gamma).$$

Hence, we have the following axioms:

$$(\mathbb{A}_{\mathbf{A1E0}})\ \frac{\Delta \not\subseteq \Gamma\&\mathrm{con}(\Gamma)}{\Gamma \mapsto \Delta}$$

Lemma 5.4.5 *Let* Γ, Δ *be sets of literals.* $\models_{\mathbf{A1E1}} \Gamma \mapsto \Delta$ *if and only if*

$$\neg\Delta \not\subseteq \Gamma\&\mathrm{con}(\Gamma).$$

□

Hence, we have the following axioms:

$$(\mathbb{A}_{\mathbf{A1E1}})\ \frac{\neg\Delta \not\subseteq \Gamma\&\mathrm{con}(\Gamma)}{\Gamma \mapsto \Delta}$$

- $\mathbf{A}0Q_2j$:

Lemma 5.4.6 *Given two sets* Γ, Δ *of literals,* $\models_{\mathbf{A0A0}} \Gamma \mapsto \Delta$ *if and only if* $\mathrm{con}(\Gamma)\&\mathrm{inval}(\Delta)\&\Gamma \cap \neg\Delta = \emptyset$. □

Hence, we have the following axiom:

$$(\mathbb{A}_{\mathbf{A0A0}})\ \frac{\mathrm{con}(\Gamma)\&\mathrm{inval}(\Delta)\&\Gamma \cap \neg\Delta = \emptyset}{\Gamma \mapsto \Delta}$$

Lemma 5.4.7 *Given two sets* Γ, Δ *of literals,* $\models_{\mathbf{A0A1}} \Gamma \mapsto \Delta$ *if and only if* $\mathrm{con}(\Gamma)\&\mathrm{inval}(\Delta)\&\Gamma \cap \Delta = \emptyset$. □

Hence, we have the following axiom:

$$(\mathbb{A}_{\mathbf{A0A1}}) \frac{\mathrm{con}(\Gamma)\&\mathrm{inval}(\Delta)\&\Gamma \cap \Delta = \emptyset}{\Gamma \mapsto \Delta}$$

Lemma 5.4.8 *Let* Γ, Δ *be sets of literals.* $\models_{\mathbf{A0E0}} \Gamma \Rightarrow \Delta$ *if and only if*

$$\neg\Delta \not\subseteq \Gamma\&\mathrm{con}(\Gamma).$$

Hence, we have the following axioms:

$$(\mathbb{A}_{\mathbf{A0E0}}) \frac{\neg\Delta \not\subseteq \Gamma\&\mathrm{con}(\Gamma)}{\Gamma \mapsto \Delta}$$

Lemma 5.4.9 *Let* Γ, Δ *be sets of literals.* $\models_{\mathbf{A0E1}} \Gamma \mapsto \Delta$ *if and only if*

$$\Delta \not\subseteq \Gamma\&\mathrm{con}(\Gamma).$$

□

Hence, we have the following axioms:

$$(\mathbb{A}_{\mathbf{A0E1}}) \frac{\Delta \not\subseteq \Gamma\&\mathrm{con}(\Gamma)}{\Gamma \mapsto \Delta}$$

- $\mathbf{E}1Q_2j$:

Lemma 5.4.10 *Let* Γ, Δ *be sets of literals.* $\Gamma \Rightarrow \Delta$ *is* $\mathbf{G}_{\mathbf{E1A0}}$*-valid if and only if* $\mathrm{inval}(\Delta)\&\neg\Gamma \not\subseteq \Delta$.

Hence, we have the following axioms:

$$(\mathbb{A}_{\mathbf{E1A0}}) \frac{\mathrm{inval}(\Delta)\&\neg\Gamma \not\subseteq \Delta}{\Gamma \mapsto \Delta}$$

Lemma 5.4.11 *Let* Γ, Δ *be sets of literals.* $\Gamma \Rightarrow \Delta$ *is* $\mathbf{G}_{\mathbf{E1A1}}$*-valid if and only if* $\mathrm{inval}(\Delta)\&\Gamma \not\subseteq \Delta$.

Hence, we have the following axioms:

$$(\mathbb{A}_{\mathbf{E1A1}}) \frac{\Gamma \not\subseteq \Delta\&\mathrm{inval}(\Delta)}{\Gamma \mapsto \Delta}$$

Lemma 5.4.12 *Given two sets* Γ, Δ *of literals,* $\models^{\mathbf{E1E0}} \Gamma \mapsto \Delta$ *if and only if* $\Gamma \neq \emptyset\&\Delta \neq \emptyset\&(\Gamma \neq \{l\} \neq \Delta)$.

Hence, we have the following axioms:

$$(\mathbb{A}_{\mathbf{E1E0}})\ \frac{\Gamma \neq \emptyset \& \Delta \neq \emptyset \& (\Gamma \neq \Delta \neq \{l\})}{\Gamma \mapsto \Delta}$$

Lemma 5.4.13 *Given two sets* Γ, Δ *of literals,* $\models_{\mathbf{E1E1}} \Gamma \mapsto \Delta$ *if and only if* $\Gamma \neq \emptyset \& \Delta \neq \emptyset \& (\Gamma \neq \{l\}$ *or* $\Delta \neq \{\neg l\})$.

Hence, we have the following axioms:

$$(\mathbb{A}_{\mathbf{E1E1}})\ \frac{\Gamma \neq \emptyset \& \Delta \neq \emptyset \& (\Gamma \neq \{l\} \text{ or } \Delta \neq \{\neg l\})}{\Gamma \mapsto \Delta}$$

- $\mathbf{E}0Q_2j$:

Lemma 5.4.14 *Let* Γ, Δ *be sets of literals.* $\Gamma \Rightarrow \Delta$ *is* $\mathbf{G}_{\mathbf{E0A0}}$*-valid if and only if* $\mathrm{inval}(\Delta)\&\Gamma \nsubseteq \Delta$.

Hence, we have the following axioms:

$$(\mathbb{A}_{\mathbf{E0A0}})\ \frac{\mathrm{inval}(\Delta)\&\Gamma \nsubseteq \Delta}{\Gamma \mapsto \Delta}$$

Lemma 5.4.15 *Let* Γ, Δ *be sets of literals.* $\Gamma \Rightarrow \Delta$ *is* $\mathbf{G}_{\mathbf{E0A1}}$*-valid if and only if* $\mathrm{inval}(\Delta)\&\neg\Gamma \nsubseteq \Delta$.

Hence, we have the following axioms:

$$(\mathbb{A}_{\mathbf{E0A1}})\ \frac{\neg\Gamma \nsubseteq \Delta \& \mathrm{inval}(\Delta)}{\Gamma \mapsto \Delta}$$

Lemma 5.4.16 *Given two sets* Γ, Δ *of literals,* $\models^{\mathbf{E0E0}} \Gamma \mapsto \Delta$ *if and only if* $\Gamma \neq \emptyset \& \Delta \neq \emptyset \& (\Gamma \neq \{l\}$ *or* $\{\neg l\} \neq \Delta)$.

Hence, we have the following axioms:

$$(\mathbb{A}_{\mathbf{E0E0}})\ \frac{\Gamma \neq \emptyset \& \Delta \neq \emptyset \& (\Gamma \neq \{l\} \text{ or } \Delta \neq \{\neg l\})}{\Gamma \mapsto \Delta}$$

Lemma 5.4.17 *Given two sets* Γ, Δ *of literals,* $\models_{\mathbf{E0E1}} \Gamma \mapsto \Delta$ *if and only if* $\Gamma \neq \emptyset \& \Delta \neq \emptyset \& (\Gamma \neq \{l\}$ *or* $\Delta \neq \{l\})$.

Hence, we have the following axioms:

$$(\mathbb{A}_{\mathbf{E0E1}})\ \frac{\Gamma \neq \emptyset \& \Delta \neq \emptyset \& (\Gamma \neq \{l\} \text{ or } \Delta \neq \{l\})}{\Gamma \mapsto \Delta}$$

Theorem 5.4.18 *Let* $\Gamma \mapsto \Delta$ *be literal.* $\Gamma \mapsto \Delta$ *is* $\mathbf{G}_{Q_1 i Q_2 j}$*-valid if and only if the precondition in* $(\mathbb{A}_{Q_1 i Q_2 j})$ *holds.* □

5.4.2 Deduction Rules

$\mathbf{G}_{L0}$ consists of the following deduction rules:

$$(\neg\neg^L)\ \frac{\Gamma, A \mapsto \Delta}{\Gamma, \neg\neg A \mapsto \Delta}$$

$$(\wedge^L)\ \frac{\left\{\begin{array}{l}\Gamma, A_1 \mapsto \Delta\\ \Gamma, A_2 \mapsto \Delta\end{array}\right.}{\Gamma, A_1 \wedge A_2 \mapsto \Delta} \qquad (\vee^L)\ \frac{\left[\begin{array}{l}\Gamma, A_1 \mapsto \Delta\\ \Gamma, A_2 \mapsto \Delta\end{array}\right.}{\Gamma, A_1 \vee A_2 \mapsto \Delta}$$

$$(\neg\wedge^L)\ \frac{\left[\begin{array}{l}\Gamma, \neg A_1 \mapsto \Delta\\ \Gamma, \neg A_2 \mapsto \Delta\end{array}\right.}{\Gamma, \neg(A_1 \vee A_2) \mapsto \Delta} \qquad (\neg\vee^L)\ \frac{\left\{\begin{array}{l}\Gamma, \neg A_1 \mapsto \Delta\\ \Gamma, \neg A_2 \mapsto \Delta\end{array}\right.}{\Gamma, \neg(A_1 \vee A_2) \mapsto \Delta}$$

$\mathbf{G}_{L1}$ consists of the following deduction rules:

$$(\neg\neg^L)\ \frac{\Gamma, A \mapsto \Delta}{\Gamma, \neg\neg A \mapsto \Delta}$$

$$(\wedge^L)\ \frac{\left[\begin{array}{l}\Gamma, A_1 \mapsto \Delta\\ \Gamma, A_2 \mapsto \Delta\end{array}\right.}{\Gamma, A_1 \wedge A_2 \mapsto \Delta} \qquad (\vee^L)\ \frac{\left\{\begin{array}{l}\Gamma, A_1 \mapsto \Delta\\ \Gamma, A_2 \mapsto \Delta\end{array}\right.}{\Gamma, A_1 \vee A_2 \mapsto \Delta}$$

$$(\neg\wedge^L)\ \frac{\left\{\begin{array}{l}\Gamma, \neg A_1 \mapsto \Delta\\ \Gamma, \neg A_2 \mapsto \Delta\end{array}\right.}{\Gamma, \neg(A_1 \vee A_2) \mapsto \Delta} \qquad (\neg\vee^L)\ \frac{\left[\begin{array}{l}\Gamma, \neg A_1 \mapsto \Delta\\ \Gamma, \neg A_2 \mapsto \Delta\end{array}\right.}{\Gamma, \neg(A_1 \vee A_2) \mapsto \Delta}$$

$\mathbf{G}_{R1}$ consists of the following deduction rules:

$$(\neg\neg^R)\ \frac{\Gamma \mapsto B, \Delta}{\Gamma \mapsto \neg\neg B, \Delta}$$

$$(\wedge^R)\ \frac{\left[\begin{array}{l}\Gamma \mapsto B_1, \Delta\\ \Gamma \mapsto B_2, \Delta\end{array}\right.}{\Gamma \mapsto B_1 \wedge B_2, \Delta} \qquad (\vee^R)\ \frac{\left\{\begin{array}{l}\Gamma \mapsto B_1, \Delta\\ \Gamma \mapsto B_2, \Delta\end{array}\right.}{\Gamma \mapsto B_1 \vee B_2, \Delta}$$

$$(\neg\wedge^R)\ \frac{\left\{\begin{array}{l}\Gamma \mapsto \neg B_1, \Delta\\ \Gamma \mapsto \neg B_2, \Delta\end{array}\right.}{\Gamma \mapsto \neg(B_1 \wedge B_2), \Delta} \qquad (\neg\vee^R)\ \frac{\left[\begin{array}{l}\Gamma \mapsto \neg B_1, \Delta\\ \Gamma \mapsto \neg B_2, \Delta\end{array}\right.}{\Gamma \mapsto \neg(B_1 \vee B_2), \Delta}$$

$\mathbf{G}_{R0}$ consists of the following deduction rules:

$$(\neg\neg^R)\ \frac{\Gamma \mapsto B, \Delta}{\Gamma \mapsto \neg\neg B, \Delta}$$

$$(\wedge^R)\ \frac{\left\{\begin{array}{l}\Gamma \mapsto B_1, \Delta\\ \Gamma \mapsto B_2, \Delta\end{array}\right.}{\Gamma \mapsto B_1 \wedge B_2, \Delta} \qquad (\vee^R)\ \frac{\left[\begin{array}{l}\Gamma \mapsto B_1, \Delta\\ \Gamma \mapsto B_2, \Delta\end{array}\right.}{\Gamma \mapsto B_1 \vee B_2, \Delta}$$

$$(\neg\wedge^R)\ \frac{\left[\begin{array}{l}\Gamma \mapsto \neg B_1, \Delta\\ \Gamma \mapsto \neg B_2, \Delta\end{array}\right.}{\Gamma \mapsto \neg(B_1 \wedge B_2), \Delta} \qquad (\neg\vee^R)\ \frac{\left\{\begin{array}{l}\Gamma \mapsto \neg B_1, \Delta\\ \Gamma \mapsto \neg B_2, \Delta\end{array}\right.}{\Gamma \mapsto \neg(B_1 \vee B_2), \Delta}$$

Then, we have the following equalities:

$$\begin{array}{ll} \mathbf{G}_{L1} = \mathbf{G}^{L1} & \mathbf{G}_{L0} = \mathbf{G}^{L0} \\ \mathbf{G}_{R1} = \mathbf{G}^{R1} & \mathbf{G}_{R0} = \mathbf{G}^{R0}. \end{array}$$

5.4.3 Deduction Systems

Let $Q_1, Q_2 \in \{\mathbf{E}, \mathbf{A}\}$ and $i, j \in \{0, 1\}$. Define

$$\mathbf{G}_{Q_1 i Q_2 j} = \mathbb{A}_{Q_1 i Q_2 j} + \mathbf{G}_{Li} + \mathbf{G}_{Rj}.$$

Definition 5.4.19 A co-sequent $\Gamma \mapsto \Delta$ is provable in $\mathbf{G}_{Q_1 i Q_2 j}$, denoted by $\vdash_{Q_1 i Q_2 j} \Gamma \mapsto \Delta$, if there is a sequence $\{\Gamma_1 \mapsto \Delta_1, ..., \Gamma_n \mapsto \Delta_n\}$ of co-sequents such that $\Gamma_n \mapsto \Delta_n = \Gamma \mapsto \Delta$, and for each $1 \leq i \leq n$, $\Gamma_i \mapsto \Delta_i$ is an axiom or is deduced from the previous co-sequents by one of the deduction rules in $\mathbf{G}_{Q_1 i Q_2 j}$.

Theorem 5.4.20 (*Soundness and completeness theorem*) *For any co-sequent* $\Gamma \mapsto \Delta$,

$$\vdash_{Q_1 i Q_2 j} \Gamma \mapsto \Delta \textit{ iff } \models_{Q_1 i Q_2 j} \Gamma \mapsto \Delta.$$

□

Theorem 5.4.21 (*Nonmonotonicity theorem*) $\mathbf{G}_{Q_1 i Q_2 j}$ *is monotonic in* Γ *if and only if* $Q_1 = \mathbf{E}$; *and nonmonotonic in* Δ *if and only if* $Q_2 = \mathbf{A}$.

5.5 R-Calculi $\mathbf{R}_{Q_1 i\, Q_2 j}$

Let $Q_1, Q_2 \in \mathbf{E}, \mathbf{A}$ and $i, j \in 0, 1$.

Given a co-sequent $\Gamma \mapsto \Delta$ and pair (A, B) of formulas, the result of $\Gamma \mapsto \Delta$ $\mathbf{G}_{Q_1 i Q_2 j}$-revising (A, B) is denoted by $\models_{Q_1 i Q_2 j} \Gamma \mapsto \Delta|(A, B) \Rightarrow \Gamma' \mapsto \Delta'$, where $\Gamma' \mapsto \Delta' =$

$$\begin{cases} \Gamma \pm_1 A \mapsto \Delta \pm_2 B & \text{if } \Gamma \pm_1 A \mapsto \Delta \pm_2 B \text{ is } \mathbf{G}_{Q_1 i Q_2 j}\text{-valid} \\ \Gamma \pm_1 A \mapsto \Delta & \text{otherwise, if } \Gamma \pm_1 A \mapsto \Delta \text{ is } \mathbf{G}_{Q_1 i Q_2 j}\text{-valid} \\ \Gamma \mapsto \Delta \pm_2 B & \text{otherwise, if } \Gamma \mapsto \Delta \pm_2 B \text{ is } \mathbf{G}_{Q_1 i Q_2 j}\text{-valid} \\ \Gamma \mapsto \Delta & \text{otherwise;} \end{cases}$$

where

$$\pm_1 A = \begin{cases} \cup\{A\} \text{ if } Q_1 = \mathbf{A} \\ -\{A\} \text{ if } Q_1 = \mathbf{E}, \end{cases} \qquad \pm_2 B = \begin{cases} \cup\{B\} \text{ if } Q_2 = \mathbf{A} \\ -\{B\} \text{ if } Q_2 = \mathbf{E}. \end{cases}$$

5.5.1 Axioms

Hence, we have the following axioms for R-calculi:

- $Q_1 i Q_2 j = \mathbf{A}0\mathbf{A}1/\mathbf{A}1\mathbf{A}0$:

$$(\mathtt{A}^{+L}_{Q_1 i Q_2 j})\ \frac{\left[\begin{array}{l}\neg l \notin \Gamma\\ l \notin \Delta\end{array}\right.}{\Gamma \mapsto \Delta|(l,m) \rightrightarrows \Gamma, l \mapsto \Delta|m}$$

$$(\mathtt{A}^{0L}_{Q_1 i Q_2 j})\ \frac{\left\{\begin{array}{l}\neg l \in \Gamma\\ l \in \Delta\end{array}\right.}{\Gamma \mapsto \Delta|(l,m) \rightrightarrows \Gamma \mapsto \Delta|m}$$

$$(\mathtt{A}^{+R}_{Q_1 i Q_2 j})\ \frac{\left[\begin{array}{l}m \notin \Gamma\\ \neg m \notin \Delta\end{array}\right.}{\Gamma' \mapsto \Delta|m \rightrightarrows \Gamma' \mapsto \Delta, m}$$

$$(\mathtt{A}^{0R}_{Q_1 i Q_2 j})\ \frac{\left\{\begin{array}{l}m \in \Gamma\\ \neg m \in \Delta\end{array}\right.}{\Gamma' \mapsto \Delta|m \rightrightarrows \Gamma' \mapsto \Delta}$$

- $Q_1 i Q_2 j = \mathbf{A}0\mathbf{A}0/\mathbf{A}1\mathbf{A}1$:

$$(\mathtt{A}^{+L}_{Q_1 i Q_2 j})\ \frac{\left[\begin{array}{l}\neg l \notin \Gamma\\ \neg l \notin \Delta\end{array}\right.}{\Gamma \mapsto \Delta|(l,m) \rightrightarrows \Gamma, l \mapsto \Delta|m}$$

$$(\mathtt{A}^{0L}_{Q_1 i Q_2 j})\ \frac{\left\{\begin{array}{l}\neg l \in \Gamma\\ \neg l \in \Delta\end{array}\right.}{\Gamma \mapsto \Delta|(l,m) \rightrightarrows \Gamma \mapsto \Delta|m}$$

$$(\mathtt{A}^{+R}_{Q_1 i Q_2 j})\ \frac{\left[\begin{array}{l}\neg m \notin \Gamma\\ \neg m \notin \Delta\end{array}\right.}{\Gamma' \mapsto \Delta|m \rightrightarrows \Gamma' \mapsto \Delta, m}$$

$$(\mathtt{A}^{0R}_{Q_1 i Q_2 j})\ \frac{\left\{\begin{array}{l}\neg m \in \Gamma\\ \neg m \in \Delta\end{array}\right.}{\Gamma' \mapsto \Delta|m \rightrightarrows \Gamma' \mapsto \Delta}$$

- $Q_1 i Q_2 j = \mathbf{A}0\mathbf{E}1/\mathbf{A}1\mathbf{E}0$:

$$(\mathtt{A}^{+L}_{Q_1 i Q_2 j})\ \frac{\neg l \notin \Gamma}{\Gamma \mapsto \Delta|(l,m) \rightrightarrows \Gamma, l \mapsto \Delta|m}$$

$$(\mathtt{A}^{0L}_{Q_1 i Q_2 j})\ \frac{\neg l \in \Gamma}{\Gamma \mapsto \Delta|(l,m) \rightrightarrows \Gamma \mapsto \Delta|m}$$

$$(\mathtt{A}^{-R}_{Q_1 i Q_2 j})\ \frac{(\Delta - \{m\}) \not\subseteq \Gamma'}{\Gamma' \mapsto \Delta|m \rightrightarrows \Gamma' \mapsto \Delta - \{m\}}$$

$$(\mathtt{A}^{0R}_{Q_1 i Q_2 j})\ \frac{(\Delta - \{m\}) \subseteq \Gamma'}{\Gamma' \mapsto \Delta|m \rightrightarrows \Gamma' \mapsto \Delta}$$

where $m \in \Delta$.

- $Q_1 i Q_2 j = \mathbf{A}0\mathbf{E}0/\mathbf{A}1\mathbf{E}1$:

$$(\mathtt{A}^{+L}_{Q_1 i Q_2 j})\ \frac{\neg l \notin \Gamma}{\Gamma \mapsto \Delta|(l,m) \rightrightarrows \Gamma, l \mapsto \Delta|m}$$

$$(\mathtt{A}^{0L}_{Q_1 i Q_2 j})\ \frac{\neg l \in \Gamma}{\Gamma \mapsto \Delta|(l,m) \rightrightarrows \Gamma \mapsto \Delta|m}$$

$$(\mathtt{A}^{-R}_{Q_1 i Q_2 j})\ \frac{\neg(\Delta - \{m\}) \not\subseteq \Gamma}{\Gamma' \mapsto \Delta|m \rightrightarrows \Gamma' \mapsto \Delta - \{m\}}$$

$$(\mathtt{A}^{0R}_{Q_1 i Q_2 j})\ \frac{\neg(\Delta - \{m\}) \subseteq \Gamma}{\Gamma' \mapsto \Delta|m \rightrightarrows \Gamma' \mapsto \Delta}$$

where $m \in \Delta$.

- $Q_1 i Q_2 j = \mathbf{E}0\mathbf{A}1/\mathbf{E}1\mathbf{A}0$:

$$(\mathtt{A}^{-L}_{Q_1 i Q_2 j})\ \frac{\neg(\Gamma - \{l\}) \not\subseteq \Delta}{\Gamma \mapsto \Delta|(l,m) \rightrightarrows \Gamma - \{l\} \mapsto \Delta|m}$$

$$(\mathtt{A}^{0L}_{Q_1 i Q_2 j})\ \frac{\neg(\Gamma - \{l\}) \subseteq \Delta}{\Gamma \mapsto \Delta|(l,m) \rightrightarrows \Gamma \mapsto \Delta|m}$$

$$(\mathtt{A}^{+R}_{Q_1 i Q_2 j})\ \frac{\left[\begin{array}{l}\neg m \notin \Delta \\ \neg\Gamma \not\subseteq \Delta, m\end{array}\right.}{\Gamma' \mapsto \Delta|m \rightrightarrows \Gamma' \mapsto \Delta, m}$$

$$(\mathtt{A}^{0R}_{Q_1 i Q_2 j})\ \frac{\left\{\begin{array}{l}\neg m \in \Delta \\ \neg\Gamma \subseteq \Delta, m\end{array}\right.}{\Gamma' \mapsto \Delta|m \rightrightarrows \Gamma' \mapsto \Delta}$$

where $l \in \Gamma$.

- $Q_1 i Q_2 j = \mathbf{E}1\mathbf{A}1/\mathbf{E}0\mathbf{A}0$:

$$(\mathtt{A}^{-L}_{Q_1 i Q_2 j})\ \frac{(\Gamma - \{l\}) \not\subseteq \Delta}{\Gamma \mapsto \Delta|(l,m) \rightrightarrows \Gamma - \{l\} \mapsto \Delta|m}$$

$$(\mathtt{A}^{0L}_{Q_1 i Q_2 j})\ \frac{(\Gamma - \{l\}) \subseteq \Delta}{\Gamma \mapsto \Delta|(l,m) \rightrightarrows \Gamma \mapsto \Delta|m}$$

$$(\mathtt{A}^{+R}_{Q_1 i Q_2 j})\ \frac{\left[\begin{array}{l}\neg m \notin \Delta \\ \Gamma \not\subseteq \Delta, m\end{array}\right.}{\Gamma' \mapsto \Delta|m \rightrightarrows \Gamma' \mapsto \Delta, m}$$

$$(\mathtt{A}^{0R}_{Q_1 i Q_2 j})\ \frac{\left\{\begin{array}{l}\neg m \in \Delta \\ \Gamma \subseteq \Delta, m\end{array}\right.}{\Gamma' \mapsto \Delta|m \rightrightarrows \Gamma' \mapsto \Delta}$$

where $l \in \Gamma$.

- $Q_1 i Q_2 j = \mathbf{E}0\mathbf{E}1/\mathbf{E}1\mathbf{E}0$:

$$(\mathtt{A}^{-L}_{Q_1 i Q_2 j}) \frac{\begin{cases} (\Gamma - \{l\}) \neq \{l'\} \\ \Delta \neq \{l'\} \end{cases}}{\Gamma \mapsto \Delta|(l,m) \rightrightarrows \Gamma - \{l\} \mapsto \Delta|m}$$

$$(\mathtt{A}^{0L}_{Q_1 i Q_2 j}) \frac{(\Gamma - \{l\}) = \Delta = \{l'\}}{\Gamma \mapsto \Delta|(l,m) \rightrightarrows \Gamma \mapsto \Delta|m}$$

$$(\mathtt{A}^{-R}_{Q_1 i Q_2 j}) \frac{\begin{cases} (\Delta - \{m\}) \neq \{m'\} \\ \Gamma \neq \{m'\} \end{cases}}{\Gamma' \mapsto \Delta|m \rightrightarrows \Gamma' \mapsto \Delta - \{m\}}$$

$$(\mathtt{A}^{0R}_{Q_1 i Q_2 j}) \frac{(\Delta - \{m\}) = \Gamma = \{m'\}}{\Gamma' \mapsto \Delta|m \rightrightarrows \Gamma' \mapsto \Delta}$$

where $l \in \Gamma$ and $m \in \Delta$.

- $Q_1 i Q_2 j = \mathbf{E}0\mathbf{E}0/\mathbf{E}1\mathbf{E}1$:

$$(\mathtt{A}^{-L}_{Q_1 i Q_2 j}) \frac{\begin{cases} (\Gamma - \{l\}) \neq \{l'\} \\ \neg\Delta \neq \{l'\} \end{cases}}{\Gamma \mapsto \Delta|(l,m) \rightrightarrows \Gamma - \{l\} \mapsto \Delta|m}$$

$$(\mathtt{A}^{0L}_{Q_1 i Q_2 j}) \frac{(\Gamma - \{l\}) = \neg\Delta = \{l'\}}{\Gamma \mapsto \Delta|(l,m) \rightrightarrows \Gamma \mapsto \Delta|m}$$

$$(\mathtt{A}^{-R}_{Q_1 i Q_2 j}) \frac{\begin{cases} \neg(\Delta - \{m\}) \neq \{m'\} \\ \Gamma \neq \{m'\} \end{cases}}{\Gamma' \mapsto \Delta|m \rightrightarrows \Gamma' \mapsto \Delta - \{m\}}$$

$$(\mathtt{A}^{0R}_{Q_1 i Q_2 j}) \frac{\neg(\Delta - \{m\}) = \Gamma = \{m'\}}{\Gamma' \mapsto \Delta|m \rightrightarrows \Gamma' \mapsto \Delta}$$

where $l \in \Gamma$ and $m \in \Delta$.

Theorem 5.5.1 *Let $\Gamma \mapsto \Delta$ be literal. $\Gamma \mapsto \Delta|(l,m) \Rightarrow \Gamma' \mapsto \Delta'$ is $\mathbf{R}_{Q_1 i Q_2 j}$-valid if and only if $\Gamma \mapsto \Delta|(l,m) \Rightarrow \Gamma' \mapsto \Delta'$ is a conclusion of some axiom $(\mathtt{A}_{Q_1 i Q_2 j})^x$, where $x \in \{+L, -L, 0L, +R, -R, 0R\}$.* □

5.5.2 Deduction Rules

Let $\mathbf{X} = \Gamma \mapsto \Delta$ and

$$\begin{aligned}\mathbf{X(A)} &= \Gamma, \mathbf{A} \mapsto \Delta\\ \mathbf{X(B)} &= \Gamma \mapsto \mathbf{B}, \Delta\\ \mathbf{X[A]} &= \Gamma - \{\mathbf{A}\} \mapsto \Delta\\ \mathbf{X[B]} &= \Gamma \mapsto \Delta - \{\mathbf{B}\}.\end{aligned}$$

Define

$$\begin{aligned}\mathbf{S}_{\mathbf{LQ_1i}} &= \mathbf{S}^{\mathbf{LQ_1i}}(\Rightarrow / \mapsto)\\ \mathbf{S}_{\mathbf{RQ_2j}} &= \mathbf{S}^{\mathbf{RQ_2j}}(\Rightarrow / \mapsto).\end{aligned}$$

5.5.3 R-Calculi

Define

$$\mathbf{R}_{Q_1 i Q_2 j} = \mathbb{A}_{Q_1 i Q_2 j} + \mathbf{S}_{\mathbf{LQ_1i}} + \mathbf{S}_{\mathbf{RQ_2j}}.$$

Definition 5.5.2 A reduction $\Gamma \mapsto \Delta|(A, B) \rightrightarrows \Gamma' \mapsto \Delta'$ is provable in $\mathbf{R}_{Q_1 i Q_2 j}$, denoted by

$$\vdash_{Q_1 i Q_2 j} \Gamma \mapsto \Delta|(A, B) \Rightarrow \Gamma' \mapsto \Delta',$$

if there is a sequence $\{\delta_1, ..., \delta_n\}$ of reductions such that $\delta_n = \Gamma \mapsto \Delta|(A, B) \Rightarrow \Gamma' \mapsto \Delta'$, and for each $1 \leq i \leq n$, δ_i is either an axiom or deduced from the previous reductions by one of the deduction rules in $\mathbf{R}_{Q_1 i Q_2 j}$.

Theorem 5.5.3 (*Soundness and completeness theorem*) *For any* $Q_1, Q_2 \in \{\mathbf{A}, \mathbf{E}\}$, $i, j \in \{0, 1\}$ *and reduction* $\delta = \Gamma \mapsto \Delta|(A, B) \Rightarrow \Gamma' \mapsto \Delta'$, δ *is* $\mathbf{R}_{Q_1 i Q_2 j}$*-valid if and only if* δ *is provable in* $\mathbf{R}_{Q_1 i Q_2 j}$. *That is,*

$$\vdash_{Q_1 i Q_2 j} \delta \textit{ iff} \models_{Q_1 i Q_2 j} \delta.$$

□

Theorem 5.5.4 (*Nonmonotonicity theorem*) $\mathbf{R}_{Q_1 i Q_2 j}$ *is composed of two parts: the enumeration/elimination part (denoted by* $\mathbf{R}^{\pm}_{Q_1 i Q_2 j}$*) and the zero part (denoted by* $\mathbf{R}^{0}_{Q_1 i Q_2 j}$*) Then,* $\mathbf{R}^{\pm}_{Q_1 i Q_2 j}$ *is nonmonotonic in* Γ *and in* Δ*; and* $\mathbf{R}^{0}_{Q_1 i Q_2 j}$ *is monotonic in* Γ *and in* Δ.

□

5.6 Conclusions

There are 16 Gentzen deduction systems:

$\mathbf{A}v$	$\mathbf{G}^{\mathbf{E}i\mathbf{E}j}$	$\mathbf{G}^{\mathbf{E}i\mathbf{A}j}$	$\mathbf{G}^{\mathbf{A}i\mathbf{E}j}$	$\mathbf{G}^{\mathbf{A}i\mathbf{A}j}$
$\mathbf{E}v$	$\mathbf{G}_{\mathbf{A}i\mathbf{A}j}$	$\mathbf{G}_{\mathbf{A}i\mathbf{E}j}$	$\mathbf{G}_{\mathbf{E}i\mathbf{A}j}$	$\mathbf{G}_{\mathbf{E}i\mathbf{E}j}$

where

$$\begin{aligned}
&\mathbf{G}^{Q_1iQ_2j} = \mathbb{A}^{Q_1iQ_2j} + \mathbf{G}^{Li} + \mathbf{G}^{Rj} \\
&\mathbf{G}_{Q_1iQ_2j} = \mathbb{A}_{Q_1iQ_2j} + \mathbf{G}_{Li} + \mathbf{G}_{Rj} \\
&\mathbf{G}^{Li} \equiv \mathbf{G}_{Li} \\
&\mathbf{G}^{Rj} \equiv \mathbf{G}_{Rj}.
\end{aligned}$$

Moreover, $\mathbf{G}^{Q_1iQ_2j}/\mathbf{G}_{Q_1iQ_2j}$ is monotonic in Γ if and only if $Q_1 = \mathbf{E}$; and nonmonotonic in Δ if and only if $Q_2 = \mathbf{A}$.

Correspondingly there are 16 R-calculi:

$\mathbf{A}v$	$\mathbf{R}^{\mathbf{E}i\mathbf{E}j}$	$\mathbf{R}^{\mathbf{E}i\mathbf{A}j}$	$\mathbf{R}^{\mathbf{A}i\mathbf{E}j}$	$\mathbf{R}^{\mathbf{A}i\mathbf{A}j}$
$\mathbf{E}v$	$\mathbf{R}_{\mathbf{A}i\mathbf{A}j}$	$\mathbf{R}_{\mathbf{A}i\mathbf{E}j}$	$\mathbf{R}_{\mathbf{E}i\mathbf{A}j}$	$\mathbf{R}_{\mathbf{E}i\mathbf{E}j}$

where

$$\begin{aligned}
&\mathbf{R}^{Q_1iQ_2j} = \mathbb{A}^{Q_1iQ_2j} + \mathbf{S}^{\mathbf{LQ_1i}} + \mathbf{S}^{\mathbf{RQ_2j}} \\
&\mathbf{R}_{Q_1iQ_2j} = \mathbb{A}_{Q_1iQ_2j} + \mathbf{S}_{\mathbf{LQ_1i}} + \mathbf{S}_{\mathbf{RQ_2j}}; \\
&\mathbf{S}^{\mathbf{LQ_1i}} \equiv \mathbf{S}_{\mathbf{LQ_1i}} \\
&\mathbf{S}^{\mathbf{RQ_2j}} \equiv \mathbf{S}_{\mathbf{RQ_2j}}.
\end{aligned}$$

Moreover, $\mathbf{R}_+^{\mathbf{E}L}, \mathbf{R}_-^{\mathbf{A}L}$ are nonmonotonic in Γ; $\mathbf{R}_0^{\mathbf{E}L}, \mathbf{R}_0^{\mathbf{A}L}$ are monotonic in Γ; $\mathbf{R}_-^{\mathbf{E}R}, \mathbf{R}_+^{\mathbf{A}R}$ are nonmonotonic in Δ; and $\mathbf{R}_0^{\mathbf{E}R}, \mathbf{R}_0^{\mathbf{A}R}$ are monotonic in Δ, where $\mathbf{R}_+^{\mathbf{E}L}/\mathbf{R}_-^{\mathbf{A}L}$ is the set of deduction rules in $\mathbf{R}^{\mathbf{E}L}/\mathbf{R}^{\mathbf{A}L}$ with mark $+/-$, respectively, and $\mathbf{R}_+^{\mathbf{E}L}/\mathbf{R}_-^{\mathbf{A}L}$ with mark 0. Similar for $\mathbf{R}_-^{\mathbf{E}R}, \mathbf{R}_+^{\mathbf{A}R}$ and $\mathbf{R}_0^{\mathbf{E}R}, \mathbf{R}_0^{\mathbf{A}R}$.

Appendix: List of all the Gentzen deduction systems $\mathbf{G}^{Q_1i\,Q_2j}/\mathbf{G}_{Q_1i\,Q_2j}$

Let

$$\begin{aligned}
&\mathbf{G}^{ij} = \mathbf{G}^{Li} + \mathbf{G}^{Rj} \\
&\mathbf{G}_{ij} = \mathbf{G}_{Li} + \mathbf{G}_{Rj}
\end{aligned}$$

Then we have the following list of all the axioms for $\mathbf{G}^{Q_1 i Q_2 j}$ and $\mathbf{G}_{Q_1 i Q_2 j}$:

system	m Γ	m Δ
$\mathbf{G}^{\mathrm{E0E1}} = \mathbf{G}^{01} + \dfrac{\Gamma \cap \Delta \neq \emptyset \text{ or incon}(\Gamma) \text{ or val}(\Delta)}{\Gamma \Rightarrow \Delta}$	Y	Y
$\mathbf{G}_{\mathrm{A1A0}} = \mathbf{G}_{10} + \dfrac{\Gamma \cap \Delta = \emptyset \& \text{con}(\Gamma) \& \text{inval}(\Delta)}{\Gamma \mapsto \Delta}$	N	N
$\mathbf{G}^{\mathrm{E0A1}} = \mathbf{G}^{01} + \dfrac{\Delta \subseteq \Gamma \text{ or incon}(\Gamma)}{\Gamma \Rightarrow \Delta}$	Y	N
$\mathbf{G}_{\mathrm{A1E0}} = \mathbf{G}_{10} + \dfrac{\Delta \not\subseteq \Gamma \& \text{con}(\Gamma)}{\Gamma \mapsto \Delta}$	N	Y

where mΓ denotes monotonic in Γ, and mΔ denotes monotonic in Δ, and

system	m Γ	m Δ
$\mathbf{G}^{\mathrm{E0E0}} = \mathbf{G}^{00} + \dfrac{\Gamma \cap \neg\Delta \neq \emptyset \text{ or incon}(\Gamma) \text{ or val}(\neg\Delta)}{\Gamma \Rightarrow \Delta}$	Y	Y
$\mathbf{G}_{\mathrm{A1A1}} = \mathbf{G}_{11} + \dfrac{\Gamma \cap \neg\Delta = \emptyset \& \text{con}(\Gamma) \& \text{inval}(\neg\Delta)}{\Gamma \mapsto \Delta}$	N	N
$\mathbf{G}^{\mathrm{E0A0}} = \mathbf{G}^{10} + \dfrac{\neg\Delta \subseteq \Gamma \text{ or incon}(\Gamma)}{\Gamma \Rightarrow \Delta}$	Y	N
$\mathbf{G}_{\mathrm{A1E1}} = \mathbf{G}_{11} + \dfrac{\neg\Delta \not\subseteq \Gamma \& \text{con}(\Gamma)}{\Gamma \mapsto \Delta}$	N	Y

and

system	m Γ	m Δ
$\mathbf{G}^{\mathrm{E1E1}} = \mathbf{G}^{11} + \dfrac{\Gamma \cap \neg\Delta \neq \emptyset \text{ or incon}(\Delta) \text{ or val}(\Delta)}{\Gamma \Rightarrow \Delta}$	Y	Y
$\mathbf{G}_{\mathrm{A0A0}} = \mathbf{G}_{00} + \dfrac{\Gamma \cap \neg\Delta = \emptyset \& \text{con}(\Gamma) \& \text{inval}(\Delta)}{\Gamma \mapsto \Delta}$	N	N
$\mathbf{G}^{\mathrm{E1A1}} = \mathbf{G}^{11} + \dfrac{\neg\Delta \subseteq \Gamma \text{ or incon}(\Gamma)}{\Gamma \Rightarrow \Delta}$	Y	N
$\mathbf{G}_{\mathrm{A0E0}} = \mathbf{G}_{00} + \dfrac{\neg\Delta \not\subseteq \Gamma \& \text{con}(\Gamma)}{\Gamma \mapsto \Delta}$	N	Y

and

system		m Γ	m Δ
$\mathbf{G}^{\mathbf{E1E0}}=\mathbf{G}^{10}+$	$\dfrac{\Gamma\cap\Delta\neq\emptyset \text{ or incon}(\Gamma)\text{ or val}(\Delta)}{\Gamma\Rightarrow\Delta}$	Y	Y
$\mathbf{G}_{\mathbf{A0A1}}=\mathbf{G}_{01}+$	$\dfrac{\Gamma\cap\Delta=\emptyset\&\text{con}(\Gamma)\&\text{inval}(\neg\Delta)}{\Gamma\mapsto\Delta}$	N	N
$\mathbf{G}^{\mathbf{E1A0}}=\mathbf{G}^{10}+$	$\dfrac{\Delta\subseteq\Gamma\text{ or incon}(\Gamma)}{\Gamma\Rightarrow\Delta}$	Y	N
$\mathbf{G}_{\mathbf{A0E1}}=\mathbf{G}_{01}+$	$\dfrac{\Delta\not\subseteq\Gamma\&\text{con}(\Gamma)}{\Gamma\mapsto\Delta}$	N	Y

and

system		m Γ	m Δ
$\mathbf{G}^{\mathbf{A0E1}}=\mathbf{G}^{01}+$	$\dfrac{\Gamma\subseteq\Delta\text{ or val}(\Delta)}{\Gamma\Rightarrow\Delta}$	N	Y
$\mathbf{G}_{\mathbf{E1A0}}=\mathbf{G}_{10}+$	$\dfrac{\Gamma\not\subseteq\Delta\&\text{con}(\Delta)}{\Gamma\mapsto\Delta}$	Y	N
$\mathbf{G}^{\mathbf{A0A1}}=\mathbf{G}^{01}+$	$\dfrac{\Delta=\Gamma=\{l\}\text{ or }\Gamma=\emptyset\text{ or }\Delta=\emptyset}{\Gamma\Rightarrow\Delta}$	N	N
$\mathbf{G}_{\mathbf{E1E0}}=\mathbf{G}_{10}+$	$\dfrac{(\Gamma\neq\emptyset\neq\Delta\&(\Gamma\neq\{l\}\text{ or }\Delta\neq\{l\})}{\Gamma\mapsto\Delta}$	Y	Y

and

system		m Γ	m Δ
$\mathbf{G}^{\mathbf{A0E0}}=\mathbf{G}^{00}+$	$\dfrac{\neg\Gamma\subseteq\Delta\text{ or val}(\neg\Delta)}{\Gamma\Rightarrow\Delta}$	N	Y
$\mathbf{G}_{\mathbf{E1A1}}=\mathbf{G}_{11}+$	$\dfrac{\neg\Gamma\not\subseteq\Delta\&\text{inval}(\neg\Delta)}{\Gamma\mapsto\Delta}$	Y	N
$\mathbf{G}^{\mathbf{A0A0}}=\mathbf{G}^{00}+$	$\dfrac{\Gamma=\neg\Delta=\{l\}\text{ or }\Gamma=\emptyset\text{ or }\Delta=\emptyset}{\Gamma\Rightarrow\Delta}$	N	N
$\mathbf{G}_{\mathbf{E1E1}}=\mathbf{G}_{11}+$	$\dfrac{(\Gamma\neq\{l\}\text{ or }\neg\Delta\neq\{l\})\&\Gamma\neq\emptyset\neq\neg\Delta}{\Gamma\mapsto\Delta}$	Y	Y

and

system		m Γ	m Δ
$\mathbf{G}^{\mathrm{A1E1}} = \mathbf{G}^{11} +$	$\dfrac{\neg\Gamma \subseteq \Delta \text{ or val}(\Delta)}{\Gamma \Rightarrow \Delta}$	N	Y
$\mathbf{G}_{\mathrm{E0A0}} = \mathbf{G}_{00} +$	$\dfrac{\neg\Gamma \nsubseteq \Delta \& \text{inval}(\Delta)}{\Gamma \mapsto \Delta}$	Y	N
$\mathbf{G}^{\mathrm{A1A1}} = \mathbf{G}^{11} +$	$\dfrac{\Delta = \neg\Gamma = \{l\} \text{ or } \neg\Gamma = \emptyset \text{ or } \Delta = \emptyset}{\Gamma \Rightarrow \Delta}$	N	N
$\mathbf{G}_{\mathrm{E0E0}} = \mathbf{G}_{00} +$	$\dfrac{(\Delta \neq \{l\} \text{ or } \neg\Gamma \neq \{l\}) \& \neg\Gamma \neq \emptyset \neq \Delta}{\Gamma \mapsto \Delta}$	Y	Y

and

system		m Γ	m Δ
$\mathbf{G}^{\mathrm{A1E0}} = \mathbf{G}^{10} +$	$\dfrac{\Gamma \subseteq \Delta \text{ or val}(\neg\Delta)}{\Gamma \Rightarrow \Delta}$	N	Y
$\mathbf{G}_{\mathrm{E0A1}} = \mathbf{G}^{01} +$	$\dfrac{\Gamma \nsubseteq \Delta \& \text{inval}(\neg\Delta)}{\Gamma \mapsto \Delta}$	Y	N
$\mathbf{G}^{\mathrm{A1A0}} = \mathbf{G}^{10} +$	$\dfrac{\neg\Gamma = \neg\Delta = \{l\} \text{ or } \neg\Gamma = \emptyset \text{ or } \neg\Delta = \emptyset}{\Gamma \Rightarrow \Delta}$	N	N
$\mathbf{G}_{\mathrm{E0E1}} = \mathbf{G}_{01} +$	$\dfrac{(\neg\Gamma \neq \{l\} \text{ or } \neg\Delta \neq \{l\}) \& \neg\Gamma \neq \emptyset \neq \Delta}{\Gamma \mapsto \Delta}$	Y	Y

References

Cao, C., Sui, Y., Wang, Y.: The nonmonotonic propositional logics. Artif. Intell. Res. **5**, 111–120 (2016)

Ginsberg, M.L. (ed.): Readings in Nonmonotonic Reasoning. Morgan Kaufmann, San Francisco (1987)

Li, W.: R-calculus: an inference system for belief revision. Comput. J. **50**, 378–390 (2007)

Li, W.: Mathematical logic, foundations for information science. In: Progress in Computer Science and Applied Logic, vol.25, Birkhäuser (2010)

Reiter, R.: A logic for default reasoning. Artif. Intell. **13**, 81–132 (1980)

Takeuti, G.: Proof Theory. In: Barwise, J. (ed.), Handbook of Mathematical Logic. Studies in Logic and the Foundations of Mathematics. North-Holland, Amsterdam, NL (1987)

Chapter 6
R-Calculi: $\mathbf{R}^{Y_1 Q_1 i Y_2 Q_2 j} / \mathbf{R}_{Y_1 Q_1 i Y_2 Q_2 j}$

Let $Q_1, Q_2 \in \{\mathbf{A}, \mathbf{E}\}, i, j \in \{0, 1\}$, and $Y_1, Y_2 \in \{\mathbf{R}, \mathbf{Q}, \mathbf{P}\}$.

Let $\mathbf{X} = \Gamma \Rightarrow \Delta$ and $X \in (A, B)$.

A reduction $\mathbf{X}|(A, B) \rightrightarrows \mathbf{X}'$ is $\mathbf{R}^{Y_1 Q_1 i Y_2 Q_2 j}$-valid (Li 2010; Takeuti and Barwise 1987), denoted by $\models^{Y_1 Q_1 i Y_2 Q_2 j} \mathbf{X}|(A, B) \rightrightarrows \mathbf{X}'$, if

$$\mathbf{X}' = \begin{cases} \mathbf{X}\{A\}\{B\} & \text{if } \mathbf{X}\{A\} \text{ and } \mathbf{X}\{A\}\{B\} \text{ are } \mathbf{G}^{Y_1 Q_1 Y_2 Q_2}-\text{ valid} \\ \mathbf{X}\{A\} & \text{if } \mathbf{X}\{A\} \text{ is } \mathbf{G}^{Y_1 Q_1 Y_2 Q_2}-\text{ valid and } \mathbf{X}\{A\}\{B\} \text{ is not} \\ \mathbf{X}\{B\} & \text{if } \mathbf{X}\{A\} \text{ is not } \mathbf{G}^{Y_1 Q_1 Y_2 Q_2}-\text{ valid and } \mathbf{X}\{B\} \text{ is} \\ \mathbf{X} & \text{otherwise} \end{cases}$$

where

$$\{A\} = \begin{cases} (A) \text{ if } Q_1 = \mathbf{A} \\ [A] \text{ if } Q_1 = \mathbf{E}\text{and } A \in \Gamma \end{cases}$$
$$\{B\} = \begin{cases} (B) \text{ if } Q_2 = \mathbf{E}\text{and } B \in \Delta \\ [B] \text{ if } Q_2 = \mathbf{A}. \end{cases}$$

Dually, a reduction $\mathbf{X}|X \rightrightarrows \mathbf{X}'$ is $\mathbf{R}_{Y_1 Q_1 i Y_2 Q_2 j}$-valid (Alchourrón et al. 1985; Darwiche and Pearl 1997; Fermé and Hansson 2011; Gärdenfors and Rott 1995), denoted by

$$\models_{Y_1 Q_1 i Y_2 Q_2 j} \mathbf{X}|X \rightrightarrows \mathbf{X}',$$

if

$$\mathbf{X}' = \begin{cases} \Gamma \pm A \mapsto \Delta|B & \text{if } \models_{Y_1 Q_1 i Y_2 Q_2 j} \Gamma \pm A \mapsto \Delta|B \\ \Gamma' \mapsto \Delta \pm B & \text{if } \models_{Y_1 Q_1 i Y_2 Q_2 j} \Gamma' \mapsto \Delta \pm B \\ \Gamma & \text{otherwise.} \end{cases}$$

W. Li and Y. Sui, *R-Calculus, IV: Propositional Logic*,
Perspectives in Formal Induction, Revision and Evolution,
https://doi.org/10.1007/978-981-19-8633-8_6

We consider the following R-calculi (Li 2007):

$$\mathbf{R}^{Y_1Q_1iY_2Q_2j}\ \mathbf{Q}^{Y_1Q_1iY_2Q_2j}\ \mathbf{P}^{Y_1Q_1iY_2Q_2j}$$

About the monotonicity (Cao et al. 2016; Reiter 1980), we have the following conclusions:

- $\mathbf{G}^{Y_1Q_1iY_2Q_2j}/\mathbf{G}_{Y_1Q_1iY_2Q_2j}$ is monotonic in Γ iff $Q_1 = \mathbf{E}$, and monotonic in Δ iff $Q_2 = \mathbf{E}$; and
- $\mathbf{R}_0^{Y_1Q_1iY_2Q_2j}/\mathbf{R}^0_{Y_1Q_1iY_2Q_2j}$ is monotonic in Γ and in Δ; and $\mathbf{R}_\pm^{Y_1Q_1iY_2Q_2j}/\mathbf{R}^\pm_{Y_1Q_1iY_2Q_2j}$ is nonmonotonic in Γ iff $Q_1 = \mathbf{A}$, and nonmonotonic in Δ iff $Q_2 = \mathbf{A}$.

6.1 Variant R-Calculi

By choosing different minimal changes, there are following R-calculi:

$$\begin{array}{c}\mathbf{R}^{\mathtt{t}}\ \mathbf{Q}^{\mathtt{t}}\ \mathbf{P}^{\mathtt{t}}\\ \mathbf{R}^{\mathtt{f}}\ \mathbf{Q}^{\mathtt{f}}\ \mathbf{P}^{\mathtt{f}}\\ \mathbf{R}_{\mathtt{t}}\ \mathbf{Q}_{\mathtt{t}}\ \mathbf{P}_{\mathtt{t}}\\ \mathbf{R}_{\mathtt{f}}\ \mathbf{Q}_{\mathtt{f}}\ \mathbf{P}_{\mathtt{f}}.\end{array}$$

We will give R-calculi $\mathbf{R}_{\mathtt{t}}, \mathbf{Q}_{\mathtt{t}}, \mathbf{P}_{\mathtt{t}}$, respectively.

6.1.1 R-Calculus $\mathbf{R}_{\mathtt{t}}$

Definition 6.1.1 Given any theory Γ and a formula γ, a theory Γ, γ' is a subset-minimal ($\subseteq$-minimal) change of Γ by γ, denoted by $\models^{\mathbf{R}}_{\mathtt{t}} \Gamma|\gamma \Rightarrow \Gamma, \gamma'$, if (i) $\gamma' = \gamma$ or $\gamma' = \lambda$; (ii) Γ, γ' is consistent, and (iii) for any set Ξ with $\Gamma, \gamma' \subset \Xi \subseteq \Gamma, \gamma$, Ξ is inconsistent.

R-calculus $\mathbf{R}_{\mathtt{t}}$ consists of the following axioms and deduction rules:

- **Axioms**:

$$(\mathtt{A}^{\mathtt{t}}_{+})\ \frac{\Gamma \nvdash \neg l}{\Gamma|l \rightrightarrows \Gamma, l}\quad (\mathtt{A}^{\mathtt{t}}_{0})\ \frac{\Gamma \vdash \neg l}{\Gamma|l \rightrightarrows \Gamma}$$

where l is a literal.

- **Deduction rules**:

$$(\neg\neg^{+})\ \frac{\Gamma|A_1 \rightrightarrows \Gamma, A_1}{\Gamma|\neg\neg A_1 \rightrightarrows \Gamma, \neg\neg A_1} \qquad (\neg\neg^{0})\ \frac{\Gamma|A_1 \rightrightarrows \Gamma}{\Gamma|\neg\neg A_1 \rightrightarrows \Gamma}$$

$$(\wedge^{+})\ \frac{\left[\begin{array}{l}\Gamma|A_1 \rightrightarrows \Gamma, A_1\\ \Gamma, A_1|A_2 \rightrightarrows \Gamma, A_1, A_2\end{array}\right.}{\Gamma|A_1 \wedge A_2 \rightrightarrows \Gamma, A_1 \wedge A_2} \qquad (\wedge^{0})\ \frac{\left\{\begin{array}{l}\Gamma|A_1 \rightrightarrows \Gamma\\ \Gamma, A_1|A_2 \rightrightarrows \Gamma, A_1\end{array}\right.}{\Gamma|A_1 \wedge A_2 \rightrightarrows \Gamma}$$

$$(\vee^{+})\ \frac{\left\{\begin{array}{l}\Gamma|A_1 \rightrightarrows \Gamma, A_1\\ \Gamma|A_2 \rightrightarrows \Gamma, A_2\end{array}\right.}{\Gamma|A_1 \vee A_2 \rightrightarrows \Gamma, A_1 \vee A_2} \qquad (\vee^{0})\ \frac{\left[\begin{array}{l}\Gamma|A_1 \rightrightarrows \Gamma\\ \Gamma|A_2 \rightrightarrows \Gamma\end{array}\right.}{\Gamma|A_1 \vee A_2 \rightrightarrows \Gamma}$$

$$(\neg\wedge^{+})\ \frac{\left\{\begin{array}{l}\Gamma|\neg A_1 \rightrightarrows \Gamma, \neg A_1\\ \Gamma|\neg A_2 \rightrightarrows \Gamma, \neg A_2\end{array}\right.}{\Gamma|\neg(A_1 \wedge A_2) \rightrightarrows \Gamma, \neg(A_1 \wedge A_2)} \qquad (\neg\wedge^{0})\ \frac{\left[\begin{array}{l}\Gamma|\neg A_1 \rightrightarrows \Gamma\\ \Gamma|\neg A_2 \rightrightarrows \Gamma\end{array}\right.}{\Gamma|\neg(A_1 \wedge A_2) \rightrightarrows \Gamma}$$

$$(\neg\vee^{+})\ \frac{\left[\begin{array}{l}\Gamma|\neg A_1 \rightrightarrows \Gamma, \neg A_1\\ \Gamma, \neg A_1|\neg A_2 \rightrightarrows \Gamma, \neg A_1, \neg A_2\end{array}\right.}{\Gamma|\neg(A_1 \vee A_2) \rightrightarrows \Gamma, \neg(A_1 \vee A_2)} \qquad (\neg\vee^{0})\ \frac{\left\{\begin{array}{l}\Gamma|\neg A_1 \rightrightarrows \Gamma\\ \Gamma, \neg A_1|\neg A_2 \rightrightarrows \Gamma, \neg A_1\end{array}\right.}{\Gamma|\neg(A_1 \vee A_2) \rightrightarrows \Gamma}$$

Definition 6.1.2 A reduction $\delta = \Gamma|A \rightrightarrows \Gamma, C$ is provable in $\mathbf{R}_{\mathrm{t}}$, denoted by $\vdash_{\mathrm{t}} \Gamma|A \rightrightarrows \Gamma, C$, if there is a sequence $\{\delta_1, \ldots, \delta_m\}$ of reductions such that $\delta_m = \delta$, and for each $i < m$, δ_{i+1} is an axiom or is deduced from the previous statements by a deduction rule in $\mathbf{R}_{\mathrm{t}}$.

Theorem 6.1.3 (Soundness and completeness theorem) *For any consistent formula set Γ and formula A, if $\Gamma|A \rightrightarrows \Gamma, A$ is provable in $\mathbf{R}_{\mathrm{t}}$ then $\Gamma \cup \{A\}$ is consistent, i.e.,*

$$\vdash_{\mathrm{t}} \Gamma|A \rightrightarrows \Gamma, A \text{ implies } \models_{\mathrm{t}} \Gamma|A \rightrightarrows \Gamma, A;$$

and if $\Gamma|A \rightrightarrows \Gamma$ is provable in $\mathbf{R}_{\mathrm{t}}$ then $\Gamma \cup \{A\}$ is inconsistent, i.e.,

$$\vdash_{\mathrm{t}} \Gamma|A \rightrightarrows \Gamma \text{ implies } \models_{\mathrm{t}} \Gamma|A \rightrightarrows \Gamma.$$

□

$\mathbf{R}_{\mathrm{t}}$ is composed of two parts: $\mathbf{R}_{\mathrm{t}}^{0}$ (consisting of axiom and deduction rules with 0) and $\mathbf{R}_{\mathrm{t}}^{+}$ (consisting of axiom and deduction rules with +). Then, $\mathbf{R}_{\mathrm{t}}^{0}$ is monotonic in Γ, and $\mathbf{R}_{\mathrm{t}}^{+}$ is nonmonotonic in Γ.

6.1.2 R-Calculus $\mathbf{R}_{\mathfrak{f}}$

Definition 6.1.4 Given any theory Γ and a formula γ, a theory Γ, γ' is a subset-minimal ($\subseteq$-minimal) change of Γ by γ, denoted by $\models_{\mathfrak{f}}^{\mathbf{R}} \Gamma|\gamma \rightrightarrows \Gamma, \gamma'$, if (i) $\gamma' = \gamma$

or $\gamma' = \lambda$; (ii) Γ, γ' is $\mathtt{f}$-consistent, and (iii) for any set Ξ with $\Gamma, \gamma' \subset \Xi \subseteq \Gamma, \gamma$, Ξ is $\mathtt{f}$-inconsistent.

R-calculus $\mathbf{R}_{\mathtt{f}}$ consists of the following axioms and deduction rules:

- **Axioms**:

$$(\mathtt{A}^{\mathtt{t}}_{+})\ \frac{\Gamma \nvdash_{\mathtt{f}} \neg l}{\Gamma|l \rightrightarrows \Gamma, l} \quad (\mathtt{A}^{\mathtt{t}}_{0})\ \frac{\Gamma \vdash_{\mathtt{f}} \neg l}{\Gamma|l \rightrightarrows \Gamma}$$

where l is a literal.

- **Deduction rules**:

$$(\neg\neg^{+})\ \frac{\Gamma|A_1 \rightrightarrows \Gamma, A_1}{\Gamma|\neg\neg A_1 \rightrightarrows \Gamma, \neg\neg A_1} \qquad (\neg\neg^{0})\ \frac{\Gamma|A_1 \rightrightarrows \Gamma}{\Gamma|\neg\neg A_1 \rightrightarrows \Gamma}$$

$$(\wedge^{+})\ \frac{\left\{\begin{array}{l}\Gamma|A_1 \rightrightarrows \Gamma, A_1\\ \Gamma|A_2 \rightrightarrows \Gamma, A_2\end{array}\right.}{\Gamma|A_1 \wedge A_2 \rightrightarrows \Gamma, A_1 \wedge A_2} \qquad (\wedge^{0})\ \frac{\left[\begin{array}{l}\Gamma|A_1 \rightrightarrows \Gamma\\ \Gamma|A_2 \rightrightarrows \Gamma\end{array}\right.}{\Gamma|A_1 \wedge A_2 \rightrightarrows \Gamma}$$

$$(\vee^{+})\ \frac{\left[\begin{array}{l}\Gamma|A_1 \rightrightarrows \Gamma, A_1\\ \Gamma, A_1|A_2 \rightrightarrows \Gamma, A_1, A_2\end{array}\right.}{\Gamma|A_1 \vee A_2 \rightrightarrows \Gamma, A_1 \vee A_2} \qquad (\vee^{0})\ \frac{\left\{\begin{array}{l}\Gamma|A_1 \rightrightarrows \Gamma\\ \Gamma, A_1|A_2 \rightrightarrows \Gamma, A_1\end{array}\right.}{\Gamma|A_1 \vee A_2 \rightrightarrows \Gamma}$$

$$(\neg\wedge^{+})\ \frac{\left[\begin{array}{l}\Gamma|\neg A_1 \rightrightarrows \Gamma, \neg A_1\\ \Gamma, \neg A_1|\neg A_2 \rightrightarrows \Gamma, \neg A_1, \neg A_2\end{array}\right.}{\Gamma|\neg(A_1 \wedge A_2) \rightrightarrows \Gamma, \neg(A_1 \wedge A_2)} \qquad (\neg\wedge^{0})\ \frac{\left\{\begin{array}{l}\Gamma|\neg A_1 \rightrightarrows \Gamma\\ \Gamma, \neg A_1|\neg A_2 \rightrightarrows \Gamma, \neg A_1\end{array}\right.}{\Gamma|\neg(A_1 \wedge A_2) \rightrightarrows \Gamma}$$

$$(\neg\vee^{+})\ \frac{\left\{\begin{array}{l}\Gamma|\neg A_1 \rightrightarrows \Gamma, \neg A_1\\ \Gamma|\neg A_2 \rightrightarrows \Gamma, \neg A_2\end{array}\right.}{\Gamma|\neg(A_1 \vee A_2) \rightrightarrows \Gamma, \neg(A_1 \vee A_2)} \qquad (\neg\vee^{0})\ \frac{\left[\begin{array}{l}\Gamma|\neg A_1 \rightrightarrows \Gamma\\ \Gamma|\neg A_2 \rightrightarrows \Gamma\end{array}\right.}{\Gamma|\neg(A_1 \vee A_2) \rightrightarrows \Gamma}$$

Definition 6.1.5 A reduction $\delta = \Gamma|A \rightrightarrows \Gamma, C$ is provable in $\mathbf{R}_{\mathtt{f}}$, denoted by $\vdash_{\mathtt{f}} \Gamma|A \rightrightarrows \Gamma, C$, if there is a sequence $\{\delta_1, \ldots, \delta_m\}$ of reductions such that $\delta_m = \delta$, and for each $i < m$, δ_{i+1} is an axiom or is deduced from the previous statements by a deduction rule in $\mathbf{R}_{\mathtt{f}}$.

Theorem 6.1.6 *(Soundness and completeness theorem) For any consistent formula set Γ and formula A,*

$$\vdash_{\mathtt{f}} \Gamma|A \rightrightarrows \Gamma' \textit{ iff } \models^{\mathtt{f}} \Gamma|A \rightrightarrows \Gamma'.$$

□

$\mathbf{R}_{\mathtt{f}}$ is composed of two parts: $\mathbf{R}^{0}_{\mathtt{f}}$ (consisting of axiom and deduction rules with 0) and $\mathbf{R}^{+}_{\mathtt{f}}$ (consisting of axiom and deduction rules with +). Then, $\mathbf{R}^{0}_{\mathtt{f}}$ is monotonic in Γ, and $\mathbf{R}^{+}_{\mathtt{f}}$ is nonmonotonic in Γ.

6.1.3 *R-Calculus* $\mathbf{Q}_{\mathtt{t}}$

Definition 6.1.7 A theory Γ, γ' is a $\preceq$-minimal change of Γ by γ, denoted by $\models^{\mathbf{Q}}_{\mathtt{t}} \Gamma|\gamma \rightrightarrows \Gamma, \gamma'$, if

(i) $\gamma' \preceq \gamma$,
(ii) $\Gamma \cup \{\gamma'\}$ is consistent, and
(iii) for any theory Ξ with $\Gamma, \gamma' \prec \Xi \preceq \Gamma, \gamma$, Ξ is inconsistent.

R-calculus $\mathbf{Q}_{\mathtt{t}}$ consists of the following axioms and deduction rules:

- **Axioms**:

$$(\mathtt{A}^{\mathtt{t}}_{+})\ \frac{\Gamma \nvdash \neg A}{\Gamma|A \rightrightarrows \Gamma, A} \qquad (\mathtt{A}^{\mathtt{t}}_{0})\ \frac{\Gamma \vdash \neg l}{\Gamma|l \rightrightarrows \Gamma, \lambda}$$

- **Deduction rules**:

$$(\wedge)\ \frac{\left[\begin{array}{l}\Gamma|A_1 \rightrightarrows \Gamma, C_1 \\ \Gamma, C_1|A_2 \rightrightarrows \Gamma, C_1, C_2\end{array}\right.}{\Gamma|A_1 \wedge A_2 \rightrightarrows \Gamma, C_1 \wedge C_2} \qquad (\vee)\ \frac{\left\{\begin{array}{l}\Gamma|A_1 \rightrightarrows \Gamma, C_1 \neq \lambda \\ \left[\begin{array}{l}\Gamma|A_1 \rightrightarrows \Gamma \\ \Gamma|A_2 \rightrightarrows \Gamma, C_2 \neq \lambda\end{array}\right. \\ \left[\begin{array}{l}\Gamma|A_1 \rightrightarrows \Gamma \\ \Gamma|A_2 \rightrightarrows \Gamma\end{array}\right.\end{array}\right.}{\Gamma|A_1 \vee A_2 \rightrightarrows \Gamma, C_1' \vee C_2'}$$

$$(\neg\wedge)\ \frac{\left\{\begin{array}{l}\Gamma|\neg A_1 \rightrightarrows \Gamma, \neg C_1 \neq \lambda \\ \left[\begin{array}{l}\Gamma|\neg A_1 \rightrightarrows \Gamma \\ \Gamma|\neg A_2 \rightrightarrows \Gamma, \neg C_2 \neq \lambda\end{array}\right. \\ \left[\begin{array}{l}\Gamma|\neg A_1 \rightrightarrows \Gamma \\ \Gamma|\neg A_2 \rightrightarrows \Gamma\end{array}\right.\end{array}\right.}{\Gamma|\neg(A_1 \wedge A_2) \rightrightarrows \Gamma, \neg(C_1' \wedge C_2')} \qquad (\neg\vee)\ \frac{\left[\begin{array}{l}\Gamma|\neg A_1 \rightrightarrows \Gamma, \neg C_1 \\ \Gamma, \neg C_1|\neg A_2 \rightrightarrows \Gamma, \neg C_1, \neg C_2\end{array}\right.}{\Gamma|\neg(A_1 \vee A_2) \rightrightarrows \Gamma, \neg(C_1 \vee C_2)}$$

where

$$C_1' \vee C_2' = \begin{cases} C_1 \vee A_2 & \text{if } C_1 \neq \lambda \\ A_1 \vee C_2 & \text{if } C_1 = \lambda \text{ and } C_2 \neq \lambda \\ \lambda & \text{otherwise} \end{cases}$$

and

$$\neg(C_1' \wedge C_2') = \begin{cases} \neg(C_1 \wedge A_2 & \text{if } C_1 \neq \lambda \\ \neg(A_1 \wedge C_2) & \text{if } C_1 = \lambda \text{ and } C_2 \neq \lambda \\ \lambda & \text{otherwise.} \end{cases}$$

Definition 6.1.8 A reduction $\delta = \Gamma|A \rightrightarrows \Gamma, C$ is provable in $\mathbf{Q}_{\mathtt{t}}$, denoted by $\vdash_{\mathtt{t}} \Gamma|A \rightrightarrows \Gamma, C$, if there is a sequence $\{\delta_1, \ldots, \delta_m\}$ of reductions such that $\delta_m = \delta$, and for each $i < m$, δ_{i+1} is an axiom or is deduced from previous reductions by a deduction rule in $\mathbf{Q}_{\mathtt{t}}$.

Theorem 6.1.9 *For any formula set Γ and formula A, there is a formula C such that $C \preceq A$ and $\Gamma|A \rightrightarrows \Gamma, C$ is provable in $\mathbf{Q}_{\mathtt{t}}$.* □

Theorem 6.1.10 *For any formula set Γ and formula A, if $\Gamma|A \rightrightarrows \Gamma, C$ is provable in $\mathbf{Q}_{\mathtt{t}}$ then $C \preceq A$ is a $\preceq$-minimal of Γ by A.* □

$\mathbf{Q}_{\mathtt{t}}$ is composed of two parts: $\mathbf{Q}_{\mathtt{t}}^0$ (axiom with 0 and deduction rules) and $\mathbf{Q}_{\mathtt{t}}^+$ (axiom with + and deduction rules). Then, $\mathbf{Q}_{\mathtt{t}}^0$ is monotonic in Γ, and $\mathbf{Q}_{\mathtt{t}}^+$ is non-monotonic in Γ.

6.1.4 R-Calculus $\mathbf{Q}_{\mathtt{f}}$

Definition 6.1.11 A theory Γ, γ' is a $\preceq$-minimal change of Γ by γ, denoted by $\models_{\mathtt{f}}^{\mathbf{Q}} \Gamma|\gamma \rightrightarrows \Gamma, \gamma'$, if

(i) $\gamma' \preceq \gamma$,
(ii) $\Gamma \cup \{\gamma'\}$ is $\mathtt{f}$-consistent, and
(iii) for any theory Ξ with $\Gamma, \gamma' \prec \Xi \preceq \Gamma, \gamma$, Ξ is $\mathtt{f}$-*inconsistent.*

R-calculus $\mathbf{Q}_{\mathtt{f}}$ consists of the following axioms and deduction rules:

- **Axioms**:

$$(\mathtt{A}_{+}^{\mathtt{f}})\ \frac{\Gamma \nvdash \neg A}{\Gamma|A \rightrightarrows \Gamma, A} \qquad (\mathtt{A}_{0}^{\mathtt{f}})\ \frac{\Gamma \vdash \neg l}{\Gamma|l \rightrightarrows \Gamma, \lambda}$$

- **Deduction rules**:

$$(\wedge)\ \frac{\left\{\begin{array}{l}\Gamma|A_1 \rightrightarrows \Gamma, C_1 \neq \lambda\\ \left[\begin{array}{l}\Gamma|A_1 \rightrightarrows \Gamma\\ \Gamma|A_2 \rightrightarrows \Gamma, C_2 \neq \lambda\end{array}\right.\\ \left[\begin{array}{l}\Gamma|A_1 \rightrightarrows \Gamma\\ \Gamma|A_2 \rightrightarrows \Gamma\end{array}\right.\end{array}\right.}{\Gamma|A_1 \wedge A_2 \rightrightarrows \Gamma, C_1' \wedge C_2'} \qquad (\vee)\ \frac{\left[\begin{array}{l}\Gamma|A_1 \rightrightarrows \Gamma, C_1\\ \Gamma, C_1|A_2 \rightrightarrows \Gamma, C_1, C_2\end{array}\right.}{\Gamma|A_1 \vee A_2 \rightrightarrows \Gamma, C_1 \vee C_2}$$

$$(\neg\wedge)\ \frac{\left[\begin{array}{l}\Gamma|\neg A_1 \rightrightarrows \Gamma, \neg C_1\\ \Gamma, \neg C_1|\neg A_2 \rightrightarrows \Gamma, \neg C_1, \neg C_2\end{array}\right.}{\Gamma|\neg(A_1 \wedge A_2) \rightrightarrows \Gamma, \neg(C_1 \wedge C_2)} \qquad (\neg\vee)\ \frac{\left\{\begin{array}{l}\Gamma|\neg A_1 \rightrightarrows \Gamma, \neg C_1 \neq \lambda\\ \left[\begin{array}{l}\Gamma|\neg A_1 \rightrightarrows \Gamma\\ \Gamma|\neg A_2 \rightrightarrows \Gamma, \neg C_2 \neq \lambda\end{array}\right.\\ \left[\begin{array}{l}\Gamma|\neg A_1 \rightrightarrows \Gamma\\ \Gamma|\neg A_2 \rightrightarrows \Gamma\end{array}\right.\end{array}\right.}{\Gamma|\neg(A_1 \vee A_2) \rightrightarrows \Gamma, \neg(C_1' \vee C_2')}$$

where

$$C_1' \wedge C_2' = \begin{cases} C_1 \wedge A_2 \text{ if } C_1 \neq \lambda \\ A_1 \wedge C_2 \text{ if } C_1 = \lambda \text{ and } C_2 \neq \lambda \\ \lambda \qquad \text{otherwise} \end{cases}$$

and

$$\neg(C_1' \vee C_2') = \begin{cases} \neg(C_1 \vee A_2 \ \text{ if } C_1 \neq \lambda \\ \neg(A_1 \vee C_2) \text{ if } C_1 = \lambda \text{ and } C_2 \neq \lambda \\ \lambda \qquad \text{otherwise.} \end{cases}$$

Definition 6.1.12 A reduction $\delta = \Gamma|A \rightrightarrows \Gamma, C$ is provable in $\mathbf{Q}_{\mathfrak{f}}$, denoted by $\vdash^{\mathbf{Q}}_{\mathfrak{f}} \Gamma|A \rightrightarrows \Gamma, C$, if there is a sequence $\{\delta_1, \ldots, \delta_m\}$ of reductions such that $\delta_m = \delta$, and for each $i < m$, δ_{i+1} is an axiom or is deduced from previous reductions by a deduction rule in $\mathbf{Q}_{\mathfrak{f}}$.

Theorem 6.1.13 (Soundness and completeness theorem) *For any formula set* Γ *and formula* A,

$$\vdash^{\mathbf{Q}}_{\mathfrak{f}} \Gamma|A \rightrightarrows \Gamma, C \text{ iff } \models^{\mathbf{Q}}_{\mathfrak{f}} \Gamma|A \rightrightarrows \Gamma, C.$$

□

$\mathbf{Q}_{\mathfrak{f}}$ is composed of two parts: $\mathbf{Q}^0_{\mathfrak{f}}$ (axiom with 0 and deduction rules) and $\mathbf{Q}^+_{\mathfrak{f}}$ (axiom with + and deduction rules). Then, $\mathbf{Q}^0_{\mathfrak{f}}$ is monotonic in Γ, and $\mathbf{Q}^+_{\mathfrak{f}}$ is non-monotonic in Γ.

6.1.5 R-Calculus $\mathbf{P}_{\mathfrak{t}}$

Definition 6.1.14 Given a theory Γ and a formula δ, theory Γ, δ' is a deduction-based minimal ($\vdash_{\preceq}$-minimal) change of δ by Γ, denoted by $\models^{\mathbf{P}}_{\mathfrak{t}} \Gamma|\delta \rightrightarrows \Gamma, \delta'$, if

(i) $\Gamma \cup \{\delta'\}$ is consistent;
(ii) $\delta' \preceq \delta$, and
(iii) for any theory Ξ with $\Gamma, \delta' \succeq \Xi \succeq \Gamma, \delta$, either $\Xi \vdash \Gamma, \delta'$ and $\Gamma, \delta' \vdash \Xi$, or Ξ is inconsistent.

R-calculus $\mathbf{P}_{\mathfrak{t}}$ consists of the following axioms and deduction rules:

- **Axiom**:

$$(\mathrm{A}^+_{\mathbf{P}}) \ \frac{\Gamma \nvdash \neg l}{\Gamma|l \rightrightarrows \Gamma, l} \qquad (\mathrm{A}^0_{\mathbf{P}}) \ \frac{\Gamma \vdash \neg l}{\Gamma|l \rightrightarrows \Gamma}$$

- **Deduction rules**:

$$(\neg\neg)\ \frac{\Gamma|A_1 \rightrightarrows \Gamma, C_1}{\Gamma|\neg\neg A_1 \rightrightarrows \Gamma, \neg\neg C_1}$$

$$(\wedge)\ \frac{\Gamma|A_1 \rightrightarrows \Gamma, C_1}{\Gamma|A_1 \wedge A_2 \rightrightarrows \Gamma, C_1|A_2} \qquad (\vee)\ \frac{\left[\begin{array}{l}\Gamma|A_1 \rightrightarrows \Gamma, C_1\\ \Gamma|A_2 \rightrightarrows \Gamma, C_2\end{array}\right.}{\Gamma|A_1 \vee A_2 \rightrightarrows \Gamma, C_1 \vee C_2}$$

$$(\neg\wedge)\ \frac{\left[\begin{array}{l}\Gamma|\neg A_1 \rightrightarrows \Gamma, \neg C_1\\ \Gamma|\neg A_2 \rightrightarrows \Gamma, \neg C_2\end{array}\right.}{\Gamma|\neg(A_1 \wedge A_2) \rightrightarrows \Gamma, \neg(C_1 \wedge C_2)} \qquad (\neg\vee)\ \frac{\Gamma|\neg A_1 \rightrightarrows \Gamma, \neg C_1}{\Gamma|\neg(A_1 \vee A_2) \rightrightarrows \Gamma, \neg C_1|\neg A_2}$$

where if C is consistent then

$$\begin{array}{l}\lambda \vee C \equiv C \vee \lambda \equiv C\\ \lambda \wedge C \equiv C \wedge \lambda \equiv C;\\ \Gamma, \lambda \equiv \Gamma\end{array}$$

and if C is inconsistent then

$$\begin{array}{l}\lambda \vee C \equiv C \vee \lambda \equiv \lambda\\ \lambda \wedge C \equiv C \wedge \lambda \equiv \lambda.\end{array}$$

Theorem 6.1.15 *For any consistent set Γ of formulas and formula A in conjunctive normal form, there is a formula C such that*

(1) $\Gamma|A \rightrightarrows \Gamma, C$ is provable in $\mathbf{P}_{\mathrm{t}}$;
(2) $C \preceq A$, and
(3) $\Gamma \cup \{C\}$ is consistent, and for any D with $C \prec D \preceq A$, either $\Gamma, C \vdash D$ and $\Gamma, D \vdash C$, or $\Gamma \cup \{D\}$ is inconsistent.

Theorem 6.1.16 *For any consistent set Γ of formulas and formula A in conjunctive normal form, if $\vdash_{\mathrm{t}} \Gamma|A \rightrightarrows \Gamma, C$ then*

(1) $C \preceq A$, and
(2) $\Gamma \cup \{C\}$ is consistent, and
(3) for any D with $C \prec D \preceq A$, either $\Gamma, C \vdash D$ and $\Gamma, D \vdash C$, or $\Gamma \cup \{D\}$ is inconsistent. □

$\mathbf{P}_{\mathrm{t}}$ is composed of two parts: $\mathbf{P}_{\mathrm{t}}^0$ (axiom with 0 and deduction rules) and $\mathbf{P}_{\mathrm{t}}^+$ (axiom with + and deduction rules). Then, $\mathbf{P}_{\mathrm{t}}^0$ is monotonic in Γ, and $\mathbf{P}_{\mathrm{t}}^+$ is nonmonotonic in Γ.

6.1.6 R-Calculus $\mathbf{P}_{\mathtt{f}}$

Definition 6.1.17 Given a theory Γ and a formula δ, theory Γ, δ' is a deduction-based minimal ($\vdash_{\preceq}$-minimal) change of δ by Γ, denoted by $\models^{\mathbf{P}}_{\mathtt{f}} \Gamma|\delta \rightrightarrows \Gamma, \delta'$, if

(i) $\Gamma \cup \{\delta'\}$ is $\mathtt{f}$-consistent;
(ii) $\delta' \preceq \delta$, and
(iii) for any theory Ξ with $\Gamma, \delta' \succeq \Xi \succeq \Gamma, \delta$, either $\Xi \vdash \Gamma, \delta'$ and $\Gamma, \delta' \vdash \Xi$, or Ξ is $\mathtt{f}$-inconsistent.

R-calculus $\mathbf{P}_{\mathtt{f}}$ consists of the following axioms and deduction rules:

- **Axiom**:

$$(\mathtt{A}^+_{\mathtt{f}})\ \frac{\Gamma \nvdash \neg l}{\Gamma|l \rightrightarrows \Gamma, l} \qquad (\mathtt{A}^0_{\mathtt{f}})\ \frac{\Gamma \vdash \neg l}{\Gamma|l \rightrightarrows \Gamma}$$

- **Deduction rules**:

$$(\neg\neg)\ \frac{\Gamma|A_1 \rightrightarrows \Gamma, C_1}{\Gamma|\neg\neg A_1 \rightrightarrows \Gamma, \neg\neg C_1}$$

$$(\wedge)\ \frac{\left[\begin{array}{l}\Gamma|A_1 \rightrightarrows \Gamma, C_1\\ \Gamma|A_2 \rightrightarrows \Gamma, C_2\end{array}\right.}{\Gamma|A_1 \wedge A_2 \rightrightarrows \Gamma, C_1 \wedge C_2} \qquad (\vee)\ \frac{\Gamma|A_1 \rightrightarrows \Gamma, C_1}{\Gamma|A_1 \vee A_2 \rightrightarrows \Gamma, C_1|A_2}$$

$$(\neg\wedge)\ \frac{\Gamma|\neg A_1 \rightrightarrows \Gamma, \neg C_1}{\Gamma|\neg(A_1 \wedge A_2) \rightrightarrows \Gamma, \neg C_1|\neg A_2} \qquad (\neg\vee)\ \frac{\left[\begin{array}{l}\Gamma|\neg A_1 \rightrightarrows \Gamma, \neg C_1\\ \Gamma|\neg A_2 \rightrightarrows \Gamma, \neg C_2\end{array}\right.}{\Gamma|\neg(A_1 \vee A_2) \rightrightarrows \Gamma, \neg(C_1 \vee C_2)}$$

Theorem 6.1.18 *For any consistent set Γ of formulas and formula A in disjunctive normal form,*

$$\vdash^{\mathbf{P}}_{\mathtt{f}} \Gamma|A \rightrightarrows \Gamma, C \text{ iff } \models^{\mathbf{P}}_{\mathtt{f}} \Gamma|A \rightrightarrows \Gamma, C.$$

□

$\mathbf{P}_{\mathtt{f}}$ is composed of two parts: $\mathbf{P}^0_{\mathtt{f}}$ (axiom with 0 and deduction rules) and $\mathbf{P}^+_{\mathtt{f}}$ (axiom with + and deduction rules). Then, $\mathbf{P}^0_{\mathtt{f}}$ is monotonic in Γ, and $\mathbf{P}^+_{\mathtt{f}}$ is non-monotonic in Γ.

6.2 R-Calculi $\mathbf{R}^{Y_1 Q_1 i Y_2 Q_2 j}$

Let $Q_1, Q_2 \in \{\mathbf{E}, \mathbf{A}\}$ and $Y_1, Y_2 \in \{S, T, P\}$.

Given a sequent $\Gamma \Rightarrow \Delta$ and pair (A, B) of formulas, the result of $\Gamma \Rightarrow \Delta$ $\mathbf{G}^{Y_1 Q_1 i Y_2 Q_2 j}$-revising (A, B) is denoted by $\models^{Y_1 Q_1 i Y_2 Q_2 j} \Gamma \Rightarrow \Delta|(A, B) \rightrightarrows \Gamma' \Rightarrow \Delta'$, where

$$\Gamma'|\Delta' = \begin{cases} \Gamma \pm_1 A \Rightarrow \Delta \pm_2 B \text{ if } \Gamma \pm_1 A \Rightarrow \Delta \pm_2 B \text{ is } \mathbf{G}^{Y_1Q_1iY_2Q_2j}\text{-valid} \\ \Gamma \pm_1 A \Rightarrow \Delta \quad \text{otherwise, if } \Gamma \pm_1 A \Rightarrow \Delta \text{ is } \mathbf{G}^{Y_1Q_1iY_2Q_2j}\text{-valid} \\ \Gamma \Rightarrow \Delta \pm_2 B \quad \text{otherwise, if } \Gamma \Rightarrow \Delta \pm_2 B \text{ is } \mathbf{G}^{Y_1Q_1iY_2Q_2j}\text{-valid} \\ \Gamma \Rightarrow \Delta \quad \text{otherwise,} \end{cases}$$

where

$$\pm_1 A = \begin{cases} \cup\{A\} \text{ if } Q_1 = \mathbf{A} \\ -\{A\} \text{ if } Q_1 = \mathbf{E}, \end{cases} \qquad \pm_2 B = \begin{cases} \cup\{B\} \text{ if } Q_2 = \mathbf{A} \\ -\{B\} \text{ if } Q_2 = \mathbf{E}. \end{cases}$$

6.2.1 Axioms

Because $\models^{Y_1Q_1iY_2Q_2j} \Gamma \Rightarrow \Delta$ iff $\models^{Q_1iQ_2j} \Gamma \Rightarrow \Delta$, axioms for $\mathbf{R}^{Y_1Q_1iY_2Q_2j}$ are same as those for $\mathbf{R}^{Q_1iQ_2j}$.

Proposition 6.2.1 *Let $\Gamma \Rightarrow \Delta$ be literal. $\models^{Y_1Q_1iY_2Q_2j} \Gamma \Rightarrow \Delta|(l,m) \Rightarrow \Gamma' \Rightarrow \Delta'$ if and only if $\models^{Q_1iQ_2j} \Gamma \Rightarrow \Delta|(l,m) \Rightarrow \Gamma' \Rightarrow \Delta'$.* □

6.2.2 Deduction Rules

There are three basic sets $\mathcal{R}, \mathcal{Q}, \mathcal{P}$ of deduction rules:

◊ Deduction class $\mathcal{R}$:

- $\mathbf{R}^{L0}$:

$$(\neg\neg^L_+)\ \frac{\mathbf{X}|A \Rightarrow \mathbf{X}(A)}{\mathbf{X}|\neg\neg A \Rightarrow \mathbf{X}(\neg\neg A_1)} \qquad (\neg\neg^L_0)\ \frac{\mathbf{X}|A \Rightarrow \mathbf{X}}{\mathbf{X}|\neg\neg A \Rightarrow \mathbf{X}}$$

$$(\wedge^L_+)\ \frac{\begin{cases}\mathbf{X}|A_1 \Rightarrow \mathbf{X}(A_1) \\ \mathbf{X}|A_2 \Rightarrow \mathbf{X}(A_2)\end{cases}}{\mathbf{X}|A_1 \wedge A_2 \Rightarrow \mathbf{X}(A_1 \wedge A_2)} \qquad (\wedge^L_0)\ \frac{\begin{bmatrix}\mathbf{X}|A_1 \Rightarrow \mathbf{X} \\ \mathbf{X}|A_2 \Rightarrow \mathbf{X}\end{bmatrix}}{\mathbf{X}|A_1 \wedge A_2 \Rightarrow \mathbf{X}}$$

$$(\vee^L_+)\ \frac{\begin{bmatrix}\mathbf{X}|A_1 \Rightarrow \mathbf{X}(A_1) \\ \mathbf{X}(A_1)|A_2 \Rightarrow \mathbf{X}(A_1, A_2)\end{bmatrix}}{\mathbf{X}|A_1 \vee A_2 \Rightarrow \mathbf{X}(A_1 \vee A_2)} \qquad (\vee^L_0)\ \frac{\begin{cases}\mathbf{X}|A_1 \Rightarrow \mathbf{X} \\ \mathbf{X}(A_1)|A_2 \Rightarrow \mathbf{X}(A_1)\end{cases}}{\mathbf{X}|A_1 \vee A_2 \Rightarrow \mathbf{X}}$$

$$(\neg\wedge^L_+)\ \frac{\begin{bmatrix}\mathbf{X}|\neg A_1 \Rightarrow \mathbf{X}(\neg A_1) \\ \mathbf{X}(\neg A_1)|\neg A_2 \Rightarrow \mathbf{X}(\neg A_1, \neg A_2)\end{bmatrix}}{\mathbf{X}|\neg(A_1 \wedge A_2) \Rightarrow \mathbf{X}(\neg(A_1 \wedge A_2))} \qquad (\neg\wedge^L_0)\ \frac{\begin{cases}\mathbf{X}|\neg A_1 \Rightarrow \mathbf{X} \\ \mathbf{X}(\neg A_1)|\neg A_2 \Rightarrow \mathbf{X}(\neg A_1)\end{cases}}{\mathbf{X}|\neg(A_1 \wedge A_2) \Rightarrow \mathbf{X}}$$

$$(\neg\vee^L_+)\ \frac{\begin{cases}\mathbf{X}|\neg A_1 \Rightarrow \mathbf{X}(\neg A_1) \\ \mathbf{X}|\neg A_2 \Rightarrow \mathbf{X}(\neg A_2)\end{cases}}{\mathbf{X}|\neg(A_1 \vee A_2) \Rightarrow \mathbf{X}(\neg(A_1 \vee A_2))} \qquad (\neg\vee^L_0)\ \frac{\begin{bmatrix}\mathbf{X}|\neg A_1 \Rightarrow \mathbf{X} \\ \mathbf{X}|\neg A_2 \Rightarrow \mathbf{X}\end{bmatrix}}{\mathbf{X}|\neg(A_1 \vee A_2) \Rightarrow \mathbf{X}}$$

- $\mathbf{R}^{R1}$:

$$(\neg\neg^R_+)\ \frac{\mathbf{X}|B \Rightarrow \mathbf{X}(B)}{\mathbf{X}|\neg\neg B \Rightarrow \mathbf{X}(\neg\neg B)}$$

$$(\wedge^R_+)\ \frac{\left[\begin{array}{l}\mathbf{X}|B_1 \Rightarrow \mathbf{X}(B_1)\\ \mathbf{X}(B_1)|B_2 \Rightarrow \mathbf{X}(B_1, B_2)\end{array}\right.}{\mathbf{X}|B_1 \wedge B_2 \Rightarrow \mathbf{X}(B_1 \wedge B_2)}$$

$$(\vee^R_+)\ \frac{\left\{\begin{array}{l}\mathbf{X}|B_1 \Rightarrow \mathbf{X}(B_1)\\ \mathbf{X}|B_2 \Rightarrow \mathbf{X}(B_2)\end{array}\right.}{\mathbf{X}|B_1 \vee B_2 \Rightarrow \mathbf{X}(B_1 \vee B_2)}$$

$$(\neg\wedge^R_+)\ \frac{\left\{\begin{array}{l}\mathbf{X}|\neg B_1 \Rightarrow \mathbf{X}(\neg B_1)\\ \mathbf{X}|\neg B_2 \Rightarrow \mathbf{X}(\neg B_2)\end{array}\right.}{\mathbf{X}|\neg(B_1 \wedge B_2) \Rightarrow \mathbf{X}(\neg(B_1 \wedge B_2))}$$

$$(\neg\vee^R_+)\ \frac{\left[\begin{array}{l}\mathbf{X}|\neg B_1 \Rightarrow \mathbf{X}(\neg B_1)\\ \mathbf{X}(\neg B_1)|\neg B_2 \Rightarrow \mathbf{X}(\neg B_1, \neg B_2)\end{array}\right.}{\mathbf{X}|\neg(B_1 \vee B_2) \Rightarrow \mathbf{X}(\neg(B_1 \vee B_2))}$$

$$(\neg\neg^R_0)\ \frac{\mathbf{X}|B \Rightarrow \mathbf{X}}{\mathbf{X}|\neg\neg B \Rightarrow \mathbf{X}}$$

$$(\wedge^R_0)\ \frac{\left\{\begin{array}{l}\mathbf{X}|B_1 \Rightarrow \mathbf{X}\\ \mathbf{X}(B_1)|B_2 \Rightarrow \mathbf{X}(B_1)\end{array}\right.}{\mathbf{X}|B_1 \wedge B_2 \Rightarrow \mathbf{X}}$$

$$(\vee^R_0)\ \frac{\left[\begin{array}{l}\mathbf{X}|B_1 \Rightarrow \mathbf{X}\\ \mathbf{X}|B_2 \Rightarrow \mathbf{X}\end{array}\right.}{\mathbf{X}|B_1 \vee B_2 \Rightarrow \mathbf{X}}$$

$$(\neg\wedge^R_0)\ \frac{\left[\begin{array}{l}\mathbf{X}|\neg B_1 \Rightarrow \mathbf{X}\\ \mathbf{X}|\neg B_2 \Rightarrow \mathbf{X}\end{array}\right.}{\mathbf{X}|\neg(B_1 \wedge B_2) \Rightarrow \mathbf{X}}$$

$$(\neg\vee^R_0)\ \frac{\left\{\begin{array}{l}\mathbf{X}|\neg B_1 \Rightarrow \mathbf{X}\\ \mathbf{X}(\neg B_1)|\neg B_2 \Rightarrow \mathbf{X}(\neg B_1)\end{array}\right.}{\mathbf{X}|\neg(B_1 \vee B_2) \Rightarrow \mathbf{X}}$$

- $\mathbf{R}^{L1}$: where $A \in \Gamma$.

$$(\neg\neg^L_0)\ \frac{\mathbf{X}|A \Rightarrow \mathbf{X}}{\mathbf{X}|\neg\neg A \Rightarrow \mathbf{X}}$$

$$(\wedge^L_0)\ \frac{\left[\begin{array}{l}\mathbf{X}|A_1 \Rightarrow \mathbf{X}\\ \mathbf{X}[A_1]|A_2 \Rightarrow \mathbf{X}[A_1]\end{array}\right.}{\mathbf{X}|A_1 \wedge A_2 \Rightarrow \mathbf{X}}$$

$$(\vee^L_0)\ \frac{\left\{\begin{array}{l}\mathbf{X}|A_1 \Rightarrow \mathbf{X}\\ \mathbf{X}|A_2 \Rightarrow \mathbf{X}\end{array}\right.}{\mathbf{X}|A_1 \vee A_2 \Rightarrow \mathbf{X}}$$

$$(\neg\wedge^L_0)\ \frac{\left\{\begin{array}{l}\mathbf{X}|\neg A_1 \Rightarrow \mathbf{X}\\ \mathbf{X}|\neg A_2 \Rightarrow \mathbf{X}\end{array}\right.}{\mathbf{X}|\neg(A_1 \wedge A_2) \Rightarrow \mathbf{X}}$$

$$(\neg\vee^L_0)\ \frac{\left\{\begin{array}{l}\mathbf{X}|\neg A_1|\mathbf{X} \Rightarrow \mathbf{X}\\ \mathbf{X}[\neg A_1]|\neg A_2 \Rightarrow \mathbf{X}[\neg A_1]\end{array}\right.}{\mathbf{X}|\neg(A_1 \vee A_2) \Rightarrow \mathbf{X}}$$

$$(\neg\neg^L_-)\ \frac{\mathbf{X}|A \Rightarrow \mathbf{X}[A]}{\mathbf{X}|\neg\neg A \Rightarrow \mathbf{X}[\neg\neg A]}$$

$$(\wedge^L_-)\ \frac{\left\{\begin{array}{l}\mathbf{X}|A_1 \Rightarrow \mathbf{X}[A_1]\\ \mathbf{X}[A_1]|A_2 \Rightarrow \mathbf{X}[A_1, A_2]\end{array}\right.}{\mathbf{X}|A_1 \wedge A_2 \Rightarrow \mathbf{X}[A_1 \wedge A_2]}$$

$$(\vee^L_-)\ \frac{\left[\begin{array}{l}\mathbf{X}|A_1 \Rightarrow \mathbf{X}[A_1]\\ \mathbf{X}|A_2 \Rightarrow \mathbf{X}[A_2]\end{array}\right.}{\mathbf{X}|A_1 \vee A_2 \Rightarrow \mathbf{X}[A_1 \vee A_2]}$$

$$(\neg\wedge^L_-)\ \frac{\left[\begin{array}{l}\mathbf{X}|\neg A_1 \Rightarrow \mathbf{X}[\neg A_1]\\ \mathbf{X}|\neg A_2 \Rightarrow \mathbf{X}[\neg A_2]\end{array}\right.}{\mathbf{X}|\neg(A_1 \wedge A_2) \Rightarrow \mathbf{X}[\neg(A_1 \wedge A_2)]}$$

$$(\neg\vee^L_-)\ \frac{\left[\begin{array}{l}\mathbf{X}|\neg A_1 \Rightarrow \mathbf{X}[\neg A_1]\\ \mathbf{X}[\neg A_1]|\neg A_2 \Rightarrow \mathbf{X}[\neg A_1, \neg A_2]\end{array}\right.}{\mathbf{X}|\neg(A_1 \vee A_2) \Rightarrow \mathbf{X}[\neg(A_1 \vee A_2)}$$

- $\mathbf{R}^{R0}$: where $B \in \Delta$

$$(\neg\neg_0^R)\ \frac{\mathbf{X}|B \Rightarrow \mathbf{X}}{\mathbf{X}|\neg\neg B \Rightarrow \mathbf{X}} \qquad (\neg\neg_-^R)\ \frac{\mathbf{X}|B \Rightarrow \mathbf{X}[B]}{\mathbf{X}|\neg\neg B \Rightarrow \mathbf{X}[\neg\neg B]}$$

$$(\wedge_0^R)\ \frac{\begin{cases}\mathbf{X}|B_1 \Rightarrow \mathbf{X}\\ \mathbf{X}|B_2 \Rightarrow \mathbf{X}\end{cases}}{\mathbf{X}|B_1 \wedge B_2 \Rightarrow \mathbf{X}} \qquad (\wedge_-^R)\ \frac{\left[\begin{array}{l}\mathbf{X}|B_1 \Rightarrow \mathbf{X}[B_1]\\ \mathbf{X}[B_1]|B_2 \Rightarrow \mathbf{X}[B_1, B_2]\end{array}\right.}{\mathbf{X}|B_1 \wedge B_2 \Rightarrow \mathbf{X}[B_1 \wedge B_2]}$$

$$(\vee_0^R)\ \frac{\left[\begin{array}{l}\mathbf{X}|B_1 \Rightarrow \mathbf{X}\\ \mathbf{X}|B_2 \Rightarrow \mathbf{X}\end{array}\right.}{\mathbf{X}|B_1 \vee B_2 \Rightarrow \mathbf{X}} \qquad (\vee_-^R)\ \frac{\begin{cases}\mathbf{X}|B_1 \Rightarrow \mathbf{X}[B_1]\\ \mathbf{X}|B_2 \Rightarrow \mathbf{X}[B_2]\end{cases}}{\mathbf{X}|B_1 \vee B_2 \Rightarrow \mathbf{X}[B_1 \vee B_2]}$$

$$(\neg\wedge_0^R)\ \frac{\left[\begin{array}{l}\mathbf{X}|\neg B_1 \Rightarrow \mathbf{X}\\ \mathbf{X}|\neg B_2 \Rightarrow \mathbf{X}\end{array}\right.}{\mathbf{X}|\neg(B_1 \wedge B_2) \Rightarrow \mathbf{X}} \qquad (\neg\wedge_-^R)\ \frac{\begin{cases}\mathbf{X}|\neg B_1 \Rightarrow \mathbf{X}[\neg B_1]\\ \mathbf{X}|\neg B_2 \Rightarrow \mathbf{X}[\neg B_2]\end{cases}}{\mathbf{X}|\neg(B_1 \wedge B_2) \Rightarrow \mathbf{X}[\neg(B_1 \wedge B_2)}$$

$$(\neg\vee_0^R)\ \frac{\begin{cases}\mathbf{X}|\neg B_1 \Rightarrow \mathbf{X}\\ \mathbf{X}|\neg B_2 \Rightarrow \mathbf{X}\end{cases}}{\mathbf{X}|\neg(B_1 \vee B_2) \Rightarrow \mathbf{X}} \qquad (\neg\vee_-^R)\ \frac{\left[\begin{array}{l}\mathbf{X}|\neg B_1 \Rightarrow \mathbf{X}[\neg B_1]\\ \mathbf{X}|\neg B_2 \Rightarrow \mathbf{X}[\neg B_2]\end{array}\right.}{\mathbf{X}|\neg(B_1 \vee B_2) \Rightarrow \mathbf{X}[\neg(B_1 \vee B_2)]}$$

◊ Deduction class $\mathcal{Q}$:

- $\mathbf{Q}^{L0}$:

$$(\neg\neg^L)\ \frac{\mathbf{X}|A \Rightarrow \mathbf{X}(A')}{\mathbf{X}|\neg\neg A \Rightarrow \mathbf{X}(\neg\neg A')}$$

$$(\wedge^L)\ \frac{\begin{cases}\left[\begin{array}{l}\mathbf{X}|A_1 \Rightarrow \mathbf{X}(C_1)\\ C_1 \neq \lambda\end{array}\right.\\ \left[\begin{array}{l}\mathbf{X}|A_1 \Rightarrow \mathbf{X}\\ \mathbf{X}|A_2 \Rightarrow \mathbf{X}(C_2)\\ C_2 \neq \lambda\end{array}\right.\\ \left[\begin{array}{l}\mathbf{X}|A_1 \Rightarrow \mathbf{X}\\ \mathbf{X}|A_2 \Rightarrow \mathbf{X}\end{array}\right.\end{cases}}{\mathbf{X}|A_1 \wedge A_2 \Rightarrow \mathbf{X}(C_1' \wedge C_2')} \qquad (\vee^L)\ \frac{\left[\begin{array}{l}\mathbf{X}|A_1 \Rightarrow \mathbf{X}(C_1)\\ \mathbf{X}(C_1)|A_2 \Rightarrow \mathbf{X}(C_1, C_2)\end{array}\right.}{\mathbf{X}|A_1 \vee A_2 \Rightarrow \mathbf{X}(C_1 \vee C_2)}$$

and

$$(\neg\wedge^L)\ \frac{\left[\begin{array}{l}\mathbf{X}|\neg A_1 \Rightarrow \mathbf{X}(\neg C_1)\\ \mathbf{X}(\neg C_1)|\neg A_2 \Rightarrow \mathbf{X}(\neg C_1, \neg C_2)\end{array}\right.}{\mathbf{X}|\neg(A_1 \wedge A_2) \Rightarrow \mathbf{X}(\neg(C_1 \wedge C_2))}$$

$$(\neg\vee^L)\ \frac{\begin{cases}\left[\begin{array}{l}\mathbf{X}|\neg A_1 \Rightarrow \mathbf{X}(\neg C_1)\\ C_1 \neq \lambda\end{array}\right.\\ \left[\begin{array}{l}\mathbf{X}|\neg A_1 \Rightarrow \mathbf{X}\\ \mathbf{X}|\neg A_2 \Rightarrow \mathbf{X}(\neg C_2)\\ C_2 \neq \lambda\end{array}\right.\\ \left[\begin{array}{l}\mathbf{X}|\neg A_1 \Rightarrow \mathbf{X}\\ \mathbf{X}|\neg A_2 \Rightarrow \mathbf{X}\end{array}\right.\end{cases}}{\mathbf{X}|\neg(A_1 \vee A_2) \Rightarrow \mathbf{X}(\neg(C_1' \vee C_2'))}$$

where

$$C_1' \wedge C_2' = \begin{cases} C_1 \wedge A_2 \text{ if } C_1 \neq \lambda \\ A_1 \wedge C_2 \text{ if } C_1 = \lambda \text{ and } C_2 \neq \lambda \\ \lambda \quad \text{otherwise} \end{cases}$$

$$\neg(C_1' \vee C_2') = \begin{cases} \neg(C_1 \vee A_2) \text{ if } C_1 \neq \lambda \\ \neg(A_1 \vee C_2) \text{ if } C_1 = \lambda \text{ and } C_2 \neq \lambda \\ \lambda \quad \text{otherwise} \end{cases}$$

- $\mathbf{Q}^{R1}$:

$$(\neg\neg^R)\ \frac{\mathbf{X}|B \Rightarrow \mathbf{X}(C)}{\mathbf{X}|\neg\neg B \Rightarrow \mathbf{X}(\neg\neg C)}$$

$$(\wedge^R)\ \frac{\begin{bmatrix} \mathbf{X}|B_1 \Rightarrow \mathbf{X}(D_1) \\ \mathbf{X}(D_1)|B_2 \Rightarrow \mathbf{X}(D_1, D_2) \end{bmatrix}}{\mathbf{X}|B_1 \wedge B_2 \Rightarrow \mathbf{X}(D_1 \wedge D_2)} \quad (\vee^R)\ \frac{\begin{cases} \begin{bmatrix} \mathbf{X}|B_1 \Rightarrow \mathbf{X}(D_1) \\ D_1 \neq \lambda \end{bmatrix} \\ \begin{bmatrix} \mathbf{X}|B_1 \Rightarrow \mathbf{X} \\ \mathbf{X}|B_2 \Rightarrow \mathbf{X}(D_2) \\ D_2 \neq \lambda \end{bmatrix} \\ \begin{bmatrix} \mathbf{X}|B_1 \Rightarrow \mathbf{X} \\ \mathbf{X}|B_2 \Rightarrow \mathbf{X} \end{bmatrix} \end{cases}}{\mathbf{X}|B_1 \vee B_2 \Rightarrow \mathbf{X}(D_1' \vee D_2')}$$

and

$$(\neg\wedge^R)\ \frac{\begin{cases} \begin{bmatrix} \mathbf{X}|\neg B_1 \Rightarrow \mathbf{X}(\neg D_1) \\ D_1 \neq \lambda \end{bmatrix} \\ \begin{bmatrix} \mathbf{X}|\neg B_1 \Rightarrow \mathbf{X} \\ \mathbf{X}|\neg B_2 \Rightarrow \mathbf{X}(\neg D_2) \\ D_2 \neq \lambda \end{bmatrix} \\ \begin{bmatrix} \mathbf{X}|\neg B_1 \Rightarrow \mathbf{X} \\ \mathbf{X}|\neg B_2 \Rightarrow \mathbf{X} \end{bmatrix} \end{cases}}{\mathbf{X}|\neg(B_1 \wedge B_2) \Rightarrow \mathbf{X}(\neg(D_1' \wedge D_2'))}$$

$$(\neg\vee^R)\ \frac{\begin{bmatrix} \mathbf{X}|\neg B_1 \Rightarrow \mathbf{X}(\neg D_1) \\ \mathbf{X}(\neg D_1)|\neg B_2 \Rightarrow \mathbf{X}(\neg D_1, \neg D_2) \end{bmatrix}}{\mathbf{X}|\neg(B_1 \vee B_2) \Rightarrow \mathbf{X}\neg(D_1 \vee D_2))}$$

where

$$D_1' \vee D_2' = \begin{cases} D_1 \vee B_2 \text{ if } D_1 \neq \lambda \\ B_1 \vee D_2 \text{ if } D_1 = \lambda \text{ and } D_2 \neq \lambda \\ \lambda \quad \text{otherwise} \end{cases}$$

$$\neg(D_1' \wedge D_2') = \begin{cases} \neg(D_1 \wedge B_2) \text{ if } D_1 \neq \lambda \\ \neg(B_1 \wedge D_2) \text{ if } D_1 = \lambda \text{ and } D_2 \neq \lambda \\ \lambda \quad \text{otherwise} \end{cases}$$

- $\mathbf{Q}^{L1}$: where $A \in \Gamma$.

$$(\neg\neg^L)\ \frac{\mathbf{X}|A \Rightarrow \mathbf{X}[C]}{\mathbf{X}|\neg\neg A \Rightarrow \mathbf{X}[\neg\neg C]}$$

$$(\wedge^L)\ \frac{\left\{\begin{array}{l}\left[\begin{array}{l}\mathbf{X}|A_1 \Rightarrow \mathbf{X}[C_1]\\ C_1 \neq \lambda\end{array}\right.\\ \left[\begin{array}{l}\mathbf{X}|A_1 \Rightarrow \mathbf{X}\\ \mathbf{X}|A_2 \Rightarrow \mathbf{X}[C_2]\\ C_2 \neq \lambda\end{array}\right.\\ \left[\begin{array}{l}\mathbf{X}|A_1 \Rightarrow \mathbf{X}\\ \mathbf{X}|A_2 \Rightarrow \mathbf{X}\end{array}\right.\end{array}\right.}{\mathbf{X}|A_1 \wedge A_2 \Rightarrow \mathbf{X}[C_1' \wedge C_2']} \qquad (\vee^L)\ \frac{\left[\begin{array}{l}\mathbf{X}|A_1 \Rightarrow \mathbf{X}[C_1]\\ \mathbf{X}[C_1]|A_2 \Rightarrow \mathbf{X}[C_1, C_2]\end{array}\right.}{\mathbf{X}|A_1 \vee A_2 \Rightarrow \mathbf{X}[C_1 \vee C_2]}$$

and

$$(\neg\wedge^L)\ \frac{\left[\begin{array}{l}\mathbf{X}|\neg A_1 \Rightarrow \mathbf{X}[\neg C_1]\\ \mathbf{X}[\neg C_1]|\neg A_2 \Rightarrow \mathbf{X}[\neg C_1, \neg C_2]\end{array}\right.}{\mathbf{X}|\neg(A_1 \wedge A_2) \Rightarrow \mathbf{X}[\neg(C_1 \wedge C_2)]}$$

$$(\neg\vee^L)\ \frac{\left\{\begin{array}{l}\left[\begin{array}{l}\mathbf{X}|\neg A_1 \Rightarrow \mathbf{X}[\neg C_1]\\ C_1 \neq \lambda\end{array}\right.\\ \left[\begin{array}{l}\mathbf{X}|\neg A_1 \Rightarrow \mathbf{X}\\ \mathbf{X}|\neg A_2 \Rightarrow \mathbf{X}[\neg C_2]\\ C_2 \neq \lambda\end{array}\right.\\ \left[\begin{array}{l}\mathbf{X}|\neg A_1 \Rightarrow \mathbf{X}\\ \mathbf{X}|\neg A_2 \Rightarrow \mathbf{X}\end{array}\right.\end{array}\right.}{\mathbf{X}|\neg(A_1 \vee A_2) \Rightarrow \mathbf{X}[\neg(C_1' \vee C_2')]}$$

where

$$C_1' \wedge C_2' = \begin{cases} C_1 \vee A_1 & \text{if } C_1 \neq \lambda \\ A_1 \vee C_1 & \text{if } C_1 = \lambda \text{ and } C_2 \neq \lambda \\ \lambda & \text{otherwise;} \end{cases}$$

$$\neg(C_1' \wedge C_2') = \begin{cases} \neg(C_1 \vee A_1) & \text{if } C_1 \neq \lambda \\ \neg(A_1 \vee C_2) & \text{if } C_1 = \lambda \text{ and } C_2 \neq \lambda \\ \lambda & \text{otherwise;} \end{cases}$$

- $\mathbf{Q}^{R0}$: where $B \in \Delta$

$$(\neg\neg^R)\ \frac{\mathbf{X}|B \Rightarrow \mathbf{X}[D]}{\mathbf{X}|\neg\neg B \Rightarrow \mathbf{X}[\neg\neg D]}$$

$$(\wedge^R)\ \frac{\left[\begin{array}{l}\mathbf{X}|B_1 \Rightarrow \mathbf{X}[D_1]\\ \mathbf{X}[D_1]|B_2 \Rightarrow \mathbf{X}[D_1, D_2]\end{array}\right.}{\mathbf{X}|B_1 \wedge B_2 \Rightarrow \mathbf{X}[D_1 \wedge D_2]}\quad (\vee^R)\ \frac{\left\{\begin{array}{l}\left[\begin{array}{l}\mathbf{X}|B_1 \Rightarrow \mathbf{X}[D_1]\\ D_1 \neq \lambda\end{array}\right.\\ \left[\begin{array}{l}\mathbf{X}|B_1 \Rightarrow \mathbf{X}\\ \mathbf{X}|B_2 \Rightarrow \mathbf{X}[D_2]\\ D_2 \neq \lambda\end{array}\right.\\ \left[\begin{array}{l}\mathbf{X}|B_1 \Rightarrow \mathbf{X}\\ \mathbf{X}|B_2 \Rightarrow \mathbf{X}\end{array}\right.\end{array}\right.}{\mathbf{X}|B_1 \vee B_2|\mathbf{X} \Rightarrow \mathbf{X}[D_1' \vee D_2']}$$

and

$$(\neg\wedge^R)\ \frac{\left\{\begin{array}{l}\left[\begin{array}{l}\mathbf{X}|\neg B_1 \Rightarrow \mathbf{X}[\neg D_1]\\ D_1 \neq \lambda\end{array}\right.\\ \left[\begin{array}{l}\mathbf{X}|\neg B_1 \Rightarrow \mathbf{X}\\ \mathbf{X}|\neg B_2 \Rightarrow \mathbf{X}[\neg D_2]\\ D_2 \neq \lambda\end{array}\right.\\ \left[\begin{array}{l}\mathbf{X}|\neg B_1 \Rightarrow \mathbf{X}\\ \mathbf{X}|\neg B_2 \Rightarrow \mathbf{X}\end{array}\right.\end{array}\right.}{\mathbf{X}|\neg(B_1 \wedge B_2) \Rightarrow \mathbf{X}[\neg(D_1' \wedge D_2')]}$$

$$(\neg\vee^R)\ \frac{\left[\begin{array}{l}\mathbf{X}|\neg B_1 \Rightarrow \mathbf{X}[\neg D_1]\\ \mathbf{X}[\neg D_1]|\neg B_2 \Rightarrow \mathbf{X}[\neg D_1, \neg D_2]\end{array}\right.}{\mathbf{X}|\neg(B_1 \vee B_2) \Rightarrow \mathbf{X}[\neg(D_1' \vee D_2')]}$$

where

$$D_1' \vee D_2' = \begin{cases} D_1 \vee B_1 \text{ if } D_1 \neq \lambda \\ B_1 \vee D_1 \text{ if } D_1 = \lambda \text{ and } D_2 \neq \lambda \\ \lambda \qquad \text{otherwise;} \end{cases}$$

$$\neg(D_1' \vee D_2') = \begin{cases} \neg(D_1 \vee B_2) \text{ if } D_1 \neq \lambda \\ \neg(B_1 \vee D_2) \text{ if } D_1 = \lambda \text{ and } D_2 \neq \lambda \\ \lambda \qquad \text{otherwise;} \end{cases}$$

◇ Deduction class $\mathcal{P}$:

- $\mathbf{P}^{L0}$:

$$(\neg\neg^L)\ \frac{\mathbf{X}|A \Rightarrow \mathbf{X}(C)}{\mathbf{X}|\neg\neg A \Rightarrow \mathbf{X}(\neg\neg C)}$$

$$(\wedge^L)\ \frac{\left[\begin{array}{l}\mathbf{X}|A_1 \Rightarrow \mathbf{X}(C_1)\\ \mathbf{X}|A_2 \Rightarrow \mathbf{X}(C_2)\end{array}\right.}{\mathbf{X}|A_1 \wedge A_2 \Rightarrow \mathbf{X}(C_1 \wedge C_2)}\quad (\vee^L)\ \frac{\left[\begin{array}{l}\mathbf{X}|A_1 \Rightarrow \mathbf{X}(C_1)\\ \mathbf{X}(C_1)|A_2 \Rightarrow \mathbf{X}(C_1, C_2)\end{array}\right.}{\mathbf{X}|A_1 \vee A_2 \Rightarrow \mathbf{X}(C_1 \vee C_2)}$$

and

$$(\neg\wedge^L)\ \frac{\left[\begin{array}{l}\mathbf{X}|\neg A_1 \Rightarrow \mathbf{X}(\neg C_1)\\ \mathbf{X}(\neg C_1)|\neg A_2 \Rightarrow \mathbf{X}(\neg C_1, \neg C_2)\end{array}\right.}{\mathbf{X}|\neg(A_1 \wedge A_2) \Rightarrow \mathbf{X}(\neg(C_1 \wedge C_2))}$$

$$(\neg\vee^L)\ \frac{\left[\begin{array}{l}\mathbf{X}|\neg A_1 \Rightarrow \mathbf{X}(\neg C_1)\\ \mathbf{X}|\neg A_2 \Rightarrow \mathbf{X}(\neg C_2)\end{array}\right.}{\mathbf{X}|\neg(A_1 \vee A_2) \Rightarrow \mathbf{X}(\neg(C_1 \vee C_2))}$$

- $\mathbf{P}^{R1}$:

$$(\neg\neg^R)\ \frac{\mathbf{X}|B \Rightarrow \mathbf{X}(D)}{\mathbf{X}|\neg\neg B \Rightarrow \mathbf{X}(\neg\neg D)}$$

$$(\wedge^R)\ \frac{\left[\begin{array}{l}\mathbf{X}|B_1 \Rightarrow \mathbf{X}(D_1)\\ \mathbf{X}(D_1)|B_2 \Rightarrow \mathbf{X}(D_1, D_2)\end{array}\right.}{\mathbf{X}|B_1 \wedge B_2 \Rightarrow \mathbf{X}(D_1 \wedge D_2)} \quad (\vee^R)\ \frac{\left[\begin{array}{l}\mathbf{X}|B_1 \Rightarrow \mathbf{X}(D_1)\\ \mathbf{X}|B_2 \Rightarrow \mathbf{X}(D_2)\end{array}\right.}{\mathbf{X}|B_1 \vee B_2 \Rightarrow \mathbf{X}(D_1 \vee D_2)}$$

and

$$(\neg\wedge^R)\ \frac{\left[\begin{array}{l}\mathbf{X}|\neg B_1 \Rightarrow \mathbf{X}(\neg D_1)\\ \mathbf{X}|\neg B_2 \Rightarrow \mathbf{X}(\neg D_2)\end{array}\right.}{\mathbf{X}|\neg(B_1 \wedge B_2) \Rightarrow \mathbf{X}(\neg(D_1 \wedge D_2))}$$

$$(\neg\vee^R)\ \frac{\left[\begin{array}{l}\mathbf{X}|\neg B_1 \Rightarrow \mathbf{X}(\neg D_1)\\ \mathbf{X}(\neg D_1)|\neg B_2 \Rightarrow \mathbf{X}(\neg D_1, \neg D_2)\end{array}\right.}{\mathbf{X}|\neg(B_1 \vee B_2) \Rightarrow \mathbf{X}(\neg(D_1 \vee D_2))}$$

- $\mathbf{P}^{L1}$: where $A \in \Gamma$.

$$(\neg\neg^L)\ \frac{\mathbf{X}|A \Rightarrow \mathbf{X}[C]}{\mathbf{X}|\neg\neg A \Rightarrow \mathbf{X}[\neg\neg C]}$$

$$(\wedge^L)\ \frac{\left[\begin{array}{l}\mathbf{X}|A_1 \Rightarrow \mathbf{X}[C_1]\\ \mathbf{X}|A_2 \Rightarrow \mathbf{X}[C_2]\end{array}\right.}{\mathbf{X}|A_1 \wedge A_2 \Rightarrow \mathbf{X}[C_1 \wedge C_2]} \quad (\vee^L)\ \frac{\left[\begin{array}{l}\mathbf{X}|A_1 \Rightarrow \mathbf{X}[C_1]\\ \mathbf{X}[C_1]|A_2 \Rightarrow \mathbf{X}[C_1, C_2]\end{array}\right.}{\mathbf{X}|A_1 \vee A_2 \Rightarrow \mathbf{X}[C_1 \vee C_2]}$$

and

$$(\neg\wedge^L)\ \frac{\left[\begin{array}{l}\mathbf{X}|\neg A_1 \Rightarrow \mathbf{X}[\neg C_1]\\ \mathbf{X}[\neg C_1]|\neg A_2 \Rightarrow \mathbf{X}[\neg C_1, \neg C_2]\end{array}\right.}{\mathbf{X}|\neg(A_1 \wedge A_2) \Rightarrow \mathbf{X}[\neg(C_1 \wedge C_2)]}$$

$$(\neg\vee^L)\ \frac{\left[\begin{array}{l}\mathbf{X}|\neg A_1 \Rightarrow \mathbf{X}[\neg C_1]\\ \mathbf{X}|\neg A_2 \Rightarrow \mathbf{X}[\neg C_2]\end{array}\right.}{\mathbf{X}|\neg(A_1 \vee A_2) \Rightarrow \mathbf{X}[\neg(C_1 \vee C_2]}$$

- $\mathbf{P}^{R0}$: where $B \in \Delta$

$$(\neg\neg^R)\ \frac{\mathbf{X}|B \Rightarrow \mathbf{X}[D]}{\mathbf{X}|\neg\neg B \Rightarrow \mathbf{X}[\neg\neg D]}$$

$$(\wedge^R)\ \frac{\left[\begin{array}{l}\mathbf{X}|B_1 \Rightarrow \mathbf{X}[D_1]\\ \mathbf{X}[D_1]|B_2 \Rightarrow \mathbf{X}[D_1, D_2]\end{array}\right.}{\mathbf{X}|B_1 \wedge B_2 \Rightarrow \mathbf{X}[D_1 \wedge D_2]}\quad (\vee^R)\ \frac{\left[\begin{array}{l}\mathbf{X}|B_1 \Rightarrow \mathbf{X}[D_1]\\ \mathbf{X}|B_2 \Rightarrow \mathbf{X}[D_2]\end{array}\right.}{\mathbf{X}|B_1 \vee B_2 \Rightarrow \mathbf{X}[D_1 \vee D_2]}$$

and

$$(\neg\wedge^R)\ \frac{\left[\begin{array}{l}\mathbf{X}|\neg B_1 \Rightarrow \mathbf{X}[\neg D_1]\\ \mathbf{X}|\neg B_2 \Rightarrow \mathbf{X}[\neg D_2]\end{array}\right.}{\mathbf{X}|\neg(B_1 \wedge B_2) \Rightarrow \mathbf{X}[\neg(D_1 \wedge D_2)]}$$

$$(\neg\vee^R)\ \frac{\left[\begin{array}{l}\mathbf{X}|\neg B_1 \Rightarrow \mathbf{X}[\neg D_1]\\ \mathbf{X}[\neg D_1]|\neg B_1 \Rightarrow \mathbf{X}[\neg D_1, \neg D_2]\end{array}\right.}{\mathbf{X}|\neg(B_1 \vee B_2) \Rightarrow \mathbf{X}[\neg(D_1 \vee D_2)]}$$

Notice that $\mathbf{Y}^{L0}$, $\mathbf{Y}^{R1}$ are for enumerating elements into Γ, and $\mathbf{Y}^{L1}$, $\mathbf{Y}^{R0}$ are for not enumerating elements into Γ.

6.2.3 R-Calculi

For $\mathbf{Y} \in \{\mathbf{R}, \mathbf{Q}, \mathbf{P}\}$, there are eight R-calculi

$$\begin{array}{cccc}\mathbf{Y}^{LE0} & \mathbf{Y}^{LA0} & \mathbf{Y}^{LE1} & \mathbf{Y}^{LA1}\\ \mathbf{Y}^{RE0} & \mathbf{Y}^{RA0} & \mathbf{Y}^{RE1} & \mathbf{Y}^{RA1}\end{array}$$

and there are four given R-calculi:

$$\begin{array}{cc}\mathbf{R}^{L0} & \mathbf{R}^{R1}\\ \mathbf{R}^{L1} & \mathbf{R}^{R0}.\end{array}$$

We define

$$\begin{array}{ll}\mathbf{Y}^{LE0} = \mathbf{Y}^{L0} & \mathbf{Y}^{RE0} = \mathbf{Y}^{R0}\\ \mathbf{Y}^{LA0} = \mathbf{Y}^{L0}((A)/[A]) & \mathbf{Y}^{RA0} = \mathbf{Y}^{R1}((B)/[B])\\ \mathbf{Y}^{LE1} = \mathbf{Y}^{L1}([A]/(A)) & \mathbf{Y}^{RE1} = \mathbf{Y}^{R1}\\ \mathbf{Y}^{LA1} = \mathbf{Y}^{L1} & \mathbf{Y}^{RA1} = \mathbf{Y}^{R1}((B)/[B]).\end{array}$$

Define

$$\mathbf{R}^{Y_1Q_1iY_2Q_2j} = \mathbb{A}^{Q_1iQ_2j} + \mathbf{Y}_1^{LQ_1i} + \mathbf{Y}_2^{RQ_2j}.$$

Definition 6.2.2 A reduction

$$\delta = \Gamma \Rightarrow \Delta|(A, B) \Rightarrow \Gamma' \Rightarrow \Delta'$$

is provable in $\mathbf{R}^{Y_1Q_1iY_2Q_2j}$, denoted by $\vdash^{Y_1Q_1iY_2Q_2j} \delta$, if there is a sequence $\{\delta_1, \ldots, \delta_n\}$ of reductions such that $\delta_n = \delta$, and for each $1 \leq i \leq n$, δ_i is either an axiom or deduced from the previous reductions by one of the deduction rules in $\mathbf{R}^{Y_1Q_1iY_2Q_2j}$.

Theorem 6.2.3 (Soundness and completeness theorem) *For any $Q_1, Q_2 \in \{\mathbf{A}, \mathbf{E}\}$, $i, j \in \{0, 1\}$, $Y_1, Y_2 \in \{\mathbf{R}, \mathbf{Q}\}$ and reduction $\delta = \Gamma \mapsto \Delta|(A, B) \Rightarrow \Gamma' \mapsto_{Y_1Q_1iY_2Q_2j} \Delta'$, δ is $\mathbf{R}^{Y_1Q_1iY_2Q_2j}$-valid if and only if δ is provable in $\mathbf{R}^{Y_1Q_1iY_2Q_2j}$. That is,*

$$\models^{Y_1Q_1iY_2Q_2j} \delta \text{ iff } \vdash^{Y_1Q_1iY_2Q_2j} \delta.$$

□

Theorem 6.2.4 (Soundness and completeness theorem) *For any $Q_1, Q_2 \in \{\mathbf{A}, \mathbf{E}\}$, $i, j \in \{0, 1\}$ and reduction $\delta = \Gamma \mapsto \Delta|(A, B) \Rightarrow \Gamma' \mapsto_{\mathbf{P}Q_1i\mathbf{P}Q_2j} \Delta'$, where A is in conjunctive normal form and B is in disjunctive normal form, δ is $\mathbf{R}^{\mathbf{P}Q_1i\mathbf{P}Q_2j}$-valid if and only if δ is provable in $\mathbf{R}^{\mathbf{P}Q_1i\mathbf{P}Q_2j}$. That is,*

$$\models^{\mathbf{P}Q_1i\mathbf{P}Q_2j} \delta \text{ iff } \vdash^{\mathbf{P}Q_1i\mathbf{P}Q_2j} \delta.$$

□

$\mathbf{R}^{Y_1Q_1iY_2Q_2j}$ is decomposed into two parts: $\mathbf{R}_{\pm}^{Y_1Q_1iY_2Q_2j}$ and $\mathbf{R}_0^{Y_1Q_1iY_2Q_2j}$.

Theorem 6.2.5 (Nonmonotonicity theorem) $\mathbf{R}_0^{Y_1Q_1iY_2Q_2j}$ *is monotonic in Γ and in Δ; and $\mathbf{R}_{\pm}^{Y_1Q_1iY_2Q_2j}$ is nonmonotonic in Γ and in Δ.* □

6.3 R-Calculi $\mathbf{R}_{Y_1Q_1iY_2Q_2j}$

Let $Q_1, Q_2 \in \{\mathbf{E}, \mathbf{A}\}$, $i, j \in \{0, 1\}$ and $Y_1, Y_2 \in \{\mathbf{R}, \mathbf{Q}, \mathbf{P}\}$.

Given a co-sequent $\Gamma \mapsto \Delta$ and pair (A, B) of formulas, the result of $\Gamma \mapsto \Delta$ $\mathbf{G}_{Y_1Q_1iY_2Q_2j}$-revising (A, B) is denoted by $\models_{Y_1Q_1iY_2Q_2j} \Gamma \mapsto \Delta|(A, B) \rightrightarrows \Gamma' \mapsto \Delta'$, if $\Gamma' \mapsto \Delta' =$

$$\begin{cases} \Gamma, \{C\} \mapsto \Delta, \{D\} & \text{if } \Gamma, \{C\} \mapsto \Delta, \{D\} \text{ is } \mathbf{G}_{Y_1Q_1iY_2Q_2j}\text{-valid} \\ \Gamma, \{C\} \mapsto \Delta & \text{if } \Gamma, \{C\} \mapsto \Delta \text{ is } \mathbf{G}_{Y_1Q_1iY_2Q_2j}\text{-valid} \\ \Gamma \mapsto \Delta, \{D\} & \text{if } \Gamma \mapsto \Delta, \{D\} \text{ is } \mathbf{G}_{Y_1Q_1iY_2Q_2j}\text{-valid} \\ \Gamma \mapsto \Delta & \text{otherwise.} \end{cases}$$

6.3.1 Axioms

Proposition 6.3.1 *Let $\Gamma \mapsto \Delta$ be literal.*

$$\models_{Y_1Q_1iY_2Q_2j} \Gamma \mapsto \Delta|(l,m) \Rightarrow \Gamma' \mapsto \Delta'$$

if and only if

$$\models_{Q_1iQ_2j} \Gamma \mapsto \Delta|(l,m) \Rightarrow \Gamma' \mapsto \Delta'$$

□

6.3.2 Deduction Rules

We define

$$\begin{array}{ll} \mathbf{Y}_{L\mathrm{E}0} = \mathbf{Y}^{L0}(\Rightarrow / \mapsto) & \mathbf{Y}_{R\mathrm{E}0} = \mathbf{Y}^{R0}(\Rightarrow / \mapsto) \\ \mathbf{Y}_{LA0} = \mathbf{Y}^{L0}(\Rightarrow / \mapsto)((A)/[A]) & \mathbf{Y}_{RA0} = \mathbf{Y}^{R1}(\Rightarrow / \mapsto)((B)/[B]) \\ \mathbf{Y}_{L\mathrm{E}1} = \mathbf{Y}^{L1}(\Rightarrow / \mapsto)([A]/(A)) & \mathbf{Y}_{R\mathrm{E}1} = \mathbf{Y}^{R1}(\Rightarrow / \mapsto) \\ \mathbf{Y}_{LA1} = \mathbf{Y}^{L1}(\Rightarrow / \mapsto) & \mathbf{Y}^{RA1} = \mathbf{Y}_{R1}(\Rightarrow / \mapsto)((B)/[B]). \end{array}$$

6.3.3 R-Calculi

Define

$$\mathbf{R}_{Y_1Q_1iY_2Q_2j} = \mathbb{A}_{Q_1iQ_2j} + \mathbf{Y}_1^{LQ_1i} + \mathbf{Y}_2^{RQ_2j}.$$

Definition 6.3.2 A reduction $\Gamma \mapsto \Delta|(A,B) \Rightarrow \Gamma' \mapsto \Delta'$ is provable in $\mathbf{R}_{Y_1Q_1iY_2Q_2j}$, denoted by

$$\vdash_{Y_1Q_1iY_2Q_2j} \Gamma \mapsto \Delta|(A,B) \Rightarrow \Gamma' \mapsto \Delta',$$

if there is a sequence $\{\delta_1, \ldots, \delta_n\}$ of reductions such that $\delta_n = \Gamma \mapsto \Delta|(A,B) \Rightarrow \Gamma' \mapsto \Delta'$, and for each $1 \leq i \leq n$, δ_i is either an axiom or deduced from the previous reductions by one of the deduction rules in $\mathbf{R}_{Y_1Q_1iY_2Q_2j}$.

Theorem 6.3.3 (Soundness and completeness theorem) *For any $Q_1, Q_2 \in \{\mathbf{A}, \mathbf{E}\}$, $i, j \in \{0, 1\}$, $Y_1, Y_2 \in \{\mathbf{R}, \mathbf{Q}\}$ and reduction $\delta = \Gamma \mapsto \Delta|(A,B) \Rightarrow \Gamma' \mapsto \Delta'$, δ is $\mathbf{R}_{Y_1Q_1iY_2Q_2j}$-valid if and only if δ is provable in $\mathbf{R}_{Y_1Q_1iY_2Q_2j}$. That is,*

$$\models_{Y_1Q_1iY_2Q_2j} \delta \text{ iff } \vdash_{Y_1Q_1iY_2Q_2j} \delta.$$

□

Theorem 6.3.4 (Soundness and completeness theorem) *For any* $Q_1, Q_2 \in \{\mathbf{A}, \mathbf{E}\}$, $i, j \in \{0, 1\}$ *and reduction* $\delta = \Gamma \mapsto \Delta|(A, B) \Rightarrow \Gamma' \mapsto \Delta'$, *where A is in disjunctive normal form and B is in conjunctive normal form,* δ *is* $\mathbf{R}_{\mathbf{P}Q_1 i \mathbf{P}Q_2 j}$*-valid if and only if* δ *is provable in* $\mathbf{R}_{\mathbf{P}Q_1 i \mathbf{P}Q_2 j}$. *That is,*

$$\models_{\mathbf{P}Q_1 i \mathbf{P}Q_2 j} \delta \text{ iff } \vdash_{\mathbf{P}Q_1 i \mathbf{P}Q_2 j} \delta. \qquad \square$$

$\mathbf{R}_{Y_1 Q_1 i Y_2 Q_2 j}$ is decomposed into two parts: $\mathbf{R}^{\pm}_{Y_1 Q_1 i Y_2 Q_2 j}$ and $\mathbf{R}^{0}_{Y_1 Q_1 i Y_2 Q_2 j}$.

Theorem 6.3.5 (Nonmonotonicity theorem) $\mathbf{R}^{0}_{Y_1 Q_1 i Y_2 Q_2 j}$ *is monotonic in* Γ *and in* Δ; *and* $\mathbf{R}^{\pm}_{Y_1 Q_1 i Y_2 Q_2 j}$ *is nonmonotonic in* Γ *and in* Δ. $\square$

6.4 Conclusions

There are the following sound and complete Gentzen deduction systems

$$\begin{aligned} \mathbf{G}^{Y_1 Q_1 i Y_2 Q_2 j} &= \mathbb{A}^{Q_1 i Q_2 j} + \mathbf{G}^{Li} + \mathbf{G}^{Rj} \\ \mathbf{G}_{Y_1 Q_1 i Y_2 Q_2 j} &= \mathbb{A}_{Q_1 i Q_2 j} + \mathbf{G}^{Li} + \mathbf{G}^{Rj}. \end{aligned}$$

Moreover, $\mathbf{G}^{Y_1 Q_1 i Y_2 Q_2 j}/\mathbf{G}_{Y_1 Q_1 i Y_2 Q_2 j}$ is monotonic in Γ if and only if $Q_1 = \mathbf{E}$; and nonmonotonic in Δ if and only if $Q_2 = \mathbf{A}$.

Correspondingly, there are the following sound and complete R-calculi:

$$\begin{aligned} \mathbf{R}^{Y_1 Q_1 i Y_2 Q_2 j} &= \mathbb{A}^{Q_1 i Q_2 j} + \mathbf{Y}_1^{L Q_1 i} + \mathbf{Y}_2^{R Q_2 j} \\ \mathbf{R}_{Y_1 Q_1 i Y_2 Q_2 j} &= \mathbb{A}_{Q_1 i Q_2 j} + \mathbf{Y}_1^{L Q_1 i} + \mathbf{Y}_2^{R Q_2 j}, \end{aligned}$$

and $\mathbf{R}_0^{Y_1 Q_1 i Y_2 Q_2 j}/\mathbf{R}^0_{Y_1 Q_1 i Y_2 Q_2 j}$ is monotonic in Γ and in Δ; and $\mathbf{R}_{\pm}^{Y_1 Q_1 i Y_2 Q_2 j}/\mathbf{R}^{\pm}_{Y_1 Q_1 i Y_2 Q_2 j}$ is nonmonotonic in Γ and in Δ.

References

Alchourrón, C.E., Gärdenfors, P., Makinson, D.: On the logic of theory change: partial meet contraction and revision functions. J. Symbol. Log. **50**, 510–530 (1985)

Cao, C., Sui, Y., Wang, Y.: The nonmonotonic propositional logics. Artif. Intell. Res. **5**, 111–120 (2016)

Darwiche, A., Pearl, J.: On the logic of iterated belief revision. Artif. Intell. **89**, 1–29 (1997)

Fermé, E., Hansson, S.O.: AGM 25 years, twenty-five years of research in belief change. J. Philosoph. Log. **40**, 295–331 (2011)

Gärdenfors, P., Rott, H.: Belief revision. In: Handbook of logic in artificial intelligence and logic programming, vol. 4, pp. 35-132. Oxford University Press (1995)

Li, W.: R-calculus: an inference system for belief revision. Comput. J. **50**, 378–390 (2007)

Li, W.: Mathematical logic, foundations for information science. In: Progress in Computer Science and Applied Logic, vol. 25, Birkhäuser (2010)

Reiter, R.: A logic for default reasoning. Artif. Intell. **13**, 81–132 (1980)
Takeuti, G.: Proof Theory. In: Barwise, J. (ed.), Handbook of Mathematical Logic. Studies in Logic and the Foundations of Mathematics. North-Holland, Amsterdam, NL (1987)

Chapter 7
R-Calculi for Supersequents

A sequent $\Gamma \Rightarrow \Delta$ is valid if for any assignment v, v satisfying each formula in Δ implies v satisfying some formula in Γ, equivalently, either v satisfies the negation of some formula in Δ or v satisfies some formula in Γ. Correspondingly, there are four kinds of tableau proof systems in propositional logic:

A multisequent $\Gamma|\Delta$ is valid (Avron 1991; Baaz and Zach 2000) if for any assignment v, either v satisfies some formula in Γ or v does not satisfy some formula in Δ. In $\mathbf{L}_3$-valued propositional logic, a multisequent is of form $\Gamma|\Theta|\Delta$, which is valid (Bochvar 1938; Fitting 1991; Gottwald 2001; Hähnle 2001) if for any assignment v, either some formula in Γ has truth-value $\mathtt{t}$, or some formula $B \in \Theta$ has truth-value $\mathtt{m}$, or some formula C in Δ has truth-value $\mathtt{f}$. In $\mathbf{B}_2$-valued propositional logic, a sequent $\Gamma \Rightarrow \Delta$ is equivalent to multisequent $\Gamma|\Delta$ (Li 2010).

A supersequent δ is of form $\Gamma|\Delta \Rightarrow \Sigma|\Pi$, where $\Gamma, \Delta, \Sigma, \Pi$ are sets of formulas. δ is valid if for any assignment v, both each formula in Γ has truth-value 1 and each formula in Δ has truth-value 0 imply either some formula in Σ has truth-value 1 or some formula in Π has truth-value 0.

In this paper, we will give a sound and complete Gentzen deduction system $\mathbf{G}^+$ for supersequents, from which four sound and complete Gentzen deduction systems for sequents are deduced (Łukasiewicz 1970; Malinowski 2009; Urquhart 2001; Wronski 1987):

- $\mathbf{G}^{\mathtt{ft}}$: a sequent $\Gamma \Rightarrow \Sigma$ is valid if for any assignment v, each formula $A \in \Gamma$ having truth-value 1 implies some formula $C \in \Sigma$ having truth-value 1;
- $\mathbf{G}^{\mathtt{ff}}$: a sequent $\Gamma \Rightarrow \Pi$ is valid if for any assignment v, each formula $A \in \Gamma$ having truth-value 1 implies some formula $D \in \Pi$ having truth-value 0;
- $\mathbf{G}^{\mathtt{tt}}$: a sequent $\Delta \Rightarrow \Sigma$ is valid if for any assignment v, each formula $B \in \Delta$ having truth-value 0 implies some formula $C \in \Sigma$ having truth-value 1;
- $\mathbf{G}^{\mathtt{tf}}$: a sequent $\Delta \Rightarrow \Pi$ is valid if for any assignment v, each formula $B \in \Delta$ having truth-value 0 implies some formula $D \in \Pi$ having truth-value 0.

W. Li and Y. Sui, *R-Calculus, IV: Propositional Logic*,
Perspectives in Formal Induction, Revision and Evolution,
https://doi.org/10.1007/978-981-19-8633-8_7

Conversely, a supersequent $\Gamma|\Delta \Rightarrow \Sigma|\Pi$ is equivalent to sequents

$$\begin{array}{l}\Gamma, \neg\Delta \Rightarrow \Sigma, \neg\Pi;\\ \Gamma, \neg\Delta \Rightarrow \neg\Sigma, \Pi;\\ \neg\Gamma, \Delta \Rightarrow \Sigma, \neg\Pi;\\ \neg\Gamma, \Delta \Rightarrow \neg\Sigma, \Pi.\end{array}$$

respectively. That is, $\Gamma|\Delta \Rightarrow \Sigma|\Pi$ is provable in $\mathbf{G}^+$ if and only if

$$\begin{array}{l}\Gamma, \neg\Delta \Rightarrow \Sigma, \neg\Pi \text{is provable in} \mathbf{G}^{\mathtt{ft}};\\ \Gamma, \neg\Delta \Rightarrow \neg\Sigma, \Pi \text{is provable in} \mathbf{G}^{\mathtt{ff}};\\ \neg\Gamma, \Delta \Rightarrow \Sigma, \neg\Pi \text{is provable in} \mathbf{G}^{\mathtt{tt}};\\ \neg\Gamma, \Delta \Rightarrow \neg\Sigma, \Pi \text{is provable in} \mathbf{G}^{\mathtt{tf}}.\end{array}$$

Also, $\mathbf{G}^+$ is taken as a combination of four tableau proof systems:

- $\mathbf{G}^{L\mathtt{f}}$: Γ is provable if and only if for any assignment v, there exists a formula $A \in \Delta$ such that $v(A) = 0$;
- $\mathbf{G}^{L\mathtt{t}}$: Δ is provable if and only if for any assignment v, there exists a formula $B \in \Delta$ such that $v(B) = 1$;
- $\mathbf{G}^{R\mathtt{t}}$: Σ is provable if and only if for any assignment v, there is a formula $C \in \Sigma$ such that $v(C) = 1$;
- $\mathbf{G}^{R\mathtt{f}}$: Π is provable if and only if for any assignment v, there is a formula $D \in \Pi$ such that $v(D) = 0$.

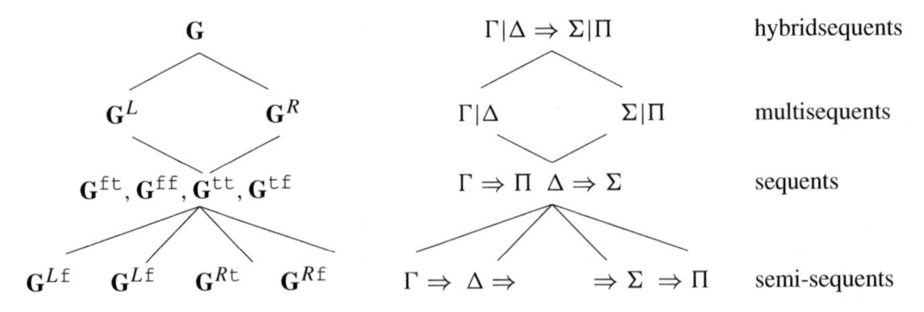

Therefore, a deduction system $\mathbf{G}^+$ for supersequents is decomposed into two deduction systems $\mathbf{G}^L$ and $\mathbf{G}^R$ for multisequents; between $\mathbf{G}^L$ and $\mathbf{G}^R$, there are four deduction systems

$$\mathbf{G}^{\mathtt{ft}}, \mathbf{G}^{\mathtt{ff}}, \mathbf{G}^{\mathtt{tt}}, \mathbf{G}^{\mathtt{tf}}$$

for sequents, which are decomposed into four tableau proof systems $\mathbf{G}^{L\mathtt{t}}$, $\mathbf{G}^{L\mathtt{f}}$, $\mathbf{G}^{R\mathtt{t}}$, $\mathbf{G}^{R\mathtt{f}}$ for theories.

Correspondingly, there are R-calculi $\mathbf{R}^+$, $\mathbf{R}_-$ for supersequents and R-calculi

$$\mathbf{R}^{\mathtt{ft}}, \mathbf{R}^{\mathtt{ff}}, \mathbf{R}^{\mathtt{tt}}, \mathbf{R}^{\mathtt{tf}}$$

for sequents, and

$$\mathbf{R}_{\mathtt{tf}}, \mathbf{R}_{\mathtt{tt}}, \mathbf{R}_{\mathtt{ff}}, \mathbf{R}_{\mathtt{ft}}$$

for co-sequents.

We will give the following deduction systems and R-calculi:

supersequents	sequents			
$\boxed{\mathbf{G}^{+}}$	$\boxed{\mathbf{G}^{\mathtt{ft}}}$	$\boxed{\mathbf{G}^{\mathtt{ff}}}$	$\mathbf{G}_{\mathtt{tt}}$	$\mathbf{G}_{\mathtt{tf}}$
$\boxed{\mathbf{G}_{-}}$	$\mathbf{G}_{\mathtt{tf}}$	$\mathbf{G}_{\mathtt{tt}}$	$\boxed{\mathbf{G}_{\mathtt{ff}}}$	$\boxed{\mathbf{G}_{\mathtt{ft}}}$
$\boxed{\mathbf{R}^{+}}$	$\boxed{\mathbf{R}^{\mathtt{ft}}}$	$\boxed{\mathbf{R}^{\mathtt{ff}}}$	$\mathbf{R}^{\mathtt{tt}}$	$\mathbf{R}^{\mathtt{tf}}$
$\boxed{\mathbf{R}_{-}}$	$\mathbf{R}_{\mathtt{tf}}$	$\mathbf{R}_{\mathtt{tt}}$	$\boxed{\mathbf{R}_{\mathtt{ff}}}$	$\boxed{\mathbf{R}_{\mathtt{ft}}}$

7.1 Supersequents

A supersequent δ is a quadruple $(\Gamma, \Delta, \Sigma, \Pi)$ of formula sets, denoted by $\Gamma|\Delta \Rightarrow \Sigma|\Pi$. And a supersequent δ is satisfied in v, denoted by $v \models \delta$, if $v \models \Gamma|\Delta$ implies $v \models \Sigma|\Pi$, where

- $v \models \Gamma|\Delta$ if for each formula $A \in \Gamma$, $v(A) = 1$; and for each formula $B \in \Delta$, $v(A) = 0$;
- $v \models \Sigma|\Pi$ if either for some formula $C \in \Sigma$, $v(C) = 1$; or for some formula $D \in \Pi$, $v(D) = 0$.

A supersequent δ is valid, denoted by $\models \delta$, if for any assignment v, $v \models \delta$.

Proposition 7.1.1 *For any supersequent* $\Gamma|\Delta \Rightarrow \Sigma|\Pi$ *and any assignment* v,

(i) $v \models \Gamma|\Delta$ *iff* $v \models \Gamma, \neg\Delta$; *and*
(ii) $v \models \Sigma|\Pi$ *iff* $v \models \Sigma, \neg\Pi$.

Therefore, $\Gamma|\Delta \Rightarrow \Sigma|\Pi$ is $\mathbf{G}^{+}$-valid iff $\Gamma, \neg\Delta \Rightarrow \Sigma, \neg\Pi$ is $\mathbf{G}^{\mathtt{ft}}$-valid, iff $\Gamma, \neg\Delta \Rightarrow \neg\Sigma, \Pi$ is $\mathbf{G}^{\mathtt{ff}}$-valid; iff $\neg\Gamma, \Delta \Rightarrow \Sigma, \neg\Pi$ is $\mathbf{G}^{\mathtt{tt}}$-valid; and iff $\neg\Gamma, \Delta \Rightarrow \neg\Sigma, \Pi$ is $\mathbf{G}^{\mathtt{tf}}$-valid.

7.1.1 *Supersequent Gentzen Deduction System* $\mathbf{G}^{+}$

Supersequent Gentzen deduction system $\mathbf{G}^{+}$ consists of the following axioms and deduction rules:

- **Axiom**:

$$(\mathrm{A}^{+}) \ \frac{\begin{array}{c}\mathrm{incon}(\Gamma \cup \neg\Delta) \text{ or } \mathrm{val}(\Sigma \cup \neg\Pi) \\ \text{or } (\Gamma \cup \neg\Delta) \cap (\Sigma \cup \neg\Pi) \neq \emptyset\end{array}}{\Gamma|\Delta \Rightarrow \Sigma|\Pi}$$

where $\Gamma, \Delta, \Sigma, \Pi$ are sets of literals.

- **Deduction rules**:

$$(\neg\neg^A)\ \frac{\Gamma, A|\Delta \Rightarrow \Sigma|\Pi}{\Gamma, \neg\neg A|\Delta \Rightarrow \Sigma|\Pi} \quad (\neg\neg^B)\ \frac{\Gamma|B, \Delta \Rightarrow \Sigma|\Pi}{\Gamma|\neg\neg B, \Delta \Rightarrow \Sigma|\Pi}$$

$$(\neg\neg^C)\ \frac{\Gamma|\Delta \Rightarrow \Sigma, C|\Pi}{\Gamma|\Delta \Rightarrow \Sigma, \neg\neg C|\Pi} \quad (\neg\neg^D)\ \frac{\Gamma|\Delta \Rightarrow \Sigma|D, \Pi}{\Gamma|\Delta \Rightarrow \Sigma|\neg\neg D, \Pi}$$

and

$$(\wedge^A)\ \frac{\left\{\begin{array}{l}\Gamma, A_1|\Delta \Rightarrow \Sigma|\Pi\\ \Gamma, A_2|\Delta \Rightarrow \Sigma|\Pi\end{array}\right.}{\Gamma, A_1 \wedge A_2|\Delta \Rightarrow \Sigma|\Pi} \quad (\wedge^B)\ \frac{\left[\begin{array}{l}\Gamma|B_1, \Delta \Rightarrow \Sigma|\Pi\\ \Gamma|B_2, \Delta \Rightarrow \Sigma|\Pi\end{array}\right.}{\Gamma|B_1 \wedge B_2, \Delta \Rightarrow \Sigma|\Pi}$$

$$(\wedge^C)\ \frac{\left[\begin{array}{l}\Gamma|\Delta \Rightarrow \Sigma, C_1|\Pi\\ \Gamma|\Delta \Rightarrow \Sigma, C_2|\Pi\end{array}\right.}{\Gamma|\Delta \Rightarrow \Sigma, C_1 \wedge C_2|\Pi} \quad (\wedge^D)\ \frac{\left\{\begin{array}{l}\Gamma|\Delta \Rightarrow \Sigma|D_1, \Pi\\ \Gamma|\Delta \Rightarrow \Sigma|D_2, \Pi\end{array}\right.}{\Gamma|\Delta \Rightarrow \Sigma|D_1 \wedge D_2, \Pi}$$

$$(\vee^A)\ \frac{\left[\begin{array}{l}\Gamma, A_1|\Delta \Rightarrow \Sigma|\Pi\\ \Gamma, A_2|\Delta \Rightarrow \Sigma|\Pi\end{array}\right.}{\Gamma, A_1 \vee A_2|\Delta \Rightarrow \Sigma|\Pi} \quad (\vee^B)\ \frac{\left\{\begin{array}{l}\Gamma|B_1, \Delta \Rightarrow \Sigma|\Pi\\ \Gamma|B_2, \Delta \Rightarrow \Sigma|\Pi\end{array}\right.}{\Gamma|B_1 \vee B_2, \Delta \Rightarrow \Sigma|\Pi}$$

$$(\vee^C)\ \frac{\left\{\begin{array}{l}\Gamma|\Delta \Rightarrow \Sigma, C_1|\Pi\\ \Gamma|\Delta \Rightarrow \Sigma, C_2|\Pi\end{array}\right.}{\Gamma|\Delta \Rightarrow \Sigma, C_1 \vee C_2|\Pi} \quad (\vee^D)\ \frac{\left[\begin{array}{l}\Gamma|\Delta \Rightarrow \Sigma|D_1, \Pi\\ \Gamma|\Delta \Rightarrow \Sigma|D_2, \Pi\end{array}\right.}{\Gamma|\Delta \Rightarrow \Sigma|D_1 \vee D_2, \Pi}$$

and

$$(\neg\wedge^A)\ \frac{\left[\begin{array}{l}\Gamma, \neg A_1|\Delta \Rightarrow \Sigma|\Pi\\ \Gamma, \neg A_2|\Delta \Rightarrow \Sigma|\Pi\end{array}\right.}{\Gamma, \neg(A_1 \wedge A_2)|\Delta \Rightarrow \Sigma|\Pi} \quad (\neg\wedge^B)\ \frac{\left\{\begin{array}{l}\Gamma|\neg B_1, \Delta \Rightarrow \Sigma|\Pi\\ \Gamma|\neg B_2, \Delta \Rightarrow \Sigma|\Pi\end{array}\right.}{\Gamma|\neg(B_1 \wedge B_2), \Delta \Rightarrow \Sigma|\Pi}$$

$$(\neg\wedge^C)\ \frac{\left\{\begin{array}{l}\Gamma|\Delta \Rightarrow \Sigma, \neg C_1|\Pi\\ \Gamma|\Delta \Rightarrow \Sigma, \neg C_2|\Pi\end{array}\right.}{\Gamma|\Delta \Rightarrow \Sigma, \neg(C_1 \wedge C_2)|\Pi} \quad (\neg\wedge^D)\ \frac{\left[\begin{array}{l}\Gamma|\Delta \Rightarrow \Sigma|\neg D_1, \Pi\\ \Gamma|\Delta \Rightarrow \Sigma|\neg D_2, \Pi\end{array}\right.}{\Gamma|\Delta \Rightarrow \Sigma|\neg(D_1 \wedge D_2), \Pi}$$

$$(\neg\vee^A)\ \frac{\left\{\begin{array}{l}\Gamma, \neg A_1|\Delta \Rightarrow \Sigma|\Pi\\ \Gamma, \neg A_2|\Delta \Rightarrow \Sigma|\Pi\end{array}\right.}{\Gamma, \neg(A_1 \vee A_2)|\Delta \Rightarrow \Sigma|\Pi} \quad (\neg\vee^B)\ \frac{\left[\begin{array}{l}\Gamma|\neg B_1, \Delta \Rightarrow \Sigma|\Pi\\ \Gamma|\neg B_2, \Delta \Rightarrow \Sigma|\Pi\end{array}\right.}{\Gamma|\neg(B_1 \vee B_2), \Delta \Rightarrow \Sigma|\Pi}$$

$$(\neg\wedge^C)\ \frac{\left[\begin{array}{l}\Gamma|\Delta \Rightarrow \Sigma, \neg C_1|\Pi\\ \Gamma|\Delta \Rightarrow \Sigma, \neg C_2|\Pi\end{array}\right.}{\Gamma|\Delta \Rightarrow \Sigma, \neg(C_1 \vee C_2)|\Pi} \quad (\neg\vee^D)\ \frac{\left\{\begin{array}{l}\Gamma|\Delta \Rightarrow \Sigma|\neg D_1, \Pi\\ \Gamma|\Delta \Rightarrow \Sigma|\neg D_2, \Pi\end{array}\right.}{\Gamma|\Delta \Rightarrow \Sigma|\neg(D_1 \vee D_2), \Pi}$$

Definition 7.1.2 A supersequent $\Gamma|\Delta \Rightarrow \Sigma|\Pi$ is provable in $\mathbf{G}^+$, denoted by $\vdash^+ \Gamma|\Delta \Rightarrow \Sigma|\Pi$, if there is a sequence $\{\Gamma_1|\Delta_1 \Rightarrow \Sigma_1|\Pi_1, \ldots, \Gamma_n|\Delta_n \Rightarrow \Sigma_n|\Pi_n\}$ of supersequents such that $\Gamma_n|\Delta_n \Rightarrow \Sigma_n|\Pi_n = \Gamma|\Delta \Rightarrow \Sigma|\Pi$, and for each $1 \leq i \leq n$, $\Gamma_i|\Delta_i \Rightarrow \Sigma_i|\Pi_i$ is either an axiom or deduced from the previous supersequents by one of the deduction rules in $\mathbf{G}^+$.

Theorem 7.1.3 (Soundness theorem) *For any supersequent* $\delta = \Gamma|\Delta \Rightarrow \Sigma|\Pi$, *if* $\vdash^+ \delta$ *then* $\models^+ \delta$.

Proof We prove that each axiom is valid and each deduction rule preserves validity.

($\mathtt{A}^+$) By Proposition 7.1.1, literal supersequent $\Gamma|\Delta \Rightarrow \Sigma|\Pi$ is $\mathbf{G}^+$-valid iff $\Gamma, \neg\Delta \Rightarrow \Sigma, \neg\Pi$ is $\mathbf{G}^{\mathtt{ft}}$-valid, iff either $\Gamma \cup \neg\Delta$ is inconsistent or $\Sigma \cup \neg\Pi$ is valid, or

$$(\Gamma \cup \neg\Delta) \cap (\Sigma \cup \neg\Pi) \neq \emptyset.$$

Hence, for any assignment v, $v \models \Gamma|\Delta \Rightarrow \Sigma|\Pi$.

($\neg\neg^A$) Assume that for any assignment v, $v \models \Gamma, A|\Delta \Rightarrow \Sigma|\Pi$. For any assignment v, assume that $v \models \Gamma, \neg\neg A|\Delta$ then $v(A) = 1 = v(\neg\neg A)$, $v \models \Gamma, A|\Delta$. By induction assumption, $v \models \Sigma|\Pi$, and hence, $v \models \Gamma, \neg\neg A|\Delta \Rightarrow \Sigma|\Pi$.

($\wedge^A$) Assume that for any assignment v, either $v \models \Gamma, A_1|\Delta \Rightarrow \Sigma|\Pi$, or $v \models \Gamma, A_2|\Delta \Rightarrow \Sigma|\Pi$. For any assignment v, if $v \models \Gamma, A_1 \wedge A_2|\Delta$ then $v \models \Gamma|\Delta$ and $v(A_1 \wedge A_2) = 1$, i.e., $v(A_1) = v(A_2) = 1$. By induction assumption, each case implies $v \models \Sigma|\Pi$.

($\wedge^B$) Assume that for any assignment v,

$$v \models \Gamma|B_1, \Delta \Rightarrow \Sigma|\Pi,$$
$$v \models \Gamma|B_2, \Delta \Rightarrow \Sigma|\Pi.$$

For any assignment v, if $v \models \Gamma|B_1 \wedge B_2, \Delta$ then $v \models \Gamma|\Delta$ and $v(B_1 \wedge B_2) = 0$, i.e., either $v(B_1) = 0$ or $v(B_2) = 0$. By induction assumption, each case implies $v \models \Sigma|\Pi$.

($\wedge^C$) Assume that for any assignment v,

$$v \models \Gamma|\Delta \Rightarrow \Sigma, C_1|\Pi,$$
$$v \models \Gamma|\Delta \Rightarrow \Sigma, C_2|\Pi.$$

For any assignment v, if $v \models \Gamma|\Delta$ then $v \models \Sigma, C_1|\Pi$ and $v \models \Sigma, C_2|\Pi$, i.e., either $v \models \Sigma|\Pi$, or $v(C_1) = v(C_2) = 1$. Hence, $v \models \Sigma, C_1 \wedge C_2|\Pi$.

($\wedge^A$) Assume that for any assignment v, either $v \models \Gamma|\Delta \Rightarrow \Sigma|D_1, \Pi$, or $v \models \Gamma|\Delta \Rightarrow \Sigma|D_2, \Pi$. For any assignment v, if $v \models \Gamma|\Delta$ then either $v \models \Sigma|D_1, \Pi$, or $v \models \Sigma|D_2, \Pi$, i.e., either $v \models \Sigma|\Pi$, or either $v(D_1) = 0$ or $v(D_2) = 0$. Hence, $v \models \Sigma|D_1 \wedge D_2, \Pi$.

Similar for other deduction rules. □

Theorem 7.1.4 (Completeness theorem) *For any supersequent $\delta = \Gamma|\Delta \Rightarrow \Sigma|\Pi$, if $\models \delta$ then $\vdash \delta$.*

Proof Given a supersequent $\delta = \Gamma|\Delta \Rightarrow \Sigma|\Pi$, we construct a tree T such that either

(i) for each branch ξ of T, some supersequent δ' at the leaf of ξ is an axiom, or
(ii) there is an assignment v such that $v \not\models \delta$.

T is constructed as follows:

- the root of T is δ;
- for a node ξ, if the supersequent δ' at ξ is literal then the node is a leaf;

- otherwise, ξ has the direct children nodes containing the following supersequents:

$$\begin{cases} \Gamma_1, A|\Delta_1 \Rightarrow \Sigma_1|\Pi_1 \text{ if } \Gamma_1, \neg\neg A|\Delta_1 \Rightarrow \Sigma_1|\Pi_1 \in \xi \\ \Gamma_1|B, \Delta_1 \Rightarrow \Sigma_1|\Pi_1 \text{ if } \Gamma_1|\neg\neg B, \Delta_1 \Rightarrow \Sigma_1|\Pi_1 \in \xi \\ \Gamma_1|\Delta_1 \Rightarrow \Sigma_1, C|\Pi_1 \text{ if } \Gamma_1|\Delta_1 \Rightarrow \Sigma_1, \neg\neg C|\Pi_1 \in \xi \\ \Gamma_1|\Delta_1 \Rightarrow \Sigma_1|D, \Pi_1 \text{ if } \Gamma_1|\Delta_1 \Rightarrow \Sigma_1|\neg\neg D, \Pi_1 \in \xi \end{cases}$$

and

$$\begin{cases} \Gamma_1, A_1, A_2|\Delta_1 \Rightarrow \Sigma_1|\Pi_1 & \text{if } \Gamma_1, A_1 \wedge A_2|\Delta_1 \Rightarrow \Sigma_1|\Pi_1 \in \xi \\ \begin{cases} \Gamma_1|B_1, \Delta_1 \Rightarrow \Sigma_1|\Pi_1 \\ \Gamma_1|B_2, \Delta_1 \Rightarrow \Sigma_1|\Pi_1 \end{cases} & \text{if } \Gamma_1|B_1 \wedge B_2, \Delta_1 \Rightarrow \Sigma_1|\Pi_1 \in \xi \\ \begin{cases} \Gamma_1|\Delta_1 \Rightarrow \Sigma_1, C_1|\Pi_1 \\ \Gamma_1|\Delta_1 \Rightarrow \Sigma_1, C_2|\Pi_1 \end{cases} & \text{if } \Gamma_1|\Delta_1 \Rightarrow \Sigma_1, C_1 \wedge C_2|\Pi_1 \in \xi \\ \Gamma_1|\Delta_1 \Rightarrow \Sigma_1|D_1, D_2, \Pi_1 & \text{if } \Gamma_1|\Delta_1 \Rightarrow \Sigma_1|D_1 \wedge D_2\Pi_1 \in \xi \\ \begin{cases} \Gamma_1, A_1|\Delta_1 \Rightarrow \Sigma_1|\Pi_1 \\ \Gamma_1, A_2|\Delta_1 \Rightarrow \Sigma_1|\Pi_1 \end{cases} & \text{if } \Gamma_1, A_1 \vee A_2|\Delta_1 \Rightarrow \Sigma_1|\Pi_1 \in \xi \\ \Gamma_1|B_1, B_2, \Delta_1 \Rightarrow \Sigma_1|\Pi_1 & \text{if } \Gamma_1|B_1 \vee B_2, \Delta_1 \Rightarrow \Sigma_1|\Pi_1 \in \xi \\ \Gamma_1|\Delta_1 \Rightarrow \Sigma_1, C_1, C_2|\Pi_1 & \text{if } \Gamma_1|\Delta_1 \Rightarrow \Sigma_1, C_1 \vee C_2|\Pi_1 \in \xi \\ \begin{cases} \Gamma_1|\Delta_1 \Rightarrow \Sigma_1|D_1, \Pi_1 \\ \Gamma_1|\Delta_1 \Rightarrow \Sigma_1|D_2, \Pi_1 \end{cases} & \text{if } \Gamma_1|\Delta_1 \Rightarrow \Sigma_1|D_1 \vee D_2\Pi_1 \in \xi \end{cases}$$

and

$$\begin{cases} \begin{cases} \Gamma_1, \neg A_1|\Delta_1 \Rightarrow \Sigma_1|\Pi_1 \\ \Gamma_1, \neg A_2|\Delta_1 \Rightarrow \Sigma_1|\Pi_1 \end{cases} & \text{if } \Gamma_1, \neg(A_1 \wedge A_2)|\Delta_1 \Rightarrow \Sigma_1|\Pi_1 \in \xi \\ \Gamma_1|\neg B_1, \neg B_2, \Delta_1 \Rightarrow \Sigma_1|\Pi_1 & \text{if } \Gamma_1|\neg(B_1 \wedge B_2), \Delta_1 \Rightarrow \Sigma_1|\Pi_1 \in \xi \\ \Gamma_1|\Delta_1 \Rightarrow \Sigma_1, \neg C_1, \neg C_2|\Pi_1 & \text{if } \Gamma_1|\Delta_1 \Rightarrow \Sigma_1, \neg(C_1 \wedge C_2)|\Pi_1 \in \xi \\ \begin{cases} \Gamma_1|\Delta_1 \Rightarrow \Sigma_1|\neg D_1, \Pi_1 \\ \Gamma_1|\Delta_1 \Rightarrow \Sigma_1|\neg D_2, \Pi_1 \end{cases} & \text{if } \Gamma_1|\Delta_1 \Rightarrow \Sigma_1|\neg(D_1 \wedge D_2), \Pi_1 \in \xi \\ \Gamma_1, \neg A_1, \neg A_2|\Delta_1 \Rightarrow \Sigma_1|\Pi_1 & \text{if } \Gamma_1, \neg(A_1 \vee A_2)|\Delta_1 \Rightarrow \Sigma_1|\Pi_1 \in \xi \\ \begin{cases} \Gamma_1|\neg B_1, \Delta_1 \Rightarrow \Sigma_1|\Pi_1 \\ \Gamma_1|\neg B_2, \Delta_1 \Rightarrow \Sigma_1|\Pi_1 \end{cases} & \text{if } \Gamma_1|\neg(B_1 \vee B_2), \Delta_1 \Rightarrow \Sigma_1|\Pi_1 \in \xi \\ \begin{cases} \Gamma_1|\Delta_1 \Rightarrow \Sigma_1, \neg C_1|\Pi_1 \\ \Gamma_1|\Delta_1 \Rightarrow \Sigma_1, \neg C_2|\Pi_1 \end{cases} & \text{if } \Gamma_1|\Delta_1 \Rightarrow \Sigma_1, \neg(C_1 \vee C_2)|\Pi_1 \in \xi \\ \Gamma_1|\Delta_1 \Rightarrow \Sigma_1|\neg D_1, \neg D_2, \Pi_1 & \text{if } \Gamma_1|\Delta_1 \Rightarrow \Sigma_1|\neg(D_1 \vee D_2), \Pi_1 \in \xi \end{cases}$$

Lemma 7.1.5 *If for each branch $\xi \subseteq T$, there is a supersequent $\delta' \in \xi$ which is an axiom in $\mathbf{G}^+$ then T is a proof of δ.*

Proof By the definition of T, T is a proof tree of δ in $\mathbf{G}^+$. □

Lemma 7.1.6 *If there is a branch $\xi \subseteq T$ such that each supersequent $\delta' \in \xi$ is not an axiom in $\mathbf{G}^+$ then there is an assignment v such that $v \not\models \delta$.*

Proof Let γ be the set of all the literal supersequents δ' in T which is not an axiom. By Proposition 7.1.1, there is an assignment v such that $v \not\models \delta'$, where δ' is at the leaf node of ξ.

We proved by induction on nodes η of ξ that each supersequent $\delta' \in \xi$ is not satisfied by v.

Case $\delta' = \Gamma'', \neg\neg A|\Delta' \Rightarrow \Sigma'|\Pi' \in \eta$. Then, η has a direct child node $\in \xi$ containing $\Gamma'', A|\Delta' \Rightarrow \Sigma'|\Pi'$. By induction assumption, $v \not\models \Gamma'', A|\Delta' \Rightarrow \Sigma'|\Pi'$, i.e., $v(A) = 1 = v(\neg\neg A)$, and $v \not\models \Gamma'', \neg\neg A|\Delta' \Rightarrow \Sigma'|\Pi'$.

Case $\delta' = \Gamma'', A_1 \wedge A_2|\Delta' \Rightarrow \Sigma'|\Pi' \in \eta$. Then, η has a direct child node $\in \xi$ containing $\Gamma'', A_1, A_2|\Delta \Rightarrow \Sigma'|\Pi'$. By induction assumption, $v \not\models \Gamma'', A_1, A_2|\Delta' \Rightarrow \Sigma'|\Pi'$, i.e., $v(A_1) = 1 = v(A_2)$, which imply $v(A_1 \wedge A_2) = 1$. Hence, $v \not\models \Gamma'', A_1 \wedge A_2|\Delta' \Rightarrow \Sigma'|\Pi'$.

Case $\delta' = \Gamma'|B_1 \wedge B_2, \Delta'' \Rightarrow \Sigma'|\Pi' \in \eta$. Then, η has a direct child node $\in \xi$ containing one of the following supersequents:

$$\begin{array}{c}\Gamma'|B_1, \Delta'' \Rightarrow \Sigma'|\Pi'; \\ \Gamma'|B_2, \Delta'' \Rightarrow \Sigma'|\Pi'.\end{array}$$

say, $\Gamma'|B_1, \Delta'' \Rightarrow \Sigma'|\Pi'$. By induction assumption, $v \not\models \Gamma'|B_1, \Delta'' \Rightarrow \Sigma'|\Pi'$, i.e., $v \not\models \Gamma'|\Delta'' \Rightarrow \Sigma'|\Pi'$ and $v(B_1) = 0$. Hence, $v(B_1 \wedge B_2) = 0$, and $v \not\models \Gamma'|B_1 \wedge B_2, \Delta'' \Rightarrow \Sigma'|\Pi'$; .

Case $\delta' = \Gamma'|\Delta' \Rightarrow \Sigma'', C_1 \wedge C_2|\Pi' \in \eta$. Then, η has a direct child node $\in \xi$ containing one of supersequents:

$$\begin{array}{c}\Gamma'|\Delta' \Rightarrow \Sigma'', C_1|\Pi'; \\ \Gamma'|\Delta' \Rightarrow \Sigma'', C_2|\Pi',\end{array}$$

say $\Gamma'|\Delta' \Rightarrow \Sigma'', C_2|\Pi'$. By induction assumption, $v \not\models \Gamma'|\Delta' \Rightarrow \Sigma'', C_2|\Pi'$; i.e., $v \not\models \Gamma'|\Delta' \Rightarrow \Sigma''|\Pi'$, and $v(C_2) = 0$. Hence, $v(C_1 \wedge C_2) = \mathtt{f}$, and $v \not\models \Gamma'|\Delta' \Rightarrow \Sigma'', C_1 \wedge C_2|\Pi'$.

Case $\delta' = \Gamma'|\Delta' \Rightarrow \Sigma'|D_1 \wedge D_2, \Pi'' \in \eta$. Then, η has a direct child node $\in \xi$ containing $\Gamma'|\Delta \Rightarrow \Sigma'|D_1, D_2, \Pi''$. By induction assumption, $v \not\models \Gamma'|\Delta' \Rightarrow \Sigma'|D_1, D_2, \Pi''$, i.e., either $v(D_1) = 0$ or $v(D_2) = 0$, which imply $v(D_1 \wedge D_2) = 0$. Hence, $v \not\models \Gamma'|\Delta' \Rightarrow \Sigma'|D_1, D_2, \Pi''$.

Similar for other cases. □

7.2 Reduction of Supersequents to Sequents

A supersequent supersequent $\Gamma|\Delta \Rightarrow \Sigma|\Xi$ is reduced into four kinds of sequents:

$$\begin{array}{l}\mathtt{ft} : \Gamma \Rightarrow \Sigma, \Delta = \Xi = \emptyset \\ \mathtt{ff} : \Gamma \Rightarrow \Xi, \Delta = \Sigma = \emptyset \\ \mathtt{tt} : \Delta \Rightarrow \Sigma, \Gamma = \Xi = \emptyset \\ \mathtt{tf} : \Delta \Rightarrow \Xi, \Gamma = \Sigma = \emptyset.\end{array}$$

7.2.1 Gentzen Deduction System $\mathbf{G}^{\mathtt{ft}}$

$\Gamma \Rightarrow \Sigma$ is $\mathbf{G}^{\mathtt{ft}}$-valid, denoted by $\models^{\mathtt{ft}} \Gamma \Rightarrow \Sigma$, if for any assignment v, $v \models \Gamma$ implies $v \models \Sigma$, where

- $v \models \Gamma$ if for each formula $A \in \Gamma$, $v(A) = 1$; and
- $v \models \Sigma$ if for some formula $C \in \Sigma$, $v(C) = 1$.
  Gentzen deduction system $\mathbf{G}^{\mathtt{ft}}$ consists of the following axioms and deduction rules:
- **Axiom**:

$$(\mathtt{A}^{\mathtt{ft}}) \frac{\left\{\begin{array}{l} \text{incon}(\Gamma) \\ \text{val}(\Sigma) \\ \Gamma \cap \Sigma \neq \emptyset \end{array}\right.}{\Gamma \Rightarrow \Sigma}$$

  where Γ, Σ are sets of literals.
- **Deduction rules**:

$$\begin{array}{ll}
(\neg\neg^{A}) \dfrac{\Gamma, A \Rightarrow \Sigma}{\Gamma, \neg\neg A \Rightarrow \Sigma} & (\neg\neg^{C}) \dfrac{\Gamma \Rightarrow \Sigma, C}{\Gamma \Rightarrow \Sigma, \neg\neg C} \\
(\wedge^{A}) \dfrac{\left\{\begin{array}{l} \Gamma, A_1 \Rightarrow \Sigma \\ \Gamma, A_2 \Rightarrow \Sigma \end{array}\right.}{\Gamma, A_1 \wedge A_2 \Rightarrow \Sigma} & (\wedge^{C}) \dfrac{\left[\begin{array}{l} \Gamma \Rightarrow \Sigma, C_1 \\ \Gamma \Rightarrow \Sigma, C_2 \end{array}\right.}{\Gamma \Rightarrow \Sigma, C_1 \wedge C_2} \\
(\vee^{A}) \dfrac{\left[\begin{array}{l} \Gamma, A_1 \Rightarrow \Sigma \\ \Gamma, A_2 \Rightarrow \Sigma \end{array}\right.}{\Gamma, A_1 \vee A_2 \Rightarrow \Sigma} & (\vee^{C}) \dfrac{\left\{\begin{array}{l} \Gamma \Rightarrow \Sigma, C_1 \\ \Gamma \Rightarrow \Sigma, C_2 \end{array}\right.}{\Gamma \Rightarrow \Sigma, C_1 \vee C_2}
\end{array}$$

and

$$\begin{array}{ll}
(\neg\wedge^{A}) \dfrac{\left[\begin{array}{l} \Gamma, \neg A_1 \Rightarrow \Sigma \\ \Gamma, \neg A_2 \Rightarrow \Sigma \end{array}\right.}{\Gamma, \neg(A_1 \wedge A_2) \Rightarrow \Sigma} & (\neg\wedge^{C}) \dfrac{\left\{\begin{array}{l} \Gamma \Rightarrow \Sigma, \neg C_1 \\ \Gamma \Rightarrow \Sigma, \neg C_2 \end{array}\right.}{\Gamma \Rightarrow \Sigma, \neg(C_1 \wedge C_2)} \\
(\neg\vee^{A}) \dfrac{\left\{\begin{array}{l} \Gamma, \neg A_1 \Rightarrow \Sigma \\ \Gamma, \neg A_2 \Rightarrow \Sigma \end{array}\right.}{\Gamma, \neg(A_1 \vee A_2) \Rightarrow \Sigma} & (\neg\wedge^{C}) \dfrac{\left[\begin{array}{l} \Gamma \Rightarrow \Sigma, \neg C_1 \\ \Gamma \Rightarrow \Sigma, \neg C_2 \end{array}\right.}{\Gamma \Rightarrow \Sigma, \neg(C_1 \vee C_2)}
\end{array}$$

Theorem 7.2.1 (Soundness and completeness theorem) *For any sequent* $\Gamma \Rightarrow \Sigma, \vdash^{\mathtt{ft}} \Gamma \Rightarrow \Sigma$ *if and only if* $\models^{\mathtt{ft}} \Gamma \Rightarrow \Sigma$. □

7.2.2 Gentzen Deduction System $\mathbf{G}^{\mathtt{ff}}$

$\Gamma \Rightarrow \Pi$ is $\mathbf{G}^{\mathtt{ff}}$-valid, denoted by $\models^{\mathtt{ff}} \Gamma \Rightarrow \Pi$, if for any assignment v, $v \models \Gamma$ implies $v \models \Pi$, where

- $v \models \Gamma$ if for each formula $A \in \Gamma$, $v(A) = 1$; and
- $v \models \Pi$ if for some formula $D \in \Sigma$, $v(D) = 0$.

  Gentzen deduction system $\mathbf{G}^{\mathtt{ff}}$ consists of the following axioms and deduction rules:
- **Axiom**:

$$(\mathtt{A}^{\mathtt{ff}})\ \frac{\begin{cases}\mathrm{incon}(\Gamma)\\ \mathrm{val}(\Pi)\\ \Gamma \cap \neg\Pi \neq \emptyset\end{cases}}{\Gamma \Rightarrow \Pi}$$

  where Γ, Π are sets of literals.
- **Deduction rules**:

$$\begin{array}{ll}
(\neg\neg^A)\ \dfrac{\Gamma, A \Rightarrow \Pi}{\Gamma, \neg\neg A \Rightarrow \Pi} & (\neg\neg^D)\ \dfrac{\Gamma \Rightarrow D, \Pi}{\Gamma \Rightarrow \neg\neg D, \Pi}\\
(\wedge^A)\ \dfrac{\begin{cases}\Gamma, A_1 \Rightarrow \Pi\\ \Gamma, A_2 \Rightarrow \Pi\end{cases}}{\Gamma, A_1 \wedge A_2 \Rightarrow \Pi} & (\wedge^D)\ \dfrac{\begin{cases}\Gamma \Rightarrow D_1, \Pi\\ \Gamma \Rightarrow D_2, \Pi\end{cases}}{\Gamma \Rightarrow D_1 \wedge D_2, \Pi}\\
(\vee^A)\ \dfrac{\left[\begin{array}{l}\Gamma, A_1 \Rightarrow \Pi\\ \Gamma, A_2 \Rightarrow \Pi\end{array}\right.}{\Gamma, A_1 \vee A_2 \Rightarrow \Pi} & (\vee^D)\ \dfrac{\left[\begin{array}{l}\Gamma \Rightarrow D_1, \Pi\\ \Gamma \Rightarrow D_2, \Pi\end{array}\right.}{\Gamma \Rightarrow D_1 \vee D_2, \Pi}
\end{array}$$

and

$$\begin{array}{ll}
(\neg\wedge^A)\ \dfrac{\left[\begin{array}{l}\Gamma, \neg A_1 \Rightarrow \Pi\\ \Gamma, \neg A_2 \Rightarrow \Pi\end{array}\right.}{\Gamma, \neg(A_1 \wedge A_2) \Rightarrow \Pi} & (\neg\wedge^D)\ \dfrac{\left[\begin{array}{l}\Gamma \Rightarrow \neg D_1, \Pi\\ \Gamma \Rightarrow \neg D_2, \Pi\end{array}\right.}{\Gamma \Rightarrow \neg(D_1 \wedge D_2), \Pi}\\
(\neg\vee^A)\ \dfrac{\begin{cases}\Gamma, \neg A_1 \Rightarrow \Pi\\ \Gamma, \neg A_2 \Rightarrow \Pi\end{cases}}{\Gamma, \neg(A_1 \vee A_2) \Rightarrow \Pi} & (\neg\vee^D)\ \dfrac{\begin{cases}\Gamma \Rightarrow \neg D_1, \Pi\\ \Gamma \Rightarrow \neg D_2, \Pi\end{cases}}{\Gamma \Rightarrow \neg(D_1 \vee D_2), \Pi}
\end{array}$$

Theorem 7.2.2 (Soundness and completeness theorem) *For any sequent* $\Gamma \Rightarrow \Pi$, $\vdash^{\mathtt{ff}} \Gamma \Rightarrow \Pi$ *if and only if* $\models^{\mathtt{ff}} \Gamma \Rightarrow \Pi$. □

7.3 R-Calculus $\mathbf{R}^+$

Given a quadruple (A, B, C, D) with $A \in \Gamma, B \in \Delta, C \in \Sigma$ and $D \in \Pi$, we use $\Gamma|\Delta \Rightarrow \Sigma|\Pi$ to revise (A, B, C, D) and obtain $\Gamma'|\Delta' \Rightarrow \Sigma'|\Pi'$, denoted by $\Gamma|\Delta \Rightarrow \Sigma|\Pi|(A, B, C, D) \rightrightarrows \Gamma'|\Delta' \Rightarrow \Sigma'|\Pi'$, if $\Gamma'|\Delta' \Rightarrow \Sigma'|\Pi' =$

$$\begin{cases}\Gamma - \{A\}|\Delta - \{B\} \Rightarrow \Sigma - \{C\}|\Pi - \{D\} \\ \qquad\qquad \text{if } \models^+ \Gamma - \{A\}|\Delta - \{B\} \Rightarrow \Sigma - \{C\}|\Pi - \{D\}\\ \Gamma|\Delta \Rightarrow \Sigma|\Pi \quad \text{otherwise.}\end{cases}$$

We denote

$$\mathbf{X} = \Gamma|\Delta \Rightarrow \Sigma|\Pi$$
$$\mathbf{X}((A, B, C, D)) = \Gamma - \{A\}|\Delta - \{B\} \Rightarrow \Sigma - \{C\}|\Pi - \{D\}.$$

R-calculus $\mathbf{R}^+$ consists of the following axioms and deduction rules:

- **Axioms**:

$$(\mathtt{A}^-)\ \frac{\left[\begin{array}{ll} \neg l \notin \Gamma & l' \notin \Gamma \\ l \notin \Delta & \neg l' \notin \Delta \\ l \notin \Sigma & \neg l' \notin \Sigma \\ \neg l \notin \Pi & l' \notin \Pi \\ \neg m \notin \Gamma & m' \notin \Gamma \\ m \notin \Delta & \neg m' \notin \Delta \\ \neg m \notin \Sigma & m' \notin \Sigma \\ m \notin \Pi & \neg m' \notin \Pi \end{array}\right.}{\mathbf{X}|(l, l', m, m') \rightrightarrows \mathbf{X}((l, l', m, m'))}$$

$$(\mathtt{A}^0)\ \frac{\left\{\begin{array}{ll} \neg l \in \Gamma & l' \in \Gamma \\ l \in \Delta & \neg l' \in \Delta \\ l \in \Sigma & \neg l' \in \Sigma \\ \neg l \in \Pi & l' \in \Pi \\ \neg m \in \Gamma & m' \in \Gamma \\ m \in \Delta & \neg m' \in \Delta \\ \neg m \in \Sigma & m' \in \Sigma \\ m \in \Pi & \neg m' \in \Pi \end{array}\right.}{\mathbf{X}|(l, l', m, m') \rightrightarrows \mathbf{X}}$$

  where $\mathbf{X}$ is literal and l, l', m, m' are literals.
- **Deduction rules**:

$$(\neg\neg_-^A)\ \frac{\mathbf{X}|(A, B, C, D) \rightrightarrows \mathbf{X}((A, B, C, D))}{\mathbf{X}|(\neg\neg A, B, C, D) \rightrightarrows \mathbf{X}((\neg\neg A, B, C, D))}$$

$$(\neg\neg_0^A)\ \frac{\mathbf{X}|(A, B, C, D) \rightrightarrows \mathbf{X}}{\mathbf{X}|(\neg\neg A, B, C, D) \rightrightarrows \mathbf{X}}$$

$$(\neg\neg_-^B)\ \frac{\mathbf{X}|(A, B, C, D) \rightrightarrows \mathbf{X}((A, B, C, D))}{\mathbf{X}|(A, \neg\neg B, C, D) \rightrightarrows \mathbf{X}((A, \neg\neg B, C, D))}$$

$$(\neg\neg_0^B)\ \frac{\mathbf{X}|(A, B, C, D) \rightrightarrows \mathbf{X}}{\mathbf{X}|(A, \neg\neg B, C, D) \rightrightarrows \mathbf{X}}$$

$$(\neg\neg_-^C)\ \frac{\mathbf{X}|(A, B, C, D) \rightrightarrows \mathbf{X}((A, B, C, D))}{\mathbf{X}|(A, B, \neg\neg C, D) \rightrightarrows \mathbf{X}((A, B, \neg\neg C, D))}$$

$$(\neg\neg_0^C)\ \frac{\mathbf{X}|(A, B, C, D) \rightrightarrows \mathbf{X}}{\mathbf{X}|(A, B, \neg\neg C, D) \rightrightarrows \mathbf{X}}$$

$$(\neg\neg_-^D)\ \frac{\mathbf{X}|(A, B, C, D) \rightrightarrows \mathbf{X}((A, B, C, D))}{\mathbf{X}|(A, B, C, \neg\neg D) \rightrightarrows \mathbf{X}((A, B, C, \neg\neg D))}$$

$$(\neg\neg_0^D)\ \frac{\mathbf{X}|(A, B, C, D) \rightrightarrows \mathbf{X}}{\mathbf{X}|(A, B, C, \neg\neg D) \rightrightarrows \mathbf{X}}$$

and

$$(\wedge^A_-)\ \frac{\left\{\begin{array}{l}\mathbf{X}|(A_1, B, C, D) \rightrightarrows \mathbf{X}((A_1, B, C, D))\\ \mathbf{X}|(A_2, B, C, D) \rightrightarrows \mathbf{X}((A_2, B, C, D))\end{array}\right.}{\mathbf{X}|(A_1 \wedge A_2, B, C, D) \rightrightarrows \mathbf{X}((A_1 \wedge A_2, B, C, D))}$$

$$(\wedge^A_0)\ \frac{\left[\begin{array}{l}\mathbf{X}|(A_1, B, C, D) \rightrightarrows \mathbf{X}\\ \mathbf{X}|(A_2, B, C, D) \rightrightarrows \mathbf{X}\end{array}\right.}{\mathbf{X}|(A_1 \wedge A_2, B, C, D) \rightrightarrows \mathbf{X}}$$

$$(\wedge^B_-)\ \frac{\left[\begin{array}{l}\mathbf{X}|(A, B_1, C, D) \rightrightarrows \mathbf{X}((A, B_1, C, D))\\ \mathbf{X}((A, B_1, C, D))|(\lambda, B_2, \lambda, \lambda) \rightrightarrows \mathbf{X}((A, B_1, C, D))((\lambda, B_2, \lambda, \lambda))\end{array}\right.}{\mathbf{X}|(A, B_1 \wedge B_2, C, D) \rightrightarrows \mathbf{X}((A, B_1 \wedge B_2, C, D))}$$

$$(\wedge^B_0)\ \frac{\left\{\begin{array}{l}\mathbf{X}|(A, B_1, C, D) \rightrightarrows \mathbf{X}\\ \mathbf{X}((A, B_1, C, D))|(\lambda, B_2, \lambda, \lambda) \rightrightarrows \mathbf{X}((A, B_1, C, D))\end{array}\right.}{\mathbf{X}|(A, B_1 \wedge B_2, C, D) \rightrightarrows \mathbf{X}}$$

and

$$(\wedge^C_-)\ \frac{\left[\begin{array}{l}\mathbf{X}|(A, B, C_1, D) \rightrightarrows \mathbf{X}((A, B, C_1, D))\\ \mathbf{X}((A, B, C_1, D))|(\lambda, \lambda, C_2, \lambda) \rightrightarrows \mathbf{X}((A, B, C_1, D))((\lambda, \lambda, C_2, \lambda))\end{array}\right.}{\mathbf{X}|(A, B, C_1 \wedge C_2, D) \rightrightarrows \mathbf{X}((A, B, C_1 \wedge C_2, D))}$$

$$(\wedge^C_0)\ \frac{\left\{\begin{array}{l}\mathbf{X}|(A, B, C_1, D) \rightrightarrows \mathbf{X}\\ \mathbf{X}((A, B, C_1, D))|(\lambda, \lambda, C_2, \lambda) \rightrightarrows \mathbf{X}((A, B, C_1, D))\end{array}\right.}{\mathbf{X}|(A, B, C_1 \wedge C_2, D) \rightrightarrows \mathbf{X}}$$

$$(\wedge^D_-)\ \frac{\left\{\begin{array}{l}\mathbf{X}|(A, B, C, D_1) \rightrightarrows \mathbf{X}((A, B, C, D_1))\\ \mathbf{X}|(A, B, C, D_2) \rightrightarrows \mathbf{X}((A, B, C, D_2))\end{array}\right.}{\mathbf{X}|(A, B, C, D_1 \wedge D_2) \rightrightarrows \mathbf{X}((A, B, C, D_1 \wedge D_2))}$$

$$(\wedge^D_0)\ \frac{\left[\begin{array}{l}\mathbf{X}|(A, B, C, D_1) \rightrightarrows \mathbf{X}\\ \mathbf{X}|(A, B, C, D_2) \rightrightarrows \mathbf{X}\end{array}\right.}{\mathbf{X}|(A, B, C, D_1 \wedge D_2) \rightrightarrows \mathbf{X}}$$

and

$$(\vee^A_-)\ \frac{\left[\begin{array}{l}\mathbf{X}|(A_1, B, C, D) \rightrightarrows \mathbf{X}((A_1, B, C, D))\\ \mathbf{X}((A_1, B, C, D))|(A_2, \lambda, \lambda, \lambda) \rightrightarrows \mathbf{X}((A_1, B, C, D))((A_2, \lambda, \lambda, \lambda))\end{array}\right.}{\mathbf{X}(A_1 \vee A_2, B, C, D) \rightrightarrows \mathbf{X}((A_1 \vee A_2, B, C, D))}$$

$$(\vee^A_0)\ \frac{\left\{\begin{array}{l}\mathbf{X}|(A_1, B, C, D) \rightrightarrows \mathbf{X}\\ \mathbf{X}((A_1, B, C, D))|(A_2, \lambda, \lambda, \lambda) \rightrightarrows \mathbf{X}((A_1, B, C, D))\end{array}\right.}{\mathbf{X}|(A_1 \vee A_2, B, C, D) \rightrightarrows \mathbf{X}}$$

$$(\vee^B_-)\ \frac{\left\{\begin{array}{l}\mathbf{X}|(A, B_1, C, D) \rightrightarrows \mathbf{X}((A, B_1, C, D))\\ \mathbf{X}|(A, B_2, C, D) \rightrightarrows \mathbf{X}((A, B_2, C, D))\end{array}\right.}{\mathbf{X}|(A, B_1 \vee B_2, C, D) \rightrightarrows \mathbf{X}((A, B_1 \vee B_2, C, D))}$$

$$(\vee^B_0)\ \frac{\left[\begin{array}{l}\mathbf{X}|(A, B_1, C, D) \rightrightarrows \mathbf{X}\\ \mathbf{X}|(A, B_2, C, D) \rightrightarrows \mathbf{X}\end{array}\right.}{\mathbf{X}|(A, B_1 \vee B_2, C, D) \rightrightarrows \mathbf{X}}$$

and

$$(\vee_-^C)\ \frac{\left\{\begin{array}{l}\mathbf{X}|(A, B, C_1, D) \rightrightarrows \mathbf{X}((A, B, C_1, D))\\ \mathbf{X}|(A, B, C_2, D) \rightrightarrows \mathbf{X}((A, B, C_2, D))\end{array}\right.}{\mathbf{X}|(A, B, C_1 \vee C_2, D) \rightrightarrows \mathbf{X}((A, B, C_1 \vee C_2, D))}$$

$$(\vee_0^C)\ \frac{\left[\begin{array}{l}\mathbf{X}|(A, B, C_1, D) \rightrightarrows \mathbf{X}\\ \mathbf{X}|(A, B, C_2, D) \rightrightarrows \mathbf{X}\end{array}\right.}{\mathbf{X}(A, B, C_1 \vee C_2, D) \rightrightarrows \mathbf{X}}$$

$$(\vee_-^D)\ \frac{\left[\begin{array}{l}\mathbf{X}|(A, B, C, D_1) \rightrightarrows \mathbf{X}((A, B, C, D_1))\\ \mathbf{X}((\lambda, \lambda, \lambda, D_1))|(A, B, C, D_2) \rightrightarrows \mathbf{X}((A, B, C, D_1))((\lambda, \lambda, \lambda, D_2))\end{array}\right.}{\mathbf{X}(A, B, C, D_1 \vee D_2) \rightrightarrows \mathbf{X}((A, B, C, D_1 \vee D_2))}$$

$$(\vee_0^D)\ \frac{\left\{\begin{array}{l}\mathbf{X}|(A, B, C, D_1) \rightrightarrows \mathbf{X}\\ \mathbf{X}((A, B, C, D_1))|(\lambda, \lambda, \lambda, D_2) \rightrightarrows \mathbf{X}((A, B, C, D_1))\end{array}\right.}{\mathbf{X}(A, B, C, D_1 \vee D_2) \rightrightarrows \mathbf{X}}$$

and

$$(\neg\wedge_-^A)\ \frac{\left[\begin{array}{l}\mathbf{X}|(\neg A_1, B, C, D) \rightrightarrows \mathbf{X}((\neg A_1, B, C, D))\\ \mathbf{X}((\neg A_1, B, C, D))|(\neg A_2, \lambda, \lambda, \lambda) \rightrightarrows \mathbf{X}((\neg A_1, B, C, D))((\neg A_2, \lambda, \lambda, \lambda))\end{array}\right.}{\mathbf{X}|(\neg(A_1 \wedge A_2), B, C, D) \rightrightarrows \mathbf{X}((\neg(A_1 \wedge A_2), B, C, D))}$$

$$(\neg\wedge_0^A)\ \frac{\left\{\begin{array}{l}\mathbf{X}|(\neg A_1, B, C, D) \rightrightarrows \mathbf{X}\\ \mathbf{X}((\neg A_1, B, C, D))|(\neg A_2, \lambda, \lambda, \lambda) \rightrightarrows \mathbf{X}((\neg A_1, B, C, D))\end{array}\right.}{\mathbf{X}|(\neg(A_1 \wedge A_2), B, C, D) \rightrightarrows \mathbf{X}}$$

$$(\neg\wedge_-^B)\ \frac{\left\{\begin{array}{l}\mathbf{X}|(A, \neg B_1, C, D) \rightrightarrows \mathbf{X}((A, \neg B_1, C, D))\\ \mathbf{X}|(A, \neg B_2, C, D) \rightrightarrows \mathbf{X}((A, \neg B_2, C, D))\end{array}\right.}{\mathbf{X}|(A, \neg(B_1 \wedge B_2), C, D) \rightrightarrows \mathbf{X}((A, \neg(B_1 \wedge B_2), C, D))}$$

$$(\neg\wedge_0^B)\ \frac{\left[\begin{array}{l}\mathbf{X}|(A, \neg B_1, C, D) \rightrightarrows \mathbf{X}\\ \mathbf{X}|(A, \neg B_2, C, D) \rightrightarrows \mathbf{X}\end{array}\right.}{\mathbf{X}|(A, \neg(B_1 \wedge B_2), C, D) \rightrightarrows \mathbf{X}}$$

and

$$(\neg\wedge_-^C)\ \frac{\left\{\begin{array}{l}\mathbf{X}|(A, B, \neg C_1, D) \rightrightarrows \mathbf{X}((A, B, \neg C_1, D))\\ \mathbf{X}|(A, B, \neg C_2, D) \rightrightarrows \mathbf{X}((A, B, \neg C_2, D))\end{array}\right.}{\mathbf{X}|(A, B, \neg(C_1 \wedge C_2), D) \rightrightarrows \mathbf{X}((A, B, \neg(C_1 \wedge C_2), D))}$$

$$(\neg\wedge_0^C)\ \frac{\left[\begin{array}{l}\mathbf{X}|(A, B, \neg C_1, D) \rightrightarrows \mathbf{X}\\ \mathbf{X}|(A, B, \neg C_2, D) \rightrightarrows \mathbf{X}\end{array}\right.}{\mathbf{X}|(A, B, \neg(C_1 \wedge C_2), D) \rightrightarrows \mathbf{X}}$$

$$(\neg\wedge_-^D)\ \frac{\left[\begin{array}{l}\mathbf{X}|(A, B, C, \neg D_1) \rightrightarrows \mathbf{X}((A, B, C, \neg D_1))\\ \mathbf{X}((A, B, C, \neg D_1))|(\lambda, \lambda, \lambda, \neg D_2) \rightrightarrows \mathbf{X}((A, B, C, \neg D_1))((\lambda, \lambda, \lambda, \neg D_2))\end{array}\right.}{\mathbf{X}|(A, B, C, \neg(D_1 \wedge D_2)) \rightrightarrows \mathbf{X}((A, B, C, \neg(D_1 \wedge D_2)))}$$

$$(\neg\wedge_0^D)\ \frac{\left\{\begin{array}{l}\mathbf{X}|(A, B, C, \neg D_1) \rightrightarrows \mathbf{X}\\ \mathbf{X}((A, B, C, \neg D_1))|(\lambda, \lambda, \lambda, \neg D_2) \rightrightarrows \mathbf{X}((A, B, C, \neg D_1))\end{array}\right.}{\mathbf{X}|(A, B, C, \neg(D_1 \wedge D_2)) \rightrightarrows \mathbf{X}}$$

and

$$(\neg\vee^A_-)\ \frac{\left\{\begin{array}{l}\mathbf{X}|(\neg A_1, B, C, D) \rightrightarrows \mathbf{X}((\neg A_1, B, C, D))\\ \mathbf{X}|(\neg A_2, B, C, D) \rightrightarrows \mathbf{X}((\neg A_2, B, C, D))\end{array}\right.}{\mathbf{X}|(\neg(A_1 \vee A_2), B, C, D) \rightrightarrows \mathbf{X}((\neg(A_1 \vee A_2), B, C, D))}$$

$$(\neg\vee^A_0)\ \frac{\left[\begin{array}{l}\mathbf{X}|(\neg A_1, B, C, D) \rightrightarrows \mathbf{X}\\ \mathbf{X}|(\neg A_2, B, C, D) \rightrightarrows \mathbf{X}\end{array}\right.}{\mathbf{X}|(\neg(A_1 \vee A_2), B, C, D) \rightrightarrows \mathbf{X}}$$

$$(\neg\vee^B_-)\ \frac{\left[\begin{array}{l}\mathbf{X}|(A, \neg B_1, C, D) \rightrightarrows \mathbf{X}((A, \neg B_1, C, D))\\ \mathbf{X}((A, \neg B_1, C, D))|(\lambda, \neg B_2, \lambda, \lambda) \rightrightarrows \mathbf{X}((A, \neg B_1, C, D))((\lambda, \neg B_2, \lambda, \lambda))\end{array}\right.}{\mathbf{X}|(A, \neg(B_1 \vee B_2), C, D) \rightrightarrows \mathbf{X}((A, \neg(B_1 \vee B_2), C, D))}$$

$$(\neg\vee^B_0)\ \frac{\left\{\begin{array}{l}\mathbf{X}|(A, \neg B_1, C, D) \rightrightarrows \mathbf{X}\\ \mathbf{X}((A, \neg B_1, C, D))|(\lambda, \neg B_2, \lambda, \lambda) \rightrightarrows \mathbf{X}((A, \neg B_1, C, D))\end{array}\right.}{\mathbf{X}|(A, \neg(B_1 \vee B_2), C, D) \rightrightarrows \mathbf{X}}$$

and

$$(\neg\vee^C_-)\ \frac{\left[\begin{array}{l}\mathbf{X}|(A, B, \neg C_1, D) \rightrightarrows \mathbf{X}((A, B, \neg C_1, D))\\ \mathbf{X}((A, B, \neg C_1, D))|(\lambda, \lambda, \neg C_2, \lambda) \rightrightarrows \mathbf{X}((A, B, \neg C_1, D))((\lambda, \lambda, \neg C_2, \lambda))\end{array}\right.}{\mathbf{X}|(A, B, \neg(C_1 \vee C_2), D) \rightrightarrows \mathbf{X}((A, B, \neg(C_1 \vee C_2), D))}$$

$$(\neg\vee^C_0)\ \frac{\left\{\begin{array}{l}\mathbf{X}|(A, B, \neg C_1, D) \rightrightarrows \mathbf{X}\\ \mathbf{X}((A, B, \neg C_1, D))|(\lambda, \lambda, \neg C_2, \lambda) \rightrightarrows \mathbf{X}((A, B, \neg C_1, D))\end{array}\right.}{\mathbf{X}|(A, B, \neg(C_1 \vee C_2), D) \rightrightarrows \mathbf{X}}$$

$$(\neg\vee^D_-)\ \frac{\left\{\begin{array}{l}\mathbf{X}|(A, B, C, \neg D_1) \rightrightarrows \mathbf{X}((A, B, C, \neg D_1))\\ \mathbf{X}|(A, B, C, \neg D_2) \rightrightarrows \mathbf{X}((A, B, C, \neg D_2))\end{array}\right.}{\mathbf{X}|(A, B, C, \neg(D_1 \vee D_2)) \rightrightarrows \mathbf{X}((A, B, C, \neg(D_1 \vee D_2)))}$$

$$(\neg\vee^D_0)\ \frac{\left[\begin{array}{l}\mathbf{X}|(A, B, C, \neg D_1) \rightrightarrows \mathbf{X}\\ \mathbf{X}|(A, B, C, \neg D_2) \rightrightarrows \mathbf{X}\end{array}\right.}{\mathbf{X}|(A, B, C, \neg(D_1 \vee D_2)) \rightrightarrows \mathbf{X}}$$

Definition 7.3.1 Given a supersequent $\mathbf{X}$ and a quadruple X with $X \in \mathbf{X}$, a reduction $\mathbf{X}|X \rightrightarrows \mathbf{X}'$ is provable in $\mathbf{R}^+$, denoted by $\vdash^+ \mathbf{X}|X \rightrightarrows \mathbf{X}'$, if there is a sequence $\{\delta_1, \ldots, \delta_n\}$ of reductions such that $\delta_n = \mathbf{X}|X \rightrightarrows \mathbf{X}'$, and for each $1 \leq i \leq n$, δ_i is an axiom or deduced from the previous reductions by one of the deduction rules in $\mathbf{R}^+$.

Theorem 7.3.2 (Soundness and completeness theorem) *For any reduction* $\mathbf{X}|X \rightrightarrows \mathbf{X}'$ *with* $X \in \mathbf{X}$,

$$\vdash^+ \mathbf{X}|X \rightrightarrows \mathbf{X}' \textit{ if and only if } \models^+ \mathbf{X}|X \rightrightarrows \mathbf{X}'.$$

□

7.3.1 $\mathbf{R}^{\mathtt{ft}}$

Given a pair (A, C) of formulas such that $A \in \Gamma$ and $C \in \Sigma$, we use $\Gamma \Rightarrow \Sigma$ to revise (A, C) and obtain a sequence $\Gamma' \Rightarrow \Sigma'$, denoted by

$$\models^{\mathtt{ft}} \Gamma \Rightarrow \Sigma|(A, C) \rightrightarrows \Gamma' \Rightarrow \Sigma',$$

if $\Gamma' \Rightarrow \Sigma' =$

$$\begin{cases} \Gamma - \{A\} \Rightarrow \Sigma - \{C\} \text{ if } \models^{\mathtt{ft}} \Gamma - \{A\} \Rightarrow \Sigma - \{C\} \\ \Gamma \Rightarrow \Sigma \quad \text{otherwise.} \end{cases}$$

Gentzen deduction system $\mathbf{G}^{\mathtt{ft}}$ consists of the following axioms and deduction rules:

- **Axioms**:

$$(\mathtt{A}_{-}^{\mathtt{ft}})\ \frac{\left[\begin{array}{l} \neg l \notin \Gamma \\ l \notin \Sigma \\ m \notin \Gamma \\ \neg m \notin \Sigma \end{array}\right.}{\Gamma \Rightarrow \Sigma|(l, m) \rightrightarrows \Gamma - \{l\} \Rightarrow \Sigma - \{m\}} \qquad (\mathtt{A}_{0}^{\mathtt{tt}})\ \frac{\left\{\begin{array}{l} \neg l \in \Gamma \\ l \in \Sigma \\ m \in \Gamma \\ \neg m \in \Sigma \end{array}\right.}{\Gamma \Rightarrow \Sigma|(l, m) \rightrightarrows \Gamma \Rightarrow \Sigma}$$

  where Γ, Σ are sets of literals, and l, m are literals such that $l \in \Gamma$ and $m \in \Sigma$.
- **Deduction rules**:

$$(\neg\neg_{-}^{A})\ \frac{\Gamma \Rightarrow \Sigma|(A, C) \rightrightarrows \Gamma - \{A\} \Rightarrow \Sigma - \{C\}}{\Gamma \Rightarrow \Sigma|(\neg\neg A, C) \rightrightarrows \Gamma - \{\neg\neg A\} \Rightarrow \Sigma - \{C\}}$$

$$(\neg\neg_{0}^{A})\ \frac{\Gamma \Rightarrow \Sigma|(A, C) \rightrightarrows \Gamma \Rightarrow \Sigma}{\Gamma \Rightarrow \Sigma|(\neg\neg A, C) \rightrightarrows \Gamma \Rightarrow \Sigma}$$

$$(\neg\neg_{-}^{C})\ \frac{\Gamma \Rightarrow \Sigma|(A, C) \rightrightarrows \Gamma - \{A\} \Rightarrow \Sigma - \{C\}}{\Gamma \Rightarrow \Sigma|(A, \neg\neg C) \rightrightarrows \Gamma - \{A\} \Rightarrow \Sigma - \{\neg\neg C\}}$$

$$(\neg\neg_{0}^{A})\ \frac{\Gamma \Rightarrow \Sigma|(A, C) \rightrightarrows \Gamma \Rightarrow \Sigma}{\Gamma \Rightarrow \Sigma|(A, \neg\neg C) \rightrightarrows \Gamma \Rightarrow \Sigma}$$

and

$$(\wedge_{-}^{A})\ \frac{\left\{\begin{array}{l} \Gamma \Rightarrow \Sigma|(A_1, C) \rightrightarrows \Gamma - \{A_1\} \Rightarrow \Sigma - \{C\} \\ \Gamma \Rightarrow \Sigma|(A_2, C) \rightrightarrows \Gamma - \{A_2\} \Rightarrow \Sigma - \{C\} \end{array}\right.}{\Gamma \Rightarrow \Sigma|(A_1 \wedge A_2, C) \rightrightarrows \Gamma - \{A_1 \wedge A_2\} \Rightarrow \Sigma - \{C\}}$$

$$(\wedge_{0}^{A})\ \frac{\left[\begin{array}{l} \Gamma \Rightarrow \Sigma|(A_1, C) \rightrightarrows \Gamma \Rightarrow \Sigma \\ \Gamma \Rightarrow \Sigma|(A_2, C) \rightrightarrows \Gamma \Rightarrow \Sigma \end{array}\right.}{\Gamma \Rightarrow \Sigma|(A_1 \wedge A_2, C) \rightrightarrows \Gamma \Rightarrow \Sigma}$$

$$(\wedge_{-}^{C})\ \frac{\left[\begin{array}{l} \Gamma \Rightarrow \Sigma|(A, C_1) \rightrightarrows \Gamma - \{A\} \Rightarrow \Sigma - \{C_1\} \\ \Gamma - \{A\} \Rightarrow \Sigma - \{C_1\}|(\lambda, C_2) \rightrightarrows \Gamma - \{A\} \Rightarrow \Sigma - \{C_1, C_2\} \end{array}\right.}{\Gamma \Rightarrow \Sigma|(A, C_1 \wedge C_2) \rightrightarrows \Gamma - \{A\} \Rightarrow \Sigma - \{C_1 \wedge C_2\}}$$

$$(\wedge_{0}^{A})\ \frac{\left\{\begin{array}{l} \Gamma \Rightarrow \Sigma|(A, C_1) \rightrightarrows \Gamma \Rightarrow \Sigma \\ \Gamma - \{A\} \Rightarrow \Sigma - \{C_1\}|(\lambda, C_2) \rightrightarrows \Gamma - \{A\} \Rightarrow \Sigma - \{C_1\} \end{array}\right.}{\Gamma \Rightarrow \Sigma|(A_1 \wedge A_2, C) \rightrightarrows \Gamma \Rightarrow \Sigma}$$

and

$$(\vee_{-}^{A})\ \dfrac{\left[\begin{array}{l}\Gamma \Rightarrow \Sigma|(A_1, C) \rightrightarrows \Gamma - \{A_1\} \Rightarrow \Sigma - \{C\}\\ \Gamma - \{A_1\} \Rightarrow \Sigma - \{C\}|(A_2, \lambda) \rightrightarrows \Gamma - \{A_1, A_2\} \Rightarrow \Sigma - \{C\}\end{array}\right.}{\Gamma \Rightarrow \Sigma|(A_1 \vee A_2, C) \rightrightarrows \Gamma - \{A_1 \vee A_2\} \Rightarrow \Sigma - \{C\}}$$

$$(\vee_{0}^{A})\ \dfrac{\left\{\begin{array}{l}\Gamma \Rightarrow \Sigma|(A_1, C) \rightrightarrows \Gamma \Rightarrow \Sigma\\ \Gamma - \{A_1\} \Rightarrow \Sigma - \{C\}|(A_2, \lambda) \rightrightarrows \Gamma - \{A_1\} \Rightarrow \Sigma - \{C\}\end{array}\right.}{\Gamma \Rightarrow \Sigma|(A_1 \vee A_2, C) \rightrightarrows \Gamma \Rightarrow \Sigma}$$

$$(\vee_{-}^{C})\ \dfrac{\left\{\begin{array}{l}\Gamma \Rightarrow \Sigma|(A, C_1) \rightrightarrows \Gamma - \{A\} \Rightarrow \Sigma - \{C_1\}\\ \Gamma \Rightarrow \Sigma|(A, C_2) \rightrightarrows \Gamma - \{A\} \Rightarrow \Sigma - \{C_2\}\end{array}\right.}{\Gamma \Rightarrow \Sigma|(A, C_1 \vee C_2) \rightrightarrows \Gamma - \{A\} \Rightarrow \Sigma - \{C_1 \vee C_2\}}$$

$$(\vee_{0}^{A})\ \dfrac{\left[\begin{array}{l}\Gamma \Rightarrow \Sigma|(A, C_1) \rightrightarrows \Gamma \Rightarrow \Sigma\\ \Gamma \Rightarrow \Sigma|(A, C_2) \rightrightarrows \Gamma \Rightarrow \Sigma\end{array}\right.}{\Gamma \Rightarrow \Sigma|(A, C_1 \vee C_2) \rightrightarrows \Gamma \Rightarrow \Sigma}$$

and

$$(\neg\wedge_{-}^{A})\ \dfrac{\left[\begin{array}{l}\Gamma \Rightarrow \Sigma|(\neg A_1, C) \rightrightarrows \Gamma - \{\neg A_1\} \Rightarrow \Sigma - \{C\}\\ \Gamma - \{\neg A_1\} \Rightarrow \Sigma - \{C\}|(\neg A_2, \lambda) \rightrightarrows \Gamma - \{\neg A_1, \neg A_2\} \Rightarrow \Sigma - \{C\}\end{array}\right.}{\Gamma \Rightarrow \Sigma|(\neg(A_1 \wedge A_2), C) \rightrightarrows \Gamma - \{\neg(A_1 \wedge A_2)\} \Rightarrow \Sigma - \{C\}}$$

$$(\neg\wedge_{0}^{A})\ \dfrac{\left\{\begin{array}{l}\Gamma \Rightarrow \Sigma|(\neg A_1, C) \rightrightarrows \Gamma \Rightarrow \Sigma\\ \Gamma - \{\neg A_1\} \Rightarrow \Sigma - \{C\}|(\neg A_2, \lambda) \rightrightarrows \Gamma - \{\neg A_1\} \Rightarrow \Sigma - \{C\}\end{array}\right.}{\Gamma \Rightarrow \Sigma|(\neg(A_1 \wedge A_2), C) \rightrightarrows \Gamma \Rightarrow \Sigma}$$

$$(\neg\wedge_{-}^{C})\ \dfrac{\left\{\begin{array}{l}\Gamma \Rightarrow \Sigma|(A, \neg C_1) \rightrightarrows \Gamma - \{A\} \Rightarrow \Sigma - \{\neg C_1\}\\ \Gamma \Rightarrow \Sigma|(A, \neg C_2) \rightrightarrows \Gamma - \{A\} \Rightarrow \Sigma - \{\neg C_2\}\end{array}\right.}{\Gamma \Rightarrow \Sigma|(A, \neg(C_1 \wedge C_2)) \rightrightarrows \Gamma - \{A\} \Rightarrow \Sigma - \{\neg(C_1 \wedge C_2)\}}$$

$$(\neg\wedge_{0}^{C})\ \dfrac{\left[\begin{array}{l}\Gamma \Rightarrow \Sigma|(A, \neg C_1) \rightrightarrows \Gamma \Rightarrow \Sigma\\ \Gamma \Rightarrow \Sigma|(A, \neg C_2) \rightrightarrows \Gamma \Rightarrow \Sigma\end{array}\right.}{\Gamma \Rightarrow \Sigma|(A, \neg(C_1 \wedge C_2)) \rightrightarrows \Gamma \Rightarrow \Sigma}$$

and

$$(\neg\vee_{-}^{A})\ \dfrac{\left\{\begin{array}{l}\Gamma \Rightarrow \Sigma|(\neg A_1, C) \rightrightarrows \Gamma - \{\neg A_1\} \Rightarrow \Sigma - \{C\}\\ \Gamma \Rightarrow \Sigma|(\neg A_2, C) \rightrightarrows \Gamma - \{\neg A_2\} \Rightarrow \Sigma - \{C\}\end{array}\right.}{\Gamma \Rightarrow \Sigma|(\neg(A_1 \vee A_2), C) \rightrightarrows \Gamma - \{\neg(A_1 \vee A_2)\} \Rightarrow \Sigma - \{C\}}$$

$$(\neg\vee_{0}^{A})\ \dfrac{\left[\begin{array}{l}\Gamma \Rightarrow \Sigma|(\neg A_1, C) \rightrightarrows \Gamma \Rightarrow \Sigma\\ \Gamma \Rightarrow \Sigma|(\neg A_2, C) \rightrightarrows \Gamma \Rightarrow \Sigma\end{array}\right.}{\Gamma \Rightarrow \Sigma|(\neg(A_1 \vee A_2), C) \rightrightarrows \Gamma \Rightarrow \Sigma}$$

$$(\neg\vee_{-}^{C})\ \dfrac{\left[\begin{array}{l}\Gamma \Rightarrow \Sigma|(A, \neg C_1) \rightrightarrows \Gamma - \{A\} \Rightarrow \Sigma - \{\neg C_1\}\\ \Gamma - \{A\} \Rightarrow \Sigma - \{\neg C_1\}|(\lambda, \neg C_2) \rightrightarrows \Gamma - \{A\} \Rightarrow \Sigma - \{\neg C_1, \neg C_2\}\end{array}\right.}{\Gamma \Rightarrow \Sigma|(A, \neg(C_1 \vee C_2)) \rightrightarrows \Gamma - \{A\} \Rightarrow \Sigma - \{\neg(C_1 \vee C_2)\}}$$

$$(\neg\vee_{0}^{C})\ \dfrac{\left\{\begin{array}{l}\Gamma \Rightarrow \Sigma|(A, \neg C_1) \rightrightarrows \Gamma \Rightarrow \Sigma\\ \Gamma - \{A\} \Rightarrow \Sigma - \{\neg C_1\}|(\lambda, \neg C_2) \rightrightarrows \Gamma - \{A\} \Rightarrow \Sigma - \{\neg C_1\}\end{array}\right.}{\Gamma \Rightarrow \Sigma|(A, \neg(C_1 \vee C_2)) \rightrightarrows \Gamma \Rightarrow \Sigma}$$

Theorem 7.3.3 (Soundness and completeness theorem) *For any sequent* $\Gamma \Rightarrow \Sigma$ *and pair* (A, C) *with* $A \in \Gamma$ *and* $C \in \Sigma$,

$$\vdash^{\mathtt{ft}} \Gamma \Rightarrow \Sigma|(A, C) \rightrightarrows \Gamma' \Rightarrow \Sigma'$$

if and only if

$$\models^{\mathtt{ft}} \Gamma \Rightarrow \Sigma|(A, C) \rightrightarrows \Gamma' \Rightarrow \Sigma',$$

where $\Gamma' = \Gamma|\Gamma - \{A\}$ *and* $\Delta' = \Delta|\Delta - \{C\}$. □

7.3.2 $\mathbf{R}^{\mathtt{ff}}$

Given a pair (A, D) of formulas such that $A \in \Gamma$ and $D \in \Pi$, we use $\Gamma \Rightarrow \Pi$ to revise (A, D) and obtain a sequence $\Gamma' \Rightarrow \Pi'$, denoted by

$$\models^{\mathtt{ff}} \Gamma \Rightarrow \Pi|(A, D) \rightrightarrows \Gamma' \Rightarrow \Pi',$$

if

$$\Gamma' \Rightarrow \Pi' = \begin{cases} \Gamma - \{A\} \Rightarrow \Pi - \{D\} & \text{if } \models^{\mathtt{ff}} \Gamma - \{A\} \Rightarrow \Pi - \{D\} \\ \Gamma \Rightarrow \Pi & \text{otherwise.} \end{cases}$$

Gentzen deduction system $\mathbf{G}^{\mathtt{ff}}$ consists of the following axioms and deduction rules:

- **Axioms**:

$$(\mathtt{A}^{\mathtt{ff}}_{-})\ \frac{\left[\begin{array}{l} \neg l \notin \Gamma \\ \neg l \notin \Pi \\ \neg m \notin \Gamma \\ \neg m \notin \Pi \end{array}\right.}{\Gamma \Rightarrow \Pi|(l, m) \rightrightarrows \Gamma - \{l\} \Rightarrow \Pi - \{m\}} \qquad (\mathtt{A}^{\mathtt{ff}}_{0})\ \frac{\left\{\begin{array}{l} \neg l \in \Gamma \\ \neg l \in \Pi \\ \neg m \in \Gamma \\ \neg m \in \Pi \end{array}\right.}{\Gamma \Rightarrow \Pi|(l, m) \rightrightarrows \Gamma \Rightarrow \Pi}$$

  where Γ, Π are sets of literals, and l, m are literals such that $l \in \Gamma$ and $m \in \Pi$.
- **Deduction rules**:

$$(\neg\neg^{A}_{-})\ \frac{\Gamma \Rightarrow \Pi|(A, D) \rightrightarrows \Gamma - \{A\} \Rightarrow \Pi - \{D\}}{\Gamma \Rightarrow \Pi|(\neg\neg A, D) \rightrightarrows \Gamma - \{\neg\neg A\} \Rightarrow \Pi - \{D\}}$$
$$(\neg\neg^{A}_{0})\ \frac{\Gamma \Rightarrow \Pi|(A, D) \rightrightarrows \Gamma \Rightarrow \Pi}{\Gamma \Rightarrow \Pi|(\neg\neg A, D) \rightrightarrows \Gamma \Rightarrow \Pi}$$
$$(\neg\neg^{D}_{-})\ \frac{\Gamma \Rightarrow \Pi|(A, D) \rightrightarrows \Gamma - \{A\} \Rightarrow \Pi - \{D\}}{\Gamma \Rightarrow \Pi|(A, \neg\neg D) \rightrightarrows \Gamma - \{A\} \Rightarrow \Pi - \{\neg\neg D\}}$$
$$(\neg\neg^{A}_{0})\ \frac{\Gamma \Rightarrow \Pi|(A, D) \rightrightarrows \Gamma \Rightarrow \Pi}{\Gamma \Rightarrow \Pi|(A, \neg\neg D) \rightrightarrows \Gamma \Rightarrow \Pi}$$

and

$$(\wedge_{-}^{A})\ \frac{\left\{\begin{array}{l}\Gamma \Rightarrow \Pi|(A_1, D) \rightrightarrows \Gamma - \{A_1\} \Rightarrow \Pi - \{D\}\\ \Gamma \Rightarrow \Pi|(A_2, D) \rightrightarrows \Gamma - \{A_2\} \Rightarrow \Pi - \{D\}\end{array}\right.}{\Gamma \Rightarrow \Pi|(A_1 \wedge A_2, D) \rightrightarrows \Gamma - \{A_1 \wedge A_2\} \Rightarrow \Pi - \{D\}}$$

$$(\wedge_{0}^{A})\ \frac{\left[\begin{array}{l}\Gamma \Rightarrow \Pi|(A_1, D) \rightrightarrows \Gamma \Rightarrow \Pi\\ \Gamma \Rightarrow \Pi|(A_2, D) \rightrightarrows \Gamma \Rightarrow \Pi\end{array}\right.}{\Gamma \Rightarrow \Pi|(A_1 \wedge A_2, D) \rightrightarrows \Gamma \Rightarrow \Pi}$$

$$(\wedge_{-}^{D})\ \frac{\left\{\begin{array}{l}\Gamma \Rightarrow \Pi|(A, D_1) \rightrightarrows \Gamma - \{A\} \Rightarrow \Pi - \{D_1\}\\ \Gamma \Rightarrow \Pi|(A, D_2) \rightrightarrows \Gamma - \{A\} \Rightarrow \Pi - \{D_2\}\end{array}\right.}{\Gamma \Rightarrow \Pi|(A, D_1 \wedge D_2) \rightrightarrows \Gamma - \{A\} \Rightarrow \Pi - \{D_1 \wedge D_2\}}$$

$$(\wedge_{0}^{A})\ \frac{\left[\begin{array}{l}\Gamma \Rightarrow \Pi|(A, D_1) \rightrightarrows \Gamma \Rightarrow \Pi\\ \Gamma \Rightarrow \Pi|(A, D_2) \rightrightarrows \Gamma \Rightarrow \Pi\end{array}\right.}{\Gamma \Rightarrow \Pi|(A, D_1 \wedge D_2) \rightrightarrows \Gamma \Rightarrow \Pi}$$

and

$$(\vee_{-}^{A})\ \frac{\left[\begin{array}{l}\Gamma \Rightarrow \Pi|(A_1, D) \rightrightarrows \Gamma - \{A_1\} \Rightarrow \Pi - \{D\}\\ \Gamma - \{A_1\} \Rightarrow \Pi - \{D\}|(A_2, \lambda) \rightrightarrows \Gamma - \{A_1, A_2\} \Rightarrow \Pi - \{D\}\end{array}\right.}{\Gamma \Rightarrow \Pi|(A_1 \vee A_2, D) \rightrightarrows \Gamma - \{A_1 \vee A_2\} \Rightarrow \Pi - \{D\}}$$

$$(\vee_{0}^{A})\ \frac{\left\{\begin{array}{l}\Gamma \Rightarrow \Pi|(A_1, D) \rightrightarrows \Gamma \Rightarrow \Pi\\ \Gamma - \{A_1\} \Rightarrow \Pi - \{D\}|(A_2, \lambda) \rightrightarrows \Gamma - \{A_1\} \Rightarrow \Pi - \{D\}\end{array}\right.}{\Gamma \Rightarrow \Pi|(A_1 \vee A_2, D) \rightrightarrows \Gamma \Rightarrow \Pi}$$

$$(\vee_{-}^{D})\ \frac{\left[\begin{array}{l}\Gamma \Rightarrow \Pi|(A, D_1) \rightrightarrows \Gamma - \{A\} \Rightarrow \Pi - \{D_1\}\\ \Gamma - \{A\} \Rightarrow \Pi - \{D_1\}|(\lambda, D_2) \rightrightarrows \Gamma - \{A\} \Rightarrow \Pi - \{D_1, D_2\}\end{array}\right.}{\Gamma \Rightarrow \Pi|(A, D_1 \vee D_2) \rightrightarrows \Gamma - \{A\} \Rightarrow \Pi - \{D_1 \vee D_2\}}$$

$$(\vee_{0}^{A})\ \frac{\left\{\begin{array}{l}\Gamma \Rightarrow \Pi|(A, D_1) \rightrightarrows \Gamma \Rightarrow \Pi\\ \Gamma - \{A\} \Rightarrow \Pi - \{D_1\}|(\lambda, D_2) \rightrightarrows \Gamma - \{A\} \Rightarrow \Pi - \{D_1\}\end{array}\right.}{\Gamma \Rightarrow \Pi|(A_1 \vee A_2, D) \rightrightarrows \Gamma \Rightarrow \Pi}$$

and

$$(\neg\wedge_{-}^{A})\ \frac{\left[\begin{array}{l}\Gamma \Rightarrow \Pi|(\neg A_1, D) \rightrightarrows \Gamma - \{\neg A_1\} \Rightarrow \Pi - \{D\}\\ \Gamma - \{\neg A_1\} \Rightarrow \Pi - \{D\}|(\neg A_2, \lambda) \rightrightarrows \Gamma - \{\neg A_1, \neg A_2\} \Rightarrow \Pi - \{D\}\end{array}\right.}{\Gamma \Rightarrow \Pi|(\neg(A_1 \wedge A_2), D) \rightrightarrows \Gamma - \{\neg(A_1 \wedge A_2)\} \Rightarrow \Pi - \{D\}}$$

$$(\neg\wedge_{0}^{A})\ \frac{\left\{\begin{array}{l}\Gamma \Rightarrow \Pi|(\neg A_1, D) \rightrightarrows \Gamma \Rightarrow \Pi\\ \Gamma - \{\neg A_1\} \Rightarrow \Pi - \{D\}|(\neg A_2, \lambda) \rightrightarrows \Gamma - \{\neg A_1\} \Rightarrow \Pi - \{D\}\end{array}\right.}{\Gamma \Rightarrow \Pi|(\neg(A_1 \wedge A_2), D) \rightrightarrows \Gamma \Rightarrow \Pi}$$

$$(\neg\wedge_{-}^{D})\ \frac{\left[\begin{array}{l}\Gamma \Rightarrow \Pi|(A, \neg D_1) \rightrightarrows \Gamma - \{A\} \Rightarrow \Pi - \{\neg D_1\}\\ \Gamma - \{A\} \Rightarrow \Pi - \{\neg D_1\}|(\lambda, \neg D_2) \rightrightarrows \Gamma - \{A\} \Rightarrow \Pi - \{\neg D_1, \neg D_2\}\end{array}\right.}{\Gamma \Rightarrow \Pi|(A, \neg(D_1 \wedge D_2)) \rightrightarrows \Gamma - \{A\} \Rightarrow \Pi - \{\neg(D_1 \wedge D_2)\}}$$

$$(\neg\wedge_{0}^{D})\ \frac{\left\{\begin{array}{l}\Gamma \Rightarrow \Pi|(A, \neg D_1) \rightrightarrows \Gamma \Rightarrow \Pi\\ \Gamma - \{A\} \Rightarrow \Pi - \{\neg D_1\}|(\lambda, \neg D_2) \rightrightarrows \Gamma - \{A\} \Rightarrow \Pi - \{\neg D_1\}\end{array}\right.}{\Gamma \Rightarrow \Pi|(A, \neg(D_1 \wedge D_2)) \rightrightarrows \Gamma \Rightarrow \Pi}$$

and

$$(\neg\vee_{-}^{A})\ \frac{\begin{cases}\Gamma\Rightarrow\Pi|(\neg A_1,D)\rightrightarrows\Gamma-\{\neg A_1\}\Rightarrow\Pi-\{D\}\\ \Gamma\Rightarrow\Pi|(\neg A_2,D)\rightrightarrows\Gamma-\{\neg A_2\}\Rightarrow\Pi-\{D\}\end{cases}}{\Gamma\Rightarrow\Pi|(\neg(A_1\vee A_2),D)\rightrightarrows\Gamma-\{\neg(A_1\vee A_2)\}\Rightarrow\Pi-\{D\}}$$

$$(\neg\vee_{0}^{A})\ \frac{\begin{bmatrix}\Gamma\Rightarrow\Pi|(\neg A_1,D)\rightrightarrows\Gamma\Rightarrow\Pi\\ \Gamma\Rightarrow\Pi|(\neg A_2,D)\rightrightarrows\Gamma\Rightarrow\Pi\end{bmatrix}}{\Gamma\Rightarrow\Pi|(\neg(A_1\vee A_2),D)\rightrightarrows\Gamma\Rightarrow\Pi}$$

$$(\neg\vee_{-}^{D})\ \frac{\begin{cases}\Gamma\Rightarrow\Pi|(A,\neg D_1)\rightrightarrows\Gamma-\{A\}\Rightarrow\Pi-\{\neg D_1\}\\ \Gamma\Rightarrow\Pi|(A,\neg D_2)\rightrightarrows\Gamma-\{A\}\Rightarrow\Pi-\{\neg D_2\}\end{cases}}{\Gamma\Rightarrow\Pi|(A,\neg(D_1\vee D_2))\rightrightarrows\Gamma-\{A\}\Rightarrow\Pi-\{\neg(D_1\vee D_2)\}}$$

$$(\neg\vee_{0}^{D})\ \frac{\begin{bmatrix}\Gamma\Rightarrow\Pi|(A,\neg D_1)\rightrightarrows\Gamma\Rightarrow\Pi\\ \Gamma\Rightarrow\Pi|(A,\neg D_2)\rightrightarrows\Gamma\Rightarrow\Pi\end{bmatrix}}{\Gamma\Rightarrow\Pi|(A,\neg(D_1\vee D_2))\rightrightarrows\Gamma\Rightarrow\Pi}$$

Theorem 7.3.4 (Soundness and completeness theorem) *For any sequent $\Gamma\Rightarrow\Pi$ and pair (A,D) with $A\in\Gamma$ and $D\in\Pi$,*

$$\vdash^{\mathtt{ff}}\Gamma\Rightarrow\Pi|(A,D)\rightrightarrows\Gamma'\Rightarrow\Pi'$$

if and only if

$$\models^{\mathtt{ff}}\Gamma\Rightarrow\Pi|(A,D)\rightrightarrows\Gamma'\Rightarrow\Pi',$$

where $\Gamma'=\Gamma|\Gamma-\{A\}$ and $\Gamma'=\Gamma|\Gamma-\{D\}$. □

7.4 Gentzen Deduction System $\mathbf{G}_{-}$

A co-supersequent $\Gamma|\Delta\mapsto\Sigma|\Pi$ is valid, denoted by $\models_{-}\Gamma|\Delta\mapsto\Sigma|\Pi$, if there is an assignment v such that $v(A)=1$ for each $A\in\Gamma$; $v(B)=0$ for each $B\in\Delta$; $v(C)=0$ for each $C\in\Sigma$, and $v(D)=1$ for each $D\in\Pi$.

$$\begin{array}{c||c}\Gamma|\Delta\Rightarrow\Sigma|\Pi & \mathtt{f}|\mathtt{t}\Rightarrow\mathtt{t}|\mathtt{f}\\ \Gamma|\Delta\mapsto\Sigma|\Pi & \mathtt{t}|\mathtt{f}\mapsto\mathtt{f}|\mathtt{t}\end{array}$$

Co-supersequent Gentzen deduction system $\mathbf{G}_{-}$ consists of the following axioms and deduction rules:

- **Axiom**:

$$(\mathtt{A}_{-})\ \frac{\begin{array}{c}\mathrm{con}(\Gamma\cup\neg\Delta)\&\mathrm{inval}(\Sigma\cup\neg\Pi)\\ (\Gamma\cup\neg\Delta)\cap(\Sigma\cup\neg\Pi)=\emptyset\end{array}}{\Gamma|\Delta\mapsto\Sigma|\Pi}$$

where Γ,Δ,Σ,Π are sets of literals.

- **Deduction rules**:

$$
(\neg\neg^{A})\ \frac{\Gamma, A|\Delta \mapsto \Sigma|\Pi}{\Gamma, \neg\neg A|\Delta \mapsto \Sigma|\Pi} \quad (\neg\neg^{B})\ \frac{\Gamma|B, \Delta \mapsto \Sigma|\Pi}{\Gamma|\neg\neg B, \Delta \mapsto \Sigma|\Pi}
$$
$$
(\neg\neg^{C})\ \frac{\Gamma|\Delta \mapsto \Sigma, C|\Pi}{\Gamma|\Delta \mapsto \Sigma, \neg\neg C|\Pi} \quad (\neg\neg^{D})\ \frac{\Gamma|\Delta \mapsto \Sigma|D, \Pi}{\Gamma|\Delta \mapsto \Sigma|\neg\neg D, \Pi}
$$

and

$$
(\wedge^{A})\ \frac{\left[\begin{array}{l}\Gamma, A_1|\Delta \mapsto \Sigma|\Pi\\ \Gamma, A_2|\Delta \mapsto \Sigma|\Pi\end{array}\right.}{\Gamma, A_1 \wedge A_2|\Delta \mapsto \Sigma|\Pi} \quad (\wedge^{B})\ \frac{\left\{\begin{array}{l}\Gamma|B_1, \Delta \mapsto \Sigma|\Pi\\ \Gamma|B_2, \Delta \mapsto \Sigma|\Pi\end{array}\right.}{\Gamma|B_1 \wedge B_2, \Delta \mapsto \Sigma|\Pi}
$$
$$
(\wedge^{C})\ \frac{\left\{\begin{array}{l}\Gamma|\Delta \mapsto \Sigma, C_1|\Pi\\ \Gamma|\Delta \mapsto \Sigma, C_2|\Pi\end{array}\right.}{\Gamma|\Delta \mapsto \Sigma, C_1 \wedge C_2|\Pi} \quad (\wedge^{D})\ \frac{\left[\begin{array}{l}\Gamma|\Delta \mapsto \Sigma|D_1, \Pi\\ \Gamma|\Delta \mapsto \Sigma|D_2, \Pi\end{array}\right.}{\Gamma|\Delta \mapsto \Sigma|D_1 \wedge D_2, \Pi}
$$
$$
(\vee^{A})\ \frac{\left\{\begin{array}{l}\Gamma, A_1|\Delta \mapsto \Sigma|\Pi\\ \Gamma, A_2|\Delta \mapsto \Sigma|\Pi\end{array}\right.}{\Gamma, A_1 \vee A_2|\Delta \mapsto \Sigma|\Pi} \quad (\vee^{B})\ \frac{\left[\begin{array}{l}\Gamma|B_1, \Delta \mapsto \Sigma|\Pi\\ \Gamma|B_2, \Delta \mapsto \Sigma|\Pi\end{array}\right.}{\Gamma|B_1 \vee B_2, \Delta \mapsto \Sigma|\Pi}
$$
$$
(\vee^{C})\ \frac{\left[\begin{array}{l}\Gamma|\Delta \mapsto \Sigma, C_1|\Pi\\ \Gamma|\Delta \mapsto \Sigma, C_2|\Pi\end{array}\right.}{\Gamma|\Delta \mapsto \Sigma, C_1 \vee C_2|\Pi} \quad (\vee^{D})\ \frac{\left\{\begin{array}{l}\Gamma|\Delta \mapsto \Sigma|D_1, \Pi\\ \Gamma|\Delta \mapsto \Sigma|D_2, \Pi\end{array}\right.}{\Gamma|\Delta \mapsto \Sigma|D_1 \vee D_2, \Pi}
$$

and

$$
(\neg\wedge^{A})\ \frac{\left\{\begin{array}{l}\Gamma, \neg A_1|\Delta \mapsto \Sigma|\Pi\\ \Gamma, \neg A_2|\Delta \mapsto \Sigma|\Pi\end{array}\right.}{\Gamma, \neg(A_1 \wedge A_2)|\Delta \mapsto \Sigma|\Pi} \quad (\neg\wedge^{B})\ \frac{\left[\begin{array}{l}\Gamma|\neg B_1, \Delta \mapsto \Sigma|\Pi\\ \Gamma|\neg B_2, \Delta \mapsto \Sigma|\Pi\end{array}\right.}{\Gamma|\neg(B_1 \wedge B_2), \Delta \mapsto \Sigma|\Pi}
$$
$$
(\neg\wedge^{C})\ \frac{\left[\begin{array}{l}\Gamma|\Delta \mapsto \Sigma, \neg C_1|\Pi\\ \Gamma|\Delta \mapsto \Sigma, \neg C_1|\Pi\end{array}\right.}{\Gamma|\Delta \mapsto \Sigma, \neg(C_1 \wedge C_2)|\Pi} \quad (\neg\wedge^{D})\ \frac{\left\{\begin{array}{l}\Gamma|\Delta \mapsto \Sigma|\neg D_1, \Pi\\ \Gamma|\Delta \mapsto \Sigma|\neg D_2, \Pi\end{array}\right.}{\Gamma|\Delta \mapsto \Sigma|\neg(D_1 \wedge D_2), \Pi}
$$
$$
(\neg\vee^{A})\ \frac{\left[\begin{array}{l}\Gamma, \neg A_1|\Delta \mapsto \Sigma|\Pi\\ \Gamma, \neg A_2|\Delta \mapsto \Sigma|\Pi\end{array}\right.}{\Gamma, \neg(A_1 \vee A_2)|\Delta \mapsto \Sigma|\Pi} \quad (\neg\vee^{B})\ \frac{\left\{\begin{array}{l}\Gamma|\neg B_1, \Delta \mapsto \Sigma|\Pi\\ \Gamma|\neg B_2, \Delta \mapsto \Sigma|\Pi\end{array}\right.}{\Gamma|\neg(B_1 \vee B_2), \Delta \mapsto \Sigma|\Pi}
$$
$$
(\neg\wedge^{C})\ \frac{\left\{\begin{array}{l}\Gamma|\Delta \mapsto \Sigma, \neg C_1|\Pi\\ \Gamma|\Delta \mapsto \Sigma, \neg C_2|\Pi\end{array}\right.}{\Gamma|\Delta \mapsto \Sigma, \neg(C_1 \vee C_2)|\Pi} \quad (\neg\vee^{D})\ \frac{\left[\begin{array}{l}\Gamma|\Delta \mapsto \Sigma|\neg D_1, \Pi\\ \Gamma|\Delta \mapsto \Sigma|\neg D_2, \Pi\end{array}\right.}{\Gamma|\Delta \mapsto \Sigma|\neg(D_1 \vee D_2), \Pi}
$$

Definition 7.4.1 A co-supersequent $\Gamma|\Delta \mapsto \Sigma|\Pi$ is provable in $\mathbf{G}_{-}$, denoted by $\vdash_{-} \Gamma|\Delta \mapsto \Sigma|\Pi$, if there is a sequence $\{\Gamma_1|\Delta_1 \mapsto \Sigma_1|\Pi_1, \ldots, \Gamma_n|\Delta_n \mapsto \Sigma_n|\Pi_n\}$ of co-supersequents such that $\Gamma_n|\Delta_n \mapsto \Sigma_n|\Pi_n = \Gamma|\Delta \mapsto \Sigma|\Pi$, and for each $1 \leq i \leq n$, $\Gamma_i|\Delta_i \mapsto \Sigma_i|\Pi_i$ is deduced from the previous co-supersequents by one of the deduction rules in $\mathbf{G}_{-}$.

Theorem 7.4.2 (Soundness and completeness theorem) *For any co-supersequent* $\delta = \Gamma|\Delta \mapsto \Sigma|\Pi$,

$$
\vdash_{-} \delta \text{ iff } \models_{-} \delta.
$$

□

7.5 Reduction of Supersequents to Sequents

A co-supersequent $\Gamma|\Delta \mapsto \Sigma|\Xi$ is reduced into four kinds of co-sequents:

$$\begin{array}{l}
\mathtt{tf}: \Gamma \mapsto \Sigma, \Delta = \Xi = \emptyset \\
\mathtt{tt}: \Gamma \mapsto \Xi, \Delta = \Sigma = \emptyset \\
\mathtt{ff}: \Delta \mapsto \Sigma, \Gamma = \Xi = \emptyset \\
\mathtt{ft}: \Delta \mapsto \Xi, \Gamma = \Sigma = \emptyset.
\end{array}$$

7.5.1 *Gentzen Deduction System* $\mathbf{G}_{\mathtt{ff}}$

A co-sequent $\Delta \mapsto \Sigma$ is $\mathbf{G}_{\mathtt{ff}}$-valid, denoted by $\models_{\mathtt{ff}} \Delta \mapsto \Sigma$, if there is an assignment v such that for every formula $B \in \Delta, v(A) = 0$, and for every formula $C \in \Sigma, v(C) = 0$.

Gentzen deduction system $\mathbf{G}_{\mathtt{ff}}$ consists of the following axiom and deduction rules:

- **Axiom**:

$$(\mathtt{A}_{\mathtt{ff}}) \frac{\left[\begin{array}{l}\mathrm{con}(\Delta) \\ \mathrm{inval}(\Sigma) \\ \neg\Delta \cap \Sigma = \emptyset\end{array}\right.}{\Delta \mapsto \Sigma}$$

where Δ, Σ are sets of literals.

- **Deduction rules**:

$$\begin{array}{ll}
(\neg\neg^B) \dfrac{B, \Delta \mapsto \Sigma}{\neg\neg B, \Delta \mapsto \Sigma} & (\neg\neg^C) \dfrac{\Delta \mapsto \Sigma, C}{\Delta \mapsto \Sigma, \neg\neg C} \\
(\wedge^B) \dfrac{\left\{\begin{array}{l}B_1, \Delta \mapsto \Sigma \\ B_2, \Delta \mapsto \Sigma\end{array}\right.}{B_1 \wedge B_2, \Delta \mapsto \Sigma} & (\wedge^C) \dfrac{\left\{\begin{array}{l}\Delta \mapsto \Sigma, C_1 \\ \Delta \mapsto \Sigma, C_2\end{array}\right.}{\Delta \mapsto \Sigma, C_1 \wedge C_2} \\
(\vee^B) \dfrac{\left[\begin{array}{l}B_1, \Delta \mapsto \Sigma \\ B_2, \Delta \mapsto \Sigma\end{array}\right.}{B_1 \vee B_2, \Delta \mapsto \Sigma} & (\vee^C) \dfrac{\left[\begin{array}{l}\Delta \mapsto \Sigma, C_1 \\ \Delta \mapsto \Sigma, C_2\end{array}\right.}{\Delta \mapsto \Sigma, C_1 \vee C_2}
\end{array}$$

and

$$\begin{array}{ll}
(\neg\wedge^B) \dfrac{\left[\begin{array}{l}\neg B_1, \Delta \mapsto \Sigma \\ \neg B_2, \Delta \mapsto \Sigma\end{array}\right.}{\neg(B_1 \wedge B_2), \Delta \mapsto \Sigma} & (\neg\wedge^C) \dfrac{\left[\begin{array}{l}\Delta \mapsto \Sigma, \neg C_1 \\ \Delta \mapsto \Sigma, \neg C_2\end{array}\right.}{\Delta \mapsto \Sigma, \neg(C_1 \wedge C_2)} \\
(\neg\vee^B) \dfrac{\left\{\begin{array}{l}\neg B_1, \Delta \mapsto \Sigma \\ \neg B_2, \Delta \mapsto \Sigma\end{array}\right.}{\neg(B_1 \vee B_2), \Delta \mapsto \Sigma} & (\neg\wedge^C) \dfrac{\left\{\begin{array}{l}\Delta \mapsto \Sigma, \neg C_1 \\ \Delta \mapsto \Sigma, \neg C_2\end{array}\right.}{\Delta \mapsto \Sigma, \neg(C_1 \vee C_2)}
\end{array}$$

Theorem 7.5.1 (Soundness and completeness theorem) *For any co-sequent* $\Delta \mapsto \Sigma, \vdash_{\mathtt{ff}} \Delta \mapsto \Sigma$ *if and only if* $\models_{\mathtt{ff}} \Delta \mapsto \Sigma$. □

7.5.2 Gentzen Deduction System $\mathbf{G}_{\mathtt{ft}}$

A co-sequent $\Delta \mapsto \Pi$ is $\mathbf{G}_{\mathtt{ft}}$-valid, denoted by $\models_{\mathtt{ft}} \Delta \mapsto \Pi$, if there is an assignment v such that for every formula $B \in \Delta, v(B) = 1$, and for every formula $D \in \Pi, v(D) = 0$.

Gentzen deduction system $\mathbf{G}_{\mathtt{ft}}$ consists of the following axiom and deduction rules:

- **Axiom**:

$$(\mathtt{A}_{\mathtt{ft}}) \frac{\left[\begin{array}{l} \mathrm{con}(\Delta) \\ \mathrm{inval}(\Pi) \\ \Delta \cap \Pi = \emptyset \end{array}\right.}{\Delta \mapsto \Pi}$$

  where Δ, Π are sets of literals.
- **Deduction rules**:

$$\begin{array}{ll}
(\neg\neg^B) \dfrac{B, \Delta \mapsto \Pi}{\neg\neg B, \Delta \mapsto \Pi} & (\neg\neg^D) \dfrac{\Delta \mapsto D, \Pi}{\Delta \mapsto \neg\neg D, \Pi} \\
(\wedge^B) \dfrac{\left\{\begin{array}{l} B_1, \Delta \mapsto \Pi \\ B_2, \Delta \mapsto \Pi \end{array}\right.}{B_1 \wedge B_2, \Delta \mapsto \Pi} & (\wedge^D) \dfrac{\left[\begin{array}{l} \Delta \mapsto D_1, \Pi \\ \Delta \mapsto D_2, \Pi \end{array}\right.}{\Delta \mapsto D_1 \wedge D_2, \Pi} \\
(\vee^B) \dfrac{\left[\begin{array}{l} B_1, \Delta \mapsto \Pi \\ B_2, \Delta \mapsto \Pi \end{array}\right.}{B_1 \vee B_2, \Delta \mapsto \Pi} & (\vee^D) \dfrac{\left\{\begin{array}{l} \Delta \mapsto D_1, \Pi \\ \Delta \mapsto D_2, \Pi \end{array}\right.}{\Delta \mapsto D_1 \vee D_2, \Pi}
\end{array}$$

and

$$\begin{array}{ll}
(\neg\wedge^B) \dfrac{\left[\begin{array}{l} \neg B_1, \Delta \mapsto \Pi \\ \neg B_2, \Delta \mapsto \Pi \end{array}\right.}{\neg(B_1 \wedge B_2), \Delta \mapsto \Pi} & (\neg\wedge^D) \dfrac{\left\{\begin{array}{l} \Delta \mapsto \neg D_1, \Pi \\ \Delta \mapsto \neg D_2, \Pi \end{array}\right.}{\Delta \mapsto \neg(D_1 \wedge D_2), \Pi} \\
(\neg\vee^B) \dfrac{\left\{\begin{array}{l} \neg B_1, \Delta \mapsto \Pi \\ \neg B_2, \Delta \mapsto \Pi \end{array}\right.}{\neg(B_1 \vee B_2), \Delta \mapsto \Pi} & (\neg\vee^D) \dfrac{\left[\begin{array}{l} \Delta \mapsto \neg D_1, \Pi \\ \Delta \mapsto \neg D_2, \Pi \end{array}\right.}{\Delta \mapsto \neg(D_1 \vee D_2), \Pi}
\end{array}$$

Theorem 7.5.2 (Soundness and completeness theorem) *For any co-sequent* $\Delta \mapsto \Pi$,

$$\vdash_{\mathtt{ft}} \Delta \mapsto \Pi \textit{ iff } \models_{\mathtt{ft}} \Delta \mapsto \Pi.$$

□

7.6 R-Calculus $\mathbf{R}_-$

Given a quadruple (A, B, C, D), we use $\Gamma|\Delta \mapsto \Sigma|\Pi$ to revise (A, B, C, D) and obtain $\Gamma'|\Delta' \mapsto \Sigma'|\Pi'$, denoted by

$$\models_- \Gamma|\Delta \mapsto \Sigma|\Pi \rightrightarrows \Gamma'|\Delta' \mapsto \Sigma'|\Pi',$$

if $\Gamma'|\Delta' \mapsto \Sigma'|\Pi' =$

$$\begin{cases} \Gamma, A|\Delta, B \mapsto \Sigma, C|\Pi, D & \text{if } \Gamma, A|\Delta, B \mapsto \Sigma, C|\Pi, D \text{ is valid} \\ \Gamma|\Delta \mapsto \Sigma|\Pi & \text{otherwise.} \end{cases}$$

We denote

$$\begin{aligned} \mathbf{X} &= \Gamma|\Delta \mapsto \Sigma|\Pi \\ \mathbf{X}[X] &= \Gamma, A|\Delta, B \mapsto \Sigma, C|\Pi, D. \end{aligned}$$

R-calculus $\mathbf{R}_-$ consists of the following axioms and deduction rules:

- **Axioms**:

$$(\mathbb{A}_+)\ \frac{\left[\begin{array}{ll} \neg l \notin \Gamma & l' \notin \Gamma \\ l \notin \Delta & \neg l' \notin \Delta \\ l \notin \Sigma & \neg l' \notin \Sigma \\ \neg l \notin \Pi & l' \notin \Pi \\ m \notin \Gamma & \neg m' \notin \Gamma \\ \neg m \notin \Delta & m' \notin \Delta \\ \neg m \notin \Sigma & m' \notin \Sigma \\ m \notin \Pi & \neg m' \notin \Pi \end{array}\right.}{\mathbf{X}|(l, l', m, m') \Rightarrow \mathbf{X}[(l, l', m, m')]}$$

$$(\mathbb{A}_0)\ \frac{\left\{\begin{array}{ll} \neg l \in \Gamma & l' \in \Gamma \\ l \in \Delta & \neg l' \in \Delta \\ l \in \Sigma & \neg l' \in \Sigma \\ \neg l \in \Pi & l' \in \Pi \\ m \in \Gamma & \neg m' \in \Gamma \\ \neg m \in \Delta & m' \in \Delta \\ \neg m \in \Sigma & m' \in \Sigma \\ m \in \Pi & \neg m' \in \Pi \end{array}\right.}{\mathbf{X}|(l, l', m, m') \Rightarrow \mathbf{X}}$$

where $\mathbf{X}$ is literal and l, l', m, m' are literals.

- **Deduction rules**:

$$(\neg\neg^{A}_{+})\ \frac{\mathbf{X}|(A, B, C, D) \Rightarrow \mathbf{X}[(A, B, C, D)]}{\mathbf{X}|(\neg\neg A, B, C, D) \Rightarrow \mathbf{X}[(\neg\neg A, B, C, D)]}$$

$$(\neg\neg^{A}_{0})\ \frac{\mathbf{X}|(A, B, C, D) \Rightarrow \mathbf{X}}{\mathbf{X}|(\neg\neg A, B, C, D) \Rightarrow \mathbf{X}}$$

$$(\neg\neg^{B}_{+})\ \frac{\mathbf{X}|(A, B, C, D) \Rightarrow \mathbf{X}[(A, B, C, D)]}{\mathbf{X}|(A, \neg\neg B, C, D) \Rightarrow \mathbf{X}[(A, \neg\neg B, C, D)]}$$

$$(\neg\neg^{B}_{0})\ \frac{\mathbf{X}|(A, B, C, D) \Rightarrow \mathbf{X}}{\mathbf{X}|(A, \neg\neg B, C, D) \Rightarrow \mathbf{X}}$$

and

$$(\neg\neg^{C}_{+})\ \frac{\mathbf{X}|(A, B, C, D) \Rightarrow \mathbf{X}[(A, B, C, D)]}{\mathbf{X}|(A, B, \neg\neg C, D) \Rightarrow \mathbf{X}[(A, B, \neg\neg C, D)]}$$

$$(\neg\neg^{C}_{0})\ \frac{\mathbf{X}|(A, B, C, D) \Rightarrow \mathbf{X}}{\mathbf{X}|(A, B, \neg\neg C, D) \Rightarrow \mathbf{X}}$$

$$(\neg\neg^{D}_{+})\ \frac{\mathbf{X}|(A, B, C, D) \Rightarrow \mathbf{X}[(A, B, C, D)]}{\mathbf{X}|(A, B, C, \neg\neg D) \Rightarrow \mathbf{X}[(A, B, C, \neg\neg D)]}$$

$$(\neg\neg^{D}_{0})\ \frac{\mathbf{X}|(A, B, C, D) \Rightarrow \mathbf{X}}{\mathbf{X}|(A, B, C, \neg\neg D) \Rightarrow \mathbf{X}}$$

and

$$(\wedge^{A}_{+})\ \frac{\left[\begin{array}{l}\mathbf{X}|(A_1, B, C, D) \Rightarrow \mathbf{X}[(A_1, B, C, D)]\\ \mathbf{X}[(A_1, B, C, D)]|(A_2, B, C, D) \Rightarrow \mathbf{X}[(A_1, B, C, D)][(A_2, B, C, D)]\end{array}\right.}{\mathbf{X}|(A_1 \wedge A_2, B, C, D) \Rightarrow \mathbf{X}[(A_1 \wedge A_2, B, C, D)]}$$

$$(\wedge^{A}_{0})\ \frac{\left\{\begin{array}{l}\mathbf{X}|(A_1, B, C, D) \Rightarrow \mathbf{X}\\ \mathbf{X}[(A_1, B, C, D)]|(A_2, B, C, D) \Rightarrow \mathbf{X}[(A_1, B, C, D)]\end{array}\right.}{\mathbf{X}|(A_1 \wedge A_2, B, C, D) \Rightarrow \mathbf{X}}$$

$$(\wedge^{B}_{+})\ \frac{\left\{\begin{array}{l}\mathbf{X}|(A, B_1, C, D) \Rightarrow \mathbf{X}[(A, B_1, C, D)]\\ \mathbf{X}|(A, B_2, C, D) \Rightarrow \mathbf{X}((A, B_2, C, D)]\end{array}\right.}{\mathbf{X}|(A, B_1 \wedge B_2, C, D) \Rightarrow \mathbf{X}[(A, B_1 \wedge B_2, C, D)]}$$

$$(\wedge^{B}_{0})\ \frac{\left[\begin{array}{l}\mathbf{X}|(A, B_1, C, D) \Rightarrow \mathbf{X}\\ \mathbf{X}|(A, B_2, C, D) \Rightarrow \mathbf{X}\end{array}\right.}{\mathbf{X}|(A, B_1 \wedge B_2, C, D) \Rightarrow \mathbf{X}}$$

and

$$(\wedge^{C}_{+})\ \frac{\left\{\begin{array}{l}\mathbf{X}|(A, B, C_1, D) \Rightarrow \mathbf{X}[(A, B, C_1, D)]\\ \mathbf{X}|(A, B, C_2, D) \Rightarrow \mathbf{X}((A, B, C_2, D)]\end{array}\right.}{\mathbf{X}|(A, B, C_1 \wedge C_2, D) \Rightarrow \mathbf{X}[(A, B, C_1 \wedge C_2, D)]}$$

$$(\wedge^{C}_{0})\ \frac{\left[\begin{array}{l}\mathbf{X}|(A, B, C_1, D) \Rightarrow \mathbf{X}\\ \mathbf{X}|(A, B, C_2, D) \Rightarrow \mathbf{X}\end{array}\right.}{\mathbf{X}|(A, B, C_1 \wedge C_2, D) \Rightarrow \mathbf{X}}$$

$$(\wedge^{D}_{+})\ \frac{\left[\begin{array}{l}\mathbf{X}|(A, B, C, D_1) \Rightarrow \mathbf{X}[(A, B, C, D_1)]\\ \mathbf{X}[(A, B, C, D_1)]|(A, B, C, D_2) \Rightarrow \mathbf{X}[(A, B, C, D_1)]((A, B, C, D_2)]\end{array}\right.}{\mathbf{X}|(A, B, C, D_1 \wedge D_2) \Rightarrow \mathbf{X}[(A, B, C, D_1 \wedge D_2)]}$$

$$(\wedge^{D}_{0})\ \frac{\left\{\begin{array}{l}\mathbf{X}|(A, B, C, D_1) \Rightarrow \mathbf{X}\\ \mathbf{X}[(A, B, C, D_1)]|(A, B, C, D_2) \Rightarrow \mathbf{X}[(A, B, C, D_1)]\end{array}\right.}{\mathbf{X}|(A, B, C, D_1 \wedge D_2) \Rightarrow \mathbf{X}}$$

and

$$(\vee^A_+)\ \frac{\left\{\begin{array}{l}\mathbf{X}|(A_1,B,C,D)\Rightarrow\mathbf{X}[(A_1,B,C,D)]\\ \mathbf{X}|(A_2,B,C,D)\Rightarrow\mathbf{X}[(A_2,B,C,D)]\end{array}\right.}{\mathbf{X}(A_1\vee A_2,B,C,D)\Rightarrow\mathbf{X}[(A_1\vee A_2,B,C,D)]}$$

$$(\vee^A_0)\ \frac{\left[\begin{array}{l}\mathbf{X}|(A_1,B,C,D)\Rightarrow\mathbf{X}\\ \mathbf{X}|(A_2,B,C,D)\Rightarrow\mathbf{X}\end{array}\right.}{\mathbf{X}(A_1\vee A_2,B,C,D)\Rightarrow\mathbf{X}}$$

$$(\vee^B_+)\ \frac{\left[\begin{array}{l}\mathbf{X}|(A,B_1,C,D)\Rightarrow\mathbf{X}[(A,B_1,C,D)]\\ \mathbf{X}[(A,B_1,C,D)]|(A,B_2,C,D)\Rightarrow\mathbf{X}[(A,B_1,C,D)][(A,B_2,C,D)]\end{array}\right.}{\mathbf{X}(A,B_1\vee B_2,C,D)\Rightarrow\mathbf{X}[(A,B_1\vee B_2,C,D)]}$$

$$(\vee^B_0)\ \frac{\left\{\begin{array}{l}\mathbf{X}|(A,B_1,C,D)\Rightarrow\mathbf{X}\\ \mathbf{X}[(A,B_1,C,D)]|(A,B_2,C,D)\Rightarrow\mathbf{X}[(A,B_1,C,D)]\end{array}\right.}{\mathbf{X}(A,B_1\vee B_2,C,D)\Rightarrow\mathbf{X}}$$

and

$$(\vee^C_+)\ \frac{\left[\begin{array}{l}\mathbf{X}|(A,B,C_1,D)\Rightarrow\mathbf{X}[(A,B,C_1,D)]\\ \mathbf{X}[(A,B,C_1,D)]|(A,B,C_2,D)\Rightarrow\mathbf{X}[(A,B,C_1,D)][(A,B,C_2,D)]\end{array}\right.}{\mathbf{X}(A,B,C_1\vee C_2,D)\Rightarrow\mathbf{X}[(A,B,C_1\vee C_2,D)]}$$

$$(\vee^C_0)\ \frac{\left\{\begin{array}{l}\mathbf{X}|(A,B,C_1,D)\Rightarrow\mathbf{X}\\ \mathbf{X}[(A,B,C_1,D)]|(A,B,C_2,D)\Rightarrow\mathbf{X}[(A,B,C_1,D)]\end{array}\right.}{\mathbf{X}(A,B,C_1\vee C_2,D)\Rightarrow\mathbf{X}}$$

$$(\vee^D_+)\ \frac{\left\{\begin{array}{l}\mathbf{X}|(A,B,C,D_1)\Rightarrow\mathbf{X}[(A,B,C,D_1)]\\ \mathbf{X}|(A,B,C,D_2)\Rightarrow\mathbf{X}[(A,B,C,D_2)]\end{array}\right.}{\mathbf{X}(A,B,C,D_1\vee D_2)\Rightarrow\mathbf{X}[(A,B,C,D_1\vee D_2)]}$$

$$(\vee^D_0)\ \frac{\left[\begin{array}{l}\mathbf{X}|(A,B,C,D_1)\Rightarrow\mathbf{X}\\ \mathbf{X}|(A,B,C,D_2)\Rightarrow\mathbf{X}\end{array}\right.}{\mathbf{X}(A,B,C,D_1\vee D_2)\Rightarrow\mathbf{X}}$$

and

$$(\neg\wedge^A_+)\ \frac{\left\{\begin{array}{l}\mathbf{X}|(\neg A_1,B,C,D)\Rightarrow\mathbf{X}[(\neg A_1,B,C,D)]\\ \mathbf{X}|(\neg A_2,B,C,D)\Rightarrow\mathbf{X}[(\neg A_2,B,C,D)]\end{array}\right.}{\mathbf{X}|(\neg(A_1\wedge A_2),B,C,D)\Rightarrow\mathbf{X}[(\neg(A_1\wedge A_2),B,C,D)]}$$

$$(\neg\wedge^A_0)\ \frac{\left[\begin{array}{l}\mathbf{X}|(\neg A_1,B,C,D)\Rightarrow\mathbf{X}\\ \mathbf{X}|(\neg A_2,B,C,D)\Rightarrow\mathbf{X}\end{array}\right.}{\mathbf{X}|(\neg(A_1\wedge A_2),B,C,D)\Rightarrow\mathbf{X}}$$

$$(\neg\wedge^B_+)\ \frac{\left[\begin{array}{l}\mathbf{X}|(A,\neg B_1,C,D)\Rightarrow\mathbf{X}[(A,\neg B_1,C,D)]\\ \mathbf{X}[(A,\neg B_1,C,D)]|(A,\neg B_2,C,D)\Rightarrow\mathbf{X}[(A,\neg B_1,C,D)][(A,\neg B_2,C,D)]\end{array}\right.}{\mathbf{X}|(A,\neg(B_1\wedge B_2),C,D)\Rightarrow\mathbf{X}[(A,\neg(B_1\wedge B_2),C,D)]}$$

$$(\neg\wedge^B_0)\ \frac{\left\{\begin{array}{l}\mathbf{X}|(A,\neg B_1,C,D)\Rightarrow\mathbf{X}\\ \mathbf{X}[(A,\neg B_1,C,D)]|(A,\neg B_2,C,D)\Rightarrow\mathbf{X}[(A,\neg B_1,C,D)]\end{array}\right.}{\mathbf{X}|(A,\neg(B_1\wedge B_2),C,D)\Rightarrow\mathbf{X}}$$

and

$$(\neg\wedge^{C}_{+})\ \frac{\left[\begin{array}{l}\mathbf{X}|(A,B,\neg C_1,D)\Rightarrow \mathbf{X}[(A,B,\neg C_1,D)]\\ \mathbf{X}[(A,B,\neg C_1,D)]|(A,B,\neg C_2,D)\Rightarrow \mathbf{X}[(A,B,\neg C_1,D)][(A,B,\neg C_2,D)]\end{array}\right.}{\mathbf{X}|(A,B,\neg(C_1\wedge C_2),D)\Rightarrow \mathbf{X}[(A,B,\neg(C_1\wedge C_2),D)]}$$

$$(\neg\wedge^{C}_{0})\ \frac{\left\{\begin{array}{l}\mathbf{X}|(A,B,\neg C_1,D)\Rightarrow \mathbf{X}\\ \mathbf{X}[(A,B,\neg C_1,D)]|(A,B,\neg C_2,D)\Rightarrow \mathbf{X}[(A,B,\neg C_1,D)]\end{array}\right.}{\mathbf{X}|(A,B,\neg(C_1\wedge C_2),D)\Rightarrow \mathbf{X}}$$

$$(\neg\wedge^{D}_{+})\ \frac{\left\{\begin{array}{l}\mathbf{X}|(A,B,C,\neg D_1)\Rightarrow \mathbf{X}[(A,B,C,\neg D_1)]\\ \mathbf{X}|(A,B,C,\neg D_2)\Rightarrow \mathbf{X}[(A,B,C,\neg D_2)]\end{array}\right.}{\mathbf{X}|(A,B,C,\neg(D_1\wedge D_2)]\Rightarrow \mathbf{X}[(A,B,C,\neg(D_1\wedge D_2)])}$$

$$(\neg\wedge^{D}_{0})\ \frac{\left[\begin{array}{l}\mathbf{X}|(A,B,C,\neg D_1)\Rightarrow \mathbf{X}\\ \mathbf{X}|(A,B,C,\neg D_2)\Rightarrow \mathbf{X}\end{array}\right.}{\mathbf{X}|(A,B,C,\neg(D_1\wedge D_2)]\Rightarrow \mathbf{X}}$$

and

$$(\neg\vee^{A}_{+})\ \frac{\left[\begin{array}{l}\mathbf{X}|(\neg A_1,B,C,D)\Rightarrow \mathbf{X}[(\neg A_1,B,C,D)]\\ \mathbf{X}[(\neg A_1,B,C,D)]|(\neg A_2,B,C,D)\Rightarrow \mathbf{X}[(\neg A_1,B,C,D)][(\neg A_2,B,C,D)]\end{array}\right.}{\mathbf{X}|(\neg(A_1\vee A_2),B,C,D)\Rightarrow \mathbf{X}[(\neg(A_1\vee A_2),B,C,D)]}$$

$$(\neg\vee^{A}_{0})\ \frac{\left\{\begin{array}{l}\mathbf{X}|(\neg A_1,B,C,D)\Rightarrow \mathbf{X}\\ \mathbf{X}[(\neg A_1,B,C,D)]|(\neg A_2,B,C,D)\Rightarrow \mathbf{X}[(\neg A_1,B,C,D)]\end{array}\right.}{\mathbf{X}|(\neg(A_1\vee A_2),B,C,D)\Rightarrow \mathbf{X}}$$

$$(\neg\vee^{B}_{+})\ \frac{\left\{\begin{array}{l}\mathbf{X}|(A,\neg B_1,C,D)\Rightarrow \mathbf{X}[(A,\neg B_1,C,D)]\\ \mathbf{X}|(A,\neg B_2,C,D)\Rightarrow \mathbf{X}[(A,\neg B_2,C,D)]\end{array}\right.}{\mathbf{X}|(A,\neg(B_1\vee B_2),C,D)\Rightarrow \mathbf{X}[(A,\neg(B_1\vee B_2),C,D)]}$$

$$(\neg\vee^{B}_{0})\ \frac{\left[\begin{array}{l}\mathbf{X}|(A,\neg B_1,C,D)\Rightarrow \mathbf{X}\\ \mathbf{X}|(A,\neg B_2,C,D)\Rightarrow \mathbf{X}\end{array}\right.}{\mathbf{X}|(A,\neg(B_1\vee B_2),C,D)\Rightarrow \mathbf{X}}$$

and

$$(\neg\vee^{C}_{+})\ \frac{\left\{\begin{array}{l}\mathbf{X}|(A,B,\neg C_1,D)\Rightarrow \mathbf{X}[(A,B,\neg C_1,D)]\\ \mathbf{X}|(A,B,\neg C_2,D)\Rightarrow \mathbf{X}[(A,B,\neg C_2,D)]\end{array}\right.}{\mathbf{X}|(A,B,\neg(C_1\vee C_2),D)\Rightarrow \mathbf{X}[(A,B,\neg(C_1\vee C_2),D)]}$$

$$(\neg\vee^{C}_{0})\ \frac{\left[\begin{array}{l}\mathbf{X}|(A,B,\neg C_1,D)\Rightarrow \mathbf{X}\\ \mathbf{X}|(A,B,\neg C_2,D)\Rightarrow \mathbf{X}\end{array}\right.}{\mathbf{X}|(A,B,\neg(C_1\vee C_2),D)\Rightarrow \mathbf{X}}$$

$$(\neg\vee^{D}_{+})\ \frac{\left[\begin{array}{l}\mathbf{X}|(A,B,C,\neg D_1)\Rightarrow \mathbf{X}[(A,B,C,\neg D_1)]\\ \mathbf{X}[(A,B,C,\neg D_1)]|(A,B,C,\neg D_2)\Rightarrow \mathbf{X}[(A,B,C,\neg D_1)][(A,B,C,\neg D_2)]\end{array}\right.}{\mathbf{X}|(A,B,C,\neg(D_1\vee D_2)]\Rightarrow \mathbf{X}[(A,B,C,\neg(D_1\vee D_2)])}$$

$$(\neg\vee^{D}_{0})\ \frac{\left\{\begin{array}{l}\mathbf{X}|(A,B,C,\neg D_1)\Rightarrow \mathbf{X}\\ \mathbf{X}[(A,B,C,\neg D_1)]|(A,B,C,\neg D_2)\Rightarrow \mathbf{X}[(A,B,C,\neg D_1)]\end{array}\right.}{\mathbf{X}|(A,B,C,\neg(D_1\vee D_2)]\Rightarrow \mathbf{X}}$$

Definition 7.6.1 Given a supersequent $\mathbf{X}$ and a quadruple X, a reduction $\mathbf{X}|X\Rightarrow \mathbf{X}'$ is provable in $\mathbf{R}_{-}$, denoted by $\vdash_{-} \mathbf{X}|X\Rightarrow \mathbf{X}'$, if there is a sequence $\{\delta_1,\ldots,\delta_n\}$ of reductions such that $\delta_n = \mathbf{X}|X \Rightarrow \mathbf{X}'$, and for each $1 \leq i \leq n$, δ_i is either an axiom or deduced from the previous reductions by one of the deduction rules in $\mathbf{R}_{-}$.

Theorem 7.6.2 (Soundness and completeness theorem) *For any reduction* $\mathbf{X}|X \Rightarrow \mathbf{X}'$,

$$\vdash_{-} \mathbf{X}|X \Rightarrow \mathbf{X}' \text{ if and only if } \models_{-} \mathbf{X}|X \Rightarrow \mathbf{X}'. \qquad \square$$

7.6.1 $\mathbf{R}_{\mathtt{ff}}$

Given a pair (B, C) of formulas, we use $\Delta \Rightarrow_{\mathtt{ff}} \Sigma$ to revise (B, C) and obtain a sequence $\Delta' \mapsto \Sigma'$, denoted by

$$\models_{\mathtt{ff}} \Delta \mapsto \Sigma|(B, C) \rightrightarrows \Delta' \mapsto \Sigma',$$

if $\Delta' \mapsto \Sigma' =$

$$\begin{cases} \Delta, B \mapsto \Sigma, C & \text{if } \Delta, B \mapsto \Sigma, C \text{ is } \mathbf{G}_{\mathtt{ff}}\text{-valid} \\ \Delta \mapsto \Sigma & \text{otherwise.} \end{cases}$$

R-calculus $\mathbf{R}_{\mathtt{ff}}$ consists of the following axioms and deduction rules:

- **Axioms**:

$$(\mathtt{A}^{+}_{\mathtt{ff}}) \frac{\begin{bmatrix} \neg l \notin \Delta \\ l \notin \Sigma \\ m \notin \Delta \\ \neg m \notin \Sigma \end{bmatrix}}{\Delta \mapsto \Sigma|(l, m) \rightrightarrows \Delta, l \mapsto \Sigma, m} \qquad (\mathtt{A}^{0}_{\mathtt{ff}}) \frac{\begin{cases} \neg l \in \Delta \\ l \in \Sigma \\ m \in \Delta \\ \neg m \in \Sigma \end{cases}}{\Delta \mapsto \Sigma|(l, m) \rightrightarrows \Delta \mapsto \Sigma}$$

where Δ, Σ are sets of literals, and l, m are literals.

- **Deduction rules**:

$$(\neg\neg^{B}_{+}) \frac{\Delta \mapsto \Sigma|(B, C) \rightrightarrows \Delta, B \mapsto \Sigma, C}{\Delta \mapsto \Sigma|(\neg\neg B, C) \rightrightarrows \Delta, \neg\neg B \mapsto \Sigma, C}$$
$$(\neg\neg^{B}_{0}) \frac{\Delta \mapsto \Sigma|(B, C) \rightrightarrows \Delta \mapsto \Sigma}{\Delta \mapsto \Sigma|(\neg\neg B, C) \rightrightarrows \Delta \mapsto \Sigma}$$
$$(\neg\neg^{C}_{+}) \frac{\Delta \mapsto \Sigma|(B, C) \rightrightarrows \Delta, B \mapsto \Sigma, C}{\Delta \mapsto \Sigma|(B, \neg\neg C) \rightrightarrows \Delta, B \mapsto \Sigma, \neg\neg C}$$
$$(\neg\neg^{C}_{0}) \frac{\Delta \mapsto \Sigma|(B, C) \rightrightarrows \Delta \mapsto \Sigma}{\Delta \mapsto \Sigma|(B, \neg\neg C) \rightrightarrows \Delta \mapsto \Sigma}$$

and

$$(\wedge_{+}^{B})\ \frac{\begin{cases}\Delta\mapsto\Sigma|(B_1,C)\rightrightarrows\Delta,B_1\mapsto\Sigma,C\\ \Delta\mapsto\Sigma|(B_2,C)\rightrightarrows\Delta,B_2\mapsto\Sigma,C\end{cases}}{\Delta\mapsto\Sigma|(B_1\wedge B_2,C)\rightrightarrows\Delta,B_1\wedge B_2\mapsto\Sigma,C}$$

$$(\wedge_{0}^{B})\ \frac{\left[\begin{array}{l}\Delta\mapsto\Sigma|(B_1,C)\rightrightarrows\Delta\mapsto\Sigma\\ \Delta\mapsto\Sigma|(B_2,C)\rightrightarrows\Delta\mapsto\Sigma\end{array}\right.}{\Delta\mapsto\Sigma|(B_1\wedge B_2,C)\rightrightarrows\Delta\mapsto\Sigma}$$

$$(\wedge_{+}^{C})\ \frac{\begin{cases}\Delta\mapsto\Sigma|(B,C_1)\rightrightarrows\Delta,B\mapsto\Sigma,C_1\\ \Delta\mapsto\Sigma|(B,C_2)\rightrightarrows\Delta,B\mapsto\Sigma,C_2\end{cases}}{\Delta\mapsto\Sigma|(B,C_1\wedge C_2)\rightrightarrows\Delta,B\mapsto\Sigma,C_1\wedge C_2}$$

$$(\wedge_{0}^{B})\ \frac{\left[\begin{array}{l}\Delta\mapsto\Sigma|(B,C_1)\rightrightarrows\Delta\mapsto\Sigma\\ \Delta\mapsto\Sigma|(B,C_2)\rightrightarrows\Delta\mapsto\Sigma\end{array}\right.}{\Delta\mapsto\Sigma|(B,C_1\wedge C_2)\rightrightarrows\Delta\mapsto\Sigma}$$

and

$$(\vee_{+}^{B})\ \frac{\left[\begin{array}{l}\Delta\mapsto\Sigma|(B_1,C)\rightrightarrows\Delta,B_1\mapsto\Sigma,C\\ \Delta,B_1\mapsto\Sigma,C|(B_2,C)\rightrightarrows\Delta,B_1,B_2\mapsto\Sigma,C\end{array}\right.}{\Delta\mapsto\Sigma|(B_1\vee B_2,C)\rightrightarrows\Delta,B_1\vee B_2\mapsto\Sigma,C}$$

$$(\vee_{0}^{B})\ \frac{\begin{cases}\Delta\mapsto\Sigma|(B_1,C)\rightrightarrows\Delta\mapsto\Sigma\\ \Delta,B_1\mapsto\Sigma,C|(B_2,C)\rightrightarrows\Delta,B_1\mapsto\Sigma,C\end{cases}}{\Delta\mapsto\Sigma|(B_1\vee B_2,C)\rightrightarrows\Delta\mapsto\Sigma}$$

$$(\vee_{+}^{C})\ \frac{\left[\begin{array}{l}\Delta\mapsto\Sigma|(B,C_1)\rightrightarrows\Delta,B\mapsto\Sigma,C_1\\ \Delta,B\mapsto\Sigma,C_1|(B,C_2)\rightrightarrows\Delta,B\mapsto\Sigma,C_1,C_2\end{array}\right.}{\Delta\mapsto\Sigma|(B,C_1\vee C_2)\rightrightarrows\Delta,B\mapsto\Sigma,C_1\vee C_2}$$

$$(\vee_{0}^{B})\ \frac{\begin{cases}\Delta\mapsto\Sigma|(B,C_1)\rightrightarrows\Delta\mapsto\Sigma\\ \Delta,B\mapsto\Sigma,C_1|(B,C_2)\rightrightarrows\Delta,B\mapsto\Sigma,C_1\end{cases}}{\Delta\mapsto\Sigma|(B,C_1\vee C_2)\rightrightarrows\Delta\mapsto\Sigma}$$

and

$$(\neg\wedge_{+}^{B})\ \frac{\left[\begin{array}{l}\Delta\mapsto\Sigma|(\neg B_1,C)\rightrightarrows\Delta,\neg B_1\mapsto\Sigma,C\\ \Delta,\neg B_1\mapsto\Sigma,C|(\neg B_2,C)\rightrightarrows\Delta,\neg B_1,\neg B_2\mapsto\Sigma,C\end{array}\right.}{\Delta\mapsto\Sigma|(\neg(B_1\wedge B_2),C)\rightrightarrows\Delta,\neg(B_1\wedge B_2)\mapsto\Sigma,C}$$

$$(\neg\wedge_{0}^{B})\ \frac{\begin{cases}\Delta\mapsto\Sigma|(\neg B_1,C)\rightrightarrows\Delta\mapsto\Sigma\\ \Delta,\neg B_1\mapsto\Sigma,C|(\neg B_2,C)\rightrightarrows\Delta,\neg B_1\mapsto\Sigma,C\end{cases}}{\Delta\mapsto\Sigma|(\neg(B_1\wedge B_2),C)\rightrightarrows\Delta\mapsto\Sigma}$$

$$(\neg\wedge_{+}^{C})\ \frac{\left[\begin{array}{l}\Delta\mapsto\Sigma|(B,\neg C_1)\rightrightarrows\Delta,B\mapsto\Sigma,\neg C_1\\ \Delta,B\mapsto\Sigma,\neg C_1|(B,\neg C_2)\rightrightarrows\Delta,B\mapsto\Sigma,\neg C_1,\neg C_2\end{array}\right.}{\Delta\mapsto\Sigma|(B,\neg(C_1\wedge C_2))\rightrightarrows\Delta,B\mapsto\Sigma,\neg(C_1\wedge C_2)}$$

$$(\neg\wedge_{0}^{C})\ \frac{\begin{cases}\Delta\mapsto\Sigma|(B,\neg C_1)\rightrightarrows\Delta\mapsto\Sigma\\ \Delta,B\mapsto\Sigma,\neg C_1|(B,\neg C_2)\rightrightarrows\Delta,B\mapsto\Sigma,\neg C_1\end{cases}}{\Delta\mapsto\Sigma|(B,\neg(C_1\wedge C_2))\rightrightarrows\Delta\mapsto\Sigma}$$

and

$$(\neg\vee_{+}^{B})\ \frac{\begin{cases}\Delta\mapsto\Sigma|(\neg B_1,C)\rightrightarrows\Delta,\neg B_1\mapsto\Sigma,C\\ \Delta\mapsto\Sigma|(\neg B_2,C)\rightrightarrows\Delta,\neg B_2\mapsto\Sigma,C\end{cases}}{\Delta\mapsto\Sigma|(\neg(B_1\vee B_2),C)\rightrightarrows\Delta,\neg(B_1\vee B_2)\mapsto\Sigma,C}$$

$$(\neg\vee_{0}^{B})\ \frac{\begin{bmatrix}\Delta\mapsto\Sigma|(\neg B_1,C)\rightrightarrows\Delta\mapsto\Sigma\\ \Delta\mapsto\Sigma|(\neg B_2,C)\rightrightarrows\Delta\mapsto\Sigma\end{bmatrix}}{\Delta\mapsto\Sigma|(\neg(B_1\vee B_2),C)\rightrightarrows\Delta\mapsto\Sigma}$$

$$(\neg\wedge_{+}^{C})\ \frac{\begin{cases}\Delta\mapsto\Sigma|(B,\neg C_1)\rightrightarrows\Delta,B\mapsto\Sigma,\neg C_1\\ \Delta\mapsto\Sigma|(B,\neg C_2)\rightrightarrows\Delta,B\mapsto\Sigma,\neg C_2\end{cases}}{\Delta\mapsto\Sigma|(B,\neg(C_1\vee C_2))\rightrightarrows\Delta,B\mapsto\Sigma,\neg(C_1\vee C_2)}$$

$$(\neg\vee_{0}^{C})\ \frac{\begin{bmatrix}\Delta\mapsto\Sigma|(B,\neg C_1)\rightrightarrows\Delta\mapsto\Sigma\\ \Delta\mapsto\Sigma|(B,\neg C_2)\rightrightarrows\Delta\mapsto\Sigma\end{bmatrix}}{\Delta\mapsto\Sigma|(B,\neg(C_1\vee C_2))\rightrightarrows\Delta\mapsto\Sigma}$$

Theorem 7.6.3 (Soundness and completeness theorem) *For any sequent* $\Delta\mapsto\Sigma$ *and pair* (B,C) *with* $B\in\Delta$ *and* $C\in\Sigma$,

$$\vdash_{\tt ff}\Delta\mapsto\Sigma|(B,C)\rightrightarrows\Delta'\mapsto\Sigma'$$

if and only if

$$\models_{\tt ff}\Delta\mapsto\Sigma|(B,C)\rightrightarrows\Delta'\mapsto\Sigma',$$

where $\Delta'=\Delta|\Delta,B$ *and* $\Sigma'=\Sigma|\Sigma,C$. □

7.6.2 $\mathbf{R}_{\tt ft}$

Given a pair (B,D) of formulas, we use $\Delta\mapsto\Pi$ to revise (B,D) and obtain a sequence $\Delta'\mapsto\Pi'$, denoted by

$$\models_{\tt ft}\Delta\mapsto\Pi|(B,D)\rightrightarrows\Delta'\mapsto\Pi',$$

if $\Delta'\mapsto\Pi'=\begin{cases}\Delta,B\mapsto\Pi,D & \text{if } \Delta,B\mapsto\Pi,\ Dis\mathbf{G}_{\tt ft}\text{–}valid\\ \Delta\mapsto\Pi & \text{otherwise.}\end{cases}$

R-calculus $\mathbf{R}_{\tt ft}$ consists of the following axioms and deduction rules:

- **Axioms**:

$$(\mathtt{A}_{\tt ft}^{+})\ \frac{\begin{bmatrix}\neg l\notin\Delta\\ \neg l\notin\Pi\\ \neg m\notin\Delta\\ \neg m\notin\Pi\end{bmatrix}}{\Delta\mapsto\Pi|(l,m)\rightrightarrows\Delta,l\mapsto\Pi,m}\qquad (\mathtt{A}_{\tt ft}^{0})\ \frac{\begin{cases}\neg l\in\Delta\\ \neg l\in\Pi\\ \neg m\in\Delta\\ \neg m\in\Pi\end{cases}}{\Delta\mapsto\Pi|(l,m)\rightrightarrows\Delta\mapsto\Pi}$$

where Δ,Π are sets of literals, and l,m are literals.

- **Deduction rules**:

$$(\neg\neg^B_+)\ \frac{\Delta \mapsto \Pi|(B, D) \rightrightarrows \Delta, B \mapsto \Pi, D}{\Delta \mapsto \Pi|(\neg\neg B, D) \rightrightarrows \Delta, \neg\neg B \mapsto \Pi, D}$$

$$(\neg\neg^B_0)\ \frac{\Delta \mapsto \Pi|(B, D) \rightrightarrows \Delta \mapsto \Pi}{\Delta \mapsto \Pi|(\neg\neg B, D) \rightrightarrows \Delta \mapsto \Pi}$$

$$(\neg\neg^D_+)\ \frac{\Delta \mapsto \Pi|(B, D) \rightrightarrows \Delta, B \mapsto \Pi, D}{\Delta \mapsto \Pi|(B, \neg\neg D) \rightrightarrows \Delta, B \mapsto \Pi, \neg\neg D}$$

$$(\neg\neg^B_0)\ \frac{\Delta \mapsto \Pi|(B, D) \rightrightarrows \Delta \mapsto \Pi}{\Delta \mapsto \Pi|(B, \neg\neg D) \rightrightarrows \Delta \mapsto \Pi}$$

and

$$(\wedge^B_+)\ \frac{\left\{\begin{array}{l}\Delta \mapsto \Pi|(B_1, D) \rightrightarrows \Delta, B_1 \mapsto \Pi, D\\ \Delta \mapsto \Pi|(B_2, D) \rightrightarrows \Delta, B_2 \mapsto \Pi, D\end{array}\right.}{\Delta \mapsto \Pi|(B_1 \wedge B_2, D) \rightrightarrows \Delta, B_1 \wedge B_2 \mapsto \Pi, D}$$

$$(\wedge^B_0)\ \frac{\left[\begin{array}{l}\Delta \mapsto \Pi|(B_1, D) \rightrightarrows \Delta \mapsto \Pi\\ \Delta \mapsto \Pi|(B_2, D) \rightrightarrows \Delta \mapsto \Pi\end{array}\right.}{\Delta \mapsto \Pi|(B_1 \wedge B_2, D) \rightrightarrows \Delta \mapsto \Pi}$$

$$(\wedge^D_+)\ \frac{\left[\begin{array}{l}\Delta \mapsto \Pi|(B, D_1) \rightrightarrows \Delta, B \mapsto \Pi, D_1\\ \Delta, B \mapsto \Pi, D_1|(B, D_2) \rightrightarrows \Delta, B \mapsto \Pi, D_1, D_2\end{array}\right.}{\Delta \mapsto \Pi|(B, D_1 \wedge D_2) \rightrightarrows \Delta, B \mapsto \Pi, D_1 \wedge D_2}$$

$$(\wedge^B_0)\ \frac{\left\{\begin{array}{l}\Delta \mapsto \Pi|(B, D_1) \rightrightarrows \Delta \mapsto \Pi\\ \Delta, B \mapsto \Pi, D_1|(B, D_2) \rightrightarrows \Delta, B \mapsto \Pi, D_1\end{array}\right.}{\Delta \mapsto \Pi|(B_1 \wedge B_2, D) \rightrightarrows \Delta \mapsto \Pi}$$

and

$$(\vee^B_+)\ \frac{\left[\begin{array}{l}\Delta \mapsto \Pi|(B_1, D) \rightrightarrows \Delta, B_1 \mapsto \Pi, D\\ \Delta, B_1 \mapsto \Pi, D|(B_2, D) \rightrightarrows \Delta, B_1, B_2 \mapsto \Pi, D\end{array}\right.}{\Delta \mapsto \Pi|(B_1 \vee B_2, D) \rightrightarrows \Delta, B_1 \vee B_2 \mapsto \Pi, D}$$

$$(\vee^B_0)\ \frac{\left\{\begin{array}{l}\Delta \mapsto \Pi|(B_1, D) \rightrightarrows \Delta \mapsto \Pi\\ \Delta, B_1 \mapsto \Pi, D|(B_2, D) \rightrightarrows \Delta, B_1 \mapsto \Pi, D\end{array}\right.}{\Delta \mapsto \Pi|(B_1 \vee B_2, D) \rightrightarrows \Delta \mapsto \Pi}$$

$$(\vee^D_+)\ \frac{\left\{\begin{array}{l}\Delta \mapsto \Pi|(B, D_1) \rightrightarrows \Delta, B \mapsto \Pi, D_1\\ \Delta \mapsto \Pi|(B, D_2) \rightrightarrows \Delta, B \mapsto \Pi, D_2\end{array}\right.}{\Delta \mapsto \Pi|(B, D_1 \vee D_2) \rightrightarrows \Delta, B \mapsto \Pi, D_1 \vee D_2}$$

$$(\vee^B_0)\ \frac{\left[\begin{array}{l}\Delta \mapsto \Pi|(B, D_1) \rightrightarrows \Delta \mapsto \Pi\\ \Delta \mapsto \Pi|(B, D_2) \rightrightarrows \Delta \mapsto \Pi\end{array}\right.}{\Delta \mapsto \Pi|(B, D_1 \vee D_2) \rightrightarrows \Delta \mapsto \Pi}$$

and

$$(\neg\wedge_{+}^{B})\ \frac{\left[\begin{array}{l}\Delta\mapsto\Pi|(\neg B_1,D)\rightrightarrows\Delta,\neg B_1\mapsto\Pi,D\\ \Delta,\neg B_1\mapsto\Pi,D|(\neg B_2,D)\rightrightarrows\Delta,\neg B_1,\neg B_2\mapsto\Pi,D\end{array}\right.}{\Delta\mapsto\Pi|(\neg(B_1\wedge B_2),D)\rightrightarrows\Delta,\neg(B_1\wedge B_2)\mapsto\Pi,D}$$

$$(\neg\wedge_{0}^{B})\ \frac{\left\{\begin{array}{l}\Delta\mapsto\Pi|(\neg B_1,D)\rightrightarrows\Delta\mapsto\Pi\\ \Delta,\neg B_1\mapsto\Pi,D|(\neg B_2,D)\rightrightarrows\Delta,\neg B_1\mapsto\Pi,D\end{array}\right.}{\Delta\mapsto\Pi|(\neg(B_1\wedge B_2),D)\rightrightarrows\Delta\mapsto\Pi}$$

$$(\neg\wedge_{+}^{D})\ \frac{\left\{\begin{array}{l}\Delta\mapsto\Pi|(B,\neg D_1)\rightrightarrows\Delta,B\mapsto\Pi,\neg D_1\\ \Delta\mapsto\Pi|(B,\neg D_2)\rightrightarrows\Delta,B\mapsto\Pi,\neg D_2\end{array}\right.}{\Delta\mapsto\Pi|(B,\neg(D_1\wedge D_2))\rightrightarrows\Delta,B\mapsto\Pi,\neg(D_1\wedge D_2)}$$

$$(\neg\wedge_{0}^{D})\ \frac{\left[\begin{array}{l}\Delta\mapsto\Pi|(B,\neg D_1)\rightrightarrows\Delta\mapsto\Pi\\ \Delta\mapsto\Pi|(B,\neg D_2)\rightrightarrows\Delta\mapsto\Pi\end{array}\right.}{\Delta\mapsto\Pi|(B,\neg(D_1\wedge D_2))\rightrightarrows\Delta\mapsto\Pi}$$

and

$$(\neg\vee_{+}^{B})\ \frac{\left\{\begin{array}{l}\Delta\mapsto\Pi|(\neg B_1,D)\rightrightarrows\Delta,\neg B_1\mapsto\Pi,D\\ \Delta\mapsto\Pi|(\neg B_2,D)\rightrightarrows\Delta,\neg B_2\mapsto\Pi,D\end{array}\right.}{\Delta\mapsto\Pi|(\neg(B_1\vee B_2),D)\rightrightarrows\Delta,\neg(B_1\vee B_2)\mapsto\Pi,D}$$

$$(\neg\vee_{0}^{B})\ \frac{\left[\begin{array}{l}\Delta\mapsto\Pi|(\neg B_1,D)\rightrightarrows\Delta\mapsto\Pi\\ \Delta\mapsto\Pi|(\neg B_2,D)\rightrightarrows\Delta\mapsto\Pi\end{array}\right.}{\Delta\mapsto\Pi|(\neg(B_1\vee B_2),D)\rightrightarrows\Delta\mapsto\Pi}$$

$$(\neg\vee_{+}^{D})\ \frac{\left[\begin{array}{l}\Delta\mapsto\Pi|(B,\neg D_1)\rightrightarrows\Delta,B\mapsto\Pi,\neg D_1\\ \Delta,B\mapsto\Pi,\neg D_1|(B,\neg D_2)\rightrightarrows\Delta,B\mapsto\Pi,\neg D_1,\neg D_2\end{array}\right.}{\Delta\mapsto\Pi|(B,\neg(D_1\vee D_2))\rightrightarrows\Delta,B\mapsto\Pi,\neg(D_1\vee D_2)}$$

$$(\neg\vee_{0}^{D})\ \frac{\left\{\begin{array}{l}\Delta\mapsto\Pi|(B,\neg D_1)\rightrightarrows\Delta\mapsto\Pi\\ \Delta,B\mapsto\Pi,\neg D_1|(B,\neg D_2)\rightrightarrows\Delta,B\mapsto\Pi,\neg D_1\end{array}\right.}{\Delta\mapsto\Pi|(B,\neg(D_1\vee D_2))\rightrightarrows\Delta\mapsto\Pi}$$

Theorem 7.6.4 (Soundness and completeness theorem) *For any sequent* $\Delta\mapsto\Pi$ *and pair* (B,D) *with* $B\in\Delta$ *and* $D\in\Pi$,

$$\vdash_{\mathtt{ft}}\Delta\mapsto\Pi|(B,D)\rightrightarrows\Delta'\mapsto\Pi'$$

if and only if

$$\models_{\mathtt{ft}}\Delta\mapsto\Pi|(B,D)\rightrightarrows\Delta'\mapsto\Pi',$$

where $\Delta'=\Delta|\Delta,B$ *and* $\Pi'=\Pi|\Pi,D$. □

7.7 Conclusions

A Gentzen deduction system is a combination of two semi-sequent deduction systems. Therefore, in each logic, there are a Gentzen deduction system and two semi-sequent deduction systems (tableau proof systems). From these deduction systems, we find that the validity/satisfiability determines the axioms; and the truth-values determine one or two sequents/semi-sequents in the premise of a deduction rule.

Gentzen deduction system $\mathbf{G}^+$ for supersequent $\Gamma|\Delta \Rightarrow \Sigma|\Pi$ is equivalent to one of the following Gentzen deduction systems for sequents:

- $\mathbf{G}^{\mathtt{ft}}$ for sequent $\Gamma, \neg\Delta \Rightarrow \Sigma, \neg\Pi$;
- $\mathbf{G}^{\mathtt{ff}}$ for sequent $\Gamma, \neg\Delta \Rightarrow \neg\Sigma, \Pi$;
- $\mathbf{G}^{\mathtt{tt}}$ for sequent $\neg\Gamma, \Delta \Rightarrow \Sigma, \neg\Pi$;
- $\mathbf{G}^{\mathtt{tf}}$ for sequent $\neg\Gamma, \Delta \Rightarrow \neg\Sigma, \Pi$.

Here, $\mathbf{G}^+$ for supersequent $\Gamma|\Delta \Rightarrow \Sigma|\Pi$ is equivalent to $\mathbf{G}^{\mathtt{tt}}$ for sequent $\Gamma, \neg\Delta \Rightarrow \Sigma, \neg\Pi$, that is, $\Gamma|\Delta \Rightarrow \Sigma|\Pi$ is provable in $\mathbf{G}^+$ iff $\Gamma, \neg\Delta \Rightarrow \Sigma, \neg\Pi$ is provable in $\mathbf{G}^{\mathtt{tt}}$.

Correspondingly, R-calculi $\mathbf{R}^+$ for supersequent $\Gamma|\Delta \Rightarrow \Sigma|\Pi$ is equivalent to one of the following R-calculi for sequents:

- $\mathbf{R}^{\mathtt{ft}}$ for sequent $\Gamma, \neg\Delta \Rightarrow \Sigma, \neg\Pi$;
- $\mathbf{R}^{\mathtt{ff}}$ for sequent $\Gamma, \neg\Delta \Rightarrow \neg\Sigma, \Pi$;
- $\mathbf{R}^{\mathtt{tt}}$ for sequent $\neg\Gamma, \Delta \Rightarrow \Sigma, \neg\Pi$;
- $\mathbf{R}^{\mathtt{tf}}$ for sequent $\neg\Gamma, \Delta \Rightarrow \neg\Sigma, \Pi$.

Dually, there is Gentzen deduction system $\mathbf{G}_-$ for co-supersequents $\Gamma|\Delta \mapsto \Sigma|\Pi$ is equivalent to one of the following Gentzen deduction systems for co-sequents:

- $\mathbf{G}_{\mathtt{tf}}$ for co-sequent $\Gamma, \neg\Delta \mapsto \Sigma, \neg\Pi$;
- $\mathbf{G}_{\mathtt{tt}}$ for co-sequent $\Gamma, \neg\Delta \mapsto \neg\Sigma, \Pi$;
- $\mathbf{G}_{\mathtt{ff}}$ for co-sequent $\neg\Gamma, \Delta \mapsto \Sigma, \neg\Pi$;
- $\mathbf{G}_{\mathtt{ft}}$ for co-sequent $\neg\Gamma, \Delta \mapsto \neg\Sigma, \Pi$.

Here, $\mathbf{G}_-$ for co-supersequent $\Gamma|\Delta \mapsto \Sigma|\Pi$ is equivalent to $\mathbf{G}_{\mathtt{tf}}$ for co-sequent $\Gamma, \neg\Delta \mapsto \Sigma, \neg\Pi$, that is, $\Gamma|\Delta \mapsto \Sigma|\Pi$ is provable in $\mathbf{G}_-$ iff $\Gamma, \neg\Delta \mapsto \Sigma, \neg\Pi$ is provable in $\mathbf{G}_{\mathtt{tf}}$.

Correspondingly, R-calculi $\mathbf{R}_-$ for co-supersequent $\Gamma|\Delta \mapsto \Sigma|\Pi$ is equivalent to one of the following R-calculi for sequents:

- $\mathbf{R}_{\mathtt{tf}}$ for co-sequent $\Gamma, \neg\Delta \mapsto \Sigma, \neg\Pi$;
- $\mathbf{R}_{\mathtt{tt}}$ for co-sequent $\Gamma, \neg\Delta \mapsto \neg\Sigma, \Pi$;
- $\mathbf{R}_{\mathtt{ff}}$ for co-sequent $\neg\Gamma, \Delta \mapsto \Sigma, \neg\Pi$;
- $\mathbf{R}_{\mathtt{ft}}$ for co-sequent $\neg\Gamma, \Delta \mapsto \neg\Sigma, \Pi$.

References

Avron, A.: Hypersequents, logical consequence and intermediate logics for concurrency. Ann. Math. Artif. Intell. **4**, 225–248 (1991)

Baaz, M., Zach, R.: Hypersequent and cut-elimination for intuitionistic fuzzy logic. In: Clote, P.G., Schwichtenberg, H., (eds.), Computer Science Logic, Proceedings of the CSL'2000, LNCS 1862, pp. 178–201. Springer

Bochvar, D.A.: On a three-valued logical calculus and its application to the analysis of the paradoxes of the classical extended functional calculus. Hist. Philos. Log. **2**, 87–112 (1938)

Fitting, M.C.: Many-valued modal logics (I,II), *Fundamenta Informaticae*, **15**(1991), 235–254; **17**(1992), 55–73

Gottwald, S.: A treatise on many-valued logics. Studies in Logic and Computation, vol. 9. Research Studies Press Ltd., Baldock (2001)

Hähnle, R.: Advanced many-valued logics. In: Gabbay, D., Guenthner, F. (eds.) Handbook of Philosophical Logic, vol. 2, pp. 297–395. Kluwer, Dordrecht (2001)

Li, W.: Mathematical logic, foundations for information science. Progress in Computer Science and Applied Logic, vol.25, Birkhäuser (2010)

Łukasiewicz, J.: Selected Works. In: Borkowski, L. (ed.) North-Holland and Warsaw: PWN, Amsterdam (1970)

Malinowski, G.: Many-valued Logic and its Philosophy. In: Gabbay, D.M., Woods, J. (eds.) Handbook of the History of Logic, vol. 8. The Many Valued and Nonmonotonic Turn in Logic, Elsevier (2009)

Urquhart, A.: Basic many-valued logic. In: Gabbay, D., Guenthner F. (eds.), Handbook of Philosophical Logic, vol. 2 (2d edn), pp. 249–295 Kluwer, Dordrecht (2001)

Wronski, A.: Remarks on a survey article on many valued logic by A. Urquhart, StudiaLogica **46**, 275–278 (1987)

Chapter 8
R-Calculi for $\rightsquigarrow$-Propositional Logic

By taking $\neg$ as a logical connective, in traditional Getzen deduction system, we have the following deduction rules

$$(\neg^L)\ \frac{\Gamma \Rightarrow A, \Delta}{\Gamma, \neg A \Rightarrow \Delta} \qquad\qquad (\neg^R)\ \frac{\Gamma, B \Rightarrow \Delta}{\Gamma \Rightarrow \neg B, \Delta}$$

to eliminate $\neg$ from the conclusion.

We take $\neg$ as a meta-logical connective, instead of as a logical connective, and obtain Gentzen deduction system $\mathbf{G}^{\mathrm{t}}$, where we use $(\neg\neg^L)/(\neg\neg^R)$ instead of $(\neg^L)/(\neg^R)$. Such a deduction system is used as the one for multivalued logic (Bolc and Borowik 2003).

Similarly we take $\to$ as a meta-logical connective, denoted by $\rightsquigarrow$, and obtain a Gentzen deduction system $\mathbf{G}^{\mathrm{t}}_{\rightsquigarrow}$ for $\rightsquigarrow$.

In this chapter, we give R-calculi $\mathbf{R}^{\mathrm{t}}_{\rightsquigarrow}, \mathbf{Q}^{\mathrm{t}}_{\rightsquigarrow}, \mathbf{P}^{\mathrm{t}}_{\rightsquigarrow}$ for $\rightsquigarrow$-propositional logic, where $\mathbf{R}^{\mathrm{t}}_{\rightsquigarrow}, \mathbf{Q}^{\mathrm{t}}_{\rightsquigarrow}, \mathbf{P}^{\mathrm{t}}_{\rightsquigarrow}$ are sound and complete with respect to $\subseteq$-minimal change, $\preceq$-minimal change and $\vdash_{\preceq}$-minimal change (Li and Sui 2014, 2013; Li et al. 2015), respectively.

By the following equivalences and inclusions: for any formulas A_1, A_2, B_1, B_2,

$$\begin{aligned}
A_1 \wedge A_2 \rightsquigarrow B &\Leftarrow A_1 \rightsquigarrow B \vee A_2 \rightsquigarrow B\\
A_1 \vee A_2 \rightsquigarrow B &\equiv A_1 \rightsquigarrow B \wedge A_2 \rightsquigarrow B\\
A \rightsquigarrow B_1 \wedge B_2 &\equiv A \rightsquigarrow B_1 \wedge A \rightsquigarrow B_2\\
A \rightsquigarrow B_1 \vee B_2 &\Leftarrow A \rightsquigarrow B_1 \vee A \rightsquigarrow B_2;
\end{aligned}$$

W. Li and Y. Sui, *R-Calculus, IV: Propositional Logic*,
Perspectives in Formal Induction, Revision and Evolution,
https://doi.org/10.1007/978-981-19-8633-8_8

and

$$\begin{array}{l} A_1 \wedge A_2 \not\rightsquigarrow B \Rightarrow A_1 \not\rightsquigarrow B \wedge A_2 \not\rightsquigarrow B \\ A_1 \vee A_2 \not\rightsquigarrow B \equiv A_1 \not\rightsquigarrow B \vee A_2 \not\rightsquigarrow B \\ A \not\rightsquigarrow B_1 \wedge B_2 \equiv A \not\rightsquigarrow B_1 \vee A \not\rightsquigarrow B_2 \\ A \not\rightsquigarrow B_1 \vee B_2 \Rightarrow A \not\rightsquigarrow B_1 \wedge A \not\rightsquigarrow B_2. \end{array}$$

we will give R-calculi $\mathbf{R}^{\mathtt{t}}_{\rightsquigarrow}, \mathbf{Q}^{\mathtt{t}}_{\rightsquigarrow}, \mathbf{P}^{\mathtt{t}}_{\rightsquigarrow}$ for $\rightsquigarrow$-propositional logic.

We will give the following deduction systems and R-calculi:

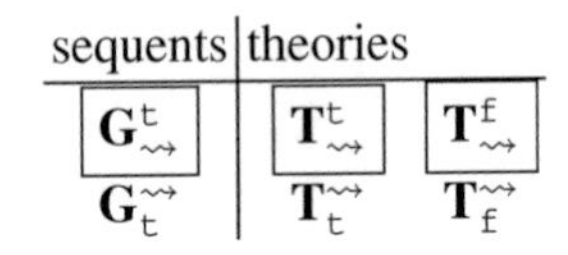

and

sequents	theories
$\mathbf{R}^{\mathtt{t}}_{\rightsquigarrow}, \mathbf{Q}^{\mathtt{t}}_{\rightsquigarrow}, \mathbf{P}^{\mathtt{t}}_{\rightsquigarrow}$	$\mathbf{R}^{\mathtt{t}}_{\rightsquigarrow}, \mathbf{Q}^{\mathtt{t}}_{\rightsquigarrow}, \mathbf{P}^{\mathtt{t}}_{\rightsquigarrow}, \mathbf{R}^{\mathtt{f}}_{\rightsquigarrow}, \mathbf{Q}^{\mathtt{f}}_{\rightsquigarrow}, \mathbf{P}^{\mathtt{f}}_{\rightsquigarrow}$
$\boxed{\mathbf{R}_{\mathtt{t}}^{\rightsquigarrow}}, \boxed{\mathbf{Q}_{\mathtt{t}}^{\rightsquigarrow}}, \boxed{\mathbf{P}_{\mathtt{t}}^{\rightsquigarrow}}$	$\boxed{\mathbf{R}_{\mathtt{t}}^{\rightsquigarrow}}, \boxed{\mathbf{Q}_{\mathtt{t}}^{\rightsquigarrow}}, \boxed{\mathbf{P}_{\mathtt{t}}^{\rightsquigarrow}}, \mathbf{R}_{\mathtt{f}}^{\rightsquigarrow}, \mathbf{Q}_{\mathtt{f}}^{\rightsquigarrow}, \mathbf{P}_{\mathtt{f}}^{\rightsquigarrow}$

8.1 $\rightsquigarrow$-Propositional Logic

$\rightsquigarrow$-Propositional logic are similar to logic programs by the correspondence given in the following.

- literals correspond to statements $l \rightsquigarrow l'$;
- clauses $l_1, \dots, l_m \leftarrow l'_1, \dots, l'_n$ correspond to sequents $l_{11} \rightsquigarrow l'_{11}, \dots, l_{1n} \rightsquigarrow l'_{1n} \Rightarrow l''_{21} \rightsquigarrow l'''_{21}, \dots, l''_{2m} \rightsquigarrow l'''_{2m}$,

where $l_1, \dots, l_m, l'_1, \dots, l'_n$ are literals and $\begin{array}{l} l_{11}, l'_{11}, \dots, l_{1n}, l'_{1n}; \\ l''_{21}, l'''_{21}, \dots, l''_{2m}, l'''_{2m} \end{array}$ are literals in propositional logic.

This section gives basic definitions and a deduction system $\mathbf{G}^{\mathtt{t}}_{\rightsquigarrow}$ for $\rightsquigarrow$-propositional logic, based on which R-calculi $\mathbf{R}^{\mathtt{t}}_{\rightsquigarrow}, \mathbf{Q}^{\mathtt{t}}_{\rightsquigarrow}, \mathbf{P}^{\mathtt{t}}_{\rightsquigarrow}$ for $\rightsquigarrow$-propositional logic will be built.

8.1.1 Basic Definitions

The logical language of $\rightsquigarrow$-propositional logic contains the following symbols:

- atomic propositional variables: $p_0, p_1, \dots$;
- formula constructors: $\neg, \vee, \wedge$; and

- formula connectives: $\rightsquigarrow, \not\rightsquigarrow$.

Formulas are defined as follows:

$$A ::= p|\neg A|A_1 \wedge A_2|A_1 \vee A_2.$$

Statements are defined as follows:

$$\delta ::= A \rightsquigarrow B|A \not\rightsquigarrow B,$$

where A, B are formulas. δ is literal (atomic) if A and B are literals (atoms).

Let v be an assignment. A statement δ is satisfied in v, denoted by $v \models \delta$, if

$$\begin{cases} v(A) = 1 \Rightarrow v(B) = 1 & \text{if } \delta = A \rightsquigarrow B \\ v(A) = 1 \& v(B) = 0 & \text{if } \delta = A \not\rightsquigarrow B. \end{cases}$$

δ is valid, denoted by $\models \delta$, if

$$\begin{cases} \mathbf{A}v(v \models \delta) \text{ if } \delta = A \rightsquigarrow B \\ \mathbf{E}v(v \models \delta) \text{ if } \delta = A \not\rightsquigarrow B. \end{cases}$$

Let Γ be a set of literal statements. A literal statement $l_1 \rightsquigarrow l_2$ is deduced from $l_1 \rightsquigarrow l$ and $l \rightsquigarrow l_2$. Define Γ^* be the transitive closure of Γ, and for each $l_1 \rightsquigarrow l_2 \in \Gamma^*$ we say that $l_1 \rightsquigarrow l_2$ is deduced from Γ, denoted by $\Gamma \vdash l_1 \rightsquigarrow l_2$. There are two deduction rules for transitivity:

$$(tran) \frac{\begin{array}{c} l_1 \rightsquigarrow l_2 \\ l_2 \rightsquigarrow l_3 \end{array}}{l_1 \rightsquigarrow l_3} \quad (neg) \frac{l_1 \rightsquigarrow l_2}{\neg l_2 \rightsquigarrow \neg l_1},$$

where $\neg l = \begin{cases} \neg p & \text{if } l = p \\ p & \text{if } l = \neg p \end{cases}$

A set Γ of statements is valid, denoted by $\models \Gamma \Rightarrow$, if for any assignment v, there is a statement $A \rightsquigarrow B/A \not\rightsquigarrow B \in \Gamma$ such that $v \models A \rightsquigarrow B/v \models A \not\rightsquigarrow B$.

Proposition 8.1.1 *For any set* Γ *of literal statements,* $\models \Gamma \Rightarrow$ *if and only if there is a statement* $l_1 \not\rightsquigarrow l_2 \in \Gamma$ *such that* $\Gamma \vdash l_1 \rightsquigarrow l_2$.

Proof Because

$$l_{11} \rightsquigarrow l_{12}, \ldots, l_{n1} \rightsquigarrow l_{n2}; l'_{11} \not\rightsquigarrow l'_{12}, \ldots, l'_{n'1} \not\rightsquigarrow l'_{n'2} \Rightarrow$$

is $\mathbf{T}^{\mathrm{f}}$-valid iff

$$\begin{array}{l} l_{11} \rightsquigarrow l_{12}, \ldots, l_{n1} \rightsquigarrow l_{n2} \\ \Rightarrow l'_{11} \rightsquigarrow l'_{12}, \ldots, l'_{n'1} \rightsquigarrow l'_{n'2} \end{array}$$

is $\mathbf{G}^{\mathtt{t}}_{\rightsquigarrow}$-valid iff there is a literal statement $l \rightsquigarrow l'$ such that

$$l_{11} \rightsquigarrow l_{12}, \ldots, l_{n1} \rightsquigarrow l_{n2} \rightsquigarrow l'''_{m'2} \vdash l \rightsquigarrow l'$$
$$l'_{11} \rightsquigarrow l'_{12}, \ldots, l'_{n'1} \rightsquigarrow l'_{n'2} \vdash l \rightsquigarrow l'$$

iff there is a statement $l \rightsquigarrow l' \in \Gamma$ such that $\Gamma \vdash l \rightsquigarrow l'$. □

Proposition 8.1.2 *For any nonempty set Δ of literal statements, $\models \Rightarrow \Delta$ if and only if there is a statement $m_1 \not\rightsquigarrow m_2 \in \Delta$ such that $\Gamma \vdash m_1 \rightsquigarrow m_2$.* □

8.2 Tableau Proof Systems

A sequent $\Rightarrow \Delta$ is $\mathbf{T}^{\mathtt{t}}_{\rightsquigarrow}$-valid, denoted by $\models^{\mathtt{t}}_{\rightsquigarrow} \Rightarrow \Delta$, if for any assignment v, there is a statement $\delta \in \Delta$ such that $v \models \delta$.

A sequent $\Gamma \Rightarrow$ is $\mathbf{T}^{\mathtt{f}}_{\rightsquigarrow}$-valid, denoted by $\models^{\mathtt{f}}_{\rightsquigarrow} \Gamma \Rightarrow$, if for any assignment v, there is a statement $\gamma \in \Gamma$ such that $v \not\models \gamma$.

A co-sequent $\mapsto \Delta$ is $\mathbf{T}^{\rightsquigarrow}_{\mathtt{f}}$-valid, denoted by $\models^{\rightsquigarrow}_{\mathtt{f}} \mapsto \Delta$, if there is an assignment v such that for any statement $\delta \in \Delta$, $v \not\models \delta$.

A co-sequent $\Gamma \mapsto$ is $\mathbf{T}^{\rightsquigarrow}_{\mathtt{t}}$-valid, denoted by $\models^{\rightsquigarrow}_{\mathtt{t}} \Gamma \mapsto$, if there is an assignment v such that for any statement $\gamma \in \Gamma$, $v \models \gamma$.

8.2.1 $\mathbf{T}^{\mathtt{t}}_{\rightsquigarrow}$

We say that a sequent $\Rightarrow \Delta$ is $\mathbf{T}^{\mathtt{t}}_{\rightsquigarrow}$-valid, denoted by $\models^{\mathtt{t}}_{\rightsquigarrow} \Rightarrow \Delta$, if for any assignment v, there is a statement $\delta \in \Delta$ such that $v \models \delta$.

Tableau proof system $\mathbf{T}^{\mathtt{t}}_{\rightsquigarrow}$ consists of the following axiom and deduction rules:

- **Axiom**:

$$(\mathtt{A}^{\mathtt{t}}_{\rightsquigarrow})\ \frac{\mathbf{E} m_1 \rightsquigarrow m_2 (m_1 \not\rightsquigarrow m_2 \in \Delta \& \Delta \vdash m_1 \rightsquigarrow m_2)}{\Rightarrow \Delta}$$

where Δ is a set of literal statements.

- **Deduction rules**:

$$(+\wedge^{RL})\ \frac{\begin{cases} \Rightarrow C_1 \rightsquigarrow D, \Delta \\ \Rightarrow C_2 \rightsquigarrow D, \Delta \end{cases}}{\Rightarrow C_1 \wedge C_2 \rightsquigarrow D, \Delta} \quad (+\wedge^{RR})\ \frac{\left[\begin{array}{l} \Rightarrow C \rightsquigarrow D_1, \Delta \\ \Rightarrow C \rightsquigarrow D_2, \Delta \end{array}\right.}{\Rightarrow C \rightsquigarrow D_1 \wedge D_2, \Delta}$$

$$(-\wedge^{RL})\ \frac{\left[\begin{array}{l} \Rightarrow C_1 \not\rightsquigarrow D, \Delta \\ \Rightarrow C_2 \not\rightsquigarrow D, \Delta \end{array}\right.}{\Rightarrow C_1 \wedge C_2 \not\rightsquigarrow D, \Delta} \quad (-\wedge^{RR})\ \frac{\begin{cases} \Rightarrow C \not\rightsquigarrow D_1, \Delta \\ \Rightarrow C \not\rightsquigarrow D_2, \Delta \end{cases}}{\Rightarrow C \not\rightsquigarrow D_1 \wedge D_2, \Delta}$$

and

$$(+\vee^{RL})\ \frac{\left[\begin{array}{l}\Rightarrow C_1 \rightsquigarrow D, \Delta\\ \Rightarrow C_2 \rightsquigarrow D, \Delta\end{array}\right.}{\Rightarrow C_1 \vee C_2 \rightsquigarrow D, \Delta} \quad (+\vee^{RR})\ \frac{\left\{\begin{array}{l}\Rightarrow C \rightsquigarrow D_1, \Delta\\ \Rightarrow C \rightsquigarrow D_2, \Delta\end{array}\right.}{\Rightarrow C \rightsquigarrow D_1 \vee D_2, \Delta}$$

$$(-\vee^{RL})\ \frac{\left\{\begin{array}{l}\Rightarrow C_1 \not\rightsquigarrow D, \Delta\\ \Rightarrow C_2 \not\rightsquigarrow D, \Delta\end{array}\right.}{\Rightarrow C_1 \vee C_2 \not\rightsquigarrow D, \Delta} \quad (-\vee^{RR})\ \frac{\left[\begin{array}{l}\Rightarrow C \not\rightsquigarrow D_1, \Delta\\ \Rightarrow C \not\rightsquigarrow D_2, \Delta\end{array}\right.}{\Rightarrow C \not\rightsquigarrow D_1 \vee D_2, \Delta}$$

and

$$(+\neg\wedge^{RL})\ \frac{\left[\begin{array}{l}\Rightarrow \neg C_1 \rightsquigarrow D, \Delta\\ \Rightarrow \neg C_2 \rightsquigarrow D, \Delta\end{array}\right.}{\Rightarrow \neg(C_1 \wedge C_2) \rightsquigarrow D, \Delta} \quad (+\neg\wedge^{RR})\ \frac{\left\{\begin{array}{l}\Rightarrow C \rightsquigarrow \neg D_1, \Delta\\ \Rightarrow C \rightsquigarrow \neg D_2, \Delta\end{array}\right.}{\Rightarrow C \rightsquigarrow \neg(D_1 \wedge D_2), \Delta}$$

$$(-\neg\wedge^{RL})\ \frac{\left\{\begin{array}{l}\Rightarrow \neg C_1 \not\rightsquigarrow D, \Delta\\ \Rightarrow \neg C_2 \not\rightsquigarrow D, \Delta\end{array}\right.}{\Rightarrow \neg(C_1 \wedge C_2) \not\rightsquigarrow D, \Delta} \quad (-\neg\wedge^{RR})\ \frac{\left[\begin{array}{l}\Rightarrow C \not\rightsquigarrow \neg D_1, \Delta\\ \Rightarrow C \not\rightsquigarrow \neg D_2, \Delta\end{array}\right.}{\Rightarrow C \not\rightsquigarrow \neg(D_1 \wedge D_2), \Delta}$$

and

$$(+\neg\vee^{RL})\ \frac{\left\{\begin{array}{l}\Rightarrow \neg C_1 \rightsquigarrow D, \Delta\\ \Rightarrow \neg C_2 \rightsquigarrow D, \Delta\end{array}\right.}{\Rightarrow \neg(C_1 \vee C_2) \rightsquigarrow D, \Delta} \quad (+\neg\vee^{RR})\ \frac{\left[\begin{array}{l}\Rightarrow C \rightsquigarrow \neg D_1, \Delta\\ \Rightarrow C \rightsquigarrow \neg D_2, \Delta\end{array}\right.}{\Rightarrow C \rightsquigarrow \neg(D_1 \vee D_2), \Delta}$$

$$(-\neg\vee^{RL})\ \frac{\left[\begin{array}{l}\Rightarrow \neg C_1 \not\rightsquigarrow D, \Delta\\ \Rightarrow \neg C_2 \not\rightsquigarrow D, \Delta\end{array}\right.}{\Rightarrow \neg(C_1 \vee C_2) \not\rightsquigarrow D, \Delta} \quad (-\neg\vee^{RR})\ \frac{\left\{\begin{array}{l}\Rightarrow C \not\rightsquigarrow \neg D_1, \Delta\\ \Rightarrow C \not\rightsquigarrow \neg D_2, \Delta\end{array}\right.}{\Rightarrow C \not\rightsquigarrow \neg(D_1 \vee D_2), \Delta}$$

Definition 8.2.1 A sequent $\Rightarrow \Delta$ is provable in $\mathbf{T}^{\mathtt{t}}_{\rightsquigarrow}$, denoted by $\vdash^{\mathtt{t}}_{\rightsquigarrow} \Rightarrow \Delta$, if there is a sequence $\{\Rightarrow \Delta_1, \ldots, \Rightarrow \Delta_n\}$ of sequents such that $\Delta_n = \Delta$, and for each $1 \leq i \leq n$, $\Rightarrow \Delta_i$ is an axiom or is deduced from the previous sequents by one deduction rule in $\mathbf{T}^{\mathtt{t}}_{\rightsquigarrow}$.

Theorem 8.2.2 (Soundness and completeness theorem) *For any sequent* $\Delta \Rightarrow$,

$$\vdash^{\mathtt{t}}_{\rightsquigarrow} \Delta \Rightarrow \textit{ if and only if } \models^{\mathtt{t}}_{\rightsquigarrow} \Delta \Rightarrow .$$

□

8.2.2 $\mathbf{T}^{\mathtt{f}}_{\rightsquigarrow}$

We say that a sequent $\Gamma \Rightarrow$ is valid, denoted by $\models^{\mathtt{f}}_{\rightsquigarrow} \Gamma \Rightarrow$, if for any assignment v, there is a statement $\gamma \in \Gamma$ such that $v \not\models \gamma$.

Tableau proof system $\mathbf{T}^{\mathtt{f}}_{\rightsquigarrow}$ consists of the following axiom and deduction rules:

- **Axiom**:

$$(\mathtt{A}^{\mathtt{f}}_{\rightsquigarrow})\ \frac{\mathbf{E} l_1 \rightsquigarrow l_2 (l_1 \not\rightsquigarrow l_2 \in \Gamma \& \Gamma \vdash l_1 \rightsquigarrow l_2)}{\Gamma \Rightarrow}$$

where Γ is a set of literal statements.

- **Deduction rules**:

$$(+\wedge^{LL})\ \frac{\left[\begin{array}{l}\Gamma, A_1 \rightsquigarrow B \Rightarrow\\ \Gamma, A_2 \rightsquigarrow B \Rightarrow\end{array}\right.}{\Gamma, A_1 \wedge A_2 \rightsquigarrow B \Rightarrow} \qquad (+\wedge^{LR})\ \frac{\left\{\begin{array}{l}\Gamma, A \rightsquigarrow B_1 \Rightarrow\\ \Gamma, A \rightsquigarrow B_2 \Rightarrow\end{array}\right.}{\Gamma, A \rightsquigarrow B_1 \wedge B_2 \Rightarrow}$$

$$(-\wedge^{LL})\ \frac{\left\{\begin{array}{l}\Gamma, A_1 \not\rightsquigarrow B \Rightarrow\\ \Gamma, A_2 \not\rightsquigarrow B \Rightarrow\end{array}\right.}{\Gamma, A_1 \wedge A_2 \not\rightsquigarrow B \Rightarrow} \qquad (-\wedge^{LR})\ \frac{\left[\begin{array}{l}\Gamma, A \not\rightsquigarrow B_1 \Rightarrow\\ \Gamma, A \not\rightsquigarrow B_2 \Rightarrow\end{array}\right.}{\Gamma, A \not\rightsquigarrow B_1 \wedge B_2 \Rightarrow}$$

and

$$(+\vee^{LL})\ \frac{\left\{\begin{array}{l}\Gamma, A_1 \rightsquigarrow B \Rightarrow\\ \Gamma, A_2 \rightsquigarrow B \Rightarrow\end{array}\right.}{\Gamma, A_1 \vee A_2 \rightsquigarrow B \Rightarrow} \qquad (+\vee^{LR})\ \frac{\left[\begin{array}{l}\Gamma, A \rightsquigarrow B_1 \Rightarrow\\ \Gamma, A \rightsquigarrow B_2 \Rightarrow\end{array}\right.}{\Gamma, A \rightsquigarrow B_1 \vee B_2 \Rightarrow}$$

$$(-\vee^{LL})\ \frac{\left[\begin{array}{l}\Gamma, A_1 \not\rightsquigarrow B \Rightarrow\\ \Gamma, A_2 \not\rightsquigarrow B \Rightarrow\end{array}\right.}{\Gamma, A_1 \vee A_2 \not\rightsquigarrow B \Rightarrow} \qquad (-\vee^{LR})\ \frac{\left\{\begin{array}{l}\Gamma, A \not\rightsquigarrow B_1 \Rightarrow\\ \Gamma, A \not\rightsquigarrow B_2 \Rightarrow\end{array}\right.}{\Gamma, A \not\rightsquigarrow B_1 \vee B_2 \Rightarrow}$$

and

$$(+\neg\wedge^{LL})\ \frac{\left\{\begin{array}{l}\Gamma, \neg A_1 \rightsquigarrow B \Rightarrow\\ \Gamma, \neg A_2 \rightsquigarrow B \Rightarrow\end{array}\right.}{\Gamma, \neg(A_1 \wedge A_2) \rightsquigarrow B \Rightarrow} \qquad (+\neg\wedge^{LR})\ \frac{\left[\begin{array}{l}\Gamma, A \rightsquigarrow \neg B_1 \Rightarrow\\ \Gamma, A \rightsquigarrow \neg B_2 \Rightarrow\end{array}\right.}{\Gamma, A \rightsquigarrow \neg(B_1 \wedge B_2) \Rightarrow}$$

$$(-\neg\wedge^{LL})\ \frac{\left[\begin{array}{l}\Gamma, \neg A_1 \not\rightsquigarrow B \Rightarrow\\ \Gamma, \neg A_2 \not\rightsquigarrow B \Rightarrow\end{array}\right.}{\Gamma, \neg(A_1 \wedge A_2) \not\rightsquigarrow B \Rightarrow} \qquad (-\neg\wedge^{LR})\ \frac{\left\{\begin{array}{l}\Gamma, A \not\rightsquigarrow \neg B_1 \Rightarrow\\ \Gamma, A \not\rightsquigarrow \neg B_2 \Rightarrow\end{array}\right.}{\Gamma, A \not\rightsquigarrow \neg(B_1 \wedge B_2) \Rightarrow}$$

and

$$(+\neg\vee^{LL})\ \frac{\left[\begin{array}{l}\Gamma, \neg A_1 \rightsquigarrow B \Rightarrow\\ \Gamma, \neg A_2 \rightsquigarrow B \Rightarrow\end{array}\right.}{\Gamma, \neg(A_1 \vee A_2) \rightsquigarrow B \Rightarrow} \qquad (+\neg\vee^{LR})\ \frac{\left\{\begin{array}{l}\Gamma, A \rightsquigarrow \neg B_1 \Rightarrow\\ \Gamma, A \rightsquigarrow \neg B_2 \Rightarrow\end{array}\right.}{\Gamma, A \rightsquigarrow \neg(B_1 \vee B_2) \Rightarrow}$$

$$(-\neg\vee^{LL})\ \frac{\left\{\begin{array}{l}\Gamma, \neg A_1 \not\rightsquigarrow B \Rightarrow\\ \Gamma, \neg A_2 \not\rightsquigarrow B \Rightarrow\end{array}\right.}{\Gamma, \neg(A_1 \vee A_2) \not\rightsquigarrow B \Rightarrow} \qquad (-\vee^{LR})\ \frac{\left[\begin{array}{l}\Gamma, A \not\rightsquigarrow \neg B_1 \Rightarrow\\ \Gamma, A \not\rightsquigarrow \neg B_2 \Rightarrow\end{array}\right.}{\Gamma, A \not\rightsquigarrow \neg(B_1 \vee B_2) \Rightarrow}$$

Definition 8.2.3 A sequent $\Gamma \Rightarrow$ is provable, denoted by $\vdash^{\text{f}}_{\rightsquigarrow} \Gamma \Rightarrow$, if there is a sequence $\{\Gamma_1 \Rightarrow, \ldots, \Gamma_n \Rightarrow\}$ of sequents such that $\Gamma_n = \Gamma$, and for each $1 \leq i \leq n$, $\Gamma_i \Rightarrow$ is an axiom or is deduced from the previous sequents by one deduction rule in $\mathbf{T}^{\text{f}}_{\rightsquigarrow}$.

Theorem 8.2.4 (Soundness and completeness theorem) *For any sequent* $\Gamma \Rightarrow$,

$$\vdash^{\text{f}}_{\rightsquigarrow} \Gamma \Rightarrow \ \textit{iff} \ \models^{\text{f}}_{\rightsquigarrow} \Gamma \Rightarrow .$$

□

8.2.3 Tableau Proof Systems $\mathbf{T}^{\rightsquigarrow}_{\mathtt{t}}/\mathbf{T}^{\rightsquigarrow}_{\mathtt{f}}$

Dually we have nonmonotonic tableau proof systems $\mathbf{T}^{\rightsquigarrow}_{\mathtt{t}}, \mathbf{T}^{\rightsquigarrow}_{\mathtt{f}}$.

Let $\mathbf{T}^*_{\rightsquigarrow}$ consist of two parts $\mathtt{A}^*_{\rightsquigarrow}$ and $\mathbf{T}^*_{\rightsquigarrow}(-)$. Then,

$$\begin{aligned}\mathbf{T}^{\rightsquigarrow}_{\mathtt{t}} &= \mathtt{A}^{\rightsquigarrow}_{\mathtt{t}} + \sigma(\mathbf{T}^{\mathtt{f}}_{\rightsquigarrow}(-)),\\ \mathbf{T}^{\rightsquigarrow}_{\mathtt{f}} &= \mathtt{A}^{\rightsquigarrow}_{\mathtt{f}} + \sigma(\mathbf{T}^{\mathtt{t}}_{\rightsquigarrow}(-)),\end{aligned}$$

where σ is a mapping mapping deduction rules in $\mathbf{T}^{\mathtt{t}}_{\rightsquigarrow}/\mathbf{T}^{\mathtt{f}}_{\rightsquigarrow}$ into $\mathbf{T}^{\rightsquigarrow}_{\mathtt{t}}/\mathbf{T}^{\rightsquigarrow}_{\mathtt{f}}$, such that

$$\begin{array}{ll}\sigma(\Rightarrow \Delta) = \Gamma \mapsto & \sigma(\Gamma \Rightarrow) = \mapsto \Delta\\ \sigma(C) = A & \sigma(A) = C,\\ \sigma(D) = B & \sigma(B) = D,\\ \sigma(x) = x, & \end{array}$$

for any other symbols x, and

$$\begin{array}{l}(\mathtt{A}^{\rightsquigarrow}_{\mathtt{t}})\ \dfrac{\sim \mathbf{E} l_1 \rightsquigarrow l_2 (l_1 \not\rightsquigarrow l_2 \in \Gamma \& \Gamma \vdash l_1 \rightsquigarrow l_2)}{\Gamma \mapsto}\\ (\mathtt{A}^{\rightsquigarrow}_{\mathtt{t}})\ \dfrac{\sim \mathbf{E} m_1 \rightsquigarrow m_2 (m_1 \not\rightsquigarrow m_2 \in \Delta \& \Delta \vdash m_1 \rightsquigarrow m_2)}{\mapsto \Delta}\end{array}$$

where Γ, Δ are sets of literal statements.

Theorem 8.2.5 (Soundness and completeness theorem) *(i) For any co-sequent* $\Gamma \mapsto$,

$$\vdash^{\rightsquigarrow}_{\mathtt{t}} \Gamma \mapsto \textit{ iff } \models^{\rightsquigarrow}_{\mathtt{t}} \Gamma \mapsto .$$

(ii) For any co-sequent $\mapsto \Delta$,

$$\vdash^{\rightsquigarrow}_{\mathtt{f}} \mapsto \Delta \textit{ iff } \models^{\rightsquigarrow}_{\mathtt{f}} \mapsto \Delta.$$

□

8.3 R-Calculi $\mathbf{R}^*_{\rightsquigarrow}$

In this section we will R-calculi $\mathbf{R}^{\mathtt{t}}_{\rightsquigarrow}$ and $\mathbf{R}^{\mathtt{f}}_{\rightsquigarrow}$.

8.3.1 R-Calculus $\mathbf{R}^{\mathtt{t}}_{\rightsquigarrow}$

Given Δ and a statement $\delta \in \Delta$, we use Δ to revise δ and obtain Δ', denoted by $\models^{\mathtt{t}}_{\rightsquigarrow} \Delta|\delta \rightrightarrows \Delta'$, if

$$\Delta' = \begin{cases} \Delta - \{\delta\} \text{ if } \models^{\mathtt{t}}_{\rightsquigarrow} \Rightarrow \Delta - \{\delta\} \\ \Delta \quad \text{otherwise.} \end{cases}$$

R-calculus $\mathbf{R}^{\mathtt{t}}_{\rightsquigarrow}$ consists of the following axioms and deduction rules:

- **Axioms**:

$$(\mathtt{A}^{\mathtt{t}}_{-\rightsquigarrow})\frac{\mathbf{E}\delta' \neq \delta(\Delta \vdash \delta' \& \sim \delta' \in \Delta)}{\Delta|\delta \rightrightarrows \Delta - \{\delta\}} \quad (\mathtt{A}^{\mathtt{t}}_{0\rightsquigarrow})\frac{\mathbf{A}\delta' \neq \delta(\Delta \nvdash \delta' \text{ or } \sim \delta' \notin \Delta)}{\Delta|\delta \rightrightarrows \Delta}$$

where $\Delta \cup \{\delta\}$ is a set of literal statements, and

$$\sim \delta = \begin{cases} A \not\rightsquigarrow B \text{ if } \delta = A \rightsquigarrow B \\ A \rightsquigarrow B \text{ if } \delta = A \not\rightsquigarrow B. \end{cases}$$

- **Deduction rules**:

$$(+\wedge^L_-)\frac{\begin{cases} \Delta|C_1 \rightsquigarrow D \rightrightarrows \Delta - \{C_1 \rightsquigarrow D\} \\ \Delta|C_2 \rightsquigarrow D \rightrightarrows \Delta - \{C_2 \rightsquigarrow D\} \end{cases}}{\Delta|C_1 \wedge C_2 \rightsquigarrow D \rightrightarrows \Delta - \{C_1 \wedge C_2 \rightsquigarrow D\}}$$

$$(+\wedge^L_0)\frac{\begin{bmatrix} \Delta|C_1 \rightsquigarrow D \rightrightarrows \Delta \\ \Delta|C_2 \rightsquigarrow D \rightrightarrows \Delta \end{bmatrix}}{\Delta|C_1 \wedge C_2 \rightsquigarrow D \rightrightarrows \Delta}$$

$$(+\wedge^R_-)\frac{\begin{bmatrix} \Delta|C \rightsquigarrow D_1 \rightrightarrows \Delta - \{C \rightsquigarrow D_1\} \\ \Delta - \{C \rightsquigarrow D_1\}|C \rightsquigarrow D_2 \rightrightarrows \Delta - \{C \rightsquigarrow D_1, C \rightsquigarrow D_2\} \end{bmatrix}}{\Delta|C \rightsquigarrow D_1 \wedge D_2 \rightrightarrows \Delta - \{C \rightsquigarrow D_1 \wedge D_2\}}$$

$$(+\wedge^R_0)\frac{\begin{cases} \Delta|C \rightsquigarrow D_1 \rightrightarrows \Delta \\ \Delta - \{C \rightsquigarrow D_1\}|C \rightsquigarrow D_2 \rightrightarrows \Delta - \{C \rightsquigarrow D_1\} \end{cases}}{\Delta|C \rightsquigarrow D_1 \wedge D_2 \rightrightarrows \Delta}$$

and

$$(-\wedge^L_-)\frac{\begin{bmatrix} \Delta|C_1 \not\rightsquigarrow D \rightrightarrows \Delta - \{C_1 \not\rightsquigarrow D\} \\ \Delta - \{C_1 \not\rightsquigarrow D\}|C_2 \not\rightsquigarrow D \rightrightarrows \Delta - \{C_1 \not\rightsquigarrow D, C_2 \not\rightsquigarrow D\} \end{bmatrix}}{\Delta|C_1 \wedge C_2 \not\rightsquigarrow D \rightrightarrows \Delta - \{C_1 \wedge C_2 \not\rightsquigarrow D\}}$$

$$(-\wedge^L_0)\frac{\begin{cases} \Delta|C_1 \not\rightsquigarrow D \rightrightarrows \Delta \\ \Delta - \{C_1 \not\rightsquigarrow D\}|C_2 \not\rightsquigarrow D \rightrightarrows \Delta - \{C_1 \not\rightsquigarrow D\} \end{cases}}{\Delta|C_1 \wedge C_2 \not\rightsquigarrow D \rightrightarrows \Delta}$$

$$(-\wedge^R_-)\frac{\begin{cases} \Delta|C \not\rightsquigarrow D_1 \rightrightarrows \Delta - \{C \not\rightsquigarrow D_1\} \\ \Delta|C \not\rightsquigarrow D_2 \rightrightarrows \Delta - \{C \not\rightsquigarrow D_2\} \end{cases}}{\Delta|C \not\rightsquigarrow D_1 \wedge D_2 \rightrightarrows \Delta - \{C \not\rightsquigarrow D_1 \wedge D_2\}}$$

$$(-\wedge^R_0)\frac{\begin{bmatrix} \Delta|C \not\rightsquigarrow D_1 \rightrightarrows \Delta \\ \Delta|C \not\rightsquigarrow D_2 \rightrightarrows \Delta \end{bmatrix}}{\Delta|C \not\rightsquigarrow D_1 \wedge D_2 \rightrightarrows \Delta}$$

and

$$(+\vee^{L}_{-})\ \frac{\left[\begin{array}{l}\Delta|C_1 \rightsquigarrow D \rightrightarrows \Delta-\{C_1 \rightsquigarrow D\}\\ \Delta-\{C_1 \rightsquigarrow D\}|C_2 \rightsquigarrow D \rightrightarrows \Delta-\{C_1 \rightsquigarrow D, C_2 \rightsquigarrow D\}\end{array}\right.}{\Delta|C_1 \vee C_2 \rightsquigarrow D \rightrightarrows \Delta-\{C_1 \vee C_2 \rightsquigarrow D\}}$$

$$(+\vee^{L}_{0})\ \frac{\left\{\begin{array}{l}\Delta|C_1 \rightsquigarrow D \rightrightarrows \Delta\\ \Delta-\{C_1 \rightsquigarrow D\}|C_2 \rightsquigarrow D \rightrightarrows \Delta-\{C_1 \rightsquigarrow D\}\end{array}\right.}{\Delta|C_1 \vee C_2 \rightsquigarrow D \rightrightarrows \Delta}$$

$$(+\vee^{R}_{-})\ \frac{\left\{\begin{array}{l}\Delta|C \rightsquigarrow D_1 \rightrightarrows \Delta-\{C_1 \rightsquigarrow D\}\\ \Delta|C \rightsquigarrow D_2 \rightrightarrows \Delta-\{C_2 \rightsquigarrow D\}\end{array}\right.}{\Delta|C \rightsquigarrow D_1 \vee D_2 \rightrightarrows \Delta-\{C_1 \vee C_2 \rightsquigarrow D\}}$$

$$(+\vee^{R}_{0})\ \frac{\left[\begin{array}{l}\Delta|C \rightsquigarrow D_1 \rightrightarrows \Delta\\ \Delta|C \rightsquigarrow D_2 \rightrightarrows \Delta\end{array}\right.}{\Delta|C \rightsquigarrow D_1 \vee D_2 \rightrightarrows \Delta}$$

and

$$(-\vee^{L}_{-})\ \frac{\left\{\begin{array}{l}\Delta|C_1 \not\mapsto D \rightrightarrows \Delta-\{C_1 \not\mapsto D\}\\ \Delta|C_2 \not\mapsto D \rightrightarrows \Delta-\{C_2 \not\mapsto D\}\end{array}\right.}{\Delta|C_1 \vee C_2 \not\mapsto D \rightrightarrows \Delta-\{C_1 \vee C_2 \not\mapsto D\}}$$

$$(-\vee^{L}_{0})\ \frac{\left[\begin{array}{l}\Delta|C_1 \not\mapsto D \rightrightarrows \Delta\\ \Delta|C_2 \not\mapsto D \rightrightarrows \Delta\end{array}\right.}{\Delta|C_1 \vee C_2 \not\mapsto D \rightrightarrows \Delta}$$

$$(-\vee^{R}_{-})\ \frac{\left[\begin{array}{l}\Delta|C \not\mapsto D_1 \rightrightarrows \Delta-\{C \not\mapsto D_1\}\\ \Delta-\{C \not\mapsto D_1\}|C \not\mapsto D_2 \rightrightarrows \Delta-\{C \not\mapsto D_1, C \not\mapsto D_2\}\end{array}\right.}{\Delta|C \not\mapsto D_1 \vee D_2 \rightrightarrows \Delta-\{C \not\mapsto D_1 \vee D_2\}}$$

$$(-\vee^{R}_{0})\ \frac{\left\{\begin{array}{l}\Delta|C \not\mapsto D_1 \rightrightarrows \Delta\\ \Delta-\{C \not\mapsto D_1\}|C \not\mapsto D_2 \rightrightarrows \Delta-\{C \not\mapsto D_1\}\end{array}\right.}{\Delta|C \not\mapsto D_1 \vee D_2 \rightrightarrows \Delta}$$

and

$$(+\neg\wedge^{L}_{-})\ \frac{\left[\begin{array}{l}\Delta|\neg C_1 \rightsquigarrow D \rightrightarrows \Delta-\{\neg C_1 \rightsquigarrow D\}\\ \Delta-\{\neg C_1 \rightsquigarrow D\}|\neg C_2 \rightsquigarrow D \rightrightarrows \Delta-\{\neg C_1 \rightsquigarrow D, \neg C_2 \rightsquigarrow D\}\end{array}\right.}{\Delta|\neg(C_1 \wedge C_2) \rightsquigarrow D \rightrightarrows \Delta-\{\neg(C_1 \wedge C_2) \rightsquigarrow D\}}$$

$$(+\neg\wedge^{L}_{0})\ \frac{\left\{\begin{array}{l}\Delta|\neg C_1 \rightsquigarrow D \rightrightarrows \Delta\\ \Delta-\{\neg C_1 \rightsquigarrow D\}|\neg C_2 \rightsquigarrow D \rightrightarrows \Delta-\{\neg C_1 \rightsquigarrow D\}\end{array}\right.}{\Delta|\neg(C_1 \wedge C_2) \rightsquigarrow D \rightrightarrows \Delta}$$

$$(+\neg\wedge^{R}_{-})\ \frac{\left\{\begin{array}{l}\Delta|C \rightsquigarrow \neg D_1 \rightrightarrows \Delta-\{C \rightsquigarrow \neg D_1\}\\ \Delta|C \rightsquigarrow \neg D_2 \rightrightarrows \Delta-\{C \rightsquigarrow \neg D_2\}\end{array}\right.}{\Delta|C \rightsquigarrow \neg(D_1 \wedge D_2) \rightrightarrows \Delta-\{C \rightsquigarrow \neg(D_1 \wedge D_2)\}}$$

$$(+\neg\wedge^{R}_{0})\ \frac{\left[\begin{array}{l}\Delta|C \rightsquigarrow \neg D_1 \rightrightarrows \Delta\\ \Delta|C \rightsquigarrow \neg D_2 \rightrightarrows \Delta\end{array}\right.}{\Delta|C \rightsquigarrow \neg(D_1 \wedge D_2) \rightrightarrows \Delta}$$

and

$$(-\neg\wedge^L_-)\ \frac{\left\{\begin{array}{l}\Delta|\neg C_1 \not\leadsto D \rightrightarrows \Delta-\{\neg C_1 \not\leadsto D\}\\ \Delta|\neg C_2 \not\leadsto D \rightrightarrows \Delta-\{\neg C_2 \not\leadsto D\}\end{array}\right.}{\Delta|\neg(C_1\wedge C_2) \not\leadsto D \rightrightarrows \Delta-\{\neg(C_1\wedge C_2) \not\leadsto D\}}$$

$$(-\neg\wedge^L_0)\ \frac{\left[\begin{array}{l}\Delta|\neg C_1 \not\leadsto D \rightrightarrows \Delta\\ \Delta|\neg C_2 \not\leadsto D \rightrightarrows \Delta\end{array}\right.}{\Delta|\neg(C_1\wedge C_2) \not\leadsto D \rightrightarrows \Delta}$$

$$(-\neg\wedge^R_-)\ \frac{\left[\begin{array}{l}\Delta|C \not\leadsto \neg D_1 \rightrightarrows \Delta-\{C \not\leadsto \neg D_1\}\\ \Delta-\{C \not\leadsto \neg D_1\}|C \not\leadsto \neg D_2 \rightrightarrows \Delta-\{C \not\leadsto \neg D_1, C \not\leadsto \neg D_2\}\end{array}\right.}{\Delta|C \not\leadsto \neg(D_1\wedge D_2) \rightrightarrows \Delta-\{C \not\leadsto \neg(D_1\wedge D_2)\}}$$

$$(-\neg\wedge^R_0)\ \frac{\left\{\begin{array}{l}\Delta|C \not\leadsto \neg D_1 \rightrightarrows \Delta\\ \Delta-\{C \not\leadsto \neg D_1\}|C \not\leadsto \neg D_2 \rightrightarrows \Delta-\{C \not\leadsto \neg D_1\end{array}\right.}{\Delta|C \not\leadsto \neg(D_1\wedge D_2) \rightrightarrows \Delta}$$

and

$$(+\neg\vee^L_-)\ \frac{\left\{\begin{array}{l}\Delta|\neg C_1 \leadsto D \rightrightarrows \Delta-\{\neg C_1 \leadsto D\}\\ \Delta|\neg C_2 \leadsto D \rightrightarrows \Delta-\{\neg C_2 \leadsto D\}\end{array}\right.}{\Delta|\neg(C_1\vee C_2) \leadsto D \rightrightarrows \Delta-\{\neg(C_1\vee C_2) \leadsto D\}}$$

$$(+\neg\vee^L_0)\ \frac{\left[\begin{array}{l}\Delta|\neg C_1 \leadsto D \rightrightarrows \Delta\\ \Delta|\neg C_2 \leadsto D \rightrightarrows \Delta\end{array}\right.}{\Delta|\neg(C_1\vee C_2) \leadsto D \rightrightarrows \Delta}$$

$$(+\neg\vee^R_-)\ \frac{\left[\begin{array}{l}\Delta|C \leadsto \neg D_1 \rightrightarrows \Delta-\{C \leadsto \neg D_1\}\\ \Delta-\{C \leadsto \neg D_1\}|C \leadsto \neg D_2 \rightrightarrows \Delta-\{C \leadsto \neg D_1, C \leadsto \neg D_2\}\end{array}\right.}{\Delta|C \leadsto \neg(D_1\vee D_2) \rightrightarrows \Delta-\{C \leadsto \neg(D_1\vee D_2)\}}$$

$$(+\neg\vee^R_0)\ \frac{\left\{\begin{array}{l}\Delta|C \leadsto \neg D_1 \rightrightarrows \Delta\\ \Delta-\{C \leadsto \neg D_1\}|C \leadsto \neg D_2 \rightrightarrows \Delta-\{C \leadsto \neg D_1\}\end{array}\right.}{\Delta|C \leadsto \neg(D_1\vee D_2) \rightrightarrows \Delta}$$

and

$$(-\neg\vee^L_-)\ \frac{\left[\begin{array}{l}\Delta|\neg C_1 \not\leadsto D \rightrightarrows \Delta-\{\neg C_1 \not\leadsto D\}\\ \Delta-\{\neg C_1 \not\leadsto D\}|\neg C_2 \not\leadsto D \rightrightarrows \Delta-\{\neg C_1 \not\leadsto D, \neg C_2 \not\leadsto D\}\end{array}\right.}{\Delta|\neg(C_1\vee C_2) \not\leadsto D \rightrightarrows \Delta-\{\neg(C_1\vee C_2) \not\leadsto D\}}$$

$$(-\neg\vee^L_0)\ \frac{\left\{\begin{array}{l}\Delta|\neg C_1 \not\leadsto D \rightrightarrows \Delta\\ \Delta-\{\neg C_1 \not\leadsto D\}|\neg C_2 \not\leadsto D \rightrightarrows \Delta-\{\neg C_1 \not\leadsto D\}\end{array}\right.}{\Delta|\neg(C_1\vee C_2) \not\leadsto D \rightrightarrows \Delta}$$

$$(-\neg\vee^R_-)\ \frac{\left\{\begin{array}{l}\Delta|C \not\leadsto \neg D_1 \rightrightarrows \Delta-\{C \not\leadsto \neg D_1\}\\ \Delta|C \not\leadsto \neg D_2 \rightrightarrows \Delta-\{C \not\leadsto \neg D_2\}\end{array}\right.}{\Delta|C \not\leadsto \neg(D_1\vee D_2) \rightrightarrows \Delta-\{C \not\leadsto \neg(D_1\vee D_2)\}}$$

$$(-\neg\vee^R_0)\ \frac{\left[\begin{array}{l}\Delta|C \not\leadsto \neg D_1 \rightrightarrows \Delta\\ \Delta|C \not\leadsto \neg D_2 \rightrightarrows \Delta\end{array}\right.}{\Delta|C \not\leadsto \neg(D_1\vee D_2) \rightrightarrows \Delta}$$

Definition 8.3.1 A reduction $\theta = \Delta|\delta \Rightarrow \Delta'$ is provable, denoted by $\vdash^{\mathtt{t}}_{\rightsquigarrow} \theta$, if there is a sequence $\{\theta_1, \ldots, \theta_n\}$ of reductions such that $\theta_n = \theta$, and for each $1 \le i \le n, \theta_i$ is an axiom or is deduced from the previous reductions by one deduction rule in $\mathbf{R}^{\mathtt{t}}_{\rightsquigarrow}$.

Theorem 8.3.2 (Soundness and completeness theorem) *For any reduction* $\theta = \Delta|\delta \Rightarrow \Delta'$,

$$\vdash^{\mathtt{t}}_{\rightsquigarrow} \theta \text{ iff } \models^{\mathtt{t}}_{\rightsquigarrow} \theta.$$

□

8.3.2 *R-Calculus* $\mathbf{R}^{\mathtt{f}}_{\rightsquigarrow}$

Given Γ and a statement $\gamma \in \Gamma$, we use Γ to revise γ and obtain Γ', denoted by $\models^{\mathtt{f}}_{\rightsquigarrow} \Gamma|\gamma \rightrightarrows \Gamma'$, if

$$\Gamma' = \begin{cases} \Gamma - \{\gamma\} & \text{if } \models^{\mathtt{f}}_{\rightsquigarrow} \Rightarrow \Gamma - \{\gamma\} \\ \Gamma & \text{otherwise.} \end{cases}$$

R-calculus $\mathbf{R}^{\mathtt{f}}_{\rightsquigarrow}$ consists of the following axioms and deduction rules:

- **Axioms**:

$$(\mathtt{A}^{\mathtt{f}}_{-})\ \frac{\mathbf{E}\gamma' \neq \gamma(\sim \gamma' \in \Gamma \& \Gamma \vdash \gamma')}{\Gamma|\gamma \rightrightarrows \Gamma - \{\gamma\}} \quad (\mathtt{A}^{\mathtt{f}}_{0})\ \frac{\sim \mathbf{E}\gamma' \neq \gamma(\sim \gamma' \in \Gamma \& \Gamma \vdash \gamma')}{\Gamma|\gamma \rightrightarrows \Gamma}$$

  where Γ is a set of literal statements.
- **Deduction rules**:

$$(+\wedge^{L}_{-})\ \frac{\left[\begin{array}{l} \Gamma|A_1 \rightsquigarrow B \rightrightarrows \Gamma - \{A_1 \rightsquigarrow B\} \\ \Gamma - \{A_1 \rightsquigarrow B\}|A_2 \rightsquigarrow B \rightrightarrows \Gamma - \{A_1 \rightsquigarrow B, A_2 \rightsquigarrow B\} \end{array}\right.}{\Gamma|A_1 \wedge A_2 \rightsquigarrow B \rightrightarrows \Gamma - \{A_1 \wedge A_2 \rightsquigarrow B\}}$$

$$(+\wedge^{L}_{0})\ \frac{\left\{\begin{array}{l} \Gamma|A_1 \rightsquigarrow B \rightrightarrows \Gamma \\ \Gamma - \{A_1 \rightsquigarrow B\}|A_2 \rightsquigarrow B \rightrightarrows \Gamma - \{A_1 \rightsquigarrow B\} \end{array}\right.}{\Gamma|A_1 \wedge A_2 \rightsquigarrow B \rightrightarrows \Gamma}$$

$$(+\wedge^{R}_{-})\ \frac{\left\{\begin{array}{l} \Gamma|A \rightsquigarrow B_1 \rightrightarrows \Gamma - \{A \rightsquigarrow B_1\} \\ \Gamma|A \rightsquigarrow B_2 \rightrightarrows \Gamma - \{A \rightsquigarrow B_2\} \end{array}\right.}{\Gamma|A \rightsquigarrow B_1 \wedge B_2 \rightrightarrows \Gamma - \{A \rightsquigarrow B_1 \wedge B_2\}}$$

$$(+\wedge^{R}_{0})\ \frac{\left[\begin{array}{l} \Gamma|A \rightsquigarrow B_1 \rightrightarrows \Gamma \\ \Gamma|A \rightsquigarrow B_2 \rightrightarrows \Gamma \end{array}\right.}{\Gamma|A \rightsquigarrow B_1 \wedge B_2 \rightrightarrows \Gamma}$$

and

$$(-\wedge^L_-)\ \frac{\left\{\begin{array}{l}\Gamma|A_1 \not\rightsquigarrow B \rightrightarrows \Gamma - \{A_1 \not\rightsquigarrow B\}\\ \Gamma|A_2 \not\rightsquigarrow B \rightrightarrows \Gamma - \{A_2 \not\rightsquigarrow B\}\end{array}\right.}{\Gamma|A_1 \wedge A_2 \not\rightsquigarrow B \rightrightarrows \Gamma - \{A_1 \wedge A_2 \not\rightsquigarrow B\}}$$

$$(-\wedge^L_0)\ \frac{\left[\begin{array}{l}\Gamma|A_1 \not\rightsquigarrow B \rightrightarrows \Gamma\\ \Gamma|A_2 \not\rightsquigarrow B \rightrightarrows \Gamma\end{array}\right.}{\Gamma|A_1 \wedge A_2 \not\rightsquigarrow B \rightrightarrows \Gamma}$$

$$(-\wedge^R_-)\ \frac{\left[\begin{array}{l}\Gamma|A \not\rightsquigarrow B_1 \rightrightarrows \Gamma - \{A \not\rightsquigarrow B_1\}\\ \Gamma - \{A \not\rightsquigarrow B_1\}|A \not\rightsquigarrow B_2 \rightrightarrows \Gamma - \{A \not\rightsquigarrow B_1, A \not\rightsquigarrow B_2\}\end{array}\right.}{\Gamma|A \not\rightsquigarrow B_1 \wedge B_2 \rightrightarrows \Gamma - \{A \not\rightsquigarrow B_1 \wedge B_2\}}$$

$$(-\wedge^R_0)\ \frac{\left\{\begin{array}{l}\Gamma|A \not\rightsquigarrow B_1 \rightrightarrows \Gamma\\ \Gamma - \{A \not\rightsquigarrow B_1\}|A \not\rightsquigarrow B_2 \rightrightarrows \Gamma - \{A \not\rightsquigarrow B_1\}\end{array}\right.}{\Gamma|A \not\rightsquigarrow B_1 \wedge B_2 \rightrightarrows \Gamma}$$

and

$$(+\vee^L_-)\ \frac{\left\{\begin{array}{l}\Gamma|A_1 \rightsquigarrow B \rightrightarrows \Gamma - \{A_1 \rightsquigarrow B\}\\ \Gamma|A_2 \rightsquigarrow B \rightrightarrows \Gamma - \{A_1 \rightsquigarrow B\}\end{array}\right.}{\Gamma|A_1 \vee A_2 \rightsquigarrow B \rightrightarrows \Gamma - \{A_1 \vee A_2 \rightsquigarrow B\}}$$

$$(+\vee^L_0)\ \frac{\left[\begin{array}{l}\Gamma|A_1 \rightsquigarrow B \rightrightarrows \Gamma\\ \Gamma|A_2 \rightsquigarrow B \rightrightarrows \Gamma\end{array}\right.}{\Gamma|A_1 \vee A_2 \rightsquigarrow B \rightrightarrows \Gamma}$$

$$(+\vee^R_-)\ \frac{\left[\begin{array}{l}\Gamma|A \rightsquigarrow B_1 \rightrightarrows \Gamma - \{A \rightsquigarrow B_1\}\\ \Gamma - \{A \rightsquigarrow B_1\}|A \rightsquigarrow B_2 \rightrightarrows \Gamma - \{A \rightsquigarrow B_1, A \rightsquigarrow B_2\}\end{array}\right.}{\Gamma|A \rightsquigarrow B_1 \vee B_2 \rightrightarrows \Gamma - \{A \rightsquigarrow B_1 \vee B_2\}}$$

$$(+\vee^R_0)\ \frac{\left\{\begin{array}{l}\Gamma|A \rightsquigarrow B_1 \rightrightarrows \Gamma\\ \Gamma - \{A \rightsquigarrow B_1\}|A \rightsquigarrow B_2 \rightrightarrows \Gamma - \{A \rightsquigarrow B_1\}\end{array}\right.}{\Gamma|A \rightsquigarrow B_1 \vee B_2 \rightrightarrows \Gamma}$$

and

$$(-\vee^L_-)\ \frac{\left[\begin{array}{l}\Gamma|A_1 \not\rightsquigarrow B \rightrightarrows \Gamma - \{A_1 \not\rightsquigarrow B\}\\ \Gamma - \{A_1 \not\rightsquigarrow B\}|A_2 \not\rightsquigarrow B \rightrightarrows \Gamma - \{A_1 \not\rightsquigarrow B, A_2 \not\rightsquigarrow B\}\end{array}\right.}{\Gamma|A_1 \vee A_2 \not\rightsquigarrow B \rightrightarrows \Gamma - \{A_1 \vee A_2 \not\rightsquigarrow B\}}$$

$$(-\vee^L_0)\ \frac{\left\{\begin{array}{l}\Gamma|A_1 \not\rightsquigarrow B \rightrightarrows \Gamma\\ \Gamma - \{A_1 \not\rightsquigarrow B\}|A_2 \not\rightsquigarrow B \rightrightarrows \Gamma - \{A_1 \not\rightsquigarrow B\}\end{array}\right.}{\Gamma|A_1 \vee A_2 \not\rightsquigarrow B \rightrightarrows \Gamma}$$

$$(-\vee^R_-)\ \frac{\left\{\begin{array}{l}\Gamma|A \not\rightsquigarrow B_1 \rightrightarrows \Gamma - \{A \not\rightsquigarrow B_1\}\\ \Gamma|A \not\rightsquigarrow B_2 \rightrightarrows \Gamma - \{A \not\rightsquigarrow B_2\}\end{array}\right.}{\Gamma|A \not\rightsquigarrow B_1 \vee B_2 \rightrightarrows \Gamma - \{A \not\rightsquigarrow B_1 \vee B_2\}}$$

$$(-\vee^R_0)\ \frac{\left[\begin{array}{l}\Gamma|A \not\rightsquigarrow B_1 \rightrightarrows \Gamma\\ \Gamma|A \not\rightsquigarrow B_2 \rightrightarrows \Gamma\end{array}\right.}{\Gamma|A \not\rightsquigarrow B_1 \vee B_2 \rightrightarrows \Gamma}$$

and

$$(+\neg\wedge^L_-)\ \frac{\left\{\begin{array}{l}\Gamma|\neg A_1 \rightsquigarrow B \rightrightarrows \Gamma-\{\neg A_1 \rightsquigarrow B\}\\ \Gamma|\neg A_2 \rightsquigarrow B \rightrightarrows \Gamma-\{\neg A_2\rightsquigarrow B\}\end{array}\right.}{\Gamma|\neg(A_1\wedge A_2)\rightsquigarrow B\rightrightarrows\Gamma-\{\neg(A_1\wedge A_2)\rightsquigarrow B\}}$$

$$(+\neg\wedge^L_0)\ \frac{\left[\begin{array}{l}\Gamma|\neg A_1 \rightsquigarrow B \rightrightarrows \Gamma\\ \Gamma|\neg A_2 \rightsquigarrow B \rightrightarrows \Gamma\end{array}\right.}{\Gamma|\neg(A_1\wedge A_2)\rightsquigarrow B\rightrightarrows\Gamma}$$

$$(+\neg\wedge^R_-)\ \frac{\left[\begin{array}{l}\Gamma|A \rightsquigarrow \neg B_1 \rightrightarrows \Gamma-\{A \rightsquigarrow \neg B_1\}\\ \Gamma-\{A \rightsquigarrow \neg B_1\}|A \rightsquigarrow \neg B_2 \rightrightarrows \Gamma-\{A\rightsquigarrow \neg B_1, A\rightsquigarrow \neg B_2\}\end{array}\right.}{\Gamma|A\rightsquigarrow \neg(B_1\wedge B_2)\rightrightarrows\Gamma-\{A\rightsquigarrow \neg(B_1\wedge B_2)\}}$$

$$(+\neg\wedge^R_0)\ \frac{\left\{\begin{array}{l}\Gamma|A \rightsquigarrow \neg B_1 \rightrightarrows \Gamma\\ \Gamma-\{A \rightsquigarrow \neg B_1\}|A \rightsquigarrow \neg B_2 \rightrightarrows \Gamma-\{A\rightsquigarrow \neg B_1\}\end{array}\right.}{\Gamma|A\rightsquigarrow \neg(B_1\wedge B_2)\rightrightarrows\Gamma}$$

and

$$(-\neg\wedge^L_-)\ \frac{\left[\begin{array}{l}\Gamma|\neg A_1 \not\rightsquigarrow B \rightrightarrows \Gamma-\{\neg A_1 \not\rightsquigarrow B\}\\ \Gamma-\{\neg A_1 \not\rightsquigarrow B\}|\neg A_2 \not\rightsquigarrow B \rightrightarrows \Gamma-\{\neg A_1\not\rightsquigarrow B, \neg A_2\not\rightsquigarrow B\}\end{array}\right.}{\Gamma|\neg(A_1\wedge A_2)\not\rightsquigarrow B\rightrightarrows\Gamma-\{\neg(A_1\wedge A_2)\not\rightsquigarrow B\}}$$

$$(-\neg\wedge^L_0)\ \frac{\left\{\begin{array}{l}\Gamma|\neg A_1 \not\rightsquigarrow B \rightrightarrows \Gamma\\ \Gamma-\{\neg A_1 \not\rightsquigarrow B\}|\neg A_2 \not\rightsquigarrow B \rightrightarrows \Gamma-\{\neg A_1\not\rightsquigarrow B\}\end{array}\right.}{\Gamma|\neg(A_1\wedge A_2)\not\rightsquigarrow B\rightrightarrows\Gamma}$$

$$(-\neg\wedge^R_-)\ \frac{\left\{\begin{array}{l}\Gamma|A \not\rightsquigarrow \neg B_1 \rightrightarrows \Gamma-\{A \not\rightsquigarrow \neg B_1\}\\ \Gamma|A \not\rightsquigarrow \neg B_2 \rightrightarrows \Gamma-\{A\not\rightsquigarrow \neg B_2\}\end{array}\right.}{\Gamma|A\not\rightsquigarrow \neg(B_1\wedge B_2)\rightrightarrows\Gamma-\{A\not\rightsquigarrow \neg(B_1\wedge B_2)\}}$$

$$(-\neg\wedge^R_0)\ \frac{\left[\begin{array}{l}\Gamma|A \not\rightsquigarrow \neg B_1 \rightrightarrows \Gamma\\ \Gamma|A \not\rightsquigarrow \neg B_2 \rightrightarrows \Gamma\end{array}\right.}{\Gamma|A\not\rightsquigarrow \neg(B_1\wedge B_2)\rightrightarrows\Gamma}$$

and

$$(+\neg\vee^L_-)\ \frac{\left[\begin{array}{l}\Gamma|\neg A_1 \rightsquigarrow B \rightrightarrows \Gamma-\{\neg A_1 \rightsquigarrow B\}\\ \Gamma-\{\neg A_1 \rightsquigarrow B\}|\neg A_2 \rightsquigarrow B \rightrightarrows \Gamma-\{\neg A_1\rightsquigarrow B, \neg A_2\rightsquigarrow B\}\end{array}\right.}{\Gamma|\neg(A_1\vee A_2)\rightsquigarrow B\rightrightarrows\Gamma-\{\neg(A_1\vee A_2)\rightsquigarrow B\}}$$

$$(+\neg\vee^L_0)\ \frac{\left\{\begin{array}{l}\Gamma|\neg A_1 \rightsquigarrow B \rightrightarrows \Gamma\\ \Gamma-\{\neg A_1 \rightsquigarrow B\}|\neg A_2 \rightsquigarrow B \rightrightarrows \Gamma-\{\neg A_1\rightsquigarrow B\}\end{array}\right.}{\Gamma|\neg(A_1\vee A_2)\rightsquigarrow B\rightrightarrows\Gamma}$$

$$(+\neg\vee^R_-)\ \frac{\left\{\begin{array}{l}\Gamma|A \rightsquigarrow \neg B_1 \rightrightarrows \Gamma-\{A \rightsquigarrow \neg B_1\}\\ \Gamma|A \rightsquigarrow \neg B_2 \rightrightarrows \Gamma-\{A\rightsquigarrow \neg B_2\}\end{array}\right.}{\Gamma|A\rightsquigarrow \neg(B_1\vee B_2)\rightrightarrows\Gamma-\{A\rightsquigarrow \neg(B_1\vee B_2)\}}$$

$$(+\neg\vee^R_0)\ \frac{\left[\begin{array}{l}\Gamma|A \rightsquigarrow \neg B_1 \rightrightarrows \Gamma\\ \Gamma|A \rightsquigarrow \neg B_2 \rightrightarrows \Gamma\end{array}\right.}{\Gamma|A\rightsquigarrow \neg(B_1\vee B_2)\rightrightarrows\Gamma}$$

and

$$(-\neg\vee^L_-)\ \frac{\begin{cases}\Gamma|\neg A_1 \not\mapsto B \rightrightarrows \Gamma - \{\neg A_1 \not\mapsto B\}\\ \Gamma|\neg A_2 \not\mapsto B \rightrightarrows \Gamma - \{\neg A_2 \not\mapsto B\}\end{cases}}{\Gamma|\neg(A_1 \vee A_2) \not\mapsto B \rightrightarrows \Gamma - \{\neg(A_1 \vee A_2) \not\mapsto B\}}$$

$$(-\neg\vee^L_0)\ \frac{\begin{bmatrix}\Gamma|\neg A_1 \not\mapsto B \rightrightarrows \Gamma\\ \Gamma|\neg A_2 \not\mapsto B \rightrightarrows \Gamma\end{bmatrix}}{\Gamma|\neg(A_1 \vee A_2) \not\mapsto B \rightrightarrows \Gamma}$$

$$(-\vee^R_-)\ \frac{\begin{bmatrix}\Gamma|A \not\mapsto \neg B_1 \rightrightarrows \Gamma - \{A \not\mapsto \neg B_1\}\\ \Gamma - \{A \not\mapsto \neg B_1\}|A \not\mapsto \neg B_2 \rightrightarrows \Gamma - \{A \not\mapsto \neg B_1, A \not\mapsto \neg B_2\}\end{bmatrix}}{\Gamma|A \not\mapsto \neg(B_1 \vee B_2) \rightrightarrows \Gamma - \{A \not\mapsto \neg(B_1 \vee B_2)\}}$$

$$(-\vee^R_0)\ \frac{\begin{cases}\Gamma|A \not\mapsto \neg B_1 \rightrightarrows \Gamma\\ \Gamma - \{A \not\mapsto \neg B_1\}|A \not\mapsto \neg B_2 \rightrightarrows \Gamma - \{A \not\mapsto \neg B_1\}\end{cases}}{\Gamma|A \not\mapsto \neg(B_1 \vee B_2) \rightrightarrows \Gamma}$$

Definition 8.3.3 A reduction $\theta = \Gamma|\gamma \Rightarrow \Gamma'$ is provable in $\mathbf{R}^{\mathrm{f}}_{\rightsquigarrow}$, denoted by $\vdash^{\mathrm{f}}_{\rightsquigarrow} \theta$, if there is a sequence $\{\theta_1, \ldots, \theta_n\}$ of reductions such that $\theta_n = \theta$, and for each $1 \leq i \leq n$, θ_i is an axiom or is deduced from the previous reductions by one deduction rule in $\mathbf{R}^{\mathrm{f}}_{\rightsquigarrow}$.

Theorem 8.3.4 (Soundness and completeness theorem) *For any reduction* $\theta = \Gamma|\gamma \Rightarrow \Gamma'$,

$$\vdash^{\mathrm{f}}_{\rightsquigarrow} \theta \text{ iff } \models^{\mathrm{f}}_{\rightsquigarrow} \theta.$$

□

8.4 Other Minimal Changes

Definition 8.4.1 Given a formula A, a formula B is a subformula of A, denoted by $B \leq A$, if either $A = B$, or

(i) if $A = \neg A_1$ then $B \leq A_1$;
(ii) if $A = A_1 \vee A_2$ or $A_1 \wedge A_2$ then either

$B \leq A_1$ or $B \leq A_2$.

For example, let $A = (p \vee q) \wedge (r \vee s)$. Then,

$$p \vee q, r \vee s \leq A;$$

and

$$p \wedge r, q \wedge r, p \wedge (r \vee s) \not\leq A.$$

It is clear that the R-calculi $\mathbf{R}^{\mathrm{t}}_{\rightsquigarrow}, \mathbf{R}^{\mathrm{f}}_{\rightsquigarrow}$ preserves the $\subseteq$-minimal change, that is, if $\Gamma|\gamma \rightrightarrows \Gamma'$ then Γ' is a $\subseteq$-minimal subtheory of Γ, γ, which is maximal consistent with Γ, that is, for any theory Ξ with $\Gamma \supseteq \Xi \supset \Gamma'$, Ξ is inconsistent with Γ.

Definition 8.4.2 Given a formula $A[B_1, \ldots, B_n]$, where $[B_1]$ is an occurrence of B_1 in A, a formula $B = A[\lambda, \ldots, \lambda] = A[B_1/\lambda, \ldots, B_n/\lambda]$, where the occurrence B_i is replaced by the empty formula λ, is called a pseudo-subformula of A, denoted by $B \preceq A$.

For example, let $A = (p \vee q) \wedge (r \vee s)$. Then,

$$p \vee q, r \vee s, p \wedge r, q \wedge r, p \wedge (r \vee s) \preceq A.$$

There are two other minimal changes, and we will consider:

(i) Pseudo-subformula-minimal ($\preceq$-minimal) change, where $\preceq$ is the pseudo-subformula relation, just as the subformula relation $\leq$, where a statement γ' is a pseudo-substatement of γ if eliminating some substatements in γ results in γ.

(ii) Deduction-based minimal ($\vdash_{\preceq}$-minimal) change, where a theory Γ' is a $\vdash_{\preceq}$-minimal change of Γ (denoted by $\models_{\mathbf{P}} \Gamma|\gamma \rightrightarrows \Gamma'$), if $\Gamma' \succeq \Gamma - \{\gamma\}$ is consistent with γ, and for any theory Ξ with $\Gamma' \succ \Xi \succeq \Gamma - \{\gamma\}$ either $\gamma, \Xi \vdash \Gamma'$ and $\gamma, \Gamma' \vdash \Xi$, or Ξ is inconsistent with γ.

We will give R-calculi $\mathbf{Q}^{\mathrm{t}}_{\rightsquigarrow}$ and $\mathbf{P}^{\mathrm{t}}_{\rightsquigarrow}$ which are sound and complete for $\preceq$-minimal change and $\vdash_{\preceq}$-minimal change, respectively.

In fact, there should be following R-calculi

$\mathbf{R}^{\mathrm{t}}_{\rightsquigarrow}$	$\mathbf{Q}^{\mathrm{t}}_{\rightsquigarrow}$	$\mathbf{P}^{\mathrm{t}}_{\rightsquigarrow}$
$\mathbf{R}^{\mathrm{f}}_{\rightsquigarrow}$	$\mathbf{Q}^{\mathrm{f}}_{\rightsquigarrow}$	$\mathbf{P}^{\mathrm{f}}_{\rightsquigarrow}$
$\mathbf{R}^{\rightsquigarrow}_{\mathrm{t}}$	$\mathbf{Q}^{\rightsquigarrow}_{\mathrm{t}}$	$\mathbf{P}^{\rightsquigarrow}_{\mathrm{t}}$
$\mathbf{R}^{\rightsquigarrow}_{\mathrm{f}}$	$\mathbf{Q}^{\rightsquigarrow}_{\mathrm{f}}$	$\mathbf{P}^{\rightsquigarrow}_{\mathrm{f}}$

and we will give three R-calculi $\mathbf{R}^{\rightsquigarrow}_{\mathrm{t}}$, $\mathbf{Q}^{\rightsquigarrow}_{\mathrm{t}}$ and $\mathbf{P}^{\rightsquigarrow}_{\mathrm{t}}$.

8.4.1 R-Calculus $\mathbf{R}^{\rightsquigarrow}_{\mathrm{t}}$

Given Γ and a statement $\gamma \in \Gamma$, we use Γ to revise γ and obtain Γ', denoted by $\models^{\rightsquigarrow}_{\mathrm{t}} \Gamma|\gamma \rightrightarrows \Gamma'$, if

$$\Gamma' = \begin{cases} \Gamma, \gamma & \text{if } \models^{\rightsquigarrow}_{\mathrm{t}} \Gamma, \gamma \mapsto \\ \Gamma & \text{otherwise.} \end{cases}$$

R-calculus $\mathbf{R}^{\rightsquigarrow}_{\mathrm{t}}$ consists of the following axioms and deduction rules:

- **Axioms**:

$$(\mathrm{A}^{+\rightsquigarrow}_{\mathrm{t}})\ \frac{\mathbf{E}\gamma' \neq \gamma(\Gamma \vdash \gamma' \& \sim\gamma' \in \Gamma)}{\Gamma|\gamma \rightrightarrows \Gamma, \gamma} \quad (\mathrm{A}^{0\rightsquigarrow}_{\mathrm{t}})\ \frac{\mathbf{A}\gamma' \neq \gamma(\Gamma \nvdash \gamma' \text{ or } \sim\gamma' \notin \Gamma)}{\Gamma|\gamma \rightrightarrows \Gamma}$$

where $\Gamma \cup \{\gamma\}$ is a set of literal statements.

- **Deduction rules**:

$$(+\wedge_{-}^{L})\ \frac{\left\{\begin{array}{l}\Gamma|A_1 \rightsquigarrow B \rightrightarrows \Gamma, A_1 \rightsquigarrow B\\ \Gamma|A_2 \rightsquigarrow B \rightrightarrows \Gamma, A_2 \rightsquigarrow B\end{array}\right.}{\Gamma|A_1 \wedge A_2 \rightsquigarrow B \rightrightarrows \Gamma, A_1 \wedge A_2 \rightsquigarrow B}$$

$$(+\wedge_{0}^{L})\ \frac{\left[\begin{array}{l}\Gamma|A_1 \rightsquigarrow B \rightrightarrows \Gamma\\ \Gamma|A_2 \rightsquigarrow B \rightrightarrows \Gamma\end{array}\right.}{\Gamma|A_1 \wedge A_2 \rightsquigarrow B \rightrightarrows \Gamma}$$

$$(+\wedge_{-}^{R})\ \frac{\left[\begin{array}{l}\Gamma|A \rightsquigarrow B_1 \rightrightarrows \Gamma, A \rightsquigarrow B_1\\ \Gamma, A \rightsquigarrow B_1|A \rightsquigarrow B_2 \rightrightarrows \Gamma, A \rightsquigarrow B_1, A \rightsquigarrow B_2\end{array}\right.}{\Gamma|A \rightsquigarrow B_1 \wedge B_2 \rightrightarrows \Gamma, A \rightsquigarrow B_1 \wedge B_2}$$

$$(+\wedge_{0}^{R})\ \frac{\left\{\begin{array}{l}\Gamma|A \rightsquigarrow B_1 \rightrightarrows \Gamma\\ \Gamma, A \rightsquigarrow B_1|A \rightsquigarrow B_2 \rightrightarrows \Gamma, A \rightsquigarrow B_1\end{array}\right.}{\Gamma|A \rightsquigarrow B_1 \wedge B_2 \rightrightarrows \Gamma, A \rightsquigarrow B_1 \wedge B_2}$$

and

$$(-\wedge_{-}^{L})\ \frac{\left[\begin{array}{l}\Gamma|A_1 \not\rightsquigarrow B \rightrightarrows \Gamma, A_1 \not\rightsquigarrow B\\ \Gamma, A_1 \not\rightsquigarrow B|A_2 \not\rightsquigarrow B \rightrightarrows \Gamma, A_1 \not\rightsquigarrow B, A_2 \not\rightsquigarrow B\end{array}\right.}{\Gamma|A_1 \wedge A_2 \not\rightsquigarrow B \rightrightarrows \Gamma, A_1 \wedge A_2 \not\rightsquigarrow B}$$

$$(-\wedge_{0}^{L})\ \frac{\left\{\begin{array}{l}\Gamma|A_1 \not\rightsquigarrow B \rightrightarrows \Gamma\\ \Gamma, A_1 \not\rightsquigarrow B|A_2 \not\rightsquigarrow B \rightrightarrows \Gamma, A_1 \not\rightsquigarrow B\end{array}\right.}{\Gamma|A_1 \wedge A_2 \not\rightsquigarrow B \rightrightarrows \Gamma, A_1 \wedge A_2 \not\rightsquigarrow B}$$

$$(-\wedge_{-}^{R})\ \frac{\left\{\begin{array}{l}\Gamma|A \not\rightsquigarrow B_1 \rightrightarrows \Gamma, A \not\rightsquigarrow B_1\\ \Gamma|A \not\rightsquigarrow B_2 \rightrightarrows \Gamma, A \not\rightsquigarrow B_2\end{array}\right.}{\Gamma|A \not\rightsquigarrow B_1 \wedge B_2 \rightrightarrows \Gamma, A \not\rightsquigarrow B_1 \wedge B_2}$$

$$(-\wedge_{0}^{R})\ \frac{\left[\begin{array}{l}\Gamma|A \not\rightsquigarrow B_1 \rightrightarrows \Gamma\\ \Gamma|A \not\rightsquigarrow B_2 \rightrightarrows \Gamma\end{array}\right.}{\Gamma|A \not\rightsquigarrow B_1 \wedge B_2 \rightrightarrows \Gamma}$$

and

$$(+\vee_{-}^{L})\ \frac{\left[\begin{array}{l}\Gamma|A_1 \rightsquigarrow B \rightrightarrows \Gamma, A_1 \rightsquigarrow B\\ \Gamma, A_1 \rightsquigarrow B|A_2 \rightsquigarrow B \rightrightarrows \Gamma, A_1 \rightsquigarrow B, A_2 \rightsquigarrow B\end{array}\right.}{\Gamma|A_1 \vee A_2 \rightsquigarrow B \rightrightarrows \Gamma, A_1 \vee A_2 \rightsquigarrow B}$$

$$(+\vee_{0}^{L})\ \frac{\left\{\begin{array}{l}\Gamma|A_1 \rightsquigarrow B \rightrightarrows \Gamma\\ \Gamma, A_1 \rightsquigarrow B|A_2 \rightsquigarrow B \rightrightarrows \Gamma, A_1 \rightsquigarrow B\end{array}\right.}{\Gamma|A_1 \vee A_2 \rightsquigarrow B \rightrightarrows \Gamma}$$

$$(+\vee_{-}^{R})\ \frac{\left\{\begin{array}{l}\Gamma|A \rightsquigarrow B_1 \rightrightarrows \Gamma, A \rightsquigarrow B_1\\ \Gamma|A \rightsquigarrow B_2 \rightrightarrows \Gamma, A \rightsquigarrow B_2\end{array}\right.}{\Gamma|A \rightsquigarrow B_1 \vee B_2 \rightrightarrows \Gamma, A \rightsquigarrow B_1 \vee B_2}$$

$$(+\vee_{0}^{R})\ \frac{\left[\begin{array}{l}\Gamma|A \rightsquigarrow B_1 \rightrightarrows \Gamma\\ \Gamma|A \rightsquigarrow B_2 \rightrightarrows \Gamma\end{array}\right.}{\Gamma|A \rightsquigarrow B_1 \vee B_2 \rightrightarrows \Gamma}$$

and

$$(-\vee^L_-)\ \frac{\begin{cases}\Gamma|A_1 \not\rightarrowtail B \rightrightarrows \Gamma, A_1 \not\rightarrowtail B\\ \Gamma|A_2 \not\rightarrowtail B \rightrightarrows \Gamma, A_2 \not\rightarrowtail B\end{cases}}{\Gamma|A_1 \vee A_2 \not\rightarrowtail B \rightrightarrows \Gamma, A_1 \vee A_2 \not\rightarrowtail B}$$

$$(-\vee^L_0)\ \frac{\left[\begin{array}{l}\Gamma|A_1 \not\rightarrowtail B \rightrightarrows \Gamma\\ \Gamma|A_2 \not\rightarrowtail B \rightrightarrows \Gamma\end{array}\right.}{\Gamma|A_1 \vee A_2 \not\rightarrowtail B \rightrightarrows \Gamma}$$

$$(-\vee^R_-)\ \frac{\left[\begin{array}{l}\Gamma|A \not\rightarrowtail B_1 \rightrightarrows \Gamma, A \not\rightarrowtail B_1\\ \Gamma, A \not\rightarrowtail B_1|A \not\rightarrowtail B_2 \rightrightarrows \Gamma, A \not\rightarrowtail B_1, A \not\rightarrowtail B_2\end{array}\right.}{\Gamma|A \not\rightarrowtail B_1 \vee B_2 \rightrightarrows \Gamma, A \not\rightarrowtail B_1 \vee B_2}$$

$$(-\vee^R_0)\ \frac{\begin{cases}\Gamma|A \not\rightarrowtail B_1 \rightrightarrows \Gamma\\ \Gamma, A \not\rightarrowtail B_1|A \not\rightarrowtail B_2 \rightrightarrows \Gamma, A \not\rightarrowtail B_1\end{cases}}{\Gamma|A \not\rightarrowtail B_1 \vee B_2 \rightrightarrows \Gamma}$$

and

$$(+\neg\wedge^L_-)\ \frac{\left[\begin{array}{l}\Gamma|\neg A_1 \leadsto B \rightrightarrows \Gamma, \neg A_1 \leadsto B\\ \Gamma, \neg A_1 \leadsto B|\neg A_2 \leadsto B \rightrightarrows \Gamma, \neg A_1 \leadsto B, \neg A_2 \leadsto B\end{array}\right.}{\Gamma|\neg(A_1 \wedge A_2) \leadsto B \rightrightarrows \Gamma, \neg(A_1 \wedge A_2) \leadsto B}$$

$$(+\neg\wedge^L_0)\ \frac{\begin{cases}\Gamma|\neg A_1 \leadsto B \rightrightarrows \Gamma\\ \Gamma, \neg A_1 \leadsto B|\neg A_2 \leadsto B \rightrightarrows \Gamma, \neg A_1 \leadsto B\end{cases}}{\Gamma|\neg(A_1 \wedge A_2) \leadsto B \rightrightarrows \Gamma}$$

$$(+\neg\wedge^R_-)\ \frac{\begin{cases}\Gamma|A \leadsto \neg B_1 \rightrightarrows \Gamma, A \leadsto \neg B_1\\ \Gamma|A \leadsto \neg B_2 \rightrightarrows \Gamma, A \leadsto \neg B_2\end{cases}}{\Gamma|A \leadsto \neg(B_1 \wedge B_2) \rightrightarrows \Gamma, A \leadsto \neg(B_1 \wedge B_2)}$$

$$(+\neg\wedge^R_0)\ \frac{\left[\begin{array}{l}\Gamma|A \leadsto \neg B_1 \rightrightarrows \Gamma\\ \Gamma|A \leadsto \neg B_2 \rightrightarrows \Gamma\end{array}\right.}{\Gamma|A \leadsto \neg(B_1 \wedge B_2) \rightrightarrows \Gamma}$$

and

$$(-\neg\wedge^L_-)\ \frac{\begin{cases}\Gamma|\neg A_1 \not\rightarrowtail B \rightrightarrows \Gamma, \neg A_1 \not\rightarrowtail B\\ \Gamma|\neg A_2 \not\rightarrowtail B \rightrightarrows \Gamma, \neg A_2 \not\rightarrowtail B\end{cases}}{\Gamma|\neg(A_1 \wedge A_2) \not\rightarrowtail B \rightrightarrows \Gamma, \neg(A_1 \wedge A_2) \not\rightarrowtail B}$$

$$(-\neg\wedge^L_0)\ \frac{\left[\begin{array}{l}\Gamma|\neg A_1 \not\rightarrowtail B \rightrightarrows \Gamma\\ \Gamma|\neg A_2 \not\rightarrowtail B \rightrightarrows \Gamma\end{array}\right.}{\Gamma|\neg(A_1 \wedge A_2) \not\rightarrowtail B \rightrightarrows \Gamma}$$

$$(-\neg\wedge^R_-)\ \frac{\left[\begin{array}{l}\Gamma|A \not\rightarrowtail \neg B_1 \rightrightarrows \Gamma, A \not\rightarrowtail \neg B_1\\ \Gamma, A \not\rightarrowtail \neg B_1|A \not\rightarrowtail \neg B_2 \rightrightarrows \Gamma, A \not\rightarrowtail \neg B_1, A \not\rightarrowtail \neg B_2\end{array}\right.}{\Gamma|A \not\rightarrowtail \neg(B_1 \wedge B_2) \rightrightarrows \Gamma, A \not\rightarrowtail \neg(B_1 \wedge B_2)}$$

$$(-\neg\wedge^R_0)\ \frac{\begin{cases}\Gamma|A \not\rightarrowtail \neg B_1 \rightrightarrows \Gamma\\ \Gamma, A \not\rightarrowtail \neg B_1|A \not\rightarrowtail \neg B_2 \rightrightarrows \Gamma, A \not\rightarrowtail \neg B_1\end{cases}}{\Gamma|A \not\rightarrowtail \neg(B_1 \wedge B_2) \rightrightarrows \Gamma}$$

and

$$(+\neg\vee^L_-)\ \frac{\left\{\begin{array}{l}\Gamma|\neg A_1 \leadsto B \rightrightarrows \Gamma, \neg A_1 \leadsto B\\ \Gamma|\neg A_2 \leadsto B \rightrightarrows \Gamma, \neg A_2 \leadsto B\end{array}\right.}{\Gamma|\neg(A_1 \vee A_2) \leadsto B \rightrightarrows \Gamma, \neg(A_1 \vee A_2) \leadsto B}$$

$$(+\neg\vee^L_0)\ \frac{\left[\begin{array}{l}\Gamma|\neg A_1 \leadsto B \rightrightarrows \Gamma\\ \Gamma|\neg A_2 \leadsto B \rightrightarrows \Gamma\end{array}\right.}{\Gamma|\neg(A_1 \vee A_2) \leadsto B \rightrightarrows \Gamma}$$

$$(+\neg\vee^R_-)\ \frac{\left[\begin{array}{l}\Gamma|A \leadsto \neg B_1 \rightrightarrows \Gamma, A \leadsto \neg B_1\\ \Gamma, A \leadsto \neg B_1|A \leadsto \neg B_2 \rightrightarrows \Gamma, A \leadsto \neg B_1, A \leadsto \neg B_2\end{array}\right.}{\Gamma|A \leadsto \neg(B_1 \vee B_2) \rightrightarrows \Gamma, A \leadsto \neg(B_1 \vee B_2)}$$

$$(+\neg\vee^R_0)\ \frac{\left[\begin{array}{l}\Gamma|A \leadsto \neg B_1 \rightrightarrows \Gamma\\ \Gamma, A \leadsto \neg B_1|A \leadsto \neg B_2 \rightrightarrows \Gamma, A \leadsto \neg B_1\end{array}\right.}{\Gamma|A \leadsto \neg(B_1 \vee B_2) \rightrightarrows \Gamma}$$

and

$$(-\neg\vee^L_-)\ \frac{\left[\begin{array}{l}\Gamma|\neg A_1 \not\leadsto B \rightrightarrows \Gamma, \neg A_1 \not\leadsto B\\ \Gamma, \neg A_1 \not\leadsto B|\neg A_2 \not\leadsto B \rightrightarrows \Gamma, \neg A_1 \not\leadsto B, \neg A_2 \not\leadsto B\end{array}\right.}{\Gamma|\neg(A_1 \vee A_2) \not\leadsto B \rightrightarrows \Gamma, \neg(A_1 \vee A_2) \not\leadsto B}$$

$$(-\neg\vee^L_0)\ \frac{\left\{\begin{array}{l}\Gamma|\neg A_1 \not\leadsto B \rightrightarrows \Gamma\\ \Gamma, \neg A_1 \not\leadsto B|\neg A_2 \not\leadsto B \rightrightarrows \Gamma, \neg A_1 \not\leadsto B\end{array}\right.}{\Gamma|\neg(A_1 \vee A_2) \not\leadsto B \rightrightarrows \Gamma}$$

$$(-\neg\vee^R_-)\ \frac{\left\{\begin{array}{l}\Gamma|A \not\leadsto \neg B_1 \rightrightarrows \Gamma, A \not\leadsto \neg B_1\\ \Gamma|A \not\leadsto \neg B_2 \rightrightarrows \Gamma, A \not\leadsto \neg B_2\end{array}\right.}{\Gamma|A \not\leadsto \neg(B_1 \vee B_2) \rightrightarrows \Gamma, A \not\leadsto \neg(B_1 \vee B_2)}$$

$$(-\neg\vee^R_0)\ \frac{\left[\begin{array}{l}\Gamma|A \not\leadsto \neg B_1 \rightrightarrows \Gamma\\ \Gamma|A \not\leadsto \neg B_2 \rightrightarrows \Gamma\end{array}\right.}{\Gamma|A \not\leadsto \neg(B_1 \vee B_2) \rightrightarrows \Gamma}$$

Definition 8.4.3 A reduction $\theta = \Gamma|\gamma \Rightarrow \Gamma'$ is provable in $\mathbf{R}_{\mathtt{t}}^{\leadsto}$, denoted by $\vdash_{\mathtt{t}}^{\leadsto} \theta$, if there is a sequence $\{\theta_1, \ldots, \theta_n\}$ of reductions such that $\theta_n = \theta$, and for each $1 \le i \le n$, θ_i is an axiom or is deduced from the previous reductions by one deduction rule in $\mathbf{R}_{\mathtt{t}}^{\leadsto}$.

Theorem 8.4.4 (Soundness and completeness theorem) *For any reduction* $\theta = \Gamma|\gamma \Rightarrow \Gamma'$,

$$\vdash_{\mathtt{t}}^{\leadsto} \theta \textit{ iff } \models_{\mathtt{t}}^{\leadsto} \theta.$$

□

8.4.2 R-Calculus $\mathbf{Q}_{\mathtt{t}}^{\leadsto}$

Given Γ and a statement γ, we use Γ to revise γ and obtain Γ', denoted by $\models_{\mathtt{t}}^{\leadsto} \Gamma|\gamma \rightrightarrows \Gamma'$, if

$$\Gamma' = \begin{cases} \Gamma, \gamma & \text{if } \models_{\mathtt{t}}^{\leadsto} \Gamma, \gamma \mapsto \\ \Gamma & \text{otherwise.} \end{cases}$$

R-calculus $\mathbf{Q}_{\mathtt{t}}^{\leadsto}$ consists of the following axioms and deduction rules:

- **Axioms**:

$$(\mathtt{A}_{+\mathtt{t}}^{\leadsto})\ \frac{\Gamma \nvdash \sim \gamma}{\Gamma|\gamma \rightrightarrows \Gamma, \gamma} \quad (\mathtt{A}_{0\mathtt{t}}^{\leadsto})\ \frac{\Gamma \vdash \sim \gamma}{\Gamma|\gamma \rightrightarrows \Gamma}$$

where Γ, γ is a set of literal statements.

- **Deduction rules**:

$$(+\wedge^L)\ \frac{\begin{cases} \begin{bmatrix} \Gamma|A_1 \leadsto B \rightrightarrows \Gamma, C_1 \leadsto B \\ C_1 \neq \lambda \end{bmatrix} \\ \begin{bmatrix} \Gamma|A_1 \leadsto B \rightrightarrows \Gamma \\ \Gamma|A_2 \leadsto B \rightrightarrows \Gamma, C_2 \leadsto B \\ C_2 \neq \lambda \end{bmatrix} \\ \begin{bmatrix} \Gamma|A_1 \leadsto B \rightrightarrows \Gamma \\ \Gamma|A_2 \leadsto B \rightrightarrows \Gamma \end{bmatrix} \end{cases}}{\Gamma|A_1 \leadsto B \rightrightarrows \Gamma, \gamma}$$

$$(+\wedge^R)\ \frac{\Gamma|A \leadsto B_1 \rightrightarrows \Gamma, A \leadsto D_1 \quad \Gamma, A \leadsto D_1|A \leadsto B_2 \rightrightarrows \Gamma, A \leadsto D_1, A \leadsto D_2}{\Gamma|A \leadsto B_1 \wedge B_2 \rightrightarrows \Gamma, A \leadsto D_1 \wedge D_2}$$

$$\text{where } \gamma = \begin{cases} C_1 \wedge A_2 \leadsto B & \text{if } C_1 \neq \lambda \\ A_1 \wedge C_2 \leadsto B & \text{if } C_1 = \lambda \neq C_2 \text{ and} \\ \lambda & \text{otherwise;} \end{cases}$$

$$(-\wedge^L)\ \frac{\Gamma|A_1 \not\leadsto B \rightrightarrows \Gamma, C_1 \not\leadsto B \quad \Gamma, C_1 \not\leadsto B|A_2 \not\leadsto B \rightrightarrows \Gamma, C_1 \not\leadsto B, C_2 \not\leadsto B}{\Gamma|A_1 \wedge A_2 \not\leadsto B \rightrightarrows \Gamma, C_1 \wedge C_2 \not\leadsto B\}}$$

$$(-\wedge^R)\ \frac{\begin{cases} \begin{bmatrix} \Gamma|A \not\leadsto B_1 \rightrightarrows \Gamma, A \not\leadsto D_1 \\ D_1 \neq \lambda \end{bmatrix} \\ \begin{bmatrix} \Gamma|A \not\leadsto B_1 \rightrightarrows \Gamma \\ \Gamma|A \not\leadsto B_2 \rightrightarrows \Gamma, A \not\leadsto D_2 \\ D_2 \neq \lambda \end{bmatrix} \\ \begin{bmatrix} \Gamma|A \not\leadsto B_1 \rightrightarrows \Gamma \\ \Gamma|A \not\leadsto B_2 \rightrightarrows \Gamma \end{bmatrix} \end{cases}}{\Gamma|A \not\leadsto B_1 \wedge B_2 \rightrightarrows \Gamma, \gamma'}$$

$$\text{where } \gamma' = \begin{cases} A \not\leadsto D_1 \wedge B_2 & \text{if } D_1 \neq \lambda \\ A \not\leadsto B_1 \wedge D_2 & \text{if } D_1 = \lambda \neq D_2 \text{ and} \\ \lambda & \text{otherwise;} \end{cases}$$

$$(+\vee^{L})\ \frac{\begin{array}{l}\Gamma|A_1 \rightsquigarrow B \rightrightarrows \Gamma, C_1 \rightsquigarrow B\\ \Gamma, C_1 \rightsquigarrow B|A_2 \rightsquigarrow B \rightrightarrows \Gamma, C_1 \rightsquigarrow B, C_2 \rightsquigarrow B\end{array}}{\Gamma|A_1 \vee A_2 \rightsquigarrow B \rightrightarrows \Gamma, C_1 \vee C_2 \rightsquigarrow B\}}$$

$$(+\vee^{R})\ \frac{\left\{\begin{array}{l}\left[\begin{array}{l}\Gamma|A \rightsquigarrow B_1 \rightrightarrows \Gamma, A \rightsquigarrow D_1\\ D_1 \neq \lambda\end{array}\right.\\ \left[\begin{array}{l}\Gamma|A \rightsquigarrow B_1 \rightrightarrows \Gamma\\ \Gamma|A \rightsquigarrow B_2 \rightrightarrows \Gamma, A \rightsquigarrow D_2\\ D_2 \neq \lambda\end{array}\right.\\ \left[\begin{array}{l}\Gamma|A \rightsquigarrow B_1 \rightrightarrows \Gamma\\ \Gamma|A \rightsquigarrow B_2 \rightrightarrows \Gamma\end{array}\right.\end{array}\right.}{\Gamma|A \rightsquigarrow B_1 \vee B_2 \rightrightarrows \Gamma, \gamma''}$$

where $\gamma'' = \begin{cases} A \rightsquigarrow D_1 \vee B_2 \text{ if } D_1 \neq \lambda \\ A \rightsquigarrow B_1 \vee D_2 \text{ if } D_1 = \lambda \neq D_2 \text{ and} \\ \lambda \qquad\qquad \text{otherwise;} \end{cases}$

$$(-\vee^{L})\ \frac{\left\{\begin{array}{l}\left[\begin{array}{l}\Gamma|A_1 \not\rightsquigarrow B \rightrightarrows \Gamma, C_1 \not\rightsquigarrow B\\ C_1 \neq \lambda\end{array}\right.\\ \left[\begin{array}{l}\Gamma|A_1 \not\rightsquigarrow B \rightrightarrows \Gamma\\ \Gamma|A_2 \not\rightsquigarrow B \rightrightarrows \Gamma, C_2 \not\rightsquigarrow B\\ C_2 \neq \lambda\end{array}\right.\\ \left[\begin{array}{l}\Gamma|A_1 \not\rightsquigarrow B \rightrightarrows \Gamma\\ \Gamma|A_2 \not\rightsquigarrow B \rightrightarrows \Gamma\end{array}\right.\end{array}\right.}{\Gamma|A_1 \vee A_2 \not\rightsquigarrow B \rightrightarrows \Gamma, \gamma'''}$$

$$(-\vee^{R})\ \frac{\begin{array}{l}\Gamma|A \not\rightsquigarrow B_1 \rightrightarrows \Gamma, A \not\rightsquigarrow D_1\\ \Gamma, A \not\rightsquigarrow D_1|A \not\rightsquigarrow B_2 \rightrightarrows \Gamma, A \not\rightsquigarrow D_1, A \not\rightsquigarrow D_2\end{array}}{\Gamma|A \not\rightsquigarrow B_1 \vee B_2 \rightrightarrows \Gamma, A \not\rightsquigarrow D_1 \vee D_2\}}$$

where $\gamma''' = \begin{cases} C_1 \vee A_2 \not\rightsquigarrow B \text{ if } C_1 \neq \lambda \\ A_1 \vee C_2 \not\rightsquigarrow B \text{ if } C_1 = \lambda \neq C_2 \text{ and} \\ \lambda \qquad\qquad \text{otherwise;} \end{cases}$

$$(+\neg\wedge^{L})\ \frac{\begin{array}{l}\Gamma|\neg A_1 \rightsquigarrow B \rightrightarrows \Gamma, \neg C_1 \rightsquigarrow B\\ \Gamma, \neg C_1 \rightsquigarrow B|\neg A_2 \rightsquigarrow B \rightrightarrows \Gamma, \neg C_1 \rightsquigarrow B, \neg C_2 \rightsquigarrow B\end{array}}{\Gamma|\neg(A_1 \wedge A_2) \rightsquigarrow B \rightrightarrows \Gamma, \neg(C_1 \wedge C_2) \rightsquigarrow B\}}$$

$$(+\neg\wedge^{R})\ \frac{\left\{\begin{array}{l}\left[\begin{array}{l}\Gamma|A \rightsquigarrow \neg B_1 \rightrightarrows \Gamma, A \rightsquigarrow \neg D_1\\ D_1 \neq \lambda\end{array}\right.\\ \left[\begin{array}{l}\Gamma|A \rightsquigarrow \neg B_1 \rightrightarrows \Gamma\\ \Gamma|A \rightsquigarrow \neg B_2 \rightrightarrows \Gamma, A \rightsquigarrow \neg D_2\\ D_2 \neq \lambda\end{array}\right.\\ \left[\begin{array}{l}\Gamma|A \rightsquigarrow \neg B_1 \rightrightarrows \Gamma\\ \Gamma|A \rightsquigarrow \neg B_2 \rightrightarrows \Gamma\end{array}\right.\end{array}\right.}{\Gamma|A \rightsquigarrow \neg(B_1 \wedge B_2) \rightrightarrows \Gamma, \gamma^{(4)}}$$

$$\text{where } \gamma^{(4)} = \begin{cases} A \rightsquigarrow \neg(D_1 \vee B_2) \text{ if } D_1 \neq \lambda \\ A \rightsquigarrow \neg(B_1 \vee D_2) \text{ if } D_1 = \lambda \neq D_2 \text{ and} \\ \lambda \quad \text{otherwise;} \end{cases}$$

$$(-\neg\wedge^L)\ \frac{\begin{cases} \begin{bmatrix} \Gamma|\neg A_1 \not\rightsquigarrow B \rightrightarrows \Gamma, \neg C_1 \not\rightsquigarrow B \\ C_1 \neq \lambda \end{bmatrix} \\ \begin{bmatrix} \Gamma|\neg A_1 \not\rightsquigarrow B \rightrightarrows \Gamma \\ \Gamma|\neg A_2 \not\rightsquigarrow B \rightrightarrows \Gamma, \neg C_2 \not\rightsquigarrow B \\ C_2 \neq \lambda \end{bmatrix} \\ \begin{bmatrix} \Gamma|\neg A_1 \not\rightsquigarrow B \rightrightarrows \Gamma \\ \Gamma|\neg A_2 \not\rightsquigarrow B \rightrightarrows \Gamma \end{bmatrix} \end{cases}}{\Gamma|\neg(A_1 \wedge A_2) \not\rightsquigarrow B \rightrightarrows \Gamma, \gamma^{(5)}}$$

$$(-\neg\wedge^R)\ \frac{\begin{array}{l} \Gamma|A \not\rightsquigarrow \neg B_1 \rightrightarrows \Gamma, A \not\rightsquigarrow \neg D_1 \\ \Gamma, A \not\rightsquigarrow \neg D_1|A \not\rightsquigarrow \neg B_2 \rightrightarrows \Gamma, A \not\rightsquigarrow \neg D_2 \end{array}}{\Gamma|A \not\rightsquigarrow \neg(B_1 \wedge B_2) \rightrightarrows \Gamma, A \not\rightsquigarrow \neg(D_1 \wedge D_2)\}}$$

$$\text{where } \gamma^{(5)} = \begin{cases} \neg(C_1 \wedge A_2) \not\rightsquigarrow B \text{ if } C_1 \neq \lambda \\ \neg(A_1 \wedge C_2) \not\rightsquigarrow B \text{ if } C_1 = \lambda \neq C_2 \text{ and} \\ \lambda \quad \text{otherwise;} \end{cases}$$

$$(+\neg\vee^L)\ \frac{\begin{cases} \begin{bmatrix} \Gamma|\neg A_1 \rightsquigarrow B \rightrightarrows \Gamma, \neg C_1 \rightsquigarrow B \\ C_1 \neq \lambda \end{bmatrix} \\ \begin{bmatrix} \Gamma|\neg A_1 \rightsquigarrow B \rightrightarrows \Gamma \\ \Gamma|\neg A_2 \rightsquigarrow B \rightrightarrows \Gamma, \neg C_2 \rightsquigarrow B \\ C_2 \neq \lambda \end{bmatrix} \\ \begin{bmatrix} \Gamma|\neg A_1 \rightsquigarrow B \rightrightarrows \Gamma \\ \Gamma|\neg A_2 \rightsquigarrow B \rightrightarrows \Gamma \end{bmatrix} \end{cases}}{\Gamma|\neg(A_1 \vee A_2) \rightsquigarrow B \rightrightarrows \Gamma, \gamma^{(6)}}$$

$$(+\neg\vee^R)\ \frac{\begin{array}{l} \Gamma|A \rightsquigarrow \neg B_1 \rightrightarrows \Gamma, A \rightsquigarrow \neg D_1 \\ \Gamma, A \rightsquigarrow \neg D_1|A \rightsquigarrow \neg B_2 \rightrightarrows \Gamma, A \rightsquigarrow \neg D_1, A \rightsquigarrow \neg D_2 \end{array}}{\Gamma|A \rightsquigarrow \neg(B_1 \vee B_2) \rightrightarrows \Gamma, A \rightsquigarrow \neg(D_1 \vee D_2)\}}$$

$$\text{where } \gamma^{(6)} = \begin{cases} \neg(C_1 \vee A_2) \rightsquigarrow B \text{ if } C_1 \neq \lambda \\ \neg(A_1 \vee C_2) \rightsquigarrow B \text{ if } C_1 = \lambda \neq C_2 \text{ and} \\ \lambda \quad \text{otherwise;} \end{cases}$$

$$(-\neg\vee^L)\ \frac{\begin{array}{l}\Gamma|\neg A_1 \not\mapsto B \rightrightarrows \Gamma, \neg C_1 \not\mapsto B\\ \Gamma, \neg C_1 \not\mapsto B|\neg A_2 \not\mapsto B \rightrightarrows \Gamma, \neg C_1 \not\mapsto B, \neg C_2 \not\mapsto B\end{array}}{\Gamma|\neg(A_1 \vee A_2) \not\mapsto B \rightrightarrows \Gamma, \neg(C_1 \vee C_2) \not\mapsto B\}}$$

$$(-\vee^R)\ \frac{\left\{\begin{array}{l}\left[\begin{array}{l}\Gamma|A \not\mapsto \neg B_1 \rightrightarrows \Gamma, A \not\mapsto \neg D_1\\ D_1 \neq \lambda\end{array}\right.\\ \left[\begin{array}{l}\Gamma|A \not\mapsto \neg B_1 \rightrightarrows \Gamma\\ \Gamma|A \not\mapsto \neg B_2 \rightrightarrows \Gamma, A \not\mapsto \neg D_2\\ D_2 \neq \lambda\end{array}\right.\\ \left[\begin{array}{l}\Gamma|A \not\mapsto \neg B_1 \rightrightarrows \Gamma\\ \Gamma|A \not\mapsto \neg B_2 \rightrightarrows \Gamma\end{array}\right.\end{array}\right.}{\Gamma|A \not\mapsto \neg(B_1 \vee B_2) \rightrightarrows \Gamma, \gamma^{(7)}}$$

where $\gamma^{(7)} = \begin{cases} A \not\mapsto \neg(D_1 \vee B_2) & \text{if } D_1 \neq \lambda \\ A \not\mapsto \neg(B_1 \vee D_2) & \text{if } D_1 = \lambda \neq D_2 \\ \lambda & \text{otherwise.} \end{cases}$

Definition 8.4.5 A reduction $\theta = \Gamma|\gamma \Rightarrow \Gamma'$ is provable in $\mathbf{Q}_{\mathtt{t}}^{\rightsquigarrow}$, denoted by $\vdash_{\mathtt{t}}^{\rightsquigarrow} \theta$, if there is a sequence $\{\theta_1, \ldots, \theta_n\}$ of reductions such that $\theta_n = \theta$, and for each $1 \leq i \leq n, \theta_i$ is an axiom or is deduced from the previous reductions by one deduction rule in $\mathbf{Q}_{\mathtt{t}}^{\rightsquigarrow}$.

Theorem 8.4.6 (Soundness and completeness theorem) *For any reduction* $\theta = \Gamma|\gamma \Rightarrow \Gamma'$,

$$\vdash_{\mathtt{t}}^{\rightsquigarrow} \theta \text{ iff } \models_{\mathtt{t}}^{\rightsquigarrow} \theta.$$

□

8.4.3 *R-Calculus* $\mathbf{P}_{\mathtt{t}}^{\rightsquigarrow}$

A statement $A \rightsquigarrow B$ is in conjunctive normal form if A is in disjunctive normal form and B is in conjunctive normal form; and $A \rightsquigarrow B$ is in disjunctive normal form if A is in conjunctive normal form and B is in disjunctive normal form.

Given Γ and a statement γ, we use Γ to revise γ and obtain Γ, γ', denoted by $\models_{\mathtt{t}}^{\rightsquigarrow} \Gamma|\gamma \rightrightarrows \Gamma, \gamma'$, if

$$\gamma' = \begin{cases} \gamma & \text{if } \models_{\mathtt{t}}^{\rightsquigarrow} \Gamma, \gamma' \mapsto \\ \lambda & \text{otherwise.} \end{cases}$$

R-calculus $\mathbf{P}_{\mathtt{t}}^{\rightsquigarrow}$ consists of the following axioms and deduction rules:

- **Axioms**:

$$(\mathtt{A}_{+\mathtt{t}}^{\rightsquigarrow})\ \frac{\neg\gamma \notin \Gamma \& \Gamma \nvdash \neg\gamma}{\Gamma|\gamma \rightrightarrows \Gamma, \gamma} \quad (\mathtt{A}_{0\mathtt{t}}^{\rightsquigarrow})\ \frac{\neg\gamma \in \Gamma \text{ or } \Gamma \vdash \neg\gamma}{\Gamma|\gamma \rightrightarrows \Gamma}$$

where Γ, γ is a set of literal statements.

- **Deduction rules**:

$$(+\wedge^L)\ \frac{\begin{array}{l}\Gamma|A_1 \rightsquigarrow B \rightrightarrows \Gamma, C_1 \rightsquigarrow B\\ \Gamma|A_2 \rightsquigarrow B \rightrightarrows \Gamma, C_2 \rightsquigarrow B\end{array}}{\Gamma|A_1 \wedge A_2 \rightsquigarrow B \rightrightarrows \Gamma, C_1 \wedge C_2 \rightsquigarrow B}$$

$$(+\wedge^R)\ \frac{\begin{array}{l}\Gamma|A \rightsquigarrow B_1 \rightrightarrows \Gamma, A \rightsquigarrow D_1\\ \Gamma, A \rightsquigarrow D_1|A \rightsquigarrow B_2 \rightrightarrows \Gamma, A \rightsquigarrow D_1, A \rightsquigarrow D_2\end{array}}{\Gamma|A \rightsquigarrow B_1 \wedge B_2 \rightrightarrows \Gamma, A \rightsquigarrow D_1 \wedge D_2}$$

$$(-\wedge^L)\ \frac{\begin{array}{l}\Gamma|A_1 \not\rightsquigarrow B \rightrightarrows \Gamma, C_1 \not\rightsquigarrow B\\ \Gamma, C_1 \not\rightsquigarrow B|A_2 \not\rightsquigarrow B \rightrightarrows \Gamma, C_1 \not\rightsquigarrow B, C_2 \not\rightsquigarrow B\end{array}}{\Gamma|A_1 \wedge A_2 \not\rightsquigarrow B \rightrightarrows \Gamma, C_1 \wedge C_2 \not\rightsquigarrow B\}}$$

$$(-\wedge^R)\ \frac{\begin{array}{l}\Gamma|A \not\rightsquigarrow B_1 \rightrightarrows \Gamma, A \not\rightsquigarrow D_1\\ \Gamma|A \not\rightsquigarrow B_2 \rightrightarrows \Gamma, A \not\rightsquigarrow D_2\end{array}}{\Gamma|A \not\rightsquigarrow B_1 \wedge B_2 \rightrightarrows \Gamma, A \not\rightsquigarrow D_1 \wedge D_2\}}$$

and

$$(+\vee^L)\ \frac{\begin{array}{l}\Gamma|A_1 \rightsquigarrow B \rightrightarrows \Gamma, C_1 \rightsquigarrow B\\ \Gamma, C_1 \rightsquigarrow B|A_2 \rightsquigarrow B \rightrightarrows \Gamma, C_1 \rightsquigarrow B, C_2 \rightsquigarrow B\end{array}}{\Gamma|A_1 \vee A_2 \rightsquigarrow B \rightrightarrows \Gamma, C_1 \vee C_2 \rightsquigarrow B\}}$$

$$(+\vee^R)\ \frac{\begin{array}{l}\Gamma|A \rightsquigarrow B_1 \rightrightarrows \Gamma, A \rightsquigarrow D_1\\ \Gamma|A \rightsquigarrow B_2 \rightrightarrows \Gamma, A \rightsquigarrow D_2\end{array}}{\Gamma|A \rightsquigarrow B_1 \vee B_2 \rightrightarrows \Gamma, A \rightsquigarrow B_1 \vee D_2\}}$$

$$(-\vee^L)\ \frac{\begin{array}{l}\Gamma|A_1 \not\rightsquigarrow B \rightrightarrows \Gamma, C_1 \not\rightsquigarrow B\\ \Gamma|A_2 \not\rightsquigarrow B \rightrightarrows \Gamma, C_2 \not\rightsquigarrow B\end{array}}{\Gamma|A_1 \vee A_2 \not\rightsquigarrow B \rightrightarrows \Gamma, C_1 \vee C_2 \not\rightsquigarrow B\}}$$

$$(-\vee^R)\ \frac{\begin{array}{l}\Gamma|A \not\rightsquigarrow B_1 \rightrightarrows \Gamma, A \not\rightsquigarrow D_1\\ \Gamma, A \not\rightsquigarrow D_1|A \not\rightsquigarrow B_2 \rightrightarrows \Gamma, A \not\rightsquigarrow D_1, A \not\rightsquigarrow D_2\end{array}}{\Gamma|A \not\rightsquigarrow B_1 \vee B_2 \rightrightarrows \Gamma, A \not\rightsquigarrow D_1 \vee D_2\}}$$

and

$$(+\neg\wedge^L)\ \frac{\begin{array}{l}\Gamma|\neg A_1 \rightsquigarrow B \rightrightarrows \Gamma, \neg C_1 \rightsquigarrow B\\ \Gamma, \neg C_1 \rightsquigarrow B|\neg A_2 \rightsquigarrow B \rightrightarrows \Gamma, \neg C_1 \rightsquigarrow B, \neg C_2 \rightsquigarrow B\end{array}}{\Gamma|\neg(A_1 \wedge A_2) \rightsquigarrow B \rightrightarrows \Gamma, \neg(C_1 \wedge C_2) \rightsquigarrow B\}}$$

$$(+\neg\wedge^R)\ \frac{\begin{array}{l}\Gamma|A \rightsquigarrow \neg B_1 \rightrightarrows \Gamma, A \rightsquigarrow \neg D_1\\ \Gamma|A \rightsquigarrow \neg B_2 \rightrightarrows \Gamma, A \rightsquigarrow \neg D_2\end{array}}{\Gamma|A \rightsquigarrow \neg(B_1 \wedge B_2) \rightrightarrows \Gamma, A \rightsquigarrow \neg(D_1 \wedge D_2)\}}$$

$$(-\neg\wedge^L)\ \frac{\begin{array}{l}\Gamma|\neg A_1 \not\rightsquigarrow B \rightrightarrows \Gamma, \neg C_1 \not\rightsquigarrow B\\ \Gamma|\neg A_2 \not\rightsquigarrow B \rightrightarrows \Gamma, \neg C_2 \not\rightsquigarrow B\end{array}}{\Gamma|\neg(A_1 \wedge A_2) \not\rightsquigarrow B \rightrightarrows \Gamma, \neg(C_1 \wedge C_2) \not\rightsquigarrow B\}}$$

$$(-\neg\wedge^R)\ \frac{\begin{array}{l}\Gamma|A \not\rightsquigarrow \neg B_1 \rightrightarrows \Gamma, A \not\rightsquigarrow \neg D_1\\ \Gamma, A \not\rightsquigarrow \neg D_1|A \not\rightsquigarrow \neg B_2 \rightrightarrows \Gamma, A \not\rightsquigarrow \neg D_2\end{array}}{\Gamma|A \not\rightsquigarrow \neg(B_1 \wedge B_2) \rightrightarrows \Gamma, A \not\rightsquigarrow \neg(D_1 \wedge D_2)\}}$$

and

$$(+\neg\vee^L)\ \frac{\Gamma|\neg A_1 \rightsquigarrow B \rightrightarrows \Gamma, \neg C_1 \rightsquigarrow B \quad \Gamma|\neg A_2 \rightsquigarrow B \rightrightarrows \Gamma, \neg C_2 \rightsquigarrow B}{\Gamma|\neg(A_1 \vee A_2) \rightsquigarrow B \rightrightarrows \Gamma, \neg(C_1 \vee C_2) \rightsquigarrow B\}}$$

$$(+\neg\vee^R)\ \frac{\Gamma|A \rightsquigarrow \neg B_1 \rightrightarrows \Gamma, A \rightsquigarrow \neg D_1 \quad \Gamma, A \rightsquigarrow \neg D_1|A \rightsquigarrow \neg B_2 \rightrightarrows \Gamma, A \rightsquigarrow \neg D_1, A \rightsquigarrow \neg D_2}{\Gamma|A \rightsquigarrow \neg(B_1 \vee B_2) \rightrightarrows \Gamma, A \rightsquigarrow \neg(D_1 \vee D_2)\}}$$

$$(-\neg\vee^L)\ \frac{\Gamma|\neg A_1 \not\rightsquigarrow B \rightrightarrows \Gamma, \neg C_1 \not\rightsquigarrow B \quad \Gamma, \neg C_1 \not\rightsquigarrow B|\neg A_2 \not\rightsquigarrow B \rightrightarrows \Gamma, \neg C_1 \not\rightsquigarrow B, \neg C_2 \not\rightsquigarrow B}{\Gamma|\neg(A_1 \vee A_2) \not\rightsquigarrow B \rightrightarrows \Gamma, \neg(C_1 \vee C_2) \not\rightsquigarrow B\}}$$

$$(-\vee^R)\ \frac{\Gamma|A \not\rightsquigarrow \neg B_1 \rightrightarrows \Gamma, A \not\rightsquigarrow \neg D_1 \quad \Gamma|A \not\rightsquigarrow \neg B_2 \rightrightarrows \Gamma, A \not\rightsquigarrow \neg D_2}{\Gamma|A \not\rightsquigarrow \neg(B_1 \vee B_2) \rightrightarrows \Gamma, A \not\rightsquigarrow \neg(D_1 \vee D_2)\}}$$

Definition 8.4.7 A reduction $\theta = \Gamma|\gamma \Rightarrow \Gamma'$ is provable in $\mathbf{P}_{\mathtt{t}}^{\rightsquigarrow}$, denoted by $\vdash_{\mathtt{t}}^{\rightsquigarrow} \theta$, if there is a sequence $\{\theta_1, \ldots, \theta_n\}$ of reductions such that $\theta_n = \theta$, and for each $1 \leq i \leq n, \theta_i$ is an axiom or is deduced from the previous reductions by one deduction rule in $\mathbf{P}_{\mathtt{t}}^{\rightsquigarrow}$.

Theorem 8.4.8 (Soundness and completeness theorem) *For any reduction* $\theta = \Gamma|\gamma \Rightarrow \Gamma'$, *where* γ *is in conjunctive normal form,*

$$\vdash_{\mathtt{t}}^{\rightsquigarrow} \theta \text{ iff } \models_{\mathtt{t}}^{\rightsquigarrow} \theta. \qquad \square$$

8.5 Gentzen Deduction Systems

There are four Gentzen deduction systems

$\mathbf{G}^{\mathtt{t}}_{\rightsquigarrow}$	$\mathbf{G}^{\mathtt{f}}_{\rightsquigarrow}$
$\mathbf{G}_{\mathtt{t}}^{\rightsquigarrow}$	$\mathbf{G}_{\mathtt{f}}^{\rightsquigarrow}$.

We will give the full description of $\mathbf{G}^{\mathtt{t}}_{\rightsquigarrow}$, and others follow from it.

8.5.1 Sequents

Given two sets Γ, Δ of statements, let $\Gamma \Rightarrow \Delta$ be a sequent, and an assignment v satisfies $\Gamma \Rightarrow \Delta$, denoted by $v \models \Gamma \Rightarrow \Delta$, if v satisfying every statement $\gamma \in \Gamma$ implies v satisfying some $\delta \in \Delta$.

$\Gamma \Rightarrow \Delta$ is $\mathbf{G}^{\mathtt{t}}_{\rightsquigarrow}$-valid, denoted by $\models^{\mathtt{t}}_{\rightsquigarrow} \Gamma \Rightarrow \Delta$, if for any assignment $v, v \models \Gamma \Rightarrow \Delta$.

A statement $l_1 \rightsquigarrow l_2$ is called a literal statement if $l_1, l_2 ::= p|\neg p$.

Proposition 8.5.1 *For any formulas* A_1, A_2, B_1, B_2,

$$\begin{aligned} A_1 \wedge A_2 \rightsquigarrow B &\Leftarrow A_1 \rightsquigarrow B \underline{\vee} A_2 \rightsquigarrow B \\ A_1 \vee A_2 \rightsquigarrow B &\equiv A_1 \rightsquigarrow B \underline{\wedge} A_2 \rightsquigarrow B \\ A \rightsquigarrow B_1 \wedge B_2 &\equiv A \rightsquigarrow B_1 \underline{\wedge} A \rightsquigarrow B_2 \\ A \rightsquigarrow B_1 \vee B_2 &\Leftarrow A \rightsquigarrow B_1 \underline{\vee} A \rightsquigarrow B_2; \end{aligned}$$

and

$$\begin{aligned} A_1 \wedge A_2 \not\rightsquigarrow B &\Rightarrow A_1 \not\rightsquigarrow B \underline{\wedge} A_2 \not\rightsquigarrow B \\ A_1 \vee A_2 \not\rightsquigarrow B &\equiv A_1 \not\rightsquigarrow B \underline{\vee} A_2 \not\rightsquigarrow B \\ A \not\rightsquigarrow B_1 \wedge B_2 &\equiv A \not\rightsquigarrow B_1 \underline{\vee} A \not\rightsquigarrow B_2 \\ A \not\rightsquigarrow B_1 \vee B_2 &\Rightarrow A \not\rightsquigarrow B_1 \underline{\wedge} A \not\rightsquigarrow B_2. \end{aligned}$$

□

For example, $\models \Rightarrow l_1 \rightsquigarrow l_2, l_3 \rightsquigarrow l_4$ iff $\models \Rightarrow \neg l_1, l_2, \neg l_3, l_4$, iff

		$\Rightarrow l_1 \rightsquigarrow l_2, l_3 \rightsquigarrow l_4$
$\neg l_1$	$l_2 = l_1$	$\Rightarrow l_1 \rightsquigarrow l_1, l_3 \rightsquigarrow l_4$
	$\neg l_3 = l_1$	$\Rightarrow l_1 \rightsquigarrow l_2, \neg l_1 \rightsquigarrow l_4$
	$l_4 = l_1$	$\Rightarrow l_1 \rightsquigarrow l_2, l_3 \rightsquigarrow l_1$
l_2	$\neg l_1 = \neg l_2$	$\Rightarrow l_2 \rightsquigarrow l_2, l_3 \rightsquigarrow l_4$
	$\neg l_3 = \neg l_2$	$\Rightarrow l_1 \rightsquigarrow l_2, l_2 \rightsquigarrow l_4$
	$l_4 = \neg l_2$	$\Rightarrow l_1 \rightsquigarrow l_2, l_3 \rightsquigarrow \neg l_2$
$\neg l_3$	$\neg l_1 = l_3$	$\Rightarrow \neg l_3 \rightsquigarrow l_2, l_3 \rightsquigarrow l_4$
	$l_2 = l_3$	$\Rightarrow l_1 \rightsquigarrow l_3, l_3 \rightsquigarrow l_4$
	$l_4 = l_3$	$\Rightarrow l_1 \rightsquigarrow l_2, l_3 \rightsquigarrow l_3$
l_4	$\neg l_1 = \neg l_4$	$\Rightarrow l_4 \rightsquigarrow l_2, l_3 \rightsquigarrow l_4$
	$l_2 = \neg l_4$	$\Rightarrow l_1 \rightsquigarrow \neg l_4, \neg l_3 \rightsquigarrow l_4$
	$\neg l_3 = \neg l_4$	$\Rightarrow l_1 \rightsquigarrow l_2, l_4 \rightsquigarrow l_4$

Proposition 8.5.2 $\models l_1 \rightsquigarrow l_2 \Rightarrow l_3 \rightsquigarrow l_4$ *if and only if* $\neg l_1, l_2 \in \{\neg l_3, l_4\}$.

Proof For any literals l_1, l_2, l_3, l_4,

$$\begin{aligned}
\models l_1 \rightsquigarrow l_2 \Rightarrow l_3 \rightsquigarrow l_4 \text{ iff } & \models \neg l_1 \Rightarrow \neg l_3, l_4 \& \models l_2 \Rightarrow \neg l_3, l_4 \\
\text{iff } & \neg l_1 \in \{\neg l_3, l_4\} \& l_2 \in \{\neg l_3, l_4\} \\
\text{iff } & (\neg l_1 = \neg l_3 \& l_2 = \neg l_3) or (\neg l_1 = \neg l_3 \& l_2 = l_4) \\
& or (\neg l_1 = l_4 \& l_2 = \neg l_3) or (\neg l_1 = l_4 \& l_2 = l_4) \\
\text{iff } & (l_1 = l_3 \& l_2 = \neg l_3) or (l_1 = l_3 \& l_2 = l_4) \\
& or (\neg l_1 = l_4 \& l_2 = \neg l_3) or (\neg l_1 = l_4 \& l_2 = l_4) \\
\text{iff } & l_1 \rightsquigarrow l_2 \Rightarrow l_3 \rightsquigarrow l_4 = \begin{cases} l_3 \rightsquigarrow \neg l_3 \Rightarrow l_3 \rightsquigarrow l_4 \\ l_3 \rightsquigarrow l_4 \Rightarrow l_3 \rightsquigarrow l_4 \\ \neg l_4 \rightsquigarrow \neg l_3 \Rightarrow l_3 \rightsquigarrow l_4 \\ \neg l_4 \rightsquigarrow l_4 \Rightarrow l_3 \rightsquigarrow l_4 \end{cases}
\end{aligned}$$

□

Proposition 8.5.3 $\models l_1 \rightsquigarrow l_2, l_3 \rightsquigarrow l_4 \Rightarrow l'_1 \rightsquigarrow l'_2, l'_3 \rightsquigarrow l'_4$ *iff*

$$\begin{aligned}
&\{\neg l_1, \neg l_3\} \cap \{\neg l'_1, l'_2, \neg l'_3 l'_4\} \neq \emptyset \& \{\neg l_1, l_4\} \cap \{\neg l'_1, l'_2, \neg l'_3 l'_4\} \neq \emptyset \\
&\& \{l_2, \neg l_3\} \cap \{\neg l'_1, l'_2, \neg l'_3 l'_4\} \neq \emptyset \& \{l_2, l_4\} \cap \{\neg l'_1, l'_2, \neg l'_3 l'_4\} \neq \emptyset.
\end{aligned}$$

Proof $\models l_1 \rightsquigarrow l_2, l_3 \rightsquigarrow l_4 \Rightarrow l'_1 \rightsquigarrow l'_2, l'_3 \rightsquigarrow l'_4$ iff

$$\begin{aligned}
&\models \neg l_1, \neg l_3 \Rightarrow l'_1 \rightsquigarrow l'_2, l'_3 \rightsquigarrow l'_4 \& \models \neg l_1, l_4 \Rightarrow l'_1 \rightsquigarrow l'_2, l'_3 \rightsquigarrow l'_4 \\
&\& \models l_2, \neg l_3 \Rightarrow l'_1 \rightsquigarrow l'_2, l'_3 \rightsquigarrow l'_4 \& \models l_2, l_4 \Rightarrow l'_1 \rightsquigarrow l'_2, l'_3 \rightsquigarrow l'_4,
\end{aligned}$$

iff

$$\begin{aligned}
&\{\neg l_1, \neg l_3\} \cap \{\neg l'_1, l'_2, \neg l'_3 l'_4\} \neq \emptyset \& \{\neg l_1, l_4\} \cap \{\neg l'_1, l'_2, \neg l'_3 l'_4\} \neq \emptyset \\
&\& \{l_2, \neg l_3\} \cap \{\neg l'_1, l'_2, \neg l'_3 l'_4\} \neq \emptyset \& \{l_2, l_4\} \cap \{\neg l'_1, l'_2, \neg l'_3 l'_4\} \neq \emptyset.
\end{aligned}$$

□

Proposition 8.5.4 *Let* Δ, Γ *be sets of literal statements such that*

$$\begin{aligned}
\Gamma &= \{l_{11} \rightsquigarrow l_{12}, \ldots, l_{n1} \rightsquigarrow l_{n2}; l'_{11} \not\rightsquigarrow l'_{12}, \ldots, l'_{n'1} \not\rightsquigarrow l'_{n'2}\} \\
\Delta &= \{l''_{11} \rightsquigarrow l''_{12}, \ldots, l''_{m1} \rightsquigarrow l''_{m2}; l'''_{11} \not\rightsquigarrow l'''_{12}, \ldots, l'''_{m'1} \not\rightsquigarrow l'''_{m'2}\}.
\end{aligned}$$

Then, $\models^{\rightsquigarrow} \Gamma \Rightarrow \Delta$ *if and only if for any* $f : \{1, \ldots, n\} \rightsquigarrow \{1, 2\}$ *and* $g : \{1, \ldots, m'\} \rightsquigarrow \{1, 2\}$ *either*

(i) $\mathbf{E}l(l, \neg l \in \sigma_f(\Gamma))$, *or*
(ii) $\mathbf{E}l(l, \neg l \in \tau_g(\Delta))$, *or*
(iii) $\sigma_f(\Gamma) \cap \tau_g(\Delta) \neq \emptyset$,

where

$$\begin{aligned}
\sigma_f(\Gamma) &= \{\neg^{f(1)} l_{1f(1)}, \ldots, \neg^{f(n)} l_{nf(n)}; l'_{11}, \neg l'_{12}, \ldots, l'_{n'1}, \neg l'_{n'2}\} \\
\tau_g(\Delta) &= \{l''_{11}, \neg l''_{12}, \ldots, l''_{m1}, \neg l''_{m2}; \neg^{g(1)} l'''_{1g(1)}, \ldots, \neg^{g(m')} l'''_{m'g(m')}\},
\end{aligned}$$

Proof Because $\Gamma \Rightarrow \Delta$

$$\begin{aligned}
&\text{iff } l_{11} \rightsquigarrow l_{12}, \ldots, l_{n1} \rightsquigarrow l_{n2}; l'_{11} \not\rightsquigarrow l'_{12}, \ldots, l'_{n'1} \not\rightsquigarrow l'_{n'2} \\
&\quad \Rightarrow l''_{11} \rightsquigarrow l''_{12}, \ldots, l''_{m1} \rightsquigarrow l''_{m2}; l'''_{11} \not\rightsquigarrow l'''_{12}, \ldots, l'''_{m'1} \not\rightsquigarrow l'''_{m'2} \\
&\text{iff } \neg l_{11} \vee l_{12}, \ldots, \neg l_{n1} \vee l_{n2}; l'_{11} \wedge \neg l'_{12}, \ldots, l'_{n'1} \wedge \neg l'_{n'2} \\
&\quad \Rightarrow \neg l''_{11} \vee l''_{12}, \ldots, \neg l''_{m1} \vee l''_{m2}; l'''_{11} \wedge \neg l'''_{12}, \ldots, l'''_{m'1} \wedge \neg l'''_{m'2} \\
&\text{iff } \neg l_{11} \vee l_{12}, \ldots, \neg l_{n1} \vee l_{n2}; l'_{11}, \neg l'_{12}, \ldots, l'_{n'1}, \neg l'_{n'2} \\
&\quad \Rightarrow \neg l''_{11}, l''_{12}, \ldots, \neg l''_{m1}, l''_{m2}; l'''_{11} \wedge \neg l'''_{12}, \ldots, l'''_{m'1} \wedge \neg l'''_{m'2} \\
&\text{iff } \mathbf{A} f : \{1, \ldots, n\} \rightsquigarrow \{1, 2\} \mathbf{A} g : \{1, \ldots, m'\} \rightsquigarrow \{1, 2\}(\\
&\quad \neg^{f(1)} l_{1f(1)}, \ldots, \neg^{f(n)} l_{nf(n)}; l'_{11}, \neg l'_{12}, \ldots, l'_{n'1}, \neg l'_{n'2} \\
&\quad \Rightarrow l''_{11}, \neg l''_{12}, \ldots, l''_{m1}, \neg l''_{m2}; \neg^{g(1)} l'''_{1g(1)}, \ldots, \neg^{g(m')} l'''_{m'g(m')}) \\
&\text{iff } \mathbf{A} f : \{1, \ldots, n\} \rightsquigarrow \{1, 2\} \mathbf{A} g : \{1, \ldots, m'\} \rightsquigarrow \{1, 2\} \\
&\quad (\text{incon}(\sigma_f(\Gamma)) \text{ or incon}(\tau_g(\Delta)) \text{ or } \Gamma \cap \Delta \neq \emptyset).
\end{aligned}$$

Assume that $\Gamma \Rightarrow \Delta$ is valid. For any assignment v, there are functions $f : \{1, \ldots, n\} \rightsquigarrow \{1, 2\}$ and $g : \{1, \ldots, m'\} \rightsquigarrow \{1, 2\}$ such that

$$v \models \neg^{f(1)} l_{1f(1)}, \ldots, \neg^{f(n)} l_{nf(n)}; l'_{11}, \neg l'_{12}, \ldots, l'_{n'1}, \neg l'_{n'2}$$

implies

$$v \models l''_{11}, \neg l''_{12}, \ldots, l''_{m1}, \neg l''_{m2}; \neg^{g(1)} l'''_{1g(1)}, \ldots, \neg^{g(m')} l'''_{m'g(m')}.$$

That is, either

(i) $\mathbf{E} l(l, \neg l \in \sigma_f(\Gamma))$, or
(ii) $\mathbf{E} l(l, \neg l \in \tau_g(\Delta))$, or
(iii) $\sigma_f(\Gamma) \cap \tau_g(\Delta) \neq \emptyset$,

Conversely, assume that there are functions $f : \{1, \ldots, n\} \rightsquigarrow \{1, 2\}$ and $g : \{1, \ldots, m'\} \rightsquigarrow \{1, 2\}$ such that

(i') $\neg \mathbf{E} l(l, \neg l \in \sigma_f(\Gamma)$, and
(ii') $\neg \mathbf{E} l(l, \neg l \in \tau_g(\Delta))$, and
(iii') $\sigma_f(\Gamma) \cap \tau_g(\Delta) = \emptyset$.

Define an assignment v such that for any variable p,

$$v(p) = 1 \text{ iff } p \in \sigma_f(\Gamma) \text{ or } \neg p \in \tau_g(\Delta).$$

Then, v is well-defined and $v \models \sigma_f(\Gamma)$, $v \not\models \tau_g(\Delta)$, which imply $v \not\models \Gamma \Rightarrow \Delta$. □

Proposition 8.5.5 *Let Δ, Γ be sets of literal statements. Then, $\models^{\mathrm{t}}_{\rightsquigarrow} \Gamma \Rightarrow \Delta$ if and only if either there is a statement $l_1 \not\rightsquigarrow l_2 \in \Gamma$ such that $\Gamma \Rightarrow \Delta \vdash l_1 \rightsquigarrow l_2$, or there is a statement $m_1 \not\rightsquigarrow m_2 \in \Delta$ such that $\Gamma \Rightarrow \Delta \vdash m_1 \rightsquigarrow m_2$.*

Proof The proposition follows from Propositions 8.1.1 and 8.1.2. □

8.5.2 Gentzen Deduction System $\mathbf{G}^{\mathtt{t}}_{\rightsquigarrow}$

Gentzen deduction system $\mathbf{G}^{\mathtt{t}}_{\rightsquigarrow}$ consists of the following axioms and deduction rules:

- **Axiom**:

$$(\mathtt{A}^{\mathtt{t}}_{\rightsquigarrow})\ \frac{\begin{cases}\mathbf{E}l_1 \not\rightsquigarrow l_2 \in \Gamma(\Gamma \Rightarrow \Delta \vdash l_1 \rightsquigarrow l_2)\\ \mathbf{E}m_1 \not\rightsquigarrow m_2 \in \Delta(\Gamma \Rightarrow \Delta \vdash m_1 \rightsquigarrow m_2)\end{cases}}{\Gamma \Rightarrow \Delta},$$

where Δ, Γ are sets of literal statements, and $\Gamma \Rightarrow \Delta \vdash l_1 \rightsquigarrow l_2$ if either $\Gamma \vdash l_1 \rightsquigarrow l_2$, or $\Delta \vdash l_1 \rightsquigarrow l_2$, or $\neg\Gamma \cup \Delta \vdash l_1 \rightsquigarrow l_2$, or $\Gamma \cup \neg\Delta \vdash l_1 \rightsquigarrow l_2$, and $\neg\delta = \sim \delta$.

- **Deduction rules**:

$$(\neg^L)\ \frac{\Gamma \Rightarrow A \rightsquigarrow B, \Delta}{\Gamma, A \not\rightsquigarrow B \Rightarrow \Delta} \qquad (\neg^R)\ \frac{\Gamma, A \rightsquigarrow B \Rightarrow \Delta}{\Gamma \Rightarrow A \not\rightsquigarrow B, \Delta}$$

$$(\wedge^{LL})\ \frac{\begin{bmatrix}\Gamma, A_1 \rightsquigarrow B \Rightarrow \Delta\\ \Gamma, A_2 \rightsquigarrow B \Rightarrow \Delta\end{bmatrix}}{\Gamma, A_1 \wedge A_2 \rightsquigarrow B \Rightarrow \Delta} \qquad (\wedge^{RL})\ \frac{\begin{cases}\Gamma \Rightarrow C_1 \rightsquigarrow D, \Delta\\ \Gamma \Rightarrow C_2 \rightsquigarrow D, \Delta\end{cases}}{\Gamma \Rightarrow C_1 \wedge C_2 \rightsquigarrow D, \Delta}$$

$$(\wedge^{LR})\ \frac{\begin{cases}\Gamma, A \rightsquigarrow B_1 \Rightarrow \Delta\\ \Gamma, A \rightsquigarrow B_2 \Rightarrow \Delta\end{cases}}{\Gamma, A \rightsquigarrow B_1 \wedge B_2, \Delta} \qquad (\wedge^{RR})\ \frac{\begin{bmatrix}\Gamma \Rightarrow C \rightsquigarrow D_1, \Delta\\ \Gamma \Rightarrow C \rightsquigarrow D_2, \Delta\end{bmatrix}}{\Gamma \Rightarrow C \rightsquigarrow D_1 \wedge D_2, \Delta}$$

and

$$(+\vee^{LL})\ \frac{\begin{cases}\Gamma, A_1 \rightsquigarrow B \Rightarrow \Delta\\ \Gamma, A_1 \rightsquigarrow B \Rightarrow \Delta\end{cases}}{\Gamma, A_1 \vee A_2 \rightsquigarrow B \Rightarrow \Delta} \qquad (+\vee^{RL})\ \frac{\begin{bmatrix}\Gamma \Rightarrow C_1 \rightsquigarrow D, \Delta\\ \Gamma \Rightarrow C_2 \rightsquigarrow D, \Delta\end{bmatrix}}{\Gamma \Rightarrow C_1 \vee C_2 \rightsquigarrow D, \Delta}$$

$$(+\vee^{LR})\ \frac{\begin{bmatrix}\Gamma, A \rightsquigarrow B_1 \Rightarrow \Delta\\ \Gamma, A \rightsquigarrow B_2 \Rightarrow \Delta\end{bmatrix}}{\Gamma, A \rightsquigarrow B_1 \vee B_2 \Rightarrow \Delta} \qquad (+\vee^{RR})\ \frac{\begin{cases}\Gamma \Rightarrow C \rightsquigarrow D_1, \Delta\\ \Gamma \Rightarrow C \rightsquigarrow D_2, \Delta\end{cases}}{\Gamma \Rightarrow C \rightsquigarrow D_1 \vee D_2, \Delta}$$

Definition 8.5.6 A sequent $\Gamma \Rightarrow \Delta$ is provable in $\mathbf{G}^{\mathtt{t}}_{\rightsquigarrow}$, denoted by $\vdash^{\mathtt{t}}_{\rightsquigarrow} \Gamma \Rightarrow \Delta$, if there is a sequence $\{\Gamma_1 \Rightarrow \Delta_1, \ldots, \Gamma_n \Rightarrow \Delta_n\}$ of sequents such that $\Gamma_n \Rightarrow \Delta_n = \Gamma \Rightarrow \Delta$, and for each $1 \leq i \leq n$, $\Gamma_i \Rightarrow \Delta_i$ is an axiom or deduced from the previous sequents by one of the deduction rules in $\mathbf{G}^{\mathtt{t}}_{\rightsquigarrow}$.

Theorem 8.5.7 (Soundness theorem) *For any sequent* $\Gamma \Rightarrow \Delta$,

$$\vdash^{\mathtt{t}}_{\rightsquigarrow} \Gamma \Rightarrow \Delta \textit{ implies } \models^{\mathtt{t}}_{\rightsquigarrow} \Gamma \Rightarrow \Delta.$$

Proof We prove that each axiom is valid and each deduction rule preserves the validity.

$(\mathtt{A}^{\mathtt{t}}_{\rightsquigarrow})$ Assume that Γ and Δ satisfy the condition in axiom $(\mathtt{A}^{\mathtt{t}}_{\rightsquigarrow})$, by Proposition 8.5.5, for any assignment v, $v \models \Gamma \Rightarrow \Delta$.

$(+\wedge^{LL})$ Assume that for any assignment v,

$$v(\Gamma, A_1 \rightsquigarrow B) = 1 \text{ implies } v(\Delta) = 1;$$
$$v(\Gamma, A_2 \rightsquigarrow B) = 1 \text{ implies } v(\Delta) = 1.$$

For any assignment v, assume that $v(\Gamma, A_1 \wedge A_2 \rightsquigarrow B) = 1$. Then, $v(A_1 \wedge A_2) = 1$ implies $v(B) = 1$. If $v(A_1) = 0$ or $v(A_2) = 0$ then $v(A_1 \rightsquigarrow B) = 1$ or $v(A_2 \rightsquigarrow B) = 1$, and by assumption, $v(\Delta) = 1$; if $v(A_1) = 1$ and $v(A_2) = 1$ then $v(B) = 1$, i.e., $v(A_1 \rightsquigarrow B) = 1, v(A_2 \rightsquigarrow B) = 1$, and by assumption, $v(\Delta) = 1$.

($+\wedge_1^{RL}$) Assume that for any assignment v,

$$v(\Gamma) = 1 \text{ implies } v(C_1 \rightsquigarrow D, \Delta) = 1.$$

For any assignment v, assume that $v(\Gamma) = 1$. If $v(\Delta) = 1$ then $v(C_1 \wedge C_2 \rightsquigarrow D, \Delta) = 1$; otherwise, by assumption, $v(C_1 \rightsquigarrow D) = 1$, and if $v(C_1 \wedge C_2) = 0$ then $v(C_1 \wedge C_2 \rightsquigarrow D) = 1$, and so $v(C_1 \wedge C_2 \rightsquigarrow D, \Delta) = 1$; if $v(C_1 \wedge C_2) = 1$ then $v(C_1) = 1$ and by assumption of $v(C_1 \rightsquigarrow D) = 1, v(D) = 1$, i.e., $v(C_1 \wedge C_2 \rightsquigarrow D) = 1$, and $v(C_1 \wedge C_2 \rightsquigarrow D, \Delta) = 1$.

($+\wedge^{LR}$) Assume that for any assignment v, either

$$v(\Gamma, A \rightsquigarrow B_1) = 1 \text{ implies } v(\Delta) = 1,$$

or

$$v(\Gamma, A \rightsquigarrow B_2) = 1 \text{ implies } v(\Delta) = 1,$$

say the former holds. For any assignment v, assume that $v(\Gamma, A \rightsquigarrow B_1 \wedge B_2) = 1$. Then, $v(A \rightsquigarrow B_1 \wedge B_2) = 1$. If $v(A) = 0$ then $v(A \rightsquigarrow B_1) = 1$, and by the assumption, $v(\Delta) = 1$; if $v(A) = 1$ then $v(B_1 \wedge B_2) = 1, v(B_1) = 1$, and by the assumption, $v(\Delta) = 1$.

($+\wedge^{RR}$) Assume that for any assignment v,

$$v(\Gamma) = 1 \text{ implies } v(C \rightsquigarrow D_1, \Delta) = 1;$$
$$v(\Gamma) = 1 \text{ implies } v(C \rightsquigarrow D_2, \Delta) = 1.$$

For any assignment v, assume that $v(\Gamma) = 1$. Then, $v(C \rightsquigarrow D_1, \Delta) = 1$ and $v(C \rightsquigarrow D_2, \Delta) = 1$. If $v(\Delta) = 1$ then $v(C \rightsquigarrow D_1 \wedge D_2, \Delta) = 1$; if $v(\Delta) = 0$ then $v(C \rightsquigarrow D_1) = 1$ and $v(C \rightsquigarrow D_2) = 1$. If $v(C) = 0$ then $v(C \rightsquigarrow D_1 \wedge D_2) = 1$, and so $v(C \rightsquigarrow D_1 \wedge D_2, \Delta) = 1$; otherwise, $v(D_1) = 1, v(D_2) = 1$, i.e., $v(C \rightsquigarrow D_1 \wedge D_2) = 1$, and so $v(C \rightsquigarrow D_1 \wedge D_2, \Delta) = 1$.

Similar for other cases. □

Theorem 8.5.8 (Completeness theorem) *For any sequent* $\Gamma \Rightarrow \Delta$,

$$\models^{\mathtt{t}}_{\rightsquigarrow} \Gamma \Rightarrow \Delta \textit{ implies } \vdash^{\mathtt{t}}_{\rightsquigarrow} \Gamma \Rightarrow \Delta.$$

Proof Given a sequent $\Gamma \Rightarrow \Delta$, we construct a tree T as follows:

- the root of T is $\Gamma \Rightarrow \Delta$;
- if $\Gamma' \Rightarrow \Delta'$ is a node such that Γ', Δ' are sets of literal statements then the node is a leaf;
- if $\Gamma' \Rightarrow \Delta'$ is a node of T which is not a leaf then $\Gamma' \Rightarrow \Delta'$ has the direct children nodes containing the following sequents:

$$\begin{cases} \Gamma_1 \Rightarrow A \rightsquigarrow B, \Delta_1 & \text{if } \Gamma' \Rightarrow \Delta' = \Gamma_1, A \not\rightsquigarrow B \Rightarrow \Delta_1 \\ \Gamma_1, A \rightsquigarrow B \Rightarrow \Delta_1 & \text{if } \Gamma' \Rightarrow \Delta' = \Gamma_1 \Rightarrow A \not\rightsquigarrow B, \Delta_1 \\ \begin{cases} \Gamma_1, A_1 \rightsquigarrow B \Rightarrow \Delta_1 \\ \Gamma_1, A_2 \rightsquigarrow B \Rightarrow \Delta_1 \end{cases} & \text{if } \Gamma' \Rightarrow \Delta' = \Gamma_1, A_1 \wedge A_2 \rightsquigarrow B \Rightarrow \Delta_1 \\ \left[\begin{array}{l} \Gamma_1 \Rightarrow C_1 \rightsquigarrow D, \Delta_1 \\ \Gamma_1 \Rightarrow C_2 \rightsquigarrow D, \Delta_1 \end{array} \right. & \text{if } \Gamma' \Rightarrow \Delta' = \Gamma_1 \Rightarrow C_1 \wedge C_2 \rightsquigarrow D, \Delta_1 \\ \left[\begin{array}{l} \Gamma_1, A \rightsquigarrow B_1 \Rightarrow \Delta_1 \\ \Gamma_1, A \rightsquigarrow B_2 \Rightarrow \Delta_1 \end{array} \right. & \text{if } \Gamma' \Rightarrow \Delta' = \Gamma_1, A \rightsquigarrow B_1 \wedge B_2 \Rightarrow \Delta_1 \\ \begin{cases} \Gamma_1 \Rightarrow C \rightsquigarrow D_1, \Delta_1 \\ \Gamma_1 \Rightarrow C \rightsquigarrow D_2, \Delta_1 \end{cases} & \text{if } \Gamma' \Rightarrow \Delta' = \Gamma_1 \Rightarrow C \rightsquigarrow D_1 \wedge D_2, \Delta_1 \end{cases}$$

and

$$\begin{cases} \left[\begin{array}{l} \Gamma_1, A_1 \rightsquigarrow B \Rightarrow \Delta_1 \\ \Gamma_1, A_2 \rightsquigarrow B \Rightarrow \Delta_1 \end{array} \right. & \text{if } \Gamma' \Rightarrow \Delta' = \Gamma_1, A_1 \vee A_2 \rightsquigarrow B \Rightarrow \Delta_1 \\ \begin{cases} \Gamma_1 \Rightarrow C_1 \rightsquigarrow D, \Delta \\ \Gamma_1 \Rightarrow C_2 \rightsquigarrow D, \Delta_1 \end{cases} & \text{if } \Gamma' \Rightarrow \Delta' = \Gamma_1 \Rightarrow C_1 \vee C_2 \rightsquigarrow D, \Delta_1 \\ \begin{cases} \Gamma_1, A \rightsquigarrow B_1 \Rightarrow \Delta \\ \Gamma_1, A \rightsquigarrow B_2 \Rightarrow \Delta_1 \end{cases} & \text{if } \Gamma' \Rightarrow \Delta' = \Gamma_1, A \rightsquigarrow B_1 \vee B_2 \Rightarrow \Delta_1 \\ \left[\begin{array}{l} \Gamma_1 \Rightarrow C \rightsquigarrow D_1, \Delta_1 \\ \Gamma_1 \Rightarrow C \rightsquigarrow D_2, \Delta_1 \end{array} \right. & \text{if } \Gamma' \Rightarrow \Delta' = \Gamma_1 \Rightarrow C \rightsquigarrow D_1 \vee D_2, \Delta_1 \end{cases}$$

Lemma 8.5.9 *If for each branch $\alpha \subseteq T$, the sequent at the leaf of α is an axiom in $\mathbf{G}^{\mathrm{t}}_{\rightsquigarrow}$ then T is a proof tree of $\Gamma \Rightarrow \Delta$ in $\mathbf{G}^{\mathrm{t}}_{\rightsquigarrow}$.*

Lemma 8.5.10 *If there is a branch $\alpha \subseteq T$ such that the leaf of α is not an axiom in $\mathbf{G}^{\mathrm{t}}_{\rightsquigarrow}$ then there is an assignment v such that $v \not\models \Gamma' \Rightarrow \Delta'$ for each $\Gamma' \Rightarrow \Delta' \in \alpha$.*

Proof Let $\Gamma' \Rightarrow \Delta'$ be the leaf of α. By Proposition 8.5.5, there is an assignment v such that $v \models \Gamma'$ and $v \not\models \Delta'$.

Fix any $\gamma \in \alpha$, and assume that $v \not\models \gamma$.

Case $(+\wedge^{LL})$. If γ is generated from $\beta \in \alpha$ by $(\wedge^{LL})$ then there are formulas A_1, A_2, B and $\beta \in \alpha$ such that

$$\begin{array}{l} \gamma = \Gamma'', A_i \rightsquigarrow B \Rightarrow \Delta', \\ \beta = \Gamma'', A_1 \wedge A_2 \rightsquigarrow B \Rightarrow \Delta', \end{array}$$

where $i \in \{1, 2\}$. By induction assumption,

$$v \models \Gamma'', A_1 \wedge A_2 \rightsquigarrow B.$$

Then

$$v \models \Gamma'', A_i \rightsquigarrow B,$$

and by induction assumption, $v \not\models \Delta'$.

Case $(+\wedge^{RL})$. If γ is generated from $\beta \in \alpha$ by $(\wedge^R)$ then there are formulas C, D_1, D_2 such that

$$\begin{array}{l}\gamma = \Gamma'' \Rightarrow C \rightsquigarrow D_1, C \rightsquigarrow D_2, \Delta'',\\ \beta = \Gamma'' \Rightarrow C \rightsquigarrow D_1 \wedge D_2, \Delta''.\end{array}$$

By induction assumption, $v \models \Gamma''$, and

$$v \not\models C \rightsquigarrow D_1, C \rightsquigarrow D_2, \Delta''.$$

Therefore,

$$v \not\models C \rightsquigarrow D_1 \wedge D_2, \Delta''.$$

Similar for other cases. □

8.6 R-Calculi

Given a co-sequent $\Gamma \mapsto \Delta$ of statements and a sequent (γ, δ) of statements, $\Gamma, \gamma' \mapsto \Delta, \delta'$ is a $\subseteq$-minimal change of $\Gamma \mapsto \Delta$ by (γ, δ), denoted by

$$\models_{\mathtt{t}} \Gamma \mapsto \Delta|(\gamma, \delta) \rightrightarrows \Gamma, \gamma' \mapsto \Delta, \delta',$$

if (γ', δ') is minimal such that (i) $(\gamma', \delta') \subseteq (\gamma, \delta)$ is consistent with $\Gamma \mapsto \Delta$, and (ii) for any sequent $(\gamma'', \delta'') \in \{(\gamma, \delta)\} - \{(\gamma', \delta')\}, \Gamma, \gamma'' \mapsto \Delta, \delta''$ is valid.

In this section, we will give an R-calculus $\mathbf{R}_{\mathtt{t}}, \mathbf{Q}_{\mathtt{t}}, \mathbf{P}_{\mathtt{t}}$ such that for any sequent $\Gamma \mapsto \Delta$ and a pair (γ, δ) of statements, $\Gamma \mapsto \Delta|(\gamma, \delta) \rightrightarrows \Gamma, \gamma' \mapsto \Delta, \delta'$ is provable in $\mathbf{R}_{\mathtt{t}}/\mathbf{Q}_{\mathtt{t}}, /\mathbf{P}_{\mathtt{t}}$ if and only if $\Gamma, \gamma' \mapsto \Delta, \delta'$ is a $\mathbf{R}/\mathbf{Q}/\mathbf{P}$-minimal change of $\Gamma \mapsto \Delta$ by (γ, δ).

8.6.1 R-Calculus $\mathbf{R}_{\mathtt{t}}$

R-calculus $\mathbf{R}_{\mathtt{t}}$ consists of the following axioms and deduction rules:

- **Axioms**:

$$(+\mathtt{A}_{\mathtt{t}}^{L+})\ \frac{\Gamma \mapsto \Delta \nvdash l \not\rightsquigarrow l'}{\Gamma \mapsto \Delta|(l \rightsquigarrow l', \delta) \rightrightarrows \Gamma, l \rightsquigarrow l' \mapsto \Delta|\delta}$$

$$(+\mathtt{A}_{\mathtt{t}}^{L0})\ \frac{\Gamma \mapsto \Delta \vdash l \not\rightsquigarrow l'}{\Gamma \mapsto \Delta|(l \rightsquigarrow l', \delta) \rightrightarrows \Gamma \mapsto \Delta|\delta}$$

$$(-\mathtt{A}_{\mathtt{t}}^{L+})\ \frac{\Gamma \mapsto \Delta \nvdash l \rightsquigarrow l'}{\Gamma \mapsto \Delta|(l \not\rightsquigarrow l', \delta) \rightrightarrows \Gamma, l \not\rightsquigarrow l' \mapsto \Delta|\delta}$$

$$(-\mathtt{A}_{\mathtt{t}}^{L0})\ \frac{\Gamma \mapsto \Delta \vdash l \rightsquigarrow l'}{\Gamma \mapsto \Delta|(l \not\rightsquigarrow l', \delta) \rightrightarrows \Gamma \mapsto \Delta|\delta}$$

$$(+\mathtt{A}_{\mathtt{t}}^{R+})\ \frac{\Gamma' \mapsto \Delta \nvdash m \not\rightsquigarrow m'}{\Gamma' \mapsto \Delta|m \rightsquigarrow m' \rightrightarrows \Gamma' \mapsto \Delta, m \rightsquigarrow m'}$$

$$(+\mathtt{A}_{\mathtt{t}}^{R0})\ \frac{\Gamma' \mapsto \Delta \vdash m \not\rightsquigarrow m'}{\Gamma' \mapsto \Delta|m \rightsquigarrow m' \rightrightarrows \Gamma' \mapsto \Delta}$$

$$(-\mathtt{A}_{\mathtt{t}}^{R+})\ \frac{\Gamma' \mapsto \Delta \nvdash m \rightsquigarrow m'}{\Gamma' \mapsto \Delta|m \not\rightsquigarrow m' \rightrightarrows \Gamma' \mapsto \Delta, m \not\rightsquigarrow m'}$$

$$(-\mathtt{A}_{\mathtt{t}}^{R0})\ \frac{\Gamma' \mapsto \Delta \vdash m \rightsquigarrow m'}{\Gamma' \mapsto \Delta|m \not\rightsquigarrow m' \rightrightarrows \Gamma' \mapsto \Delta}$$

- **Deduction rules**:

$$(\neg^{L}+)\ \frac{\Gamma \mapsto \Delta|(, A \rightsquigarrow B; C \rightsquigarrow D) \rightrightarrows \Gamma \mapsto \Delta, A \rightsquigarrow B|C \rightsquigarrow D}{\Gamma \mapsto \Delta|(A \not\rightsquigarrow B, C \rightsquigarrow D) \rightrightarrows \Gamma, A \not\rightsquigarrow B \mapsto \Delta|C \rightsquigarrow D}$$

$$(\neg_{0}^{L})\ \frac{\Gamma \mapsto \Delta|(, A \rightsquigarrow B; C \rightsquigarrow D) \rightrightarrows \Gamma \mapsto \Delta|C \rightsquigarrow D}{\Gamma \mapsto \Delta|(A \not\rightsquigarrow B, C \rightsquigarrow D) \rightrightarrows \Gamma \mapsto \Delta|C \rightsquigarrow D}$$

$$(\neg^{R}+)\ \frac{\Gamma \mapsto \Delta|(C \rightsquigarrow D, \lambda) \rightrightarrows \Gamma, C \rightsquigarrow D \mapsto \Delta}{\Gamma \mapsto \Delta|C \not\rightsquigarrow D \rightrightarrows \Gamma \mapsto \Delta, C \not\rightsquigarrow D}$$

$$(\neg_{0}^{R})\ \frac{\Gamma \mapsto \Delta|(C \rightsquigarrow D, \lambda) \rightrightarrows \Gamma \mapsto \Delta}{\Gamma \mapsto \Delta|C \not\rightsquigarrow D \rightrightarrows \Gamma \mapsto \Delta}$$

and

$$(\wedge_{+}^{LL})\ \frac{\left[\begin{array}{l}\Gamma \mapsto \Delta|(A_1 \rightsquigarrow B, C \rightsquigarrow D) \rightrightarrows \Gamma, A_1 \rightsquigarrow B \mapsto \Delta|C \rightsquigarrow D\\ \Gamma, A_1 \rightsquigarrow B \mapsto \Delta|(A_1 \rightsquigarrow B, C \rightsquigarrow D) \rightrightarrows \Gamma, A_1 \rightsquigarrow B, A_2 \rightsquigarrow B \mapsto \Delta|C \rightsquigarrow D\end{array}\right.}{\Gamma \mapsto \Delta|(A_1 \wedge A_2 \rightsquigarrow B, C \rightsquigarrow D) \rightrightarrows \Gamma, A_1 \wedge A_2 \rightsquigarrow B \mapsto \Delta|C \rightsquigarrow D}$$

$$(\wedge_{0}^{LL})\ \frac{\left\{\begin{array}{l}\Gamma \mapsto \Delta|(A_1 \rightsquigarrow B, C \rightsquigarrow D) \rightrightarrows \Gamma \mapsto \Delta|C \rightsquigarrow D\\ \Gamma, A_1 \rightsquigarrow B \mapsto \Delta|(A_2 \rightsquigarrow B, C \rightsquigarrow D) \rightrightarrows \Gamma, A_1 \rightsquigarrow B \mapsto \Delta|C \rightsquigarrow D\end{array}\right.}{\Gamma \mapsto \Delta|(A_1 \wedge A_2 \rightsquigarrow B, C \rightsquigarrow D) \rightrightarrows \Gamma \mapsto \Delta|C \rightsquigarrow D}$$

$$(\vee_{+}^{LL})\ \frac{\left\{\begin{array}{l}\Gamma \mapsto \Delta|(A_1 \rightsquigarrow B, C \rightsquigarrow D) \rightrightarrows \Gamma, A_1 \rightsquigarrow B \mapsto \Delta|C \rightsquigarrow D\\ \Gamma \mapsto \Delta|(A_2 \rightsquigarrow B, C \rightsquigarrow D) \rightrightarrows \Gamma, A_2 \rightsquigarrow B \mapsto \Delta|C \rightsquigarrow D\end{array}\right.}{\Gamma \mapsto \Delta|(A_1 \vee A_2 \rightsquigarrow B, C \rightsquigarrow D) \rightrightarrows \Gamma, A_1 \vee A_2 \rightsquigarrow B \mapsto \Delta|C \rightsquigarrow D}$$

$$(\vee_{0}^{LL})\ \frac{\left[\begin{array}{l}\Gamma \mapsto \Delta|(A_1 \rightsquigarrow B, C \rightsquigarrow D) \rightrightarrows \Gamma \mapsto \Delta|C \rightsquigarrow D\\ \Gamma \mapsto \Delta|(A_2 \rightsquigarrow B, C \rightsquigarrow D) \rightrightarrows \Gamma \mapsto \Delta|C \rightsquigarrow D\end{array}\right.}{\Gamma \mapsto \Delta|(A_1 \vee A_2 \rightsquigarrow B, C \rightsquigarrow D) \rightrightarrows \Gamma \mapsto \Delta|C \rightsquigarrow D}$$

and

$$(\wedge_{+}^{LR})\ \frac{\left\{\begin{array}{l}\Gamma \mapsto \Delta|(A \rightsquigarrow B_1, C \rightsquigarrow D) \rightrightarrows \Gamma, A \rightsquigarrow B_1 \mapsto \Delta|C \rightsquigarrow D\\ \Gamma \mapsto \Delta|(A \rightsquigarrow B_2, C \rightsquigarrow D) \rightrightarrows \Gamma, A \rightsquigarrow B_2 \mapsto \Delta|C \rightsquigarrow D\end{array}\right.}{\Gamma \mapsto \Delta|(A \rightsquigarrow B_1 \wedge B_2, C \rightsquigarrow D) \rightrightarrows \Gamma, A \rightsquigarrow B_1 \wedge B_2 \mapsto \Delta|C \rightsquigarrow D}$$

$$(\wedge_{0}^{LR})\ \frac{\left[\begin{array}{l}\Gamma \mapsto \Delta|(A \rightsquigarrow B_1, C \rightsquigarrow D) \rightrightarrows \Gamma \mapsto \Delta|C \rightsquigarrow D\\ \Gamma \mapsto \Delta|(A \rightsquigarrow B_2, C \rightsquigarrow D) \rightrightarrows \Gamma \mapsto \Delta|C \rightsquigarrow D\end{array}\right.}{\Gamma \mapsto \Delta|(A \rightsquigarrow B_1 \wedge B_2, C \rightsquigarrow D) \rightrightarrows \Gamma \mapsto \Delta|C \rightsquigarrow D}$$

$$(\vee_{+}^{LR})\ \frac{\left[\begin{array}{l}\Gamma \mapsto \Delta|(A \rightsquigarrow B_1, C \rightsquigarrow D) \rightrightarrows \Gamma, A \rightsquigarrow B_1 \mapsto \Delta|C \rightsquigarrow D\\ \Gamma, A \rightsquigarrow B_1 \mapsto \Delta|(A \rightsquigarrow B_2, C \rightsquigarrow D) \rightrightarrows \Gamma, A \rightsquigarrow B_1, A \rightsquigarrow B_2 \mapsto \Delta|C \rightsquigarrow D\end{array}\right.}{\Gamma \mapsto \Delta|(A \rightsquigarrow B_1 \vee B_2, C \rightsquigarrow D) \rightrightarrows \Gamma, A \rightsquigarrow B_1 \vee B_2 \mapsto \Delta|C \rightsquigarrow D}$$

$$(\vee_{0}^{LR})\ \frac{\left\{\begin{array}{l}\Gamma \mapsto \Delta|(A \rightsquigarrow B_1, C \rightsquigarrow D) \rightrightarrows \Gamma \mapsto \Delta|C \rightsquigarrow D\\ \Gamma, A \rightsquigarrow B_1 \mapsto \Delta|(A \rightsquigarrow B_2, C \rightsquigarrow D) \rightrightarrows \Gamma, A \rightsquigarrow B_1 \mapsto \Delta|C \rightsquigarrow D\end{array}\right.}{\Gamma \mapsto \Delta|(A \rightsquigarrow B_1 \vee B_2, C \rightsquigarrow D) \rightrightarrows \Gamma \mapsto \Delta|C \rightsquigarrow D}$$

and

$$(\wedge_{+}^{RL})\ \frac{\left\{\begin{array}{l}\Gamma' \mapsto \Delta|C_1 \rightsquigarrow D \rightrightarrows \Gamma' \mapsto \Delta, C_1 \rightsquigarrow D\\ \Gamma' \mapsto \Delta|C_2 \rightsquigarrow D \rightrightarrows \Gamma' \mapsto \Delta, C_2 \rightsquigarrow D\end{array}\right.}{\Gamma' \mapsto \Delta|C_1 \wedge C_2 \rightsquigarrow D \rightrightarrows \Gamma' \mapsto \Delta, C_1 \wedge C_2 \rightsquigarrow D}$$

$$(\wedge_{0}^{RL})\ \frac{\left[\begin{array}{l}\Gamma' \mapsto \Delta|C_1 \rightsquigarrow D \rightrightarrows \Gamma' \mapsto \Delta\\ \Gamma' \mapsto \Delta|C_2 \rightsquigarrow D \rightrightarrows \Gamma' \mapsto \Delta\end{array}\right.}{\Gamma' \mapsto \Delta|C_1 \wedge C_2 \rightsquigarrow D \rightrightarrows \Gamma' \mapsto \Delta}$$

$$(\vee_{+}^{RL})\ \frac{\left[\begin{array}{l}\Gamma' \mapsto \Delta|C_1 \rightsquigarrow D \rightrightarrows \Gamma' \mapsto \Delta, C_1 \rightsquigarrow D\\ \Gamma' \mapsto \Delta, C_1 \rightsquigarrow D|C_2 \rightsquigarrow D \rightrightarrows \Gamma' \mapsto \Delta, C_1 \rightsquigarrow D, C_2 \rightsquigarrow D\end{array}\right.}{\Gamma' \mapsto \Delta|C_1 \vee C_2 \rightsquigarrow D \rightrightarrows \Gamma' \mapsto \Delta, C_1 \vee C_2 \rightsquigarrow D}$$

$$(\vee_{0}^{RL})\ \frac{\left\{\begin{array}{l}\Gamma' \mapsto \Delta|C_1 \rightsquigarrow D \rightrightarrows \Gamma' \mapsto \Delta\\ \Gamma' \mapsto \Delta, C_1 \rightsquigarrow D|C_2 \rightsquigarrow D \rightrightarrows \Gamma' \mapsto \Delta, C_1 \rightsquigarrow D\end{array}\right.}{\Gamma' \mapsto \Delta|C_1 \vee C_2 \rightsquigarrow D \rightrightarrows \Gamma' \mapsto \Delta, C_1 \vee C_2 \rightsquigarrow D}$$

and

$$(\wedge_{+}^{RR})\ \frac{\left[\begin{array}{l}\Gamma' \mapsto \Delta|C \rightsquigarrow D_1 \rightrightarrows \Gamma' \mapsto \Delta, C \rightsquigarrow D_1\\ \Gamma' \mapsto \Delta, C \rightsquigarrow D_1|C \rightsquigarrow D_2 \rightrightarrows \Gamma' \mapsto \Delta, C \rightsquigarrow D_1, C \rightsquigarrow D_2\end{array}\right.}{\Gamma' \mapsto \Delta|C \rightsquigarrow D_1 \wedge D_2 \rightrightarrows \Gamma' \mapsto \Delta, C \rightsquigarrow D_1 \wedge D_2}$$

$$(\wedge_{0}^{RR})\ \frac{\left\{\begin{array}{l}\Gamma' \mapsto \Delta|C \rightsquigarrow D_1 \rightrightarrows \Gamma' \mapsto \Delta\\ \Gamma' \mapsto \Delta, C \rightsquigarrow D_1|C \rightsquigarrow D_2 \rightrightarrows \Gamma' \mapsto \Delta, C \rightsquigarrow D_1, C \rightsquigarrow D_2\end{array}\right.}{\Gamma' \mapsto \Delta|C \rightsquigarrow D_1 \wedge D_2 \rightrightarrows \Gamma' \mapsto \Delta}$$

$$(\vee_{+}^{RR})\ \frac{\left\{\begin{array}{l}\Gamma' \mapsto \Delta|C \rightsquigarrow D_1 \rightrightarrows \Gamma' \mapsto \Delta, C \rightsquigarrow D_1\\ \Gamma' \mapsto \Delta|C \rightsquigarrow D_2 \rightrightarrows \Gamma' \mapsto \Delta, C \rightsquigarrow D_2\end{array}\right.}{\Gamma' \mapsto \Delta|C \rightsquigarrow D_1 \vee D_2 \rightrightarrows \Gamma' \mapsto \Delta, C \rightsquigarrow D_1 \vee D_2}$$

$$(\vee_{0}^{RR})\ \frac{\left[\begin{array}{l}\Gamma' \mapsto \Delta|C \rightsquigarrow D_1 \rightrightarrows \Gamma' \mapsto \Delta\\ \Gamma' \mapsto \Delta|C \rightsquigarrow D_2 \rightrightarrows \Gamma' \mapsto \Delta\end{array}\right.}{\Gamma' \mapsto \Delta|C \rightsquigarrow D_1 \vee D_2 \rightrightarrows \Gamma' \mapsto \Delta}$$

Definition 8.6.1 A reduction $\theta = \Gamma \mapsto \Delta|(\gamma, \delta) \rightrightarrows \Gamma, \gamma' \mapsto \Delta, \theta'$ is provable in $\mathbf{R}_{\mathtt{t}}$, denoted by $\vdash_{\mathtt{t}} \theta$, if there is a sequence $\{\theta_1, \ldots, \theta_m\}$ of reductions such that $\theta_m =$

θ, and for each $i < m$, θ_{i+1} is an axiom or is deduced from the previous reductions by a deduction rule in $\mathbf{R}_{\mathrm{t}}$.

Theorem 8.6.2 (Soundness and completeness theorem) *For any sequent* $\Gamma \mapsto \Delta$ *and a statement pair* (γ, δ),

$$\vdash_{\mathrm{t}} \delta \text{ iff } \models \delta,$$

where $\theta = \Gamma \mapsto \Delta|(\gamma, \delta) \rightrightarrows \Gamma, \gamma' \mapsto \Delta, \delta'$. □

8.6.2 *R-Calculus* $\mathbf{Q}_{\mathrm{t}}$

Given a sequent $\Gamma \mapsto \Delta$ of statements and a sequent (γ, δ) of statements, $\Gamma, \gamma' \mapsto \Delta, \delta'$ is a $\preceq$-minimal change of $\Gamma \mapsto \Delta$ by (γ, δ), denoted by

$$\models_{\mathrm{t}} \Gamma \mapsto \Delta|(\gamma, \delta) \rightrightarrows \Gamma, \gamma' \mapsto \Delta, \delta',$$

if $\Gamma, \gamma' \mapsto \Delta, \delta'$ is minimal such that (i) $(\gamma', \delta') \preceq (\gamma, \delta)$ is consistent with $\Gamma \mapsto \Delta$, and (ii) for any (γ'', δ'') with $(\gamma', \delta') \prec (\gamma'', \delta'') \preceq (\gamma, \delta)$, $\Gamma, \gamma'' \mapsto \Delta, \delta''$ is invalid.

R-calculus $\mathbf{Q}_{\mathrm{t}}$ consists of the following axioms and deduction rules:

- **Axioms**: same as in $\mathbf{R}_{\mathrm{t}}$.
- **Deduction rules**:

$$(\neg^L +)\ \frac{\Gamma \mapsto \Delta|(, A \rightsquigarrow B; C \rightsquigarrow D) \rightrightarrows \Gamma \mapsto \Delta, A \rightsquigarrow B|C \rightsquigarrow D}{\Gamma \mapsto \Delta|(A \not\rightsquigarrow B, C \rightsquigarrow D) \rightrightarrows \Gamma, A \not\rightsquigarrow B \mapsto \Delta|C \rightsquigarrow D}$$

$$(\neg^L_0)\ \frac{\Gamma \mapsto \Delta|(, A \rightsquigarrow B; C \rightsquigarrow D) \rightrightarrows \Gamma \mapsto \Delta|C \rightsquigarrow D}{\Gamma \mapsto \Delta|(A \not\rightsquigarrow B, C \rightsquigarrow D) \rightrightarrows \Gamma \mapsto \Delta|C \rightsquigarrow D}$$

$$(\neg^R +)\ \frac{\Gamma \mapsto \Delta|(C \rightsquigarrow D, \lambda) \rightrightarrows \Gamma, C \rightsquigarrow D \mapsto \Delta}{\Gamma \mapsto \Delta|C \not\rightsquigarrow D \rightrightarrows \Gamma \mapsto \Delta, C \not\rightsquigarrow D}$$

$$(\neg^R_0)\ \frac{\Gamma \mapsto \Delta|(C \rightsquigarrow D, \lambda) \rightrightarrows \Gamma \mapsto \Delta}{\Gamma \mapsto \Delta|C \not\rightsquigarrow D \rightrightarrows \Gamma \mapsto \Delta}$$

and

$$(\wedge^{LL})\ \frac{\left[\begin{array}{l}\Gamma \mapsto \Delta|(A_1 \rightsquigarrow B, C \rightsquigarrow D) \rightrightarrows \Gamma, A_1' \rightsquigarrow B \mapsto \Delta|C \rightsquigarrow D \\ \Gamma, A_1' \rightsquigarrow B \mapsto \Delta|(A_2 \rightsquigarrow B, C \rightsquigarrow D) \rightrightarrows \Gamma, A_1' \rightsquigarrow B, A_2' \rightsquigarrow B \mapsto \Delta|C \rightsquigarrow D\end{array}\right.}{\Gamma \mapsto \Delta|(A_1 \wedge A_2 \rightsquigarrow B, C \rightsquigarrow D) \rightrightarrows \Gamma, A_1' \wedge A_2' \rightsquigarrow B \mapsto \Delta|C \rightsquigarrow D}$$

$$(\vee^{LL})\ \frac{\left\{\begin{array}{l}\left[\begin{array}{l}\Gamma \mapsto \Delta|(A_1 \rightsquigarrow B, C \rightsquigarrow D) \rightrightarrows \Gamma, A_1' \rightsquigarrow B \mapsto \Delta|C \rightsquigarrow D \\ A_1' \neq \lambda\end{array}\right. \\ \left[\begin{array}{l}\Gamma \mapsto \Delta|(A_1 \rightsquigarrow B, C \rightsquigarrow D) \rightrightarrows \Gamma \mapsto \Delta|C \rightsquigarrow D \\ \Gamma \mapsto \Delta|(A_1 \rightsquigarrow B, C \rightsquigarrow D) \rightrightarrows \Gamma, A_2' \rightsquigarrow B \mapsto \Delta|C \rightsquigarrow D \\ A_2' \neq \lambda\end{array}\right. \\ \left[\begin{array}{l}\Gamma \mapsto \Delta|(A_1 \rightsquigarrow B, C \rightsquigarrow D) \rightrightarrows \Gamma \mapsto \Delta|C \rightsquigarrow D \\ \Gamma \mapsto \Delta|(A_2 \rightsquigarrow B, C \rightsquigarrow D) \rightrightarrows \Gamma \mapsto \Delta|C \rightsquigarrow D\end{array}\right.\end{array}\right.}{\Gamma \mapsto \Delta|(A_1 \vee A_2 \rightsquigarrow B, C \rightsquigarrow D) \rightrightarrows \Gamma, A_1'' \vee A_2'' \rightsquigarrow B \mapsto \Delta|C \rightsquigarrow D}$$

$$\text{where } A_1'' \vee A_2'' = \begin{cases} A_1' \vee A_2 & \text{if } A_1' \neq \lambda \\ A_1 \vee A_2' & \text{if } A_1' = \lambda \neq A_2' \text{ and} \\ \lambda & \text{otherwise,} \end{cases}$$

$$(\wedge^{LR})\ \frac{\left\{\begin{array}{l} \left[\begin{array}{l} \Gamma \mapsto \Delta|(A \rightsquigarrow B_1, C \rightsquigarrow D) \rightrightarrows \Gamma, A \rightsquigarrow B_1' \mapsto \Delta|C \rightsquigarrow D \\ B_1' \neq \lambda \end{array}\right. \\ \left[\begin{array}{l} \Gamma \mapsto \Delta|(A \rightsquigarrow B_1, C \rightsquigarrow D) \rightrightarrows \Gamma \mapsto \Delta|C \rightsquigarrow D \\ \Gamma \mapsto \Delta|(A \rightsquigarrow B_2, C \rightsquigarrow D) \rightrightarrows \Gamma, A \rightsquigarrow B_2' \mapsto \Delta|C \rightsquigarrow D \\ B_2' \neq \lambda \end{array}\right. \\ \left[\begin{array}{l} \Gamma \mapsto \Delta|(A \rightsquigarrow B_1, C \rightsquigarrow D) \rightrightarrows \Gamma \mapsto \Delta|C \rightsquigarrow D \\ \Gamma \mapsto \Delta|(A \rightsquigarrow B_2, C \rightsquigarrow D) \rightrightarrows \Gamma \mapsto \Delta|C \rightsquigarrow D \end{array}\right. \end{array}\right.}{\Gamma \mapsto \Delta|(A \rightsquigarrow B_1 \wedge B_2, C \rightsquigarrow D) \rightrightarrows \Gamma, A \rightsquigarrow B_1'' \wedge B_2'' \mapsto \Delta|C \rightsquigarrow D}$$

$$(\vee^{LR})\ \frac{\begin{array}{l} \Gamma \mapsto \Delta|(A \rightsquigarrow B_1, C \rightsquigarrow D) \rightrightarrows \Gamma, A \rightsquigarrow B_1' \mapsto \Delta|C \rightsquigarrow D \\ \Gamma, A \rightsquigarrow B_1' \mapsto \Delta|(A \rightsquigarrow B_2, C \rightsquigarrow D) \rightrightarrows \Gamma, A \rightsquigarrow B_1', A \rightsquigarrow B_2' \mapsto \Delta|C \rightsquigarrow D \end{array}}{\Gamma \mapsto \Delta|(A \rightsquigarrow B_1 \vee B_2, C \rightsquigarrow D) \rightrightarrows \Gamma, A \rightsquigarrow B_1' \vee B_2' \mapsto \Delta|C \rightsquigarrow D}$$

$$\text{where } B_1'' \vee B_2'' = \begin{cases} B_1' \vee B_2 & \text{if } B_1' \neq \lambda \\ B_1 \vee B_2' & \text{if } B_1' = \lambda \neq B_2' \text{ and} \\ \lambda & \text{otherwise,} \end{cases}$$

$$(\wedge^{RL})\ \frac{\left\{\begin{array}{l} \left[\begin{array}{l} \Gamma' \mapsto \Delta|C_1 \rightsquigarrow D \rightrightarrows \Gamma' \mapsto \Delta, C_1' \rightsquigarrow D \\ C' \neq \lambda \end{array}\right. \\ \left[\begin{array}{l} \Gamma' \mapsto \Delta|C_1 \rightsquigarrow D \rightrightarrows \Gamma' \mapsto \Delta \\ \Gamma' \mapsto \Delta|C_2 \rightsquigarrow D \rightrightarrows \Gamma' \mapsto \Delta, C_2' \rightsquigarrow D \\ C_2' \neq \lambda \end{array}\right. \\ \left[\begin{array}{l} \Gamma' \mapsto \Delta|C_1 \rightsquigarrow D \rightrightarrows \Gamma' \mapsto \Delta \\ \Gamma' \mapsto \Delta|C_2 \rightsquigarrow D \rightrightarrows \Gamma' \mapsto \Delta \end{array}\right. \end{array}\right.}{\Gamma' \mapsto \Delta|C_1 \wedge C_2 \rightsquigarrow D \rightrightarrows \Gamma' \mapsto \Delta, C_1'' \wedge C_2'' \rightsquigarrow D}$$

$$(\vee^{RL})\ \frac{\begin{array}{l} \Gamma' \mapsto \Delta|C_1 \rightsquigarrow D \rightrightarrows \Gamma' \mapsto \Delta, C_1' \rightsquigarrow D \\ \Gamma' \mapsto \Delta, C_1' \rightsquigarrow D|C_2 \rightsquigarrow D \rightrightarrows \Gamma' \mapsto \Delta, C_1' \rightsquigarrow D, C_2' \rightsquigarrow D \end{array}}{\Gamma' \mapsto \Delta|C_1 \vee C_2 \rightsquigarrow D \rightrightarrows \Gamma' \mapsto \Delta, C_1' \vee C_2' \rightsquigarrow D}$$

$$\text{where } C_1'' \vee C_2'' = \begin{cases} C_1' \vee C_2 & \text{if } C_1' \neq \lambda \\ C_1 \vee C_2' & \text{if } C_1' = \lambda \neq C_2' \text{ and} \\ \lambda & \text{otherwise,} \end{cases}$$

$$(\wedge^{RR})\ \frac{\begin{array}{l} \Gamma' \mapsto \Delta|C \rightsquigarrow D_1 \rightrightarrows \Gamma' \mapsto \Delta, C \rightsquigarrow D_1' \\ \Gamma' \mapsto \Delta, C \rightsquigarrow D_1'|C \rightsquigarrow D_2 \rightrightarrows \Gamma' \mapsto \Delta, C \rightsquigarrow D_1', C \rightsquigarrow D_2' \end{array}}{\Gamma' \mapsto \Delta|C \rightsquigarrow D_1 \wedge D_2 \rightrightarrows \Gamma' \mapsto \Delta, C \rightsquigarrow D_1' \wedge D_2'}$$

$$(\vee^{RR})\ \frac{\left\{\begin{array}{l} \left[\begin{array}{l} \Gamma' \mapsto \Delta|C \rightsquigarrow D_1 \rightrightarrows \Gamma' \mapsto \Delta, C \rightsquigarrow D_1' \\ D_1' \neq \lambda \end{array}\right. \\ \left[\begin{array}{l} \Gamma' \mapsto \Delta|C \rightsquigarrow D_1 \rightrightarrows \Gamma' \mapsto \Delta \\ \Gamma' \mapsto \Delta|C \rightsquigarrow D_2 \rightrightarrows \Gamma' \mapsto \Delta, C \rightsquigarrow D_2' \\ D_1' \neq \lambda \end{array}\right. \\ \left[\begin{array}{l} \Gamma' \mapsto \Delta|C \rightsquigarrow D_1 \rightrightarrows \Gamma' \mapsto \Delta \\ \Gamma' \mapsto \Delta|C \rightsquigarrow D_2 \rightrightarrows \Gamma' \mapsto \Delta \end{array}\right. \end{array}\right.}{\Gamma' \mapsto \Delta|C \rightsquigarrow D_1 \vee D_2 \rightrightarrows \Gamma' \mapsto \Delta, C \rightsquigarrow D_1'' \vee D_2''}$$

where $D_1'' \vee D_2'' = \begin{cases} D_1' \vee D_2 \text{ if } D_1' \neq \lambda \\ D_1 \vee D_2' \text{ if } D_1' = \lambda \neq D_2' \\ \lambda \quad\quad\quad \text{otherwise.} \end{cases}$

Definition 8.6.3 A reduction $\theta = \Gamma \mapsto \Delta|(\gamma, \delta) \rightrightarrows \Gamma, \gamma' \mapsto \Delta, \theta'$ is provable in $\mathbf{Q}_{\mathtt{t}}$, denoted by $\vdash_{\mathtt{t}} \theta$, if there is a sequence $\{\theta_1, \ldots, \theta_m\}$ of reductions such that $\theta_m = \theta$, and for each $i < m$, θ_{i+1} is an axiom or is deduced from the previous reductions by a deduction rule in $\mathbf{Q}_{\mathtt{t}}$.

Theorem 8.6.4 (Soundness and completeness theorem) *For any sequent* $\Gamma \mapsto \Delta$ *and a statement pair* (γ, δ),

$$\vdash_{\mathtt{t}} \theta \text{ iff } \models_{\mathtt{t}} \theta,$$

where $\theta = \Gamma \mapsto \Delta|(\gamma, \delta) \rightrightarrows \Gamma, \gamma' \mapsto \Delta, \delta'$. □

8.6.3 *R-Calculus* $\mathbf{P}_{\mathtt{t}}$

Given a sequent $\Gamma \mapsto \Delta$ of statements and a sequent (γ, δ) of statements, $\Gamma, \gamma' \mapsto \Delta, \delta'$ is a $\vdash_{\preceq}$-minimal change of $\Gamma \mapsto \Delta$ by (γ, δ), denoted by

$$\models_{\mathtt{t}} \Gamma \mapsto \Delta|(\gamma, \delta) \rightrightarrows \Gamma, \gamma' \mapsto \Delta, \delta',$$

if (γ', δ') is minimal such that (i) $(\gamma', \delta') \preceq (\gamma, \delta)$ is consistent with $\Gamma \mapsto \Delta$, and (ii) for any pair (γ'', δ'') with $(\gamma', \delta') \prec (\gamma'', \delta'') \preceq (\gamma, \delta)$, either

$$\Gamma, \gamma' \mapsto \Delta, \delta' \vdash \Gamma, \gamma'' \mapsto \Delta, \delta''$$

and

$$\Gamma, \gamma'' \mapsto \Delta, \delta'' \vdash \Gamma, \gamma' \mapsto \Delta, \delta',$$

or $\Gamma, \gamma'' \mapsto \Delta, \delta''$ is consistent.

R-calculus $\mathbf{P}_{\mathtt{t}}$ consists of the following axioms and deduction rules:

- **Axioms**: same as in $\mathbf{R}_{\mathtt{t}}$.
- **Deduction rules**:

$$(\neg^L +) \frac{\Gamma \mapsto \Delta|(, A \rightsquigarrow B; C \rightsquigarrow D) \rightrightarrows \Gamma \mapsto \Delta, A \rightsquigarrow B|C \rightsquigarrow D}{\Gamma \mapsto \Delta|(A \not\rightsquigarrow B, C \rightsquigarrow D) \rightrightarrows \Gamma, A \not\rightsquigarrow B \mapsto \Delta|C \rightsquigarrow D}$$

$$(\neg_0^L) \frac{\Gamma \mapsto \Delta|(, A \rightsquigarrow B; C \rightsquigarrow D) \rightrightarrows \Gamma \mapsto \Delta|C \rightsquigarrow D}{\Gamma \mapsto \Delta|(A \not\rightsquigarrow B, C \rightsquigarrow D) \rightrightarrows \Gamma \mapsto \Delta|C \rightsquigarrow D}$$

$$(\neg^R +) \frac{\Gamma \mapsto \Delta|(C \rightsquigarrow D, \lambda) \rightrightarrows \Gamma, C \rightsquigarrow D \mapsto \Delta}{\Gamma \mapsto \Delta|(\lambda, C \not\rightsquigarrow D) \rightrightarrows \Gamma \mapsto \Delta, C \not\rightsquigarrow D}$$

$$(\neg_0^R) \frac{\Gamma \mapsto \Delta|(C \rightsquigarrow D, \lambda) \rightrightarrows \Gamma \mapsto \Delta}{\Gamma \mapsto \Delta|(\lambda, C \not\rightsquigarrow D) \rightrightarrows \Gamma \mapsto \Delta}$$

where $\Gamma \mapsto \Delta|(, A \rightsquigarrow B; C \rightsquigarrow D) = (\Gamma \mapsto \Delta|(\lambda, A \rightsquigarrow B))|(\lambda, C \rightsquigarrow D)$, and

$$(\wedge^{LL})\ \frac{\begin{array}{l}\Gamma \mapsto \Delta|(A_1 \rightsquigarrow B, C \rightsquigarrow D) \rightrightarrows \Gamma, A_1' \rightsquigarrow B \mapsto \Delta|C \rightsquigarrow D\\ \Gamma, A_1' \rightsquigarrow B \mapsto \Delta|(A_2 \rightsquigarrow B, C \rightsquigarrow D) \rightrightarrows \Gamma, A_1' \rightsquigarrow B, A_2' \rightsquigarrow B \mapsto \Delta|C \rightsquigarrow D\end{array}}{\Gamma \mapsto \Delta|(A_1 \wedge A_2 \rightsquigarrow B, C \rightsquigarrow D) \rightrightarrows \Gamma, A_1' \wedge A_2' \rightsquigarrow B \mapsto \Delta|C \rightsquigarrow D}$$

$$(\vee^{LL})\ \frac{\begin{array}{l}\Gamma \mapsto \Delta|(A_1 \rightsquigarrow B, C \rightsquigarrow D) \rightrightarrows \Gamma, A_1' \rightsquigarrow B \mapsto \Delta|C \rightsquigarrow D\\ \Gamma \mapsto \Delta|(A_2 \rightsquigarrow B, C \rightsquigarrow D) \rightrightarrows \Gamma, A_2' \rightsquigarrow B \mapsto \Delta|C \rightsquigarrow D\end{array}}{\Gamma \mapsto \Delta|(A_1 \vee A_2 \rightsquigarrow B, C \rightsquigarrow D) \rightrightarrows \Gamma, A_1' \vee A_2' \rightsquigarrow B \mapsto \Delta|C \rightsquigarrow D}$$

$$(\wedge^{LR})\ \frac{\begin{array}{l}\Gamma \mapsto \Delta|(A \rightsquigarrow B_1, C \rightsquigarrow D) \rightrightarrows \Gamma, A \rightsquigarrow B_1' \mapsto \Delta|C \rightsquigarrow D\\ \Gamma \mapsto \Delta|(A \rightsquigarrow B_2, C \rightsquigarrow D) \rightrightarrows \Gamma, A \rightsquigarrow B_2' \mapsto \Delta|C \rightsquigarrow D\end{array}}{\Gamma \mapsto \Delta|(A \rightsquigarrow B_1 \wedge B_2, C \rightsquigarrow D) \rightrightarrows \Gamma, A \rightsquigarrow B_1' \wedge B_2' \mapsto \Delta|C \rightsquigarrow D}$$

$$(\vee^{LR})\ \frac{\begin{array}{l}\Gamma \mapsto \Delta|(A \rightsquigarrow B_1, C \rightsquigarrow D) \rightrightarrows \Gamma, A \rightsquigarrow B_1' \mapsto \Delta|C \rightsquigarrow D\\ \Gamma, A \rightsquigarrow B_1' \mapsto \Delta|(A \rightsquigarrow B_2, C \rightsquigarrow D) \rightrightarrows \Gamma, A \rightsquigarrow B_1', A \rightsquigarrow B_2' \mapsto \Delta|C \rightsquigarrow D\end{array}}{\Gamma \mapsto \Delta|(A \rightsquigarrow B_1 \vee B_2, C \rightsquigarrow D) \rightrightarrows \Gamma, A \rightsquigarrow B_1' \vee B_2' \mapsto \Delta|C \rightsquigarrow D}$$

and

$$(\wedge^{RL})\ \frac{\begin{array}{l}\Gamma' \mapsto \Delta|C_1 \rightsquigarrow D \rightrightarrows \Gamma' \mapsto \Delta, C_1' \rightsquigarrow D\\ \Gamma' \mapsto \Delta|C_2 \rightsquigarrow D \rightrightarrows \Gamma' \mapsto \Delta, C_2' \rightsquigarrow D\end{array}}{\Gamma' \mapsto \Delta|C_1 \wedge C_2 \rightsquigarrow D \rightrightarrows \Gamma' \mapsto \Delta, C_1' \wedge C_2' \rightsquigarrow D}$$

$$(\vee^{RL})\ \frac{\begin{array}{l}\Gamma' \mapsto \Delta|C_1 \rightsquigarrow D \rightrightarrows \Gamma' \mapsto \Delta, C_1' \rightsquigarrow D\\ \Gamma' \mapsto \Delta, C_1' \rightsquigarrow D|C_2 \rightsquigarrow D \rightrightarrows \Gamma' \mapsto \Delta, C_1' \rightsquigarrow D, C_2' \rightsquigarrow D\end{array}}{\Gamma' \mapsto \Delta|C_1 \vee C_2 \rightsquigarrow D \rightrightarrows \Gamma' \mapsto \Delta, C_1' \vee C_2' \rightsquigarrow D}$$

$$(\wedge^{RR})\ \frac{\begin{array}{l}\Gamma' \mapsto \Delta|C \rightsquigarrow D_1 \rightrightarrows \Gamma' \mapsto \Delta, C \rightsquigarrow D_1'\\ \Gamma' \mapsto \Delta, C \rightsquigarrow D_1'|C \rightsquigarrow D_2 \rightrightarrows \Gamma' \mapsto \Delta, C \rightsquigarrow D_1', C \rightsquigarrow D_2'\end{array}}{\Gamma' \mapsto \Delta|C \rightsquigarrow D_1 \wedge D_2 \rightrightarrows \Gamma' \mapsto \Delta, C \rightsquigarrow D_1' \wedge D_2'}$$

$$(\vee^{RR})\ \frac{\begin{array}{l}\Gamma' \mapsto \Delta|C \rightsquigarrow D_1 \rightrightarrows \Gamma' \mapsto \Delta, C \rightsquigarrow D_1'\\ \Gamma' \mapsto \Delta|C \rightsquigarrow D_2 \rightrightarrows \Gamma' \mapsto \Delta, C \rightsquigarrow D_2'\end{array}}{\Gamma' \mapsto \Delta|C \rightsquigarrow D_1 \vee D_2 \rightrightarrows \Gamma' \mapsto \Delta, C \rightsquigarrow D_1' \vee D_2'}$$

Definition 8.6.5 A reduction $\theta = \Gamma \mapsto \Delta|(\gamma, \delta) \rightrightarrows \Gamma, \gamma' \mapsto \Delta, \theta'$ is provable in $\mathbf{P}_{\mathtt{t}}$, denoted by $\vdash_{\mathtt{t}} \theta$, if there is a sequence $\{\theta_1, \ldots, \theta_m\}$ of reductions such that $\theta_m = \theta$, and for each $i < m$, θ_{i+1} is an axiom or is deduced from the previous reductions by a deduction rule in $\mathbf{P}_{\mathtt{t}}$.

Theorem 8.6.6 (*Soundness and completeness theorem*) *For any sequent $\Gamma \mapsto \Delta$ and a statement pair (γ, δ), where γ is in disjunctive normal form and δ is in conjunctive normal form,*

$$\vdash_{\mathtt{t}} \theta \textit{ iff } \models_{\mathtt{t}} \theta,$$

where $\theta = \Gamma \mapsto \Delta|(\gamma, \delta) \rightrightarrows \Gamma, \gamma' \mapsto \Delta, \delta'$. □

8.7 Conclusions

We considered the R-calculi based on three minimal changes and have the following R-calculi for theories:

Tableau proof system	$\mathbf{T}^{\mathtt{t}}_{\leadsto}, \mathbf{T}^{\mathtt{f}}_{\leadsto}, \mathbf{T}^{\leadsto}_{\mathtt{t}}, \mathbf{T}^{\leadsto}_{\mathtt{f}}$
R-calculus	$\mathbf{R}^{\mathtt{t}}_{\leadsto}, \mathbf{R}^{\mathtt{f}}_{\leadsto}$
minimal change	$\mathbf{R}^{\leadsto}_{\mathtt{t}}, \mathbf{Q}^{\leadsto}_{\mathtt{t}}, \mathbf{P}^{\leadsto}_{\mathtt{t}}$

and for Gentzen deduction system

Gentzen deduction system	$\mathbf{G}^{\mathtt{t}}_{\leadsto}$
minimal change	$\mathbf{R}_{\mathtt{t}}, \mathbf{Q}_{\mathtt{t}}, \mathbf{P}_{\mathtt{t}}.$

References

Bolc, L., Borowik, P.: Many-Valued Logics (2 Automated Reasoning and Practical Applications). Springer, Berlin (2003)

Li, W., Sui, Y.: A sound and complete R-calculi with respect to contraction and minimal change. Frontiers Comput. Sci. **8**, 184–191 (2014)

Li, W., Sui, Y.: The sound and complete R-calculi with respect to pseudo-revision and pre-revision, International Journal of Intelligence. Science **3**, 110–117 (2013)

Li, W., Sui, Y., Sun, M.: The sound and complete R-calculus for revising propositional theories. Sci. China: Inf. Sci. **58**, 092101:1-092101:12 (2015)

Printed in the United States
by Baker & Taylor Publisher Services